Contact Line Dynamics and Droplet Spreading

Contact Line Dynamics and Droplet Spreading

Guest Editors

Alireza Mohammad Karim
Koji Hasegawa
Maurizio Santini

MDPI

Basel • Beijing • Wuhan • Barcelona • Belgrade • Novi Sad • Cluj • Manchester

Guest Editors

Alireza Mohammad Karim
Department of Engineering
University of Cambridge
Cambridge
United Kingdom

Koji Hasegawa
Department of Mechanical Engineering
Kogakuin University of Technology & Engineering
Tokyo
Japan

Maurizio Santini
Department of Engineering and Applied Sciences
University of Bergamo
Dalmine
Italy

Editorial Office
MDPI AG
Grosspeteranlage 5
4052 Basel, Switzerland

This is a reprint of the Special Issue, published open access by the journal *Fluids* (ISSN 2311-5521), freely accessible at: https://www.mdpi.com/journal/fluids/special_issues/43141S7937.

For citation purposes, cite each article independently as indicated on the article page online and as indicated below:

Lastname, A.A.; Lastname, B.B. Article Title. *Journal Name* **Year**, *Volume Number*, Page Range.

ISBN 978-3-7258-5087-7 (Hbk)
ISBN 978-3-7258-5088-4 (PDF)
https://doi.org/10.3390/books978-3-7258-5088-4

Contents

About the Editors

Alireza Mohammad Karim

Alireza Mohammad Karim is a postdoctoral fellow in Pediatric Cardiology in the Department of Pediatrics in the Division of Cardiology at the Children's Hospital of Philadelphia. He has a Bachelor of Science in Mechanical Engineering from UCLA with the highest honor distinction (summa cum laude) alongside the honor of being the recipient of the Boeing Outstanding Undergraduate Student Award. He has two PhDs. He has obtained a PhD in Aerospace Engineering with a specialization in fluid mechanics and a minor in applied mathematics from UCLA and a PhD in Engineering with a specialization in nanotherapeutics from the University of Cambridge as a Cambridge Trust Scholar. He was named the Featured Researcher in the Collaborative Center for Aerospace Sciences in the Department of Mechanical and Aerospace Engineering at UCLA. He has earned two Master of Science degrees. He has a Master of Science in Aerospace Engineering from Caltech with the Space Fellowship Award from the Jet Propulsion Laboratory (JPL) in NASA, and a Master of Science in Mechanical Engineering from UCLA. He is the author/co-author of over 30 scientific publications. His research centers in microfluidics, nanotherapeutics, and heart valve disease. He has served as the Lead Guest Editor for Special Issues of the journals Fluids and Micromachines. He has also served as the Lead Guest Editor for a Special Collection of the journal Advances in Mechanical Engineering. He has been a member of the Topical Advisory Panel of the journal Micromachines since 2022, the Section Topical Advisory Panel Member of Non-Newtonian and Complex Fluids of the journal Fluids since 2023, and the Section Topical Advisory Panel Member of Nanotechnology Applications in Bioengineering of the journal Bioengineering since 2024.

Koji Hasegawa

Koji Hasegawa is an Associate Professor at Kogakuin University of Technology & Engineering, Japan. He earned his Ph.D. in Engineering from the University of Tsukuba in 2013, specializing in multiphase flow. From 2024 to 2025, he served as a visiting researcher at the University of Stuttgart, Germany. His research focuses on droplet dynamics, encompassing interfacial transport phenomena, wetting, spreading, and phase-change processes such as evaporation and freezing. A distinctive aspect of his work is the use of advanced optical diagnostics, including high-speed imaging and infrared thermography, to investigate multiphase flow phenomena. Dr. Hasegawa has received numerous honors, including the Young Scientists' Award from Japan's Ministry of Education, Culture, Sports, Science and Technology, as well as awards from the Japan Society of Mechanical Engineers (JSME), the Japanese Society for Multiphase Flow (JSMF), and the Japan Society of Microgravity Application (JASMA).

Maurizio Santini

Maurizio Santini is a full professor of technical physics (thermodynamic, heat and mass transfer, applied acoustic) at the University of Bergamo (Italy). His primary research interests include atomization processes (droplet impact on dry walls, liquid films and pools); liquid/solid interfaces (contact lines) and the characterization of porous media and transport phenomena in biological materials. His various research activities are predominantly experimental, particularly in the application and development of multispectral techniques for the characterization of multiphase flows at the microscale (optical microscopy, phase-Doppler anemometry, X-ray micro-tomography including phase-contrast and photon counting techniques).

Preface

Contact line appears once a liquid–air interface is in contact with a solid surface and involves interfacial physics and fluid mechanics. A moving contact line and droplet spreading have important roles in a broad spectrum of technologies including healthcare, electronics, and energy. This Special Issue Reprint provides advances in moving contact lines and droplet impact and explores their applications in biotechnology, microfluidics, nanofluidics, printings/coatings, and heat transfer.

Alireza Mohammad Karim, Koji Hasegawa, and Maurizio Santini
Guest Editors

fluids

MDPI

Editorial

Contact Line Dynamics and Droplet Spreading

Alireza Mohammad Karim

Department of Pediatrics, Division of Cardiology, Children's Hospital of Philadelphia, Philadelphia, PA 19104, USA; am2633@cantab.ac.uk

Contact line motion occurs when a liquid encounters a solid surface. The physics of liquid contact line movement and droplet spreading involves interfacial science and fluid dynamics. Contact line dynamics and droplet spreading significantly impact a broad range of scientific fields and technologies, including microfluidics, nanofluidics, printed electronics, nano-printing, 3D bioprinting, coatings, and phase-change heat transfer. The contact line is a three-phase region (gas/liquid/solid or liquid/liquid/solid) with an intricate configuration. Its structure depends on several parameters, namely the solid surface topography, liquid rheology, gas environment, and other physical properties of each phase. In this Special Issue, we aim to explore the recent developments in contact line physics and droplet impact mechanisms, encompassing theory, state-of-the-art experiments, and computations while considering their applications in various scientific and technological fields, including biotechnology, microfluidics, nanofluidics, printings, and coatings, as well as heat transfer.

In spray paintings, Ye et al. [1] computationally explored the mechanism of viscous droplet impact on dry and wetted solid walls, considering air entrapment, film structure, and flake flat pigment orientation using a computational fluid dynamics (CFD) program. The numerical results were compared with observations using high-speed visualization [1]. In printings, Shah and Mohan [2] presented a model for droplet evolution using the volume of fluid (VOF) approach to solve Navier–Stokes and continuity equations for incompressible flow, with various immiscible phases in a finite volume. Their computational model agreed well with observations of the impact, spread, and recoil of droplets on a substrate [2]. In their study, they considered three main non-dimensional parameters, including the Reynolds number, Weber number, and Ohnesorge number [2]. Water droplet impact dynamics on a two-layered pool composed of an ultrathin suspended layer of silicone oil on top of a deep water pool was explored by Dehghanghadikolaei et al. [3]. They found that the viscosity of the suspended oil layer is a major parameter to regulate droplet impact dynamics [3]. In spray coatings, Stober et al. [4] studied the asymmetric crown morphology caused by an oblique impact ($\alpha = \alpha = 60°$) of a single droplet on a horizontal wetted wall with the same liquid film. They showed that the Weber number variation in the splashing stage causes distinct crown morphologies [4].

Droplet impact and coalescence physics was investigated by Chashechkin and Ilinykh [5] via probing the flow pattern, cavity formation, and distribution of a free-falling droplet onto a liquid pool at rest. They also considered the droplet coalescence based on the observed energy conservation due to the droplet's kinetic energy and surface potential energy. Beloborodov and Vishnyakov [6] studied the coalescence dynamics of Lennard–Jones nano-droplets as a function of droplet size and temperature using molecular simulations. Comparing the duration of coalescence with the droplet lifespan in a suspension revealed that for dense suspensions and small droplets, the coalescence timescale is enhanced and needs to be taken into account in aggregation theoretical modeling [6].

In organized packings, Croce and Suzzi [7] computationally examined liquid film instability in a falling down motion on a vertical plate with lateral walls, mimicking the structured packing configuration by applying the lubrication assumption. Their numerical findings, based on a linear stability analysis and considering substrate wettability, suggested that the formed rivulets at a wavelength larger than the minimum threshold promote partial dewetting [7]. In the interfacial instability of a complex multi-phase flow, Hasegawa and Kishimoto [8] experimentally explored the contact line and interfacial instability dynamics of a binary droplet (a water and 2-propanol (IPA) mixture) on an immiscible (sunflower oil) pool. Their experimental observations showed that fingering instability is driven by the capillary effect for a liquid–liquid system and Plateau–Rayleigh instability [8].

The impact of biochemically active heterogeneous droplets on a solid surface plays a key role in the food and pharmaceutical industry. In an advanced drug delivery system, suspended therapeutic particles in an aqueous environment (i.e., heterogeneous droplets) are deposited onto a solid surface. It is critical to print uniform drug-loaded droplets on a substrate surface for sustained and regulated drug release in patients upon administration. The bioactive heterogeneous droplet impact mechanism deserves much more attention in future research. In bacterial encapsulation process, motivated by the food and pharmaceutical industry, Mohd Isa et al. [9] studied the encapsulation of bacteria in single water-in-oil (W/O) and double water-in-oil-in-water ($W_1/O/W_2$) emulsions under cold storage and temperature-regulated release conditions. In their study, storage conditions at freezing temperatures (-20 °C and -80 °C) caused an extensive droplet destabilization with a rapid release of encapsulated bacteria upon thawing, where the temperature-regulated release of encapsulated bacteria was reached [9]. Their study provides an excellent review for utilizing emulsion droplets for temperature-regulated bacterial encapsulation, which is a crucial factor to consider for food and pharmaceutical storage and usage.

The dropwise condensation of steam over a hybrid hydrophobic–hydrophilic surface was numerically explored by Croce and Suzzi [10] through a non-dimensional analysis of the heat transfer model using a Lagrangian approach. Their computational findings agreed well with observations, and they showed that heat flux was enhanced under various operating conditions by the hybrid surface [10]. In the development of energy systems and thermal management processes, Lakew et al. [11] explored a bubble growth mechanism for modeling boiling heat transfer to reach the optimum heat flux needed for its operation. They presented analyses of evaporating ultra-thin (i.e., microlayer) liquid underneath steam bubbles growing at the heated surface in the atmospheric pressure nucleate of boiling water [11].

The contact line dynamics of ionic liquids at the mesoscopic scale was investigated by Carvalho et al. [12]. They examined the effect of the deposition speed and substrate temperature on the nucleation, droplet formation, and droplet spread of the ionic liquid films formed by thermal evaporation [12]. They considered four ionic liquids consisting of an alkylimidazolium cation (C_nC_1im) and either bis(trifluoromethylsulfonyl)imide (NTf_2) or triflate (OTf) as the anion [12]. Each ionic liquid droplet was simultaneously deposited on surfaces of indium tin oxide (ITO) and silver (Ag) [12]. They found that the substrate's wettability by the ionic liquids was remarkably influenced by changes in the mass flow rate and substrate temperature [12]. Taitelbaum [13] provided a review on the top-view reactive contact line dynamics based on statistical physics, with significant attention on the spread of mercury droplets over metal-on-glass at room temperature, and compared the findings with the case of reactive droplets spreading at high temperatures. Mohammad Karim and Suszynski [14] provided a comprehensive review of the conventional physical models applied for describing the contact line dynamics, using hydrodynamics and molecular kinetics.

Three-dimensional (3D) bioprinting is a rapidly advanced technology in regenerative medicine such as cardiac tissue engineering and 3D cancer modeling [15–54]. However, more studies still need to be performed to precisely mimic the complex microstructures and nanostructures of biological tissues and accurately control the distribution of cells, biochemical factors, and biomaterials in a layer-by-layer approach.

With the emerging advancements in technology ranging from machine/deep learning, quantum power, and nanoscience, it is highly recommended to utilize these state-of-the-art tools to conduct research in the field of droplet impact for complex systems due to droplet biochemical heterogeneity and the heterogeneous topography of substrate surfaces. A recent study utilized machine learning along with experimental observation to study the spreading and freezing process of water droplet impact upon a supercooled substrate [55]. There has been significant attention on bioprinting for healthcare and medicine to print live biological cells and be used for artificial organ synthesis and tissue engineering [18,56–90]. Over the last decade, a remarkable advancement has been achieved in the field of printed electronics for the application of wearable and stretchable sensors [56].

Finally, we thank the authors and reviewers for their contributions to this Special Issue by sharing their research or their comments in the rigorous review process to provide a highly selective collection of articles in the field of contact line physics and droplet impact dynamics.

Conflicts of Interest: The author declares no conflict of interest.

References

1. Ye, Q.; Tiedje, O.; Shen, B.; Domnick, J. Impact of Viscous Droplets on Dry and Wet Substrates for Spray Painting Processes. *Fluids* **2025**, *10*, 131. [CrossRef]
2. Shah, R.; Mohan, R.V. Computational Modeling of Droplet-Based Printing Using Multiphase Volume of Fluid (VOF) Method: Prediction of Flow, Spread Behavior, and Printability. *Fluids* **2025**, *10*, 123. [CrossRef]
3. Dehghanghadikolaei, A.; Abdul Halim, B.; Khoshbakhtnejad, E.; Sojoudi, H. Dynamics of a Water Droplet Impacting an Ultrathin Layer of Oil Suspended on a Pool of Water. *Fluids* **2024**, *9*, 82. [CrossRef]
4. Stober, J.L.; Santini, M.; Schulte, K. Influence of Weber Number on Crown Morphology during an Oblique Droplet Impact on a Thin Wall Film. *Fluids* **2023**, *8*, 301. [CrossRef]
5. Chashechkin, Y.D.; Ilinykh, A.Y. Fine Flow Structure at the Miscible Fluids Contact Domain Boundary in the Impact Mode of Free-Falling Drop Coalescence. *Fluids* **2023**, *8*, 269. [CrossRef]
6. Beloborodov, D.; Vishnyakov, A. Molecular Dynamics of Nanodroplet Coalescence in Quasi-Saturated Vapor. *Fluids* **2023**, *8*, 77. [CrossRef]
7. Croce, G.; Suzzi, N. Instability of a Film Falling Down a Bounded Plate and Its Application to Structured Packing. *Fluids* **2025**, *10*, 30. [CrossRef]
8. Hasegawa, K.; Kishimoto, Y. Fingering Instability of Binary Droplets on Oil Pool. *Fluids* **2023**, *8*, 138. [CrossRef]
9. Mohd Isa, N.S.; El Kadri, H.; Vigolo, D.; Mohamed Zakhari, N.F.A.; Gkatzionis, K. Understanding the Application of Emulsion Systems for Bacterial Encapsulation and Temperature-Modulated Release. *Fluids* **2024**, *9*, 274. [CrossRef]
10. Croce, G.; Suzzi, N. Numerical Simulation of Dropwise Condensation of Steam over Hybrid Surfaces via New Non-Dimensional Heat Transfer Model. *Fluids* **2023**, *8*, 300. [CrossRef]
11. Lakew, E.; Sarchami, A.; Giustini, G.; Kim, H.; Bellur, K. Thin Film Evaporation Modeling of the Liquid Microlayer Region in a Dewetting Water Bubble. *Fluids* **2023**, *8*, 126. [CrossRef]
12. Carvalho, R.M.; Neto, C.; Santos, L.M.N.B.F.; Bastos, M.; Costa, J.C.S. Influence of the Deposition Rate and Substrate Temperature on the Morphology of Thermally Evaporated Ionic Liquids. *Fluids* **2023**, *8*, 105. [CrossRef]
13. Taitelbaum, H. Statistical Physics Perspective on Droplet Spreading in Reac-2 tive Wetting Interfaces. *Fluids* **2025**, *10*, 170. [CrossRef]
14. Mohammad Karim, A.; Suszynski, W.J. Physics of Dynamic Contact Line: Hydrodynamics Theory versus Molecular Kinetic Theory. *Fluids* **2022**, *7*, 318. [CrossRef]
15. Wang, Z.; Wang, L.; Li, T.; Liu, S.; Guo, B.; Huang, W.; Wu, Y. 3D bioprinting in cardiac tissue engineering. *Theranostics* **2021**, *11*, 7948–7969. [CrossRef]

16. Matai, I.; Kaur, G.; Seyedsalehi, A.; McClinton, A.; Laurencin, C.T. Progress in 3D bioprinting technology for tissue/organ regenerative engineering. *Biomaterials* **2020**, *226*, 119536. [CrossRef] [PubMed]
17. Zhang, J.; Wehrle, E.; Rubert, M.; Muller, R. 3D Bioprinting of Human Tissues: Biofabrication, Bioinks, and Bioreactors. *Int. J. Mol. Sci.* **2021**, *22*, 3971. [CrossRef]
18. Shukla, P.; Yeleswarapu, S.; Heinrich, M.A.; Prakash, J.; Pati, F. Mimicking tumor microenvironment by 3D bioprinting: 3D cancer modeling. *Biofabrication* **2022**, *14*, 032002. [CrossRef]
19. Lee, A.; Hudson, A.R.; Shiwarski, D.J.; Tashman, J.W.; Hinton, T.J.; Yerneni, S.; Bliley, J.M.; Campbell, P.G.; Feinberg, A.W. 3D bioprinting of collagen to rebuild components of the human heart. *Science* **2019**, *365*, 482–487. [CrossRef]
20. Abbasov, I.B. Three-Dimensional Bioprinting of Organs: Modern Trends. *Crit. Rev. Biomed. Eng.* **2022**, *50*, 19–34. [CrossRef] [PubMed]
21. Samandari, M.; Mostafavi, A.; Quint, J.; Memic, A.; Tamayol, A. In situ bioprinting: Intraoperative implementation of regenerative medicine. *Trends Biotechnol.* **2022**, *40*, 1229–1247. [CrossRef]
22. Knowlton, S.; Anand, S.; Shah, T.; Tasoglu, S. Bioprinting for Neural Tissue Engineering. *Trends Neurosci.* **2018**, *41*, 31–46. [CrossRef] [PubMed]
23. Yu, X.; Zhang, T.; Li, Y. 3D Printing and Bioprinting Nerve Conduits for Neural Tissue Engineering. *Polymers* **2020**, *12*, 1637. [CrossRef]
24. de Melo, B.A.G.; Cruz, E.M.; Ribeiro, T.N.; Mundim, M.V.; Porcionatto, M.A. 3D Bioprinting of Murine Cortical Astrocytes for Engineering Neural-Like Tissue. *J. Vis. Exp.* **2021**, *173*, e62691. [CrossRef]
25. Bedir, T.; Ulag, S.; Ustundag, C.B.; Gunduz, O. 3D bioprinting applications in neural tissue engineering for spinal cord injury repair. *Mater. Sci. Eng. C Mater. Biol. Appl.* **2020**, *110*, 110741. [CrossRef]
26. Warren, D.; Tomaskovic-Crook, E.; Wallace, G.G.; Crook, J.M. Engineering in vitro human neural tissue analogs by 3D bioprinting and electrostimulation. *APL Bioeng.* **2021**, *5*, 020901. [CrossRef] [PubMed]
27. Yeh, Y.C.; Chen, P.Y.; Chen, K.T.; Lee, I.C. Innovative MXene/SilMA-Based Conductive Bioink for Three Dimensional Bioprinting of Neural Stem Cell Spheroids in Neural Tissue Engineering. *ACS Appl. Mater. Interfaces* **2025**, *17*, 10402–10416. [CrossRef] [PubMed]
28. Sharma, R.; Restan Perez, M.; da Silva, V.A.; Thomsen, J.; Bhardwaj, L.; Andrade, T.A.M.; Alhussan, A.; Willerth, S.M. 3D bioprinting complex models of cancer. *Biomater. Sci.* **2023**, *11*, 3414–3430. [CrossRef]
29. Olejnik, A.; Semba, J.A.; Kulpa, A.; Danczak-Pazdrowska, A.; Rybka, J.D.; Gornowicz-Porowska, J. 3D Bioprinting in Skin Related Research: Recent Achievements and Application Perspectives. *ACS Synth. Biol.* **2022**, *11*, 26–38. [CrossRef]
30. Wu, Y.; Ravnic, D.J.; Ozbolat, I.T. Intraoperative Bioprinting: Repairing Tissues and Organs in a Surgical Setting. *Trends Biotechnol.* **2020**, *38*, 594–605. [CrossRef]
31. Quadri, F.; Soman, S.S.; Vijayavenkataraman, S. Progress in cardiovascular bioprinting. *Artif. Organs.* **2021**, *45*, 652–664. [CrossRef]
32. Moldovan, N.I. Progress in scaffold-free bioprinting for cardiovascular medicine. *J. Cell. Mol. Med.* **2018**, *22*, 2964–2969. [CrossRef]
33. Sun, Z.; Zhao, J.; Leung, E.; Flandes-Iparraguirre, M.; Vernon, M.; Silberstein, J.; De-Juan-Pardo, E.M.; Jansen, S. Three-Dimensional Bioprinting in Cardiovascular Disease: Current Status and Future Directions. *Biomolecules* **2023**, *13*, 1180. [CrossRef]
34. Correia Carreira, S.; Begum, R.; Perriman, A.W. 3D Bioprinting: The Emergence of Programmable Biodesign. *Adv. Healthc. Mater.* **2020**, *9*, e1900554. [CrossRef] [PubMed]
35. Ravanbakhsh, H.; Karamzadeh, V.; Bao, G.; Mongeau, L.; Juncker, D.; Zhang, Y.S. Emerging Technologies in Multi-Material Bioprinting. *Adv. Mater.* **2021**, *33*, e2104730. [CrossRef] [PubMed]
36. Sabzevari, A.; Rayat Pisheh, H.; Ansari, M.; Salati, A. Progress in bioprinting technology for tissue regeneration. *J. Artif. Organs.* **2023**, *26*, 255–274. [CrossRef]
37. Chae, S.; Cho, D.W. Biomaterial-based 3D bioprinting strategy for orthopedic tissue engineering. *Acta Biomater.* **2023**, *156*, 4–20. [CrossRef] [PubMed]
38. Jiu, J.; Liu, H.; Li, D.; Li, J.; Liu, L.; Yang, W.; Yan, L.; Li, S.; Zhang, J.; Li, X.; et al. 3D bioprinting approaches for spinal cord injury repair. *Biofabrication* **2024**, *16*, 032003. [CrossRef]
39. Herzog, J.; Franke, L.; Lai, Y.; Gomez Rossi, P.; Sachtleben, J.; Weuster-Botz, D. 3D bioprinting of microorganisms: Principles and applications. *Bioprocess Biosyst. Eng.* **2024**, *47*, 443–461. [CrossRef]
40. Kang, M.S.; Jang, J.; Jo, H.J.; Kim, W.H.; Kim, B.; Chun, H.J.; Lim, D.; Han, D.W. Advances and Innovations of 3D Bioprinting Skin. *Biomolecules* **2022**, *13*, 55. [CrossRef]
41. Hong, N.; Yang, G.H.; Lee, J.; Kim, G. 3D bioprinting and its in vivo applications. *J. Biomed. Mater. Res. B Appl. Biomater.* **2018**, *106*, 444–459. [CrossRef]
42. Chen, E.P.; Toksoy, Z.; Davis, B.A.; Geibel, J.P. 3D Bioprinting of Vascularized Tissues for in vitro and in vivo Applications. *Front. Bioeng. Biotechnol.* **2021**, *9*, 664188. [CrossRef]
43. Caceres-Alban, J.; Sanchez, M.; Casado, F.L. Bioprinting: A Strategy to Build Informative Models of Exposure and Disease. *IEEE Rev. Biomed. Eng.* **2023**, *16*, 594–610. [CrossRef]

44. Wu, Y.; Kennedy, P.; Bonazza, N.; Yu, Y.; Dhawan, A.; Ozbolat, I. Three-Dimensional Bioprinting of Articular Cartilage: A Systematic Review. *Cartilage* **2021**, *12*, 76–92. [CrossRef] [PubMed]
45. Ballard, A.; Patush, R.; Perez, J.; Juarez, C.; Kirillova, A. Bioprinting: Mechanical Stabilization and Reinforcement Strategies in Regenerative Medicine. *Tissue Eng. Part A* **2024**, *30*, 387–408. [CrossRef] [PubMed]
46. Knowlton, S.; Onal, S.; Yu, C.H.; Zhao, J.J.; Tasoglu, S. Bioprinting for cancer research. *Trends Biotechnol.* **2015**, *33*, 504–513. [CrossRef] [PubMed]
47. De Stefano, P.; Briatico-Vangosa, F.; Bianchi, E.; Pellegata, A.F.; Hartung de Hartungen, A.; Corti, P.; Dubini, G. Bioprinting of Matrigel Scaffolds for Cancer Research. *Polymers* **2021**, *13*, 2026. [CrossRef]
48. Kang, Y.; Datta, P.; Shanmughapriya, S.; Ozbolat, I.T. 3D Bioprinting of Tumor Models for Cancer Research. *ACS Appl. Bio Mater.* **2020**, *3*, 5552–5573. [CrossRef]
49. Khorsandi, D.; Rezayat, D.; Sezen, S.; Ferrao, R.; Khosravi, A.; Zarepour, A.; Khorsandi, M.; Hashemian, M.; Iravani, S.; Zarrabi, A. Application of 3D, 4D, 5D, and 6D bioprinting in cancer research: What does the future look like? *J. Mater. Chem. B* **2024**, *12*, 4584–4612. [CrossRef]
50. Leonard, F.; Godin, B. 3D In Vitro Model for Breast Cancer Research Using Magnetic Levitation and Bioprinting Method. *Methods Mol. Biol.* **2016**, *1406*, 239–251. [CrossRef]
51. Bhardwaj, N.; Yadav, S.K.; Saneja, A. Recent advances in 3D bioprinting for cancer research: From precision models to personalized therapies. *Drug Discov. Today* **2024**, *29*, 103924. [CrossRef]
52. Sun, H.; Wang, Y.; Yang, H. Revolutionizing preclinical research for pancreatic cancer: The potential of 3D bioprinting technology for personalized therapy. *Hepatobiliary Surg. Nutr.* **2023**, *12*, 616–618. [CrossRef]
53. Yang, H.; Yang, K.H.; Narayan, R.J.; Ma, S. Laser-based bioprinting for multilayer cell patterning in tissue engineering and cancer research. *Essays Biochem.* **2021**, *65*, 409–416. [CrossRef] [PubMed]
54. Goodarzi Hosseinabadi, H.; Dogan, E.; Miri, A.K.; Ionov, L. Digital Light Processing Bioprinting Advances for Microtissue Models. *ACS Biomater. Sci. Eng.* **2022**, *8*, 1381–1395. [CrossRef] [PubMed]
55. Yang, S.; Hou, Y.; Shang, Y.; Zhong, X. BPNN and CNN-based AI modeling of spreading and icing pattern of a water droplet impact on a supercooled surface. *AIP Adv.* **2022**, *12*, 045209. [CrossRef]
56. Modak, C.D.; Kumar, A.; Tripathy, A.; Sen, P. Drop impact printing. *Nat. Commun.* **2020**, *11*, 4327. [CrossRef] [PubMed]
57. Qu, J.; Dou, C.; Xu, B.; Li, J.; Rao, Z.; Tsin, A. Printing quality improvement for laser-induced forward transfer bioprinting: Numerical modeling and experimental validation. *Phys. Fluids* **2021**, *33*, 071906. [CrossRef]
58. Dasgupta, Q.; Black, L.D. A fresh slate for 3D bioprinting. *Science* **2019**, *365*, 446–447. [CrossRef]
59. Mandrycky, C.; Wang, Z.; Kim, K.; Kim, D.H. 3D bioprinting for engineering complex tissues. *Biotechnol. Adv.* **2016**, *34*, 422–434. [CrossRef]
60. Starly, B.; Shirwaiker, R. 3D bioprinting techniques. In *3D Bioprinting Nanotechnology in Tissue Engineering Regenerative Medicine*; Zhang, L.G., Fisher, J.P., Leong, K.W., Eds.; Elsevier: Amsterdam, The Netherlands, 2015.
61. Choi, Y.; Jin, P.; Lee, S.; Song, Y.; Tay, R.Y.; Kim, G.; Yoo, J.; Han, H.; Yeom, J.; Cho, J.H.; et al. All-printed chip-less wearable neuromorphic system for multimodal physiochemical health monitoring. *Nat. Commun.* **2025**, *16*, 5689. [CrossRef]
62. Dey, M.; Ozbolat, I.T. 3D bioprinting of cells, tissues and organs. *Sci. Rep.* **2020**, *10*, 14023. [CrossRef] [PubMed]
63. Van Ombergen, A.; Chalupa-Gantner, F.; Chansoria, P.; Colosimo, B.M.; Costantini, M.; Domingos, M.; Dufour, A.; De Maria, C.; Groll, J.; Jungst, T.; et al. 3D Bioprinting in Microgravity: Opportunities, Challenges, and Possible Applications in Space. *Adv. Healthc. Mater.* **2023**, *12*, e2300443. [CrossRef] [PubMed]
64. Rezapour Sarabi, M.; Yetisen, A.K.; Tasoglu, S. Bioprinting in Microgravity. *ACS Biomater. Sci. Eng.* **2023**, *9*, 3074–3083. [CrossRef]
65. Gungor-Ozkerim, P.S.; Inci, I.; Zhang, Y.S.; Khademhosseini, A.; Dokmeci, M.R. Bioinks for 3D bioprinting: An overview. *Biomater. Sci.* **2018**, *6*, 915–946. [CrossRef] [PubMed]
66. Wang, H.; Yu, H.; Zhou, X.; Zhang, J.; Zhou, H.; Hao, H.; Ding, L.; Li, H.; Gu, Y.; Ma, J.; et al. An Overview of Extracellular Matrix-Based Bioinks for 3D Bioprinting. *Front. Bioeng. Biotechnol.* **2022**, *10*, 905438. [CrossRef]
67. Kravchenko, S.V.; Sakhnov, S.N.; Myasnikova, V.V.; Trofimenko, A.I.; Buzko, V.Y. Bioprinting technologies in ophthalmology. *Vestn. Oftalmol.* **2023**, *139*, 105–112. [CrossRef]
68. Sun, W.; Starly, B.; Daly, A.C.; Burdick, J.A.; Groll, J.; Skeldon, G.; Shu, W.; Sakai, Y.; Shinohara, M.; Nishikawa, M.; et al. The bioprinting roadmap. *Biofabrication* **2020**, *12*, 022002. [CrossRef]
69. Ikram, M.; Mahmud, M.A.P.; Kalyar, A.A.; Alomayri, T.; Almahri, A.; Hussain, D. 3D-bioprinting of MXenes: Developments, medical applications, challenges, and future roadmap. *Colloids Surf. B Biointerfaces* **2025**, *251*, 114568. [CrossRef]
70. Wu, Y.; Qin, M.; Yang, X. Organ bioprinting: Progress, challenges and outlook. *J. Mater. Chem. B* **2023**, *11*, 10263–10287. [CrossRef]
71. Lam, E.H.Y.; Yu, F.; Zhu, S.; Wang, Z. 3D Bioprinting for Next-Generation Personalized Medicine. *Int. J. Mol. Sci.* **2023**, *24*, 6357. [CrossRef]
72. McMillan, A.; McMillan, N.; Gupta, N.; Kanotra, S.P.; Salem, A.K. 3D Bioprinting in Otolaryngology: A Review. *Adv. Healthc. Mater.* **2023**, *12*, e2203268. [CrossRef]

73. Mierke, C.T. Bioprinting of Cells, Organoids and Organs-on-a-Chip Together with Hydrogels Improves Structural and Mechanical Cues. *Cells* **2024**, *13*, 1638. [CrossRef]
74. Banerjee, D.; Singh, Y.P.; Datta, P.; Ozbolat, V.; O'Donnell, A.; Yeo, M.; Ozbolat, I.T. Strategies for 3D bioprinting of spheroids: A comprehensive review. *Biomaterials* **2022**, *291*, 121881. [CrossRef] [PubMed]
75. Memic, A.; Navaei, A.; Mirani, B.; Cordova, J.A.V.; Aldhahri, M.; Dolatshahi-Pirouz, A.; Akbari, M.; Nikkhah, M. Bioprinting technologies for disease modeling. *Biotechnol. Lett.* **2017**, *39*, 1279–1290. [CrossRef] [PubMed]
76. Genova, T.; Roato, I.; Carossa, M.; Motta, C.; Cavagnetto, D.; Mussano, F. Advances on Bone Substitutes through 3D Bioprinting. *Int. J. Mol. Sci.* **2020**, *21*, 7012. [CrossRef]
77. Fransen, M.F.J.; Addario, G.; Bouten, C.V.C.; Halary, F.; Moroni, L.; Mota, C. Bioprinting of kidney in vitro models: Cells, biomaterials, and manufacturing techniques. *Essays Biochem.* **2021**, *65*, 587–602. [CrossRef] [PubMed]
78. Chawla, S.; Midha, S.; Sharma, A.; Ghosh, S. Silk-Based Bioinks for 3D Bioprinting. *Adv. Healthc. Mater.* **2018**, *7*, e1701204. [CrossRef]
79. Samadi, A.; Moammeri, A.; Pourmadadi, M.; Abbasi, P.; Hosseinpour, Z.; Farokh, A.; Shamsabadipour, A.; Heydari, M.; Mohammadi, M.R. Cell Encapsulation and 3D Bioprinting for Therapeutic Cell Transplantation. *ACS Biomater. Sci. Eng.* **2023**, *9*, 1862–1890. [CrossRef]
80. Di Piazza, E.; Pandolfi, E.; Cacciotti, I.; Del Fattore, A.; Tozzi, A.E.; Secinaro, A.; Borro, L. Bioprinting Technology in Skin, Heart, Pancreas and Cartilage Tissues: Progress and Challenges in Clinical Practice. *Int. J. Environ. Res. Public Health* **2021**, *18*, 10806. [CrossRef]
81. Li, P.; Zhang, W.; Smith, L.J.; Ayares, D.; Cooper, D.K.C.; Ekser, B. The potential role of 3D-bioprinting in xenotransplantation. *Curr. Opin. Organ Transplant.* **2019**, *24*, 547–554. [CrossRef]
82. Murphy, S.V.; De Coppi, P.; Atala, A. Opportunities and challenges of translational 3D bioprinting. *Nat. Biomed. Eng.* **2020**, *4*, 370–380. [CrossRef]
83. Skeldon, G.; Lucendo-Villarin, B.; Shu, W. Three-dimensional bioprinting of stem-cell derived tissues for human regenerative medicine. *Philos. Trans. R. Soc. B Biol. Sci.* **2018**, *373*, 20170224. [CrossRef]
84. Tharakan, S.; Khondkar, S.; Ilyas, A. Bioprinting of Stem Cells in Multimaterial Scaffolds and Their Applications in Bone Tissue Engineering. *Sensors* **2021**, *21*, 7477. [CrossRef]
85. Ali, M.; PR, A.; Lee, S.J.; Jackson, J.D. Three-dimensional bioprinting for organ bioengineering: Promise and pitfalls. *Curr. Opin. Organ Transplant.* **2018**, *23*, 649–656. [CrossRef]
86. Sabetkish, S.; Currie, P.; Meagher, L. Recent trends in 3D bioprinting technology for skeletal muscle regeneration. *Acta Biomater.* **2024**, *181*, 46–66. [CrossRef]
87. Holland, I. Extrusion bioprinting: Meeting the promise of human tissue biofabrication? *Prog. Biomed. Eng.* **2025**, *7*, 023001. [CrossRef]
88. Kim, D.Y.; Liu, Y.; Kim, G.; An, S.B.; Han, I. Innovative Strategies in 3D Bioprinting for Spinal Cord Injury Repair. *Int. J. Mol. Sci.* **2024**, *25*, 9592. [CrossRef]
89. Li, Y.; Wu, J.; He, C.; He, H.; Xie, M.; Yao, K.; He, J.; Duan, Y.; Zhaung, L.; Wang, P.; et al. 3D Prestress Bioprinting of Directed Tissues. *Adv. Healthc. Mater.* **2023**, *12*, e2301487. [CrossRef]
90. Germain, N.; Dhayer, M.; Dekiouk, S.; Marchetti, P. Current Advances in 3D Bioprinting for Cancer Modeling and Personalized Medicine. *Int. J. Mol. Sci.* **2022**, *23*, 3432. [CrossRef]

fluids

MDPI

Review

Physics of Dynamic Contact Line: Hydrodynamics Theory versus Molecular Kinetic Theory

Alireza Mohammad Karim [1,*] and Wieslaw J. Suszynski [2]

1 Nanoscience Centre, Department of Engineering, University of Cambridge, 11 J. J. Thomson Avenue, Cambridge CB3 0FF, UK
2 Department of Chemical Engineering and Materials Science, University of Minnesota, 421 Washington Ave. SE, Minneapolis, MN 55455, USA
* Correspondence: am2633@cam.ac.uk

Abstract: The dynamic contact line plays a key role in various fields of interfacial physics, including bioprinting, nano-scale printing, three-dimensional printing, biomaterials, tissue engineering, smart materials, flexible printed electronics, biomedicine, and healthcare. However, there is still a lack of thorough physical understanding of its real behavior in numerous complex problems in nature and technology. The dynamic contact line exhibits a complex conformation in real-life fluid dynamics problems. Therefore, this review presents two main long-standing models that describe the physics of the dynamic contact line: hydrodynamics theory and molecular kinetics theory. Next, the role of the dynamic contact line in current advanced technologies is discussed. Finally, this review discusses future research directions to enhance the power of current physical models of the dynamic contact line.

Keywords: hydrodynamics; molecular kinetics; contact angle; contact line velocity; flexible substrate; superhydrophobic substrate

1. Introduction

When a liquid is in contact with a substrate in a gaseous environment, a contact line is formed. The contact line, also known as the three-phase contact line, is a line of intersection of three phases: solid, liquid, and gas. The contact line moves on the substrate until the total energy of the contact line reaches its minimum value, which is the thermodynamic equilibrium state. In this dynamic condition, the contact line is defined as the dynamic contact line [1–22].

The physics of the dynamic contact line has been an important subject in the field of interfacial science. The physics of the dynamic contact line is a multidisciplinary topic, and it involves numerous scientific areas, including anti-icing of air planes, various conventional coatings, smart coatings, functional coatings, inkjet printing, bio-printing, tissue engineering, functional organs, flexible printed electronics, biomaterials, smart materials, microfluidics, nanofluidics, phase-change heat exchangers, oil recovery, power plants, and various biological systems [2–14,23–60].

The angle between the liquid–air interface and liquid–substrate interface at the dynamic contact line is known as the dynamic contact angle (θ_D). When the dynamic contact line reaches the equilibrium state, the dynamic contact angle becomes the static contact angle (θ_0) [61,62]. The static contact angle is defined by the balance of the interfacial tensions at the contact line, as demonstrated by Figure 1 and formulated by Equation (1) [63]:

$$\sigma_{LG} \cos \theta_0 = \sigma_{SG} - \sigma_{SL} \quad (1)$$

in which σ_{LG} is the interfacial tension at the liquid–gas interface, σ_{SG} is the interfacial tension at the solid–gas interface, and σ_{SL} is the interfacial tension at the solid–liquid interface.

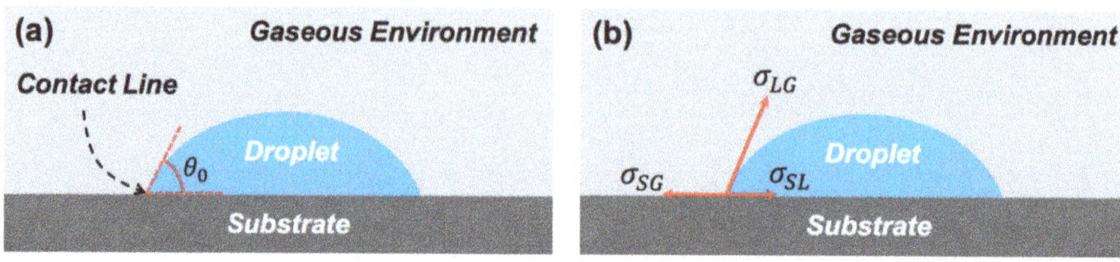

Figure 1. (**a**) Schematic representation of the contact line and the static contact angle. (**b**) A droplet, deposited on a substrate, is in the equilibrium state with the balance of the interfacial tensions along the contact line.

There are two mechanisms for contact line dynamics: spontaneous and forced [64–66]. In the spontaneous mechanism, the contact line moves on a substrate to reach the equilibrium state without any external driving force. In spontaneous mode, the capillary force and gravity play key roles in inducing the motion of the contact line on the substrate to reduce the dynamic contact angle to achieve the minimum energy value in the equilibrium state (static contact angle). For the case of a droplet spreading on a substrate, the gravitational effect is negligible as the droplet size is less than the capillary length ($l_{cap} = \sqrt{\sigma/(\rho g)}$), where σ is the droplet surface tension, ρ is the droplet density, and g is the gravitational acceleration [67].

In the forced mechanism, the contact line dynamics are governed by an external driving force (or forces). In forced contact line dynamics, the balance of the forces is between the tendency of the contact line to achieve the equilibrium condition and the external force(s) to induce the contact line to deviate from the equilibrium condition [68]. There are two modes of the forced contact line dynamics: advancing and receding. In advancing contact line dynamics, the dynamic contact angle is denoted as the advancing dynamic contact angle (θ_A). In receding contact line dynamics, the dynamic contact angle is denoted as the receding dynamic contact angle (θ_R).

The physics of the dynamic contact line is conventionally defined in terms of the relation between the dynamic contact angle (θ_D) and the contact line velocity (U), as represented in Figure 2 [19,69–72]. It is important to note that the contact line velocity is not necessarily constant in the contact line dynamics, as shown in Figure 2. It depends on the mode of contact line motion: spontaneous or forced. In forced contact line motion, the contact line velocity can be controlled to be constant. In spontaneous contact line motion, the contact line velocity might not be constant, such as a droplet spreading on a smooth glass surface in which the contact line velocity reduces as the dynamic contact angle decreases until the droplet reaches the equilibrium state. In the physical relation between the dynamic contact angle and the contact line velocity, the equilibrium advancing contact angle (θ_{0A}) and equilibrium receding contact angle (θ_{0R}) are also important [68–72]. Various experimental techniques have been used to measure the contact angle, including capillary rise, the Wilhelmy plate, a continuous vertical fiber, and a continuous moving tape [73–79].

The difference in value between the equilibrium of the advancing and receding contact angles is known as the contact angle hysteresis, as shown in Figure 2. The contact angle hysteresis signifies the heterogeneity in substrate roughness and substrate chemistry [80–90]. The contact angle hysteresis causes the pinning event and stick–slip behavior of the dynamic contact line on the substrate.

The stick–slip behavior is the phenomenon in which the dynamic contact line remains stationary for most of the time, and only from time to time does it move rapidly and abruptly [91]. The pinning event is a phenomenon in which the dynamic contact line does not move until the contact angle reaches the critical values (critical advancing and receding dynamic contact angles). The difference between the critical advancing and receding dynamic contact angles is defined as the contact angle hysteresis, which features the pinning effect. The pinning effect prevents the dynamic contact line from moving over the substrate [92].

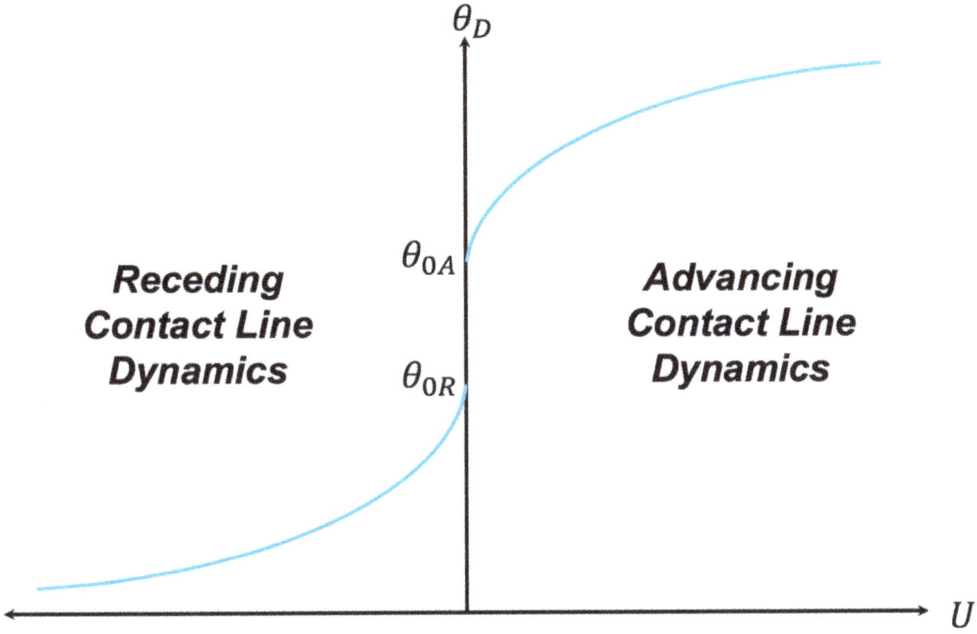

Figure 2. Relation between contact line velocity and dynamic contact angle for advancing contact line dynamics and receding contact line dynamics.

Numerous studies have attempted to reach a physical understanding of the underlying mechanism of the dynamic contact line. Several theories have been proposed to define the physics of the moving contact line by determining the relation between the dynamic contact angle and the contact line velocity [21,22,64–66,68–72,93–122]. Among them, two long-standing theories have been largely accepted in the interfacial physics community: hydrodynamics and molecular kinetics.

Regarding the significance of the dynamic contact line in various scientific fields and technologies, a recent comprehensive review outlined the challenges of the computational approach to describe the physics of the dynamic contact line [123]. There are several well-established computational models that numerically describe the flow physics of the dynamic contact line [123]. These computational models are based on direct numerical simulations of the dynamic contact line using the volume of the fluid technique [123]. There have been advancements in the volume-of-fluid-based computations to describe the dynamics of a moving contact line [123]. One of the advancements in the field of numerical simulation of the dynamic contact line was the combination of the simulation of the flow physics in a macroscopic region with the theory of hydrodynamics in the proximity of the dynamic contact line. The other main advancement in this field was the computational explanation of the dynamic contact line in a microscopic region by considering the van der Waals force [123].

The physics of the dynamic contact line involves multidisciplinary research. Physically understanding the dynamic contact line problems, such as flexible printed electronics, biosensors, functional smart nano-bio-materials, and printable solar cells, requires collaboration between numerous fields, including electrical engineering, materials science, mechanical engineering, applied physics, applied mathematics, biology, chemistry, and nanoscience [50,124]. Moreover, the dynamic contact line is a multi-scale physical problem that requires deep insights from all scales to define the true physics of the dynamic contact line (Figure 3). At the micro scale, a thin liquid layer known as the precursor film forms on a substrate, which advances the dynamic contact line [124]. At the macro scale, the hydrostatic pressure from the bulk of the liquid affects the physics of the dynamic contact line to achieve equilibrium [124]. At the mesoscopic scale, the physics of the dynamic contact line are governed by hydrostatic pressure and disjoining pressure [124].

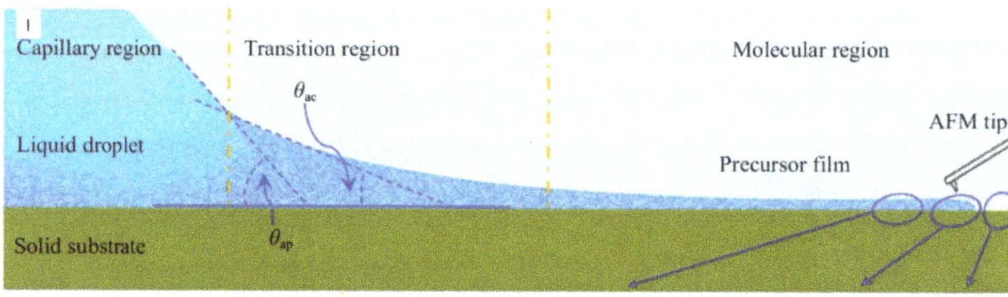

Figure 3. Multi-scale representation of the physics of contact line dynamics [124].

In the following sections, the role of the dynamic contact line in some current advanced technologies, including printings/coatings and biotechnology/healthcare, is demonstrated. After, the two long-standing physical theories that have been frequently used to describe the physics of the dynamic contact line are thoroughly reviewed. These two theories are hydrodynamics and molecular–kinetic. Moreover, the drawbacks and advantages of both theories are discussed. Finally, future research directions to enhance the power of these two theories to explain the physics of the dynamic contact line in real-life problems are discussed.

2. Applications of the Dynamic Contact Line

2.1. *Coating and Printing*

The dynamic contact line has a remarkable role in numerous areas of printing, covering conventional inkjet printing, bioprinting, three-dimensional printing, soft materials, nanoprinting, tissue engineering, functional organs, and smart materials. Furthermore, the dynamic contact line has a prominent role in different coatings modes in various areas of technology, including anti-icing of airplanes, smart materials, and the automobile industry [27,125–131].

Droplet-based printing on soft materials is of significant interest in leading-edge state-of-the-art technologies, including modern, flexible printed electronics; robotics; tissue engineering; functional organs; and soft biomaterials, as demonstrated in Figure 4 [132]. In droplet-based printing for biotechnology/healthcare and medical purposes, droplets contain biological solutions such as DNA samples, living cells, viruses, bacteria, and proteins. In droplet-based printing for electronics purposes, droplets consist of electrical-conducting solutions, such as silver inks and polymeric samples.

Coatings and printings of evaporative droplets on various substrates show remarkable applications in numerous technologies, such as self-assembly of volatile suspensions of nanoparticles, flexible electronics products, soft materials, photovoltaics products, and biosensors [133–137].

Contact line dynamics has remarkable usefulness in advanced state-of-the-art biotechnology for the production of new-generation biosensors [138,139]. Moreover, contact line dynamics plays an important role in the diagnosis of various diseases via the imaging of a spreading blood droplet and its drying process at various steps, including flow, adhesion, gelation, and fracturing [126,137].

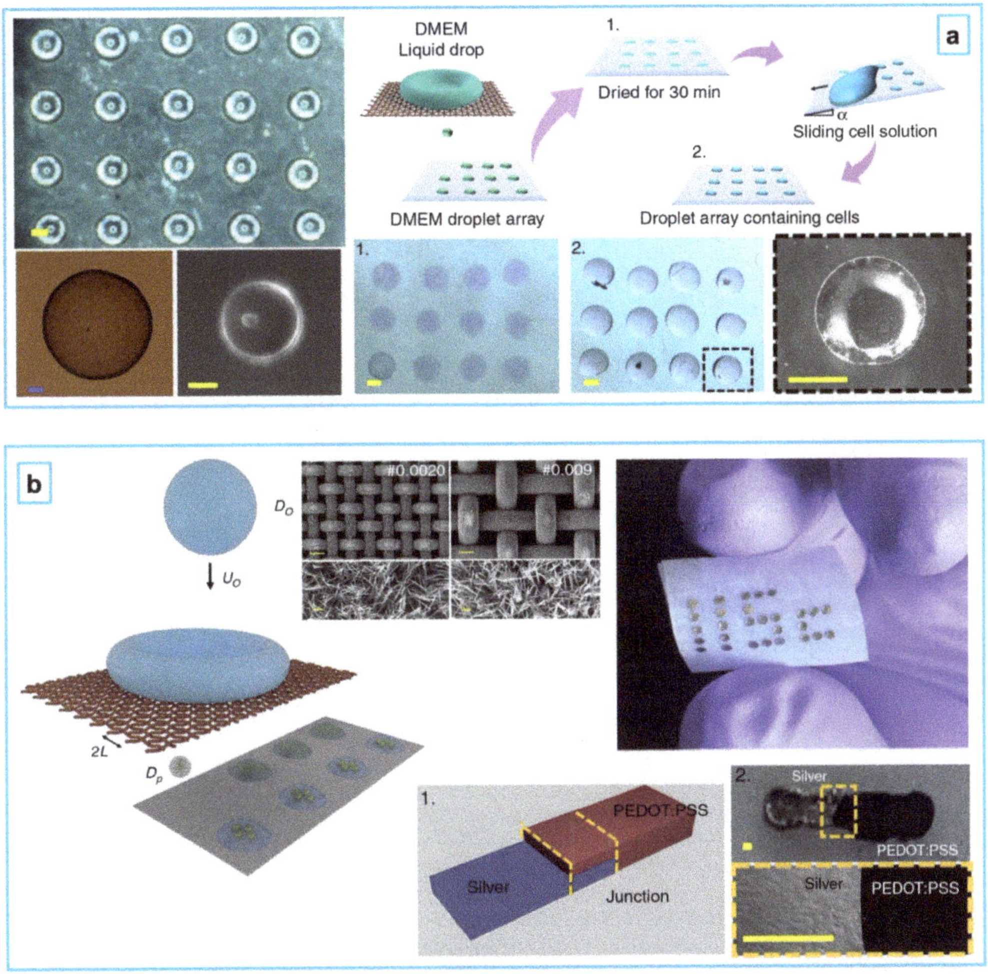

Figure 4. (**a**) Droplet-based bioprinting of droplets containing biological solutions to produce biomaterials [132]. (**b**) Droplet-based printing of electrically conducting droplets containing silver inks on soft materials to produce soft elastic printed electronics [132].

2.2. Healthcare

Due to the seriousness of the COVID-19 pandemic, the dynamic contact line of respiratory droplets on various substrates and soft materials such as face masks has a remarkable role in the fight against the spread of the coronavirus in the public (Figure 5). Currently, the scientific community in various fields is trying to develop a physical understanding of the interactions of SARS-CoV-2-virus-laden droplets with diverse materials. The interfacial interactions between virus-laden droplets on substrates with various wetting properties, such as hydrophilic, hydrophobic, and superhydrophobic surfaces, have been investigated. Moreover, recent studies have attempted to investigate the interactions of virus-contained droplets on materials with different mechanical properties, such as rigid, soft, and elastic substrates [140–142].

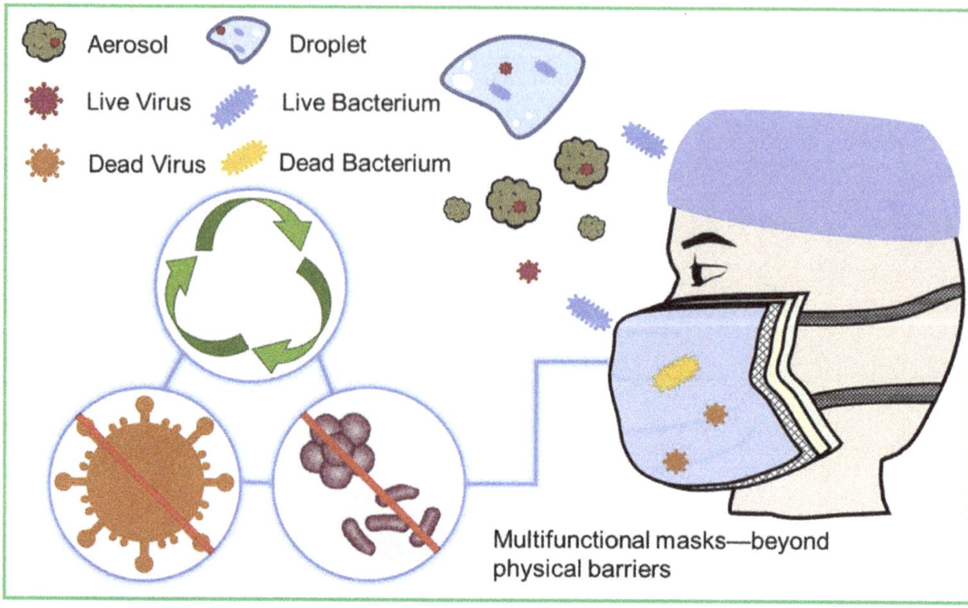

Figure 5. Significance of SARS-CoV-2-virus-containing droplet interfacial interaction with a soft superhydrophobic material such as a face mask [142].

It is important to note that extensive, full-scale computational efforts have been made to clarify the underlying physical process of droplet motion in various droplet-based applications such as droplet-based microfluidics in Polymerase Chain Reaction continuous flow and single-cell DNA analysis [143].

3. Molecular Kinetic Theory

Molecular kinetic theory explains the molecular dynamics of the liquid–air interface at the vicinity of the dynamic contact line in terms of surface diffusion and adjacent bulk phases [68,95,101]. Molecular kinetic theory considers a solid surface containing many adsorption sites over which the liquid molecules move. The liquid molecular motion near the dynamic contact line is dependent on potential barriers. Molecular kinetic theory explains the physics of the dynamic contact line according to the strength of the potential barriers at the vicinity of the contact line.

Molecular kinetic theory outlines the physical process of contact line motion on a substrate in terms of the attachment of the liquid molecules on the adsorption sites of the solid surface as the gas molecules of the environment detach from the adsorption sites, as represented in Figure 6 [95,101]. Molecular kinetic theory regards the balance of interfacial tensions with friction (F_f) at the dynamic contact line (Equation (2)):

$$F_f = \sigma\,(\cos\theta_0 - \cos\theta_D) \tag{2}$$

Therefore, molecular kinetic theory describes the physics of contact line dynamics on a substrate in terms of molecular motion near the contact line as a result of energy dissipation due to friction [101]. It is assumed that the substrate has n number of adsorption sites per unit surface area over which the molecules move. The adsorption sites are equally distant by λ. The number density of adsorption sites (n) and the corresponding distance between them (λ) are related by $n = \frac{1}{\lambda^2}$. The frequency of molecular motion near the dynamic contact line is denoted by K_0.

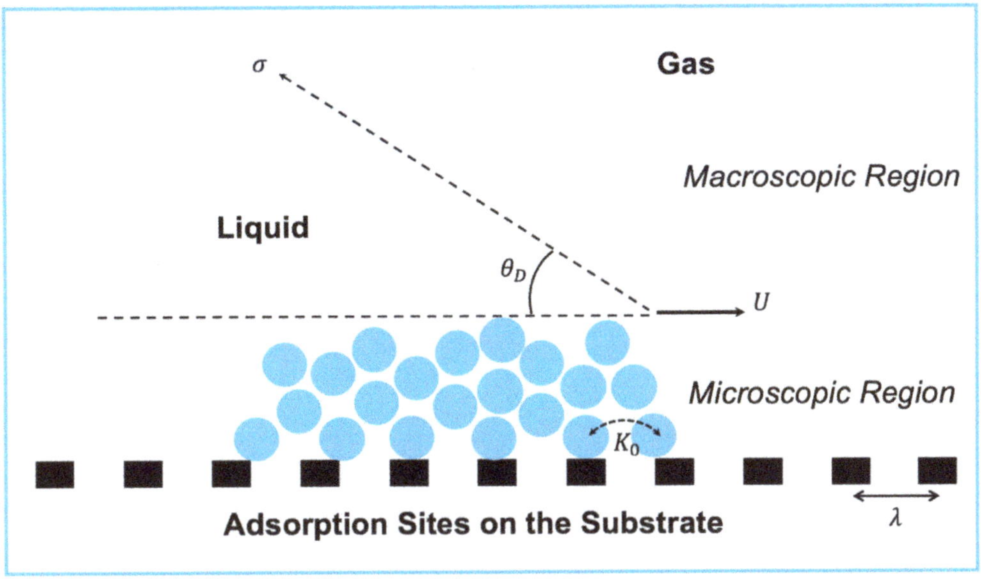

Figure 6. Schematic representation of molecular kinetic theory to explain the physics of dynamic contact line based on two parameters: λ and K_0.

Molecular kinetic theory shows the relation between the velocity (U) of the dynamic contact line and the dynamic contact angle (θ_D) in terms of the molecular kinetic parameters λ and K_0, as represented by (Equation (3)) [95,101]:

$$U = 2\,\lambda\,K_0 \sinh\left(\frac{\pm\sigma\,\lambda^2\,(\cos\theta_0 - \cos\theta_D)}{2\,k_B\,T}\right) \tag{3}$$

in which T presents the temperature and k_B represents the Boltzmann constant. In Equation (3), the positive sign stands for the dynamic contact line advancement on a substrate, and the negative sign denotes that the dynamic contact line recedes on the substrate.

The frequency (K_0) of molecular motion at the vicinity of the dynamic contact line is described by Eyring's theory of absolute reaction rates [144], as shown in Equation (4):

$$K_0 = \frac{k_B\,T}{h} exp\left(\frac{-\Delta G}{N_A\,k_B\,T}\right) \tag{4}$$

in which ΔG stands for activation free energy, related to adsorption energy; N_A denotes Avogadro's number; and h represents Planck's constant. It is necessary to know that there is no definitive method to specify the molecular kinetic parameters (λ and K_0) to describe the physics of the dynamic contact line. As a result, their values are normally estimated by fitting the experimental data with the molecular kinetic model (Equation (3)) based on known values for surface tension (σ), temperature (T), and the static contact angle (θ_0).

4. Hydrodynamics Theory

Hydrodynamics theory characterizes the physics of flow near the dynamic contact line. However, there has been a contradiction between the motion of the liquid at the dynamic contact line and the long-established no-slip boundary condition at the liquid–substrate interface. This is due to the fact that the shear stress and pressure diverge to infinity as the flow approaches the dynamic contact line [70]. Moreover, the total force applied on the substrate due to liquid flow becomes unbounded as the flow nears the dynamic contact line [70].

This issue in hydrodynamics theory was resolved by relaxing the well-established no-slip boundary condition, along the liquid–substrate interface, at the proximity of the dynamic contact line over the distance called the slip length [70,93,103–105]. This method resolves the issue of divergence of the force by the flow on the solid surface at the dynamic contact line. However, the shear stresses are still not finite at the dynamic contact line [68]. This approach adapts hydrodynamics theory to explain the physics of the dynamic contact line without divergence of the force applied on the substrate at the contact line. It is important to note that adapted hydrodynamics theory is only suitable for flow at low capillary numbers so that the liquid-free surface geometry maintains its static condition away from the dynamic contact line [68].

Hydrodynamics theory explains the physics of the dynamic contact line based on viscous bending of the liquid-free surface geometry, which is formed at the mesoscopic region of the flow. The dynamic contact angle in the macroscopic region of the flow is determined by the extrapolation of the static liquid-free surface all the way to the substrate [68]. Therefore, hydrodynamics theory defines the physics of the dynamic contact line using the matched asymptotic expansions approach, as represented in Figure 7 [68,122].

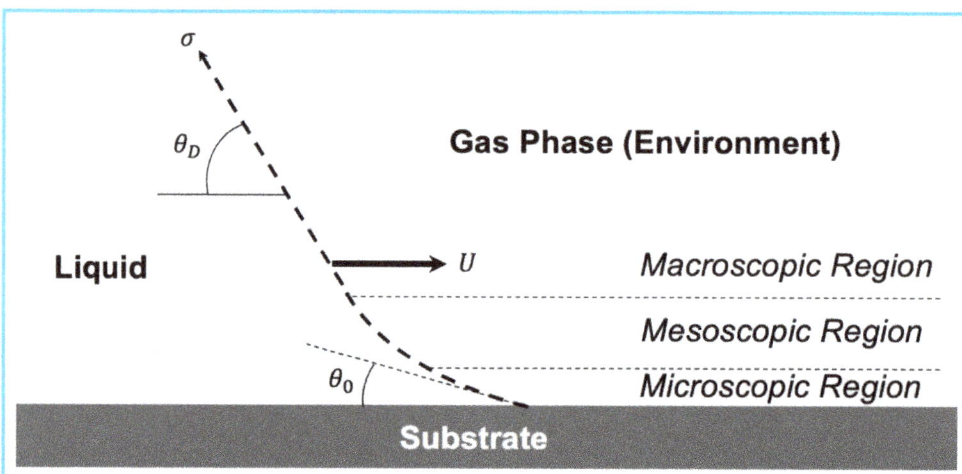

Figure 7. Schematics of hydrodynamics theory to explain the physics of dynamic contact line due to viscous bending near the contact line.

The relation between the capillary number and the dynamic contact angle, θ_D, in the macroscopic region, during the motion of a liquid-free interface on a substrate, is shown in Equation (5) [71]:

$$\frac{1}{2}\int_{\theta_0}^{\theta_D} \frac{\beta - \sin\beta \ \cos\beta}{\sin\beta} d\beta = \pm\, Ca \ \ln\left(\frac{L}{L_s}\right) \tag{5}$$

Equation (5) is valid for a steady state two-dimensional flow of a viscous liquid on a smooth horizontal solid surface at a low Reynolds number. In Equation (5), L denotes the macroscopic characteristic length and L_s represents the microscopic characteristic length, also called the slip length. θ_0 presents the static contact angle in the microscopic region, and θ_D represents the dynamic contact angle in the macroscopic region. Ca denotes the capillary number ($Ca = \mu\, U/\sigma$), in which σ is the surface tension of the liquid, μ is the shear viscosity of the liquid, and U presents the dynamic contact line velocity.

Lubrication assumption is assumed for the Navier–Stokes equations for the flow of the viscous liquid on the solid surface [71]. A steady-state wedge shape of the viscous liquid at the proximity of the dynamic contact line is also assumed. Therefore, Voinov (1976) formulated hydrodynamics theory to describe the physics of the dynamic contact line using Equations (6) and (7) [71]:

$$\theta_D{}^3 - \theta_0{}^3 = \pm 9\, Ca\, \ln\left(\frac{L}{L_s}\right);\ \theta_D \leq \frac{3\pi}{4} \tag{6}$$

$$\frac{9\pi}{4}\ln\left(\frac{1-\cos\theta_D}{1+\cos\theta_D}\right) + (\pi-\theta_D)^3 - \theta_0{}^3 = \pm 9\, Ca\, \ln\left(\frac{L}{L_s}\right);\ \theta_D \geq \frac{3\pi}{4} \tag{7}$$

Here, the positive sign stands for the advancement of the dynamic contact line on a smooth substrate, and the negative sign denotes the receding of the dynamic contact line on the substrate. Similar findings were obtained by Cox's theoretical study [69] in which gas viscosity was assumed to be negligible. Therefore, Equations (6) and (7) are called "Cox-Voinov" equations of hydrodynamics theory and are used for describing the physics of the dynamic contact line.

The characteristic length for the macroscopic region, L, is directly related to the capillary length (L_{cap}): $L \propto L_{cap}$. The proportionality constant depends on several factors, including the flow condition, fluid boundary conditions, and capillary number of the flow [8].

The microscopic characteristic length (slip length) is used to remove the long-established stress singularity at the dynamic contact line. It is shown to be dependent on the speed of the dynamic contact line and the liquid physical characteristics, and its magnitude is estimated on the order of several angstroms [2]. The magnitude of the microscopic characteristic length also depends on other parameters, including surface roughness, substrate geometry, solid surface chemistry, flow condition, and shear rate [145,146]. The prediction of the physics of the dynamic contact line by hydrodynamics theory is strongly influenced by the microscopic characteristic length [145,146].

Numerous studies on the development of hydrodynamics theory to describe the physics of the dynamic contact line have provided experimental, theoretical, and numerical attempts. Table 1 summarizes the list of alternative forms of hydrodynamics theory that explain the physics of the dynamic contact line for various conditions of flow and substrate.

Table 1. Summary of alternative forms of hydrodynamics theory in experimental, theoretical, and numerical studies.

Model	Equation	Condition(s)	Reference
Hoffman	$\theta_D = \cos^{-1}\left[1 - 2\tanh\left(5.16\left(\frac{Ca}{1+1.31\,Ca^{0.99}}\right)^{0.706}\right)\right]$	Experimental study; forced motion of silicone oil on a glass capillary; applicable for nonvolatile liquids with zero static contact angle; negligible inertia and gravity; flow with low capillary number	[106]
Tanner	$\theta_D{}^3 \propto Ca;\ R(t) \sim t^{1/10};\ \theta_D(t) \sim t^{3/10};$	Experimental study and scaling analysis of spontaneous motion of a viscous liquid droplet on a horizontal substrate; steady-state flow; proportionality constant depends on physical characteristics of liquids and flow; based on lubrication assumption; $R(t)$ is the time-varying droplet radius of the droplet and $\theta_D(t)$ is the time-varying dynamic contact angle; suitable for small dynamic contact angles	[107]
Hoffman–Voinov–Tanner	$\theta_D{}^3 \propto Ca$	Dynamics of contact line motion at low capillary numbers on a substrate with zero static contact angle; microscopic region of the flow is excluded; also suitable for flow with finite (non-zero) static contact angle; derived by numerical, experimental, and theoretical studies; also applicable for physics of viscous evaporative droplet spreading on a substrate	[71,106,107,147–152]

Table 1. *Cont.*

Model	Equation	Condition(s)	Reference
Generalized Hoffman–Voinov–Tanner	${\theta_D}^3 - {\theta_0}^3 \propto Ca$	Generalized hydrodynamics theory by Hoffman study; considers non-zero static contact angle; proportionality constant depends on physical characteristics of liquid	[71,106,107]
de Gennes	$\theta_D({\theta_0}^2 - {\theta_D}^2) = 6\,Ca\,\ln\left(\frac{L}{L_s}\right)$	Experimental/theoretical work on liquid movement on a smooth substrate; non-zero finite static contact angle; similar to generalized "Hoffman-Voinov-Tanner" model; slip length is the length of microscopic part of the dynamic contact line in which slip condition is valid; lubrication assumption is applied in the analysis; describes the physics of dynamic contact line for flow at low capillary numbers in the mesoscopic region	[72]
Cox	$\mathcal{G}(\theta_D,\zeta) - \mathcal{G}(\theta_0,\zeta) = \pm Ca\,\ln\left(\frac{L}{L_s}\right); \mathcal{G}(\theta,\zeta) = \int_0^\theta \frac{d\vartheta}{\mathcal{F}(\vartheta,\zeta)};$ $\mathcal{F}(\vartheta,\zeta) =$ $\frac{2\sin\vartheta\left[\zeta^2\left(\vartheta^2-(\sin\vartheta)^2\right)+2\,\zeta\left[\vartheta(\pi-\vartheta)+(\sin\vartheta)^2\right]+\left[(\pi-\vartheta)^2+(\sin\vartheta)^2\right]\right]}{\left[\zeta\left(\vartheta^2-(\sin\vartheta)^2\right)[(\pi-\vartheta)+(\sin\vartheta\cos\vartheta)]+\left[\left[(\pi-\vartheta)^2-(\sin\vartheta)^2\right](\vartheta-\sin\vartheta\cos\vartheta)\right]\right]};$ ζ denotes the shear viscosity ratio; positive sign is for advancing dynamic contact line; and negative sign is for receding dynamic contact line	Theoretical analysis; applies slip condition to resolve stress singularity at the dynamic contact line; slip length is the length of microscopic part of the dynamic contact line in which slip condition is valid	[69,108]
Combined hydrodynamics–molecular kinetic	${\theta_D}^3 - (\theta_0(U))^3 = \pm\,9\,Ca\,\ln\left(\frac{L}{L_s}\right)$: positive sign stands for advancing dynamic contact line, and negative sign presents receding dynamic contact line $\theta_0(U) = \cos^{-1}\left[\cos\theta_Y \pm \left(\frac{2\,k_B\,T}{\sigma\,\lambda^2}\right)\sinh^{-1}\left(\frac{U}{2\,\kappa^0\,\lambda}\right)\right]$: θ_Y is Young's static contact angle for stationary condition; positive sign is for receding liquid motion, and negative sign denotes advancing flow $R(t) \sim t^{1/7}; \theta_D(t) \sim t^{3/7}$ $R(t)$ is the time-varying droplet radius, and $\theta_D(t)$ denotes transient dynamic contact angle in the case of droplet spreading on a substrate	Modified hydrodynamics theory proposed by Voinov (1976) with contact line velocity-dependent static contact angle in which static contact angle is defined by molecular kinetic theory proposed by Blake et al. (1969) and Blake et al. (1993); three parameters need to be adjusted to describe the physics of dynamic contact line: λ, κ^0, and L_s the combined model is only valid for slow dynamic contact line; not an appropriate model for describing the physics of dynamic contact line for high-speed coatings; a promising model for physics of dynamic contact line on real materials with variable hydrophobicity and diverse contact angle hysteresis	[68,109–113,153]
Brochard-Wyart and de Gennes	$U = \frac{\sigma\,(\cos\theta_0 - \cos\theta_D)}{\frac{k_B\,T}{\kappa^0\,\lambda^3} + \frac{6\,\mu}{\theta_D}\ln\left(\frac{L}{L_s}\right)}$	Similar to combined hydrodynamics–molecular kinetic theory; receding dynamic contact line is an irreversible process with energy dissipation caused by out-of-balance surface tension and contact line speed; the equation is obtained considering energy loss due to viscous flow near the contact line and friction force; the equation is valid for flow at low capillary numbers	[110,111]

5. Molecular Kinetic Theory: Advantages and Limitations

Molecular kinetic theory is a qualitative approach to describe the physics of the dynamic contact line by examining the energy dissipation at the proximity of the dynamic contact line due to friction force. Molecular kinetic theory is less suitable to precisely describe the physics of dynamic contact line problems compared to hydrodynamics theory. Furthermore, molecular kinetic theory is qualitatively a perfect theory to outline the physics of the dynamic contact line on a solid surface with heterogeneity in chemistry or roughness [50,124,154]. Molecular kinetic theory is normally applied for a dynamic contact line with large contact angles. Molecular kinetic theory was reported to be appropriate for explaining the physics of forced movement of the dynamic contact line on rough substrates and smooth inclined substrates [55,155–158]. Molecular kinetic theory was also shown to be a promising model for describing the physics of the dynamic contact line in the area of electrowetting and cell-adhesion problems [50,124,154].

It is important to note that molecular kinetic theory normally hypothesizes the physics of the dynamic contact line. However, it was shown that in the case of contact line dynamics of a superfluid on a substrate, molecular kinetic theory successfully describes the relationship between the dynamic contact angle and contact line velocity at various temperatures [101,102]. This result demonstrates the strength of molecular kinetic theory in defining the physics of the dynamic contact line by considering the role of the environment temperature. Moreover, it was reported that molecular kinetic theory is a suitable theory for defining the dynamic contact line on a substrate with nano-size heterogeneities that cause pinning–depinning events [101,102]. This result highlights the possibility of the successful role of molecular kinetic theory in describing the physics of the dynamic contact line on a substrate with orderly or non-orderly nano-size pores.

To apply molecular kinetic theory to define the physics of the dynamic contact line, obtaining extensive experimental results without any noise is required so that the fitting of the data with the molecular kinetic model is successful in perfectly estimating the molecular kinetic parameters. This requirement is a significant challenge for many experimental studies of the dynamic contact line. Therefore, two factors are important in applying molecular kinetic theory to explain the physics of the dynamic contact line: reproducibility and reliability of the estimated molecular kinetic parameters and level of uncertainty in the fitting of data with the molecular kinetic model. Another drawback of molecular kinetic theory is its limitation to the microscopic region. Molecular kinetic theory does not consider the mesoscopic and macroscopic regions of the dynamic contact line [95]. Molecular kinetic theory is not suitable for predicting the physics of the dynamic contact line of non-Newtonian liquids [16,18,19,57,68,159–161].

6. Hydrodynamics Theory: Advantages and Limitations

On the other hand, hydrodynamics theory normally quantitatively explains the physics of the dynamic contact line, and it is mostly used for describing the physics of dynamic contact lines with small contact angles. Hydrodynamics theory assumes lubrication in the Navier–Stokes equations. Hydrodynamics theory concludes that dissipation mainly occurs in the bulk of flow due to viscosity and presumes that dissipation at the contact line is not significant. Hydrodynamics theory was shown to be a suitable theory for describing the physics of forced motion of the dynamic contact line on a chemically treated substrate [1].

In general, hydrodynamics was demonstrated to be a suitable theory for defining the physics of spontaneous liquid motion (i.e., Newtonian and non-Newtonian) on a substrate [22,46,47,162–168]. It was reported that hydrodynamics is an appropriate theory for describing the physics of motion of a liquid on biological materials, synthetic superhydrophobic surfaces, bio-inspired materials, superoleophobic materials, biomaterials, and smart substrates [50,122,160,169,170].

Hydrodynamics theory also has some weaknesses. Hydrodynamics theory is generally applicable to the physics of Newtonian liquid motion on homogeneous surfaces [114,171–176]. Hydrodynamics theory does not perfectly describe the physics of the dynamic contact line for the forced motion of a liquid on heterogeneous materials with roughness and heterogeneous surface chemistry [50,93,114,160,169–183]. Additionally, hydrodynamics theory is not suitable for explaining the physics of the dynamic contact line for the motion of complex fluids on real substrates with natural heterogeneous roughness and complex surface chemistry [114,122,171–178].

Furthermore, hydrodynamics theory does not consider the role of micro- or nano-scale pores on substrates when describing the physics of the dynamic contact line. These drawbacks significantly limit the applicability of hydrodynamics theory in describing the physics of the dynamic contact line on real interfacial problems [15,20,46,47,50,60,89,114,140,141,160,169–178,184–198].

7. Future Perspectives

Molecular kinetic theory and hydrodynamics theory necessitate estimating physical parameters: the rate of molecular activity near the contact line, the distance between adsorptions sites of the substrate, and the microscopic characteristics length. These parameters are only estimated and not exact. Therefore, both theories cannot provide a real physical understanding of the mechanism of the dynamic contact line.

As hydrodynamics theory is mostly suitable for the macroscopic and mesoscopic regions of the flow and molecular kinetic theory is generally applicable for describing the microscopic region of the dynamic contact line, it is beneficial to apply both theories simultaneously to describe dynamic contact line problems. However, the parameters that define the dynamic contact line in these two well-established theories (hydrodynamics and molecular kinetic) can only currently be estimated through fitting analyses. Therefore, the combined model currently is not applicable to precisely describe real-life dynamic contact line problems. For this reason, it is vital to pay significant attention to precisely determining physical parameters by applying current advanced technologies, including quantum computation, imaging techniques such as atomic force microscopy and cryogenic electron microscopy, and the power of machine learning.

Nanofluidics is the current advanced technology used for the production of nanofilms. The physics of the dynamic contact line in nanofluidics needs to be investigated by considering the complex fluids. The physical properties of nanofluids, such as rheology, change rapidly in the proximity to the dynamic contact line. The change in physical properties of nanofluids can influence the physical parameters that define the dynamic contact line via hydrodynamics theory and molecular kinetic theory. These physical dependencies need to be thoroughly investigated. For instance, as the slip condition strongly depends on the rheology of fluids, in nanofluidics, the slip condition can abruptly change and consequently affects hydrodynamics theory.

Another issue is the real characteristics of the substrates, which complicate the relationship between the physical parameters that define the hydrodynamics and molecular kinetic theories and the heterogenous roughness, heterogeneous surface chemistry, and flexibility of the substrates. Therefore, it is significantly important to enhance the physical understanding of this subject by applying state-of-the-art nanotechnologies such as atomic force microscopy and cryogenic electron microscopy to look closely at the influence of the substrate features on the physics of the dynamic contact line for real-life natural substrates.

Both theories are not suitable for application to real interfacial-related problems. Problems in healthcare and biomedicine that deal with interfacial physics urgently require enhanced knowledge of the dynamic contact line and its real underlying physical mechanism. The physics of the dynamic contact line is a multi-scale phenomenon that requires understanding the details of flow in microscopic, mesoscopic, and macroscopic regions simultaneously. Moreover, a dynamic contact line problem is an interdisciplinary physical event that requires thorough knowledge in various fields, such as chemistry, materials science, electrical engineering, mechanical engineering, physics, mathematics, and biology.

The dynamic contact line in real-life problems involves complexity in the liquid environment and substrate. The liquid, environment, and substrate can have complicated mixtures of phases in real-life situations. Most problems, such as bioprinting; tissue engineering; and designing smart biomaterials, which involve DNA molecules, cells, peptides, and viruses; and state-of-the-art biosensors, deal with very complex liquid models and substrates [46,47,165–168,197,198].

The present models of hydrodynamics and molecular kinetic do not consider such complexities. Therefore, it is highly recommended that the interfacial physics community directs experimental and theoretical efforts to more complex systems of solid–liquid–gas systems for an enhanced physical understanding of the dynamic contact line in such complicated real-life conditions.

Author Contributions: Conceptualization, A.M.K.; methodology, A.M.K.; software, A.M.K.; validation, A.M.K., W.J.S.; formal analysis, A.M.K.; investigation, A.M.K.; resources, A.M.K.; data curation, A.M.K.; writing—original draft preparation, A.M.K.; writing—review and editing, A.M.K.; visualization, A.M.K.; supervision, A.M.K.; project administration, A.M.K. All authors have read and agreed to the published version of the manuscript.

Funding: This research received no external funding.

Institutional Review Board Statement: This study did not require ethical approval.

Informed Consent Statement: This study did not involve humans.

Data Availability Statement: Data sharing is not applicable to this article as no new data were created or analyzed in this study.

Conflicts of Interest: The authors declare no conflict of interest.

References

1. Ren, W.; Weinan, E. Contact Line Dynamics on Heterogeneous Surfaces. *Phys. Fluids* **2011**, *23*, 072103. [CrossRef]
2. Eggers, J.; Stone, H.A. Characteristic Lengths at Moving Contact Lines for a Perfectly Wetting Fluid: The Influence of Speed on The Dynamic Contact Angle. *J. Fluid Mech.* **2004**, *505*, 309–321. [CrossRef]
3. Diaz, M.E.; Cerro, R.L. A General Solution of Dewetting Flow with a Moving Contact Line. *Phys. Fluids* **2021**, *33*, 103601. [CrossRef]
4. Dodds, S.; Carvalho, M.D.S.; Kumar, S. Stretching and Slipping of Liquid Bridges Near Plates and Cavities. *Phys. Fluids* **2009**, *21*, 092103. [CrossRef]
5. Dodds, S.; Carvalho, M.; Kumar, S. Stretching Liquid Bridges with Moving Contact Lines: The Role of Inertia. *Phys. Fluids* **2011**, *23*, 092101. [CrossRef]
6. Dodds, S.; Carvalho, M.S.; Kumar, S. The Dynamics of Three-Dimensional Liquid Bridges with Pinned and Moving Contact Lines. *J. Fluid Mech.* **2012**, *707*, 521–540. [CrossRef]
7. Ghosh, A.; Bandyopadhyay, D.; Sharma, A. Micro-Patterning of Coatings on A Fiber Surface Exploiting the Contact Instabilities of Thin Viscoelastic Films. *Phys. Fluids* **2018**, *30*, 114101. [CrossRef]
8. Harikrishnan, A.R.; Dhar, P.; Agnihotri, P.K.; Gedupudi, S.; Das, S.K. Correlating Contact Line Capillarity and Dynamic Contact Angle Hysteresis in Surfactant-Nanoparticle Based Complex Fluids. *Phys. Fluids* **2018**, *30*, 042006. [CrossRef]
9. Kumar, S. Liquid Transfer in Printing Processes: Liquid Bridges with Moving Contact Lines. *Annu. Rev. Fluid Mech.* **2015**, *47*, 67–94. [CrossRef]
10. Liu, C.-Y.; Vandre, E.; Carvalho, M.S.; Kumar, S. Dynamic Wetting Failure and Hydrodynamic Assist in Curtain Coating. *J. Fluid Mech.* **2016**, *808*, 290–315. [CrossRef]
11. Liu, C.-Y.; Vandre, E.; Carvalho, M.S.; Kumar, S. Dynamic wetting failure in surfactant solutions. *J. Fluid Mech.* **2016**, *789*, 285–309. [CrossRef]
12. Liu, C.-Y.; Carvalho, M.S.; Kumar, S. Mechanisms of Dynamic Wetting Failure in the Presence of Soluble Surfactants. *J. Fluid Mech.* **2017**, *825*, 677–703. [CrossRef]
13. Liu, C.-Y.; Carvalho, M.S.; Kumar, S. Dynamic Wetting Failure in Curtain Coating: Comparison of Model Predictions and Experimental Observations. *Chem. Eng. Sci.* **2018**, *195*, 74–82. [CrossRef]
14. Liu, M.; Chen, X.-P. Numerical Study on the Stick-Slip Motion of Contact Line Moving on Heterogeneous Surfaces. *Phys. Fluids* **2017**, *29*, 082102. [CrossRef]
15. Ding, H.; Spelt, P.D.M. Inertial Effects in Droplet Spreading: A Comparison Between Diffuse-Interface and Level-Set Simulations. *J. Fluid Mech.* **2007**, *576*, 287–296. [CrossRef]
16. Biance, A.-L.; Clanet, C.; Quéré, D. First Steps in the Spreading of a Liquid Droplet. *Phys. Rev. E* **2004**, *69*, 016301. [CrossRef]
17. Foister, R.T. The Kinetics of Displacement Wetting in Liquid/Liquid/Solid Systems. *J. Colloid Interface Sci.* **1990**, *136*, 266–282. [CrossRef]
18. Spelt, P.D. A level-Set Approach for Simulations of Flows with Multiple Moving Contact Lines with Hysteresis. *J. Comput. Phys.* **2005**, *207*, 389–404. [CrossRef]
19. Wang, H. From Contact Line Structures to Wetting Dynamics. *Langmuir* **2019**, *35*, 10233–10245. [CrossRef]
20. Popescu, M.N.; Oshanin, G.; Dietrich, S.; Cazabat, A.-M. Precursor Films in Wetting Phenomena. *J. Phys. Condens. Matter* **2012**, *24*, 243102. [CrossRef]
21. Pham, T.; Kumar, S. Imbibition and Evaporation of Droplets of Colloidal Suspensions on Permeable Substrates. *Phys. Rev. Fluids* **2019**, *4*, 034004. [CrossRef]
22. Charitatos, V.; Kumar, S. A Thin-Film Model for Droplet Spreading on Soft Solid Substrates. *Soft Matter* **2020**, *16*, 8284–8298. [CrossRef]
23. Raske, N.; Hewson, R.W.; Kapur, N.; De Boer, G. A Predictive Model for Discrete Cell Gravure Roll Coating. *Phys. Fluids* **2017**, *29*, 062101. [CrossRef]
24. Karim, A.M.; Suszynski, W.J.; Griffith, W.B.; Pujari, S.; Francis, L.F.; Carvalho, M.S. Effect of Viscoelasticity on Stability of Liquid Curtain. *J. Non-Newton. Fluid Mech.* **2018**, *257*, 83–94. [CrossRef]
25. Karim, A.M.; Suszynski, W.J.; Griffith, W.B.; Pujari, S.; Francis, L.F.; Carvalho, M.S. Effect of Rheological Properties of Shear Thinning Liquids on Curtain Stability. *J. Non-Newton. Fluid Mech.* **2018**, *263*, 69–76. [CrossRef]
26. Karim, A.M.; Suszynski, W.J.; Pujari, S.; Francis, L.F.; Carvalho, M.S. Delaying Breakup and Avoiding Air Entrainment in Curtain Coating Using a Two-Layer Liquid Structure. *Chem. Eng. Sci.* **2019**, *213*, 115376. [CrossRef]
27. Karim, A.M. Experimental Dynamics of Newtonian and Non-Newtonian Droplets Impacting Liquid Surface with Different Rheology. *Phys. Fluids* **2020**, *32*, 043102. [CrossRef]

28. Karim, A.M.; Suszynski, W.J.; Pujari, S.; Francis, L.F.; Carvalho, M.S. Contact Line Dynamics in Curtain Coating of Non-Newtonian Liquids. *Phys. Fluids* **2021**, *33*, 103103. [CrossRef]
29. Karim, A.M.; Suszynski, W.J.; Pujari, S. Liquid Film Stability and Contact Line Dynamics of Emulsion Liquid Films in Curtain Coating Process. *J. Coat. Technol. Res.* **2021**, *18*, 1531–1541. [CrossRef]
30. Mohammad Karim, A.; Pujari, S.; Suszynski, W.J.; Francis, L.F.; Carvalho, M.S.; Yadav, V. Methods for Curtain Coating Substrates. U.S. Patent No: 11,369,988 B2, 28 June 2022.
31. Mohammad Karim, A.; Pujari, S.; Suszynski, W.J.; Francis, L.F.; Carvalho, M.S.; Yadav, V. Methods for Curtain Coating Substrates. Application Publication No: WO 2019/190623 A1, 3 October 2019.
32. Omori, T.; Kajishima, T. Apparent and Microscopic Dynamic Contact Angles in Confined Flows. *Phys. Fluids* **2017**, *29*, 112107. [CrossRef]
33. Patil, N.D.; Shaikh, J.; Sharma, A.; Bhardwaj, R. Droplet Impact Dynamics Over a Range of Capillary Numbers and Surface Wettability: Assessment of Moving Contact Line Models and Energy Budget Analysis. *Phys. Fluids* **2022**, *34*, 052119. [CrossRef]
34. Peschka, D. Variational Approach to Dynamic Contact Angles for Thin Films. *Phys. Fluids* **2018**, *30*, 082115. [CrossRef]
35. Park, J.; Kumar, S. Droplet Sliding on an Inclined Substrate with a Topographical Defect. *Langmuir* **2017**, *33*, 7352–7363. [CrossRef]
36. Vandre, E.; Carvalho, M.S.; Kumar, S. Delaying the Onset of Dynamic Wetting Failure through Meniscus Confinement. *J. Fluid Mech.* **2012**, *707*, 496–520. [CrossRef]
37. Vandre, E.; Carvalho, M.S.; Kumar, S. On the Mechanism of Wetting Failure during Fluid Displacement along a Moving Substrate. *Phys. Fluids* **2013**, *25*, 102103. [CrossRef]
38. Vandre, E.; Carvalho, M.S.; Kumar, S. Characteristics of Air Entrainment during Dynamic Wetting Failure along a Planar Substrate. *J. Fluid Mech.* **2014**, *747*, 119–140. [CrossRef]
39. Xia, Y.; Qin, J. Dynamics of Moving Contact Line on a Transversely Patterned Inclined Surface. *Phys. Fluids* **2020**, *32*, 042101. [CrossRef]
40. Xu, X.; Wang, X. Theoretical Analysis for Dynamic Contact Angle Hysteresis on Chemically Patterned Surfaces. *Phys. Fluids* **2020**, *32*, 112102. [CrossRef]
41. Zhang, J.; Wang, P.; Borg, M.K.; Reese, J.; Wen, D. A Critical Assessment of the Line Tension Determined by the Modified Young's Equation. *Phys. Fluids* **2018**, *30*, 082003. [CrossRef]
42. Zheng, W.; Wen, B.; Sun, C.; Bai, B. Effects of Surface Wettability on Contact Line Motion in Liquid–Liquid Displacement. *Phys. Fluids* **2021**, *33*, 082101. [CrossRef]
43. Bergeron, V.; Bonn, D.; Martin, J.Y.; Vovelle, L. Controlling Droplet Deposition with Polymer Additives. *Nature* **2000**, *405*, 772–775. [CrossRef]
44. Sansom, M.S.; Biggin, P.C. Water at the Nanoscale. *Nature* **2001**, *414*, 157–159. [CrossRef]
45. Xu, J.; Yan, X.; Liu, G.; Xie, J. The Critical Nanofluid Concentration as the Crossover between Changed and Unchanged Solar-Driven Droplet Evaporation Rates. *Nano Energy* **2019**, *57*, 791–803. [CrossRef]
46. Wang, H.; Garimella, S.V.; Murthy, J.Y. Characteristics of an Evaporating Thin Film in a Microchannel. *Int. J. Heat Mass Transf.* **2007**, *50*, 3933–3942. [CrossRef]
47. Wang, X.D.; Lee, D.J.; Peng, A.X.F.; Lai, J.Y. Spreading Dynamics and Dynamic Contact Angle of Non-Newtonian Fluids. *Langmuir* **2007**, *23*, 8042–8047. [CrossRef]
48. Wang, X.D.; Zhang, Y.; Lee, D.J.; Peng, X.F. Spreading of Completely Wetting or Partially Wetting Power-Law Fluid on Solid Surface. *Langmuir* **2007**, *23*, 9258–9262. [CrossRef]
49. Raghupathi, P.A.; Kandlikar, S.G. Contact Line Region Heat Transfer Mechanisms for an Evaporating Interface. *Int. J. Heat Mass Transf.* **2016**, *95*, 296–306. [CrossRef]
50. Lu, G.; Wang, X.D.; Duan, Y.Y. A Critical Review of Dynamic Wetting by Complex Fluids: From Newtonian Fluids to Non-Newtonian Fluids and Nanofluids. *Adv. Colloid Interface Sci.* **2016**, *236*, 43–62. [CrossRef]
51. Sui, Y.; Ding, H.; Spelt, P.D. Numerical Simulations of Flows with Moving Contact Lines. *Annu. Rev. Fluid Mech.* **2014**, *46*, 97–119. [CrossRef]
52. Abe, Y.; Zhang, B.; Gordillo, L.; Mohammad Karim, A.; Francis, L.F.; Cheng, X. Dynamic Self-Assembly of Charged Colloidal Strings and Walls in Simple Fluid Flows. *Soft Matter* **2017**, *13*, 1681–1692. [CrossRef]
53. Karim, A.M.; Suszynski, W.J.; Francis, L.F.; Carvalho, M.S. Effect of Viscosity on Liquid Curtain Stability. *AIChE J.* **2017**, *64*, 1448–1457. [CrossRef]
54. Dwivedi, R.K.; Jain, V.; Muralidhar, K. Dynamic Contact Angle Model for Resolving Low-Viscosity Droplet Oscillations During Spreading Over a Surface with Varying Wettability. *Phys. Rev. Fluids* **2022**, *7*, 034002. [CrossRef]
55. Kim, J.-H.; Kavehpour, H.P.; Rothstein, J.P. Dynamic Contact Angle Measurements on Superhydrophobic Surfaces. *Phys. Fluids* **2015**, *27*, 032107. [CrossRef]
56. Fernández-Toledano, J.C.; Blake, T.D.; De Coninck, J. Taking a Closer Look: A Molecular-Dynamics Investigation of Microscopic and Apparent Dynamic Contact Angles. *J. Colloid Interface Sci.* **2021**, *587*, 311–323. [CrossRef]
57. Qian, T.; Wang, X.-P.; Sheng, P. Molecular Scale Contact Line Hydrodynamics of Immiscible Flows. *Phys. Rev. E* **2003**, *68*, 016306. [CrossRef]
58. Wang, X.-P.; Qian, T.; Sheng, P. Moving Contact Line on Chemically Patterned Surfaces. *J. Fluid Mech.* **2008**, *605*, 59–78. [CrossRef]

59. Chen, X.; Wang, X.; Xu, X. Analysis of the Cahn–Hilliard Equation with a Relaxation Boundary Condition Modeling the Contact Angle Dynamics. *Arch. Ration. Mech. Anal.* **2014**, *213*, 1–24. [CrossRef]
60. Savva, N.; Kalliadasis, S. Dynamics of Moving Contact Lines: A Comparison between Slip and Precursor Film Models. *Eur. Lett.* **2011**, *94*, 64004. [CrossRef]
61. Nosonovsky, M.; Ramachandran, R. Geometric Interpretation of Surface Tension Equilibrium in Superhydrophobic Systems. *Entropy* **2015**, *17*, 4684–4700. [CrossRef]
62. Nikolayev, V.S. Evaporation Effect on the Contact Angle and Contact Line Dynamics. In *The Surface Wettability Effect on Phase Change*; Marengo, M., De Coninck, J., Eds.; Springer Nature: Berlin/Heidelberg, Germany, 2022.
63. Young, T. III. An essay on the cohesion of fluids. *Philos. Trans. R. Soc. Lond.* **1805**, *95*, 65–87. [CrossRef]
64. Shikhmurzaev, Y. The Moving Contact Line on a Smooth Solid Surface. *Int. J. Multiph. Flow* **1993**, *19*, 589–610. [CrossRef]
65. Shikhmurzaev, Y.D. Mathematical Modeling of Wetting Hydrodynamics. *Fluid Dyn. Res.* **1994**, *13*, 45–64. [CrossRef]
66. Shikhmurzaev, Y.D. Moving Contact Lines in Liquid/Liquid/Solid Systems. *J. Fluid Mech.* **1997**, *334*, 211–249. [CrossRef]
67. De Ruijter, M.J.; De Coninck, J.; Blake, T.D.; Clarke, A.A.; Rankin, A. Contact Angle Relaxation during the Spreading of Partially Wetting Drops. *Langmuir* **1997**, *13*, 7293–7298. [CrossRef]
68. Blake, T.D. The Physics of Moving Wetting Lines. *J. Colloid Interface Sci.* **2006**, *299*, 1–13. [CrossRef]
69. Cox, R.G. The Dynamics of the Spreading of Liquids on a Solid Surface. Part 1. Viscous Flow. *J. Fluid Mech.* **1986**, *168*, 169–194. [CrossRef]
70. Huh, C.; Scriven, J.L.E. Hydrodynamic Model of Steady Movement of a Solid/Liquid/Fluid Contact Line. *J. Colloid Interface Sci.* **1971**, *35*, 85–101. [CrossRef]
71. Voinov, O.V. Hydrodynamics of Wetting. *Izv. Akad. Nauk. SSSR Mekhanika Zhidkosti I Gaza* **1976**, *5*, 714–721. [CrossRef]
72. De Gennes, P.G. Deposition of Langmuir-Blodgett Layers. *Colloid Polym. Sci.* **1986**, *264*, 463–465. [CrossRef]
73. Butt, H.; Liu, J.; Koynov, K.; Straub, B.; Hinduja, C.; Roismann, I.; Berger, R.; Li, X.; Vollmer, D.; Steffen, W.; et al. Contact Angle Hysteresis. *Curr. Opin. Colloid Interface Sci.* **2022**, *59*, 101574. [CrossRef]
74. Bartell, F.E.; Wooley, A.D. Solid-Liquid-Air Contact Angles and Their Dependence upon the Surface Condition of the Solid. *J. Am. Chem. Soc.* **1933**, *55*, 3518–3527. [CrossRef]
75. Schäffer, E.; Wong, P.-Z. Contact Line Dynamics Near the Pinning Threshold: A Capillary Rise and Fall Experiment. *Phys. Rev. E* **2000**, *61*, 5257–5277. [CrossRef]
76. Johnson, R.E.; Dettre, R.H. Wettability and Contact Angles. In *Surface and Colloid Science*; Matijevic, E., Ed.; Wiley-Interscience: New York, NY, USA, 1969; pp. 85–153.
77. Wang, J.-H.; Claesson, P.M.; Parker, J.L.; Yasuda, H. Dynamic Contact Angles and Contact Angle Hysteresis of Plasma Polymers. *Langmuir* **1994**, *10*, 3887–3897. [CrossRef]
78. Vega, M.J.; Gouttiere, C.; Seveno, D.; Blake, T.D.; Voue, M.; de Coninck, J. Experimental Investigation of the Link Be-Tween Static and Dynamic Wetting by Forced Wetting of Nylon Filament. *Langmuir* **2007**, *23*, 10628–10634. [CrossRef]
79. Karim, A.M.; Kavehpour, H.P. Effect of Viscous Force on Dynamic Contact Angle Measurement Using Wilhelmy Plate Method. *Colloids Surfaces A Physicochem. Eng. Asp.* **2018**, *548*, 54–60. [CrossRef]
80. Bartell, F.E.; Shepard, J.W. The Effect of Surface Roughness on Apparent Contact Angles and on Contact Angle Hysteresis. I. The System Paraffin—Water—Air. *J. Phys. Chem.* **1953**, *57*, 211–215. [CrossRef]
81. Eick, J.; Good, R.; Neumann, A. Thermodynamics of Contact Angles. II. Rough Solid Surfaces. *J. Colloid Interface Sci.* **1975**, *53*, 235–248. [CrossRef]
82. Oliver, J.F.; Mason, S.G. Liquid Spreading on Rough Metal Surfaces. *J. Mater. Sci.* **1980**, *15*, 431–437. [CrossRef]
83. Good, R.J.A. Thermodynamic Derivation of Wenzel's Modification of Young's Equation for Contact Angles, Together with a Theory of Hysteresis. *J. Am. Chem. Soc.* **1952**, *74*, 5041–5042. [CrossRef]
84. Neumann, A.; Good, R. Thermodynamics of Contact Angles. I. Heterogeneous Solid Surfaces. *J. Colloid Interface Sci.* **1972**, *38*, 341–358. [CrossRef]
85. Schwartz, L.W.; Garoff, S. Contact Angle Hysteresis on Heterogeneous surfaces. *Langmuir* **1985**, *1*, 219–230. [CrossRef]
86. Marmur, A. Contact Angle Hysteresis on Heterogeneous Smooth Surfaces. *J. Colloid Interface Sci.* **1994**, *168*, 40–46. [CrossRef]
87. Decker, E.L.; Garoff, S. Contact Line Structure and Dynamics on Surfaces with Contact Angle Hysteresis. *Langmuir* **1997**, *13*, 6321–6332. [CrossRef]
88. Brandon, S.; Marmur, A. Simulation of Contact Angle Hysteresis on Chemically Heterogeneous Surfaces. *J. Colloid Interface Sci.* **1996**, *183*, 351–355. [CrossRef] [PubMed]
89. Yue, P.; Zhou, C.; Feng, J.J. Sharp-Interface Limit of the Cahn–Hilliard Model for Moving Contact Lines. *J. Fluid Mech.* **2010**, *645*, 279–294. [CrossRef]
90. Yue, P.; Zhou, C.; Feng, J.J.; Ollivier-Gooch, C.F.; Hu, H.H. Phase-Field Simulations of Interfacial Dynamics in Viscoelastic Fluids Using Finite Elements with Adaptive Meshing. *J. Comput. Phys.* **2006**, *219*, 47–67. [CrossRef]
91. Shanahan, M.E.R. Simple Theory of "Stick-Slip" Wetting Hysteresis. *Langmuir* **1995**, *11*, 1041–1043. [CrossRef]
92. Kalinin, Y.V.; Berejnov, V.; Thorne, R.E. Contact Line Pinning by Microfabricated Patterns: Effects of Microscale topography. *Langmuir* **2009**, *25*, 5391–5397. [CrossRef]
93. Dussan, V.E.B. On the Spreading of Liquids on Solid Surfaces: Static and Dynamic Contact Lines. *Annu. Rev. Fluid Mech.* **1979**, *11*, 371–400. [CrossRef]

94. Van Der Waals, J.D. The Thermodynamic Theory of Capillarity under the Hypothesis of a Continuous Variation of Density. *J. Stat. Phys.* **1979**, *20*, 200–244. [CrossRef]
95. Blake, T.D.; Berg, J.C. *Wettability: Dynamic Contact Angles and Wetting Kinetics*; Marcel Dekker: New York, NY, USA, 1993; Volume 49, pp. 251–309.
96. Blake, T.; Shikhmurzaev, Y. Dynamic Wetting by Liquids of Different Viscosity. *J. Colloid Interface Sci.* **2002**, *253*, 196–202. [CrossRef]
97. Blake, T.D.; Bracke, M.; Shikhmurzaev, Y.D. Experimental Evidence of Non-Local Hydrodynamic Influence on the Dynamic Contact Angle. *Phys. Fluids* **1999**, *11*, 1995. [CrossRef]
98. Cahn, J.W. Critical Point Wetting. *J. Chem. Phys.* **1977**, *66*, 3667–3672. [CrossRef]
99. Jacqmin, D. Contact-Line Dynamics of a Diffuse Fluid Interface. *J. Fluid Mech.* **2000**, *402*, 57–88. [CrossRef]
100. Cahn, J.W.; Hilliard, J.E. Free Energy of a Non-Uniform System. Part I. Interfacial Free Energy. *J. Chem. Phys.* **1958**, *28*, 258–267. [CrossRef]
101. Blake, T.D.; Haynes, J.M. Kinetics of Liquid/Liquid Displacement. *J. Colloid Interf. Sci.* **1969**, *30*, 421–423. [CrossRef]
102. Prevost, A.; Rolley, E.; Guthmann, C. Thermally Activated Motion of the Contact Line of a Liquid 4 He Meniscus on a Cesium Substrate. *Phys. Rev. Lett.* **1999**, *83*, 348–351. [CrossRef]
103. Huh, C.; Mason, S.G. The Steady Movement of a Liquid Meniscus in a Capillary Tube. *J. Fluid Mech.* **1977**, *81*, 401–419. [CrossRef]
104. Dussan, V.E.B. The Moving Contact Line: The Slip Boundary Condition. *J. Fluid Mech.* **1976**, *77*, 665–684. [CrossRef]
105. Ngan, C.G.; Dussan, V.E.B. On the Dynamics of Liquid Spreading on Solid Surfaces. *J. Fluid Mech.* **1989**, *209*, 191–226. [CrossRef]
106. Hoffman, R.L. A Study of the Advancing Interface. I. Interface Shape in Liquid—Gas Systems. *J. Colloid Interface Sci.* **1975**, *50*, 228–241. [CrossRef]
107. Tanner, L.H. The Spreading of Silicone Oil Drops on Horizontal Surfaces. *J. Phys. D Appl. Phys.* **1979**, *12*, 1473–1484. [CrossRef]
108. Hocking, L.M. A Moving Fluid Interface. Part 2. The Removal of the Force Singularity by A Slip Flow. *J. Fluid Mech.* **1977**, *79*, 209–229. [CrossRef]
109. Petrov, P.G.; Petrov, J.G. A Combined Molecular-Hydrodynamic Approach to Wetting Kinetics. *Langmuir* **1992**, *8*, 1762–1767. [CrossRef]
110. Brochard-Wyart, F.; de Gennes, P. Dynamics of Partial Wetting. *Adv. Colloid Interface Sci.* **1992**, *39*, 1–11. [CrossRef]
111. De Ruijter, M.J.; De Coninck, J.; Oshanin, G. Droplet Spreading: Partial Wetting Regime Revisited. *Langmuir* **1999**, *15*, 2209–2216. [CrossRef]
112. Blake, T.D.; Ruschak, K.J. *Liquid Film Coating*; Schweizer, P.M., Kistler, S.F., Eds.; Chapman & Hall: London, UK, 1997; Volume 63.
113. Petrov, J.G.; Ralston, J.; Schneemilch, A.M.; Hayes, R.A. Dynamics of Partial Wetting and Dewetting in Well-Defined Systems. *J. Phys. Chem. B* **2003**, *107*, 1634–1645. [CrossRef]
114. Yuan, Q.; Zhao, Y.-P. Precursor Film in Dynamic Wetting, Electrowetting, and Electro-Elasto-Capillarity. *Phys. Rev. Lett.* **2010**, *104*, 246101. [CrossRef]
115. Hardy, W.B., III. The Spreading of Fluids on Glass. *Lond. Edinb. Dublin Philos. Mag. J. Sci.* **1919**, *38*, 49–55. [CrossRef]
116. Espín, L.; Kumar, S. Sagging of Evaporating Droplets of Colloidal Suspensions on Inclined Substrates. *Langmuir* **2014**, *30*, 11966–11974. [CrossRef]
117. Karapetsas, G.; Sahu, K.C.; Matar, O.K. Evaporation of Sessile Droplets Laden with Particles and Insoluble Surfactants. *Langmuir* **2016**, *32*, 6871–6881. [CrossRef]
118. Maki, K.L.; Kumar, S. Fast Evaporation of Spreading Droplets of Colloidal Suspensions. *Langmuir* **2011**, *27*, 11347–11363. [CrossRef]
119. Wray, A.W.; Papageorgiou, D.T.; Craster, R.V.; Sefiane, K.; Matar, O.K. Electrostatic Suppression of the Coffee Stain Effect. *Langmuir* **2014**, *30*, 5849–5858. [CrossRef]
120. Charitatos, V.; Kumar, S. Droplet Evaporation on Soft Solid Substrates. *Soft Matter* **2021**, *17*, 9339–9352. [CrossRef]
121. Derjaguin, B.; Churaev, N. Structural Component of Disjoining Pressure. *J. Colloid Interface Sci.* **1974**, *49*, 249–255. [CrossRef]
122. Karim, A.M. A Review of Physics of Moving Contact Line Dynamics Models and Its Applications in Interfacial Science. *J. Appl. Phys.* **2022**, *132*, 080701. [CrossRef]
123. Afkhami, S. Challenges of Numerical Simulation of Dynamic Wetting Phenomena: A Review. *Curr. Opin. Colloid Interface Sci.* **2022**, *57*, 101523. [CrossRef]
124. Zhao, Y. Moving Contact Line Problem: Advances and Perspectives. *Theor. Appl. Mech. Lett.* **2014**, *4*, 034002. [CrossRef]
125. Soltman, D.; Smith, B.; Kang, H.; Morris, S.J.S.; Subramanian, V. Methodology for Inkjet Printing of Partially Wetting Films. *Langmuir* **2010**, *26*, 15686–15693. [CrossRef]
126. Tarasevich, Y.Y.; Pravoslavnova, D.M. Segregation in Desiccated Sessile Drops of Biological Fluids. *Eur. Phys. J. E* **2007**, *22*, 311–314. [CrossRef]
127. Karim, A.M. Experimental Dynamics of Newtonian Non-Elastic and Viscoelastic Droplets Impacting Immiscible Liquid Surface. *AIP Adv.* **2019**, *9*, 125141. [CrossRef]
128. Backholm, M.; Molpeceres, D.; Vuckovac, M.; Nurmi, H.; Hokkanen, M.J.; Jokinen, V.; Timonen, J.V.I.; Ras, R.H.A. Water droplet friction and rolling dynamics on superhydrophobic surfaces. *Commun. Mater.* **2020**, *1*, 64. [CrossRef]
129. De Coninck, J. An Introduction to Wettability and Wetting Phenomena. In *The Surface Wettability Effect on Phase Change*; Springer: Berlin/Heidelberg, Germany, 2022.

130. Mugele, F.; Baret, J.-C. Electrowetting: From Basics to Applications. *J. Phys. Condens. Matter* **2005**, *17*, R705–R774. [CrossRef]
131. Jarosz, J.; Molliex, N.; Chenon, G.; Berge, B. Adaptive Eyeglasses for Presbyopia Correction: An Original Variable-Focus Technology. *Opt. Express* **2019**, *27*, 10533–10552. [CrossRef] [PubMed]
132. Modak, C.D.; Kumar, A.; Tripathy, A.; Sen, P. Drop Impact Printing. *Nat. Commun.* **2020**, *11*, 4327. [CrossRef] [PubMed]
133. Biswas, S.; Gawande, S.; Bromberg, V.; Sun, Y. Effects of Particle Size and Substrate Surface Properties on Deposition Dynamics of Inkjet-Printed Colloidal Drops for Printable Photovoltaics Fabrication. *J. Sol. Energy Eng.* **2010**, *132*, 021010. [CrossRef]
134. Joshi, A.S.; Sun, Y. Numerical Simulation of Colloidal Drop Deposition Dynamics on Patterned Substrates for Printable Electronics Fabrication. *J. Disp. Technol.* **2010**, *6*, 579–585. [CrossRef]
135. Chhasatia, V.H.; Joshi, A.S.; Sun, Y. Effect of Relative Humidity on Contact Angle and Particle Deposition Morphology of An Evaporating Colloidal Drop. *Appl. Phys. Lett.* **2010**, *97*, 231909. [CrossRef]
136. Bigioni, T.P.; Lin, X.-M.; Nguyen, T.; Corwin, E.I.; Witten, T.A.; Jaeger, H.M. Kinetically Driven Self Assembly of Highly Ordered Nanoparticle Monolayers. *Nat. Mater.* **2006**, *5*, 265–270. [CrossRef]
137. Sobac, B.; Brutin, D. Structural and Evaporative Evolutions in Desiccating Sessile Drops of Blood. *Phys. Rev. E* **2011**, *84*, 011603. [CrossRef]
138. Gosselin, E.; Eynde, J.J.V.; Petit, A.; Conti, J.; De Coninck, J. Designing a High Performance, Stable Spectroscopic Biosensor for the Binding of Large and Small Molecules. *J. Colloid Interface Sci.* **2017**, *508*, 443–454. [CrossRef]
139. Yao, M.; Tijing, L.D.; Naidu, G.; Kim, S.-H.; Matsuyama, H.; Fane, A.G.; Shon, H.K. A Review of Membrane Wettability for the Treatment of Saline Water Deploying Membrane Distillation. *Desalination* **2020**, *479*, 114312. [CrossRef]
140. Melayil, K.R.; Mitra, S.K. Wetting, Adhesion, and Droplet Impact on Face Masks. *Langmuir* **2021**, *37*, 2810–2815. [CrossRef]
141. Sharma, S.; Pinto, R.; Saha, A.; Chaudhuri, S.; Basu, S. Video: On Secondary Atomization and Blockage of Surrogate Cough Droplets in Single and Multi-Layer Face Masks. *Sci. Adv.* **2020**, *7*, eabf0452. [CrossRef]
142. Liao, M.; Liu, H.; Wang, X.; Hu, X.; Huang, Y.; Liu, X.; Brenan, K.; Mecha, J.; Nirmalan, M.; Lu, J.R. A Technical Review of Face Mask Wearing in Preventing Respiratory COVID-19 Transmission. *Curr. Opin. Colloid Interface Sci.* **2021**, *52*, 101417. [CrossRef]
143. Montessori, A.; Lauricella, M.; Tiribocchi, A. Computational Droplets: Where We Stand and How Far We Can Go. *Eur. Lett.* **2022**, *138*, 67001. [CrossRef]
144. Powell, R.E.; Roseveare, W.E.; Eyring, H. The Theory of Rate Processes. *Ind. Eng. Chem.* **1941**, *33*, 430. [CrossRef]
145. Niavarani, A.; Priezjev, N. The Effective Slip Length and Vortex Formation in Laminar Flow over a Rough Surface. *Phys. Fluids* **2009**, *21*, 052105. [CrossRef]
146. Niavarani, A.; Priezjev, N. Modeling the Combined Effect of Surface Roughness and Shear Rate on Slip Flow of Simple Fluids. *Phys. Rev. E* **2010**, *81*, 011606. [CrossRef]
147. Xiaodong, W.; Xiaofeng, P.; Yuanyuan, D.; Buxuan, W. Dynamics of Spreading of Liquid on Solid Surface. *Chin. J. Chem. Eng.* **2007**, *15*, 730–737. [CrossRef]
148. Cazabat, A.-M. How Does a Droplet Spread? *Contemp. Phys.* **1987**, *28*, 347–364. [CrossRef]
149. Roques-Carmes, T.; Mathieu, V.; Gigante, A. Experimental Contribution to the Understanding of the Dynamics of Spreading of Newtonian Fluids: Effect of Volume, Viscosity and Surfactant. *J. Colloid Interface Sci.* **2010**, *344*, 180–197. [CrossRef]
150. Eddi, A.; Winkels, K.G.; Snoeijer, J.H. Short time dynamics of viscous drop spreading. *Phys. Fluids* **2013**, *25*, 013102. [CrossRef]
151. Legendre, D.; Maglio, M. Numerical Simulation of Spreading Drops. *Colloids Surf. A Physicochem. Eng. Asp.* **2013**, *432*, 29–37. [CrossRef]
152. Zeid, W.B.; Brutin, D. Beyond Tanner's Law: Role of Contact Line Evaporation on the Spreading of Viscous Droplet. *Interfacial Phenom. Heat Transf.* **2015**, *3*, 221–229. [CrossRef]
153. Blake, T.D.; Dobson, R.A.; Ruschak, K.J. Wetting at High Capillary Numbers. *J. Colloid Interface Sci.* **2004**, *279*, 198–205. [CrossRef]
154. Sedev, R. The Molecular-Kinetic Approach to Wetting Dynamics: Achievements and Limitations. *Adv. Colloid Interface Sci.* **2015**, *222*, 661–669. [CrossRef]
155. Mohammad Karim, A. Parametric Study of Liquid Contact Line Dynamics: Adhesion vs. Hydrodynamics. Ph.D. Thesis, University of California, Los Angeles, CA, USA, 2015.
156. Karim, A.M.; Kavehpour, H.P. Dynamics of Spreading on Ultra-Hydrophobic Surfaces. *J. Coat. Technol. Res.* **2015**, *12*, 959–964. [CrossRef]
157. Karim, A.M.; Rothstein, J.P.; Kavehpour, H.P. Experimental Study of Dynamic Contact Angles on Rough Hydrophobic Surfaces. *J. Colloid Interface Sci.* **2018**, *513*, 658–665. [CrossRef]
158. Karim, A.M.; Fujii, K.; Kavehpour, H.P. Contact Line Dynamics of Gravity Driven Spreading of Liquids. *Fluid Dyn. Res.* **2021**, *53*, 035503. [CrossRef]
159. Seppecher, P. Moving Contact Lines in the Cahn-Hilliard Theory. *Int. J. Eng. Sci.* **1996**, *34*, 977–992. [CrossRef]
160. Bonn, D.; Eggers, J.; Indekeu, J.; Meunier, J.; Rolley, E. Wetting and Spreading. *Rev. Mod. Phys.* **2009**, *81*, 739–805. [CrossRef]
161. Zheng, L.; Wang, Y.-X.; Plawsky, J.L.; Wayner, P.C. Accuracy of Measurements of Curvature and Apparent Contact Angle in a Constrained Vapor Bubble Heat Exchanger. *Int. J. Heat Mass Transf.* **2002**, *45*, 2021–2030. [CrossRef]
162. Karim, A.M.; Davis, S.H.; Kavehpour, H.P. Forced versus Spontaneous Spreading of Liquids. *Langmuir* **2016**, *32*, 10153–10158. [CrossRef]

163. Karim, A.M.; Kavehpour, H.P. Spreading of Emulsions on a Solid Substrate. *J. Coat. Technol. Res.* **2013**, *11*, 103–108. [CrossRef]
164. Carré, A.; Eustache, F. Spreading Kinetics of Shear-Thinning Fluids in Wetting and Dewetting Modes. *Langmuir* **2000**, *16*, 2936–2941. [CrossRef]
165. Betelu, S.; Fontelos, M. Capillarity Driven Spreading of Power-Law Fluids. *Appl. Math. Lett.* **2003**, *16*, 1315–1320. [CrossRef]
166. Betelú, S.; Fontelos, M. Capillarity Driven Spreading of Circular Drops of Shear-Thinning Fluid. *Math. Comput. Model.* **2004**, *40*, 729–734. [CrossRef]
167. Starov, V.M.; Tyatyushkin, A.N.; Velarde, M.G.; Zhdanov, S.A. Spreading of Non-Newtonian Liquids over Solid Substrates. *J. Colloid Interface Sci.* **2003**, *257*, 284–290. [CrossRef]
168. Liang, Z.-P.; Wang, X.-D.; Duan, Y.-Y.; Min, Q. Energy-Based Model for Capillary Spreading of Power-Law Liquids on a Horizontal Plane. *Colloids Surf. A Physicochem. Eng. Asp.* **2012**, *403*, 155–163. [CrossRef]
169. Liang, Z.-P.; Wang, X.-D.; Lee, D.-J.; Peng, X.-F.; Su, A. Spreading Dynamics of Power-Law Fluid Droplets. *J. Phys. Condens. Matter* **2009**, *21*, 464117. [CrossRef]
170. Liu, K.; Tian, Y.; Jiang, L. Bio-Inspired Superoleophobic and Smart Materials: Design, Fabrication, and Application. *Prog. Mater. Sci.* **2013**, *58*, 503–564. [CrossRef]
171. Blake, T.; De Coninck, J. The Influence of Solid–Liquid Interactions on Dynamic Wetting. *Adv. Colloid Interface Sci.* **2002**, *96*, 21–36. [CrossRef]
172. Ferrari, M.; Ravera, F. Surfactants and Wetting at Superhydrophobic Surfaces: Water Solutions and Non Aqueous Liquids. *Adv. Colloid Interface Sci.* **2010**, *161*, 22–28. [CrossRef]
173. Yan, Y.; Gao, N.; Barthlott, W. Mimicking Natural Superhydrophobic Surfaces and Grasping the Wetting Process: A Review on Recent Progress in Preparing Superhydrophobic Surfaces. *Adv. Colloid Interface Sci.* **2011**, *169*, 80–105. [CrossRef]
174. Bhushan, B.; Jung, Y.C. Natural and Biomimetic Artificial Surfaces for Superhydrophobicity, Self-Cleaning, Low Adhesion, and Drag Reduction. *Prog. Mater. Sci.* **2011**, *56*, 1–108. [CrossRef]
175. Oberli, L.; Caruso, D.; Hall, C.; Fabretto, M.; Murphy, P.J.; Evans, D. Condensation and Freezing of Droplets on Superhydrophobic Surfaces. *Adv. Colloid Interface Sci.* **2014**, *210*, 47–57. [CrossRef]
176. Webb, H.K.; Crawford, R.J.; Ivanova, E.P. Wettability of Natural Superhydrophobic Surfaces. *Adv. Colloid Interface Sci.* **2014**, *210*, 58–64. [CrossRef]
177. Ma, M.; Hill, R.M. Superhydrophobic Surfaces. *Curr. Opin. Colloid Interface Sci.* **2006**, *11*, 193–202. [CrossRef]
178. Koch, K.; Bhushan, B.; Barthlott, W. Multifunctional Surface Structures of Plants: An Inspiration for Biomimetics. *Prog. Mater. Sci.* **2009**, *54*, 137–178. [CrossRef]
179. Savva, N.; Kalliadasis, S. Two-Dimensional Droplet Spreading Over Topographical Substrates. *Phys. Fluids* **2009**, *21*, 092102. [CrossRef]
180. Savva, N.; Pavliotis, G.A.; Kalliadasis, S. Contact Lines Over Random Topographical Substrates. Part 2. Dynamics. *J. Fluid Mech.* **2011**, *672*, 384–410. [CrossRef]
181. Pham, T.; Kumar, S. Drying of Droplets of Colloidal Suspensions on Rough Substrates. *Langmuir* **2017**, *33*, 10061–10076. [CrossRef]
182. Charitatos, V.; Pham, T.; Kumar, S. Droplet Evaporation on Inclined Substrates. *Phys. Rev. Fluids* **2021**, *6*, 084001. [CrossRef]
183. Espín, L.; Kumar, S. Droplet Spreading and Absorption on Rough, Permeable Substrates. *J. Fluid Mech.* **2015**, *784*, 465–486. [CrossRef]
184. Ren, W.; Hu, D.; Weinan, E. Continuum Models for the Contact Line Problem. *Phys. Fluids* **2010**, *22*, 102103. [CrossRef]
185. Lukyanov, A.V. Non-Locality of the Contact Line in Dynamic Wetting Phenomena. *J. Colloid Interface Sci.* **2021**, *608*, 2131–2141. [CrossRef]
186. Li, X.; Pozrikidis, C. Shear Flow Over a Liquid Drop Adhering to a Solid Surface. *J. Fluid Mech.* **1996**, *307*, 167–190. [CrossRef]
187. Moriarty, J.A.; Schwartz., L.W. Unsteady Spreading of Thin Liquid Films with Small Surface Tension. *Phys. Fluids A Fluid Dyn.* **1991**, *3*, 733–742. [CrossRef]
188. Myers, T. Thin Films with High Surface Tension. *SIAM Rev.* **1998**, *40*, 441–462. [CrossRef]
189. Wang, X.; Min, Q.; Zhang, Z.; Duan, Y. Effect of Moving Contact Line's Curvature on Dynamic Wetting of non-Newtonian Fluids. *Langmuir* **2018**, *34*, 15612–15620. [CrossRef] [PubMed]
190. Thompson, P.A.; Robbins, M.O. Simulations of Contact-Line Motion: Slip and the Dynamic Contact Angle. *Phys. Rev. Lett.* **1989**, *63*, 766–769. [CrossRef] [PubMed]
191. Thompson, P.A.; Brinckerhoff, W.; Robbins, M.O. Microscopic Studies of Static and Dynamic Contact Angles. *J. Adhes. Sci. Technol.* **1993**, *7*, 535–554. [CrossRef]
192. Romero, O.J.; Suszynski, W.J.; Scriven, L.E.; Carvalho, M.S. Low-Flow Limit in Slot Coating of Dilute Solutions of High Molecular Weight Polymer. *J. Non-Newton. Fluid Mech.* **2004**, *118*, 137–156. [CrossRef]
193. Wei, Y.; Ramé, E.; Walker, L.; Garoff, S. Dynamic wetting with viscous Newtonian and non-Newtonian Fluids. *J. Phys. Condens. Matter* **2009**, *21*, 464126. [CrossRef] [PubMed]
194. Cohu, O.; Benkreira, H. Entrainment of Air by a Solid Surface Plunging into a Non-Newtonian Liquid. *AIChE J.* **1998**, *44*, 2360–2368. [CrossRef]
195. Charitatos, V.; Suszynski, W.J.; Carvalho, M.S.; Kumar, S. Dynamic Wetting Failure in Shear-Thinning and Shear-Thickening Liquids. *J. Fluid Mech.* **2020**, *892*, A1. [CrossRef]

196. Liang, Z.-P.; Wang, X.-D.; Duan, Y.; Min, Q.; Wang, C.; Lee, D.-J. Dynamic Wetting of Non-Newtonian Fluids: Multicomponent Molecular-Kinetic Approach. *Langmuir* **2010**, *26*, 14594–14599. [CrossRef]
197. Carré, A.; Woehl, P. Hydrodynamic Behavior at the Triple Line of Spreading Liquids and the Divergence Problem. *Langmuir* **2002**, *18*, 3600–3603. [CrossRef]
198. Carré, A.; Woehl, P. Spreading of Silicone Oils on Glass in Two Geometries. *Langmuir* **2005**, *22*, 134–139. [CrossRef]

fluids

MDPI

Review

Statistical Physics Perspective on Droplet Spreading in Reactive Wetting Interfaces

Haim Taitelbaum

Department of Physics, Bar-Ilan University, Ramat-Gan 52900, Israel; haimt@mail.biu.ac.il

Abstract: Droplet spreading is a fascinating phenomenon. Especially when the droplet spreads, reacts, and dissolves on and into metal substrates. This reactive wetting mainly occurs at high temperatures, with a vast number of applications in industry and material science. It is common to monitor and study the process using a side-view projection of the droplet, focusing on the dynamics and shape of its contact line. However, when the spreading is monitored *top*-view, rich and non-trivial spatio-temporal patterns are revealed during different stages of the process. These patterns call for a different type of study of the perimeter of the entire droplet. Statistical physics is the natural candidate to perform such tasks, using tools developed for the study of kinetic roughening of advancing interfaces. In this review, we demonstrate the use of these tools, the growth, roughness, and persistence exponents, to study the spreading of mercury droplets on metal-on-glass at room temperature, which by itself is a unique experimental system at this range of temperatures. The universality of the results is discussed in comparison with similar patterns of reactive wetting at *high* temperatures.

Keywords: reactive-wetting; metal-on-glass; contact angle; surface tension; kinetic roughening; growth; roughness; persistence

1. Introduction

Droplet spreading is a fascinating phenomenon [1–8]. The dynamics of droplet spreading is traditionally described by the change in time of the droplet radius $R(t)$ and its contact angle $\theta(t)$ [9–12] (See Figure 1). According to the well-known Tanner's Law [9], $R(t) \sim t^{1/10}$ and $\theta(t) \sim t^{-3/10}$ for a spherical cap shape droplet. When gravity is non-negligible, Lopez et al. [10] showed that the radius grows slightly faster, as $t^{1/8}$. Nevertheless, these dynamics laws deal with a *side*-view projection of the spreading droplet in a *classical* wetting process. When other mechanisms are involved, e.g., dissolution, compound formation, chemical reaction, and more, the wetting is *reactive* [13–22]. No theoretical prediction for the dynamic radius and contact angle is known. In turn, based on many empirical results (see, e.g., [22]), it is commonly believed that in these cases the spreading is much faster, behaving like $R(t) \sim t$ for the radius. Such reactive wetting systems have many applications in industry and material science.

In this review, we describe the dynamics of droplet spreading in reactive wetting interfaces, as obtained in a series of papers by our group [23–35] and references cited therein. Several aspects make our experimental system unique. (A) It is metal on *metal-on-glass*; (B) The spreading droplet material is *mercury* [36]. (C) The experiment is performed at *room temperature*. (D) We follow the *top* view of the entire perimeter of the advancing droplet.

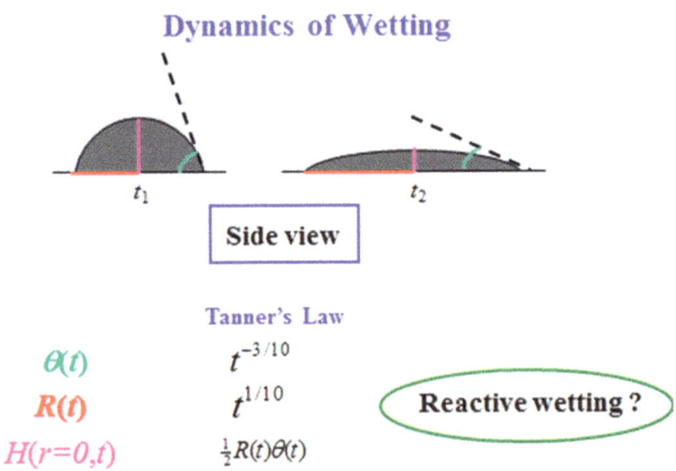

Figure 1. Wetting dynamics (side view).

Most of reactive wetting experiments are metal-on-metal at *high* temperatures (see e.g., [22]). This requires the heating of the upper metal until it turns into a liquid, which can spread. The only relevant metal that is liquid already in *room* temperature is the notorious mercury. Hence, it is the only candidate for room temperature reactive-wetting experiments. In addition, from the positive aspect, mercury is used for measurements in medical devices, dental amalgams, and more, and it is important to explore its behavior. In terms of data analysis, our study exploits *statistical physics* tools to identify the driving mechanisms in the system. The statistical physics point of view enriches the insights into the droplet spreading process. The use of these tools has the advantage of universal applicability of the results to other systems, materials, and experimental conditions, such as high temperatures.

The paper starts with a description of the experimental system and methodology. We then move on to present the two main regimes of the spreading process—bulk spreading and kinetic roughening. The next chapter is devoted to the statistical physics tools, i.e., the scaling exponents, the roughness, growth, and persistence exponents. These enable us to sketch the big picture of the entire process, and to discuss the universality of the results as reflected in high-temperature systems. We then discuss additional topics, such as the kinetic roughening in the QKPZ (Quenched Kardar-Parisi-Zhang) equation and the Ising model, and the effect of temperature on the kinetic roughening exponents.

2. The Experimental System and Methodology

The experimental system is described in detail in [23–26]. In short, silver (or gold) thin films of various thicknesses (2000–6000 Å) were evaporated, sputtered (or polished) on top of a glass substrate, forming a metal-on-glass substrate. Mercury droplets (100–150 µm diameter) were deposited on this substrate at room temperature (Figure 2). The spreading process was monitored top view using an optical microscope equipped with DIC (Differential Interference Contrast) lenses and a CCD (Charge Coupled Device) camera. The top-view images have been translated into side-view profiles [26], based on the various colors obtained using reflection-DIC light microscopy. The idea is that the object reflects light from different points of its surface with different colors that are indicative of its surface slope at each point of the mercury droplet. The measurements were taken in a diametric cross section of the droplet, parallel to the beam-splitting direction.

Some typical snapshots are shown in Figure 3 for mercury spreading on top of a silver film (4200 Å) on glass, taken at times t = 5, 10, and 15 s after the droplet deposition on the film. These snapshots demonstrate that after several seconds, the droplet perimeter changes from a seemingly smooth curve into a roughened pattern. This reflects the two main regimes in the droplet spreading process, "bulk spreading" and "kinetic roughening". In short, the transition between these stages is driven by the dissolution of the upper liquid metal into the lower solid one, followed by the capillary effect at the metal-glass interface, as will be described in detail below.

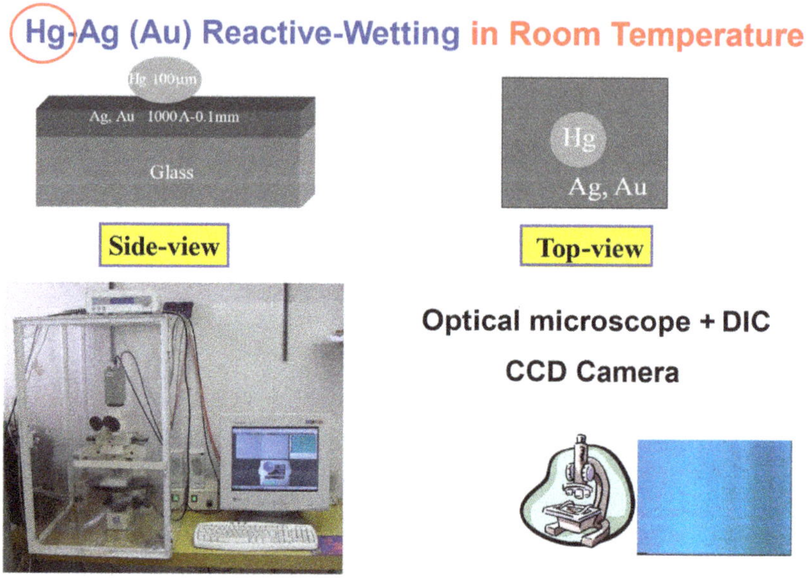

Figure 2. The experimental system of liquid metal on solid metal-on-glass at room temperature.

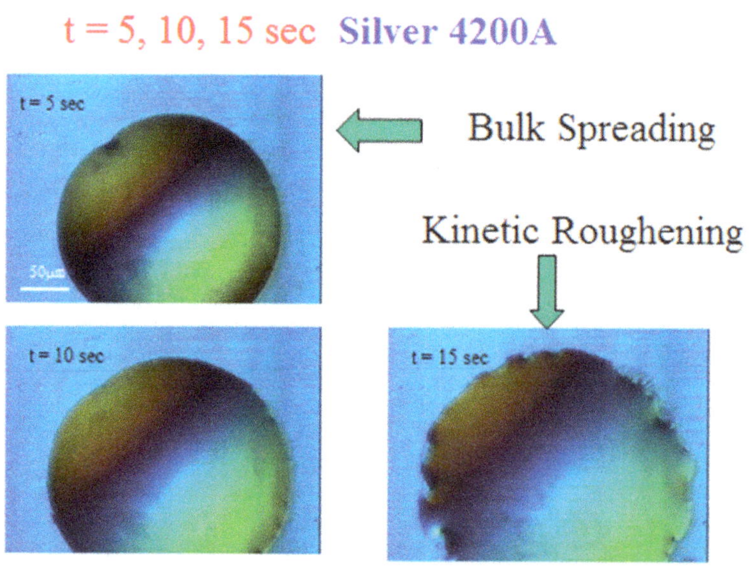

Figure 3. Typical snapshots of mercury spreading on silver film on glass under the microscope. The transition from bulk spreading to kinetic roughening is demonstrated.

The top-view images are used to reconstruct the *side-view* profiles of the spreading droplet using the technique described in detail in [26]. This is shown in Figure 4 for the same system, where the vertical scale is stretched in the main graph in order to demonstrate the spherical cap shape of the advancing droplet. The bottom sketch is the real scale. Such reconstructed profiles are used to study the spreading kinetics.

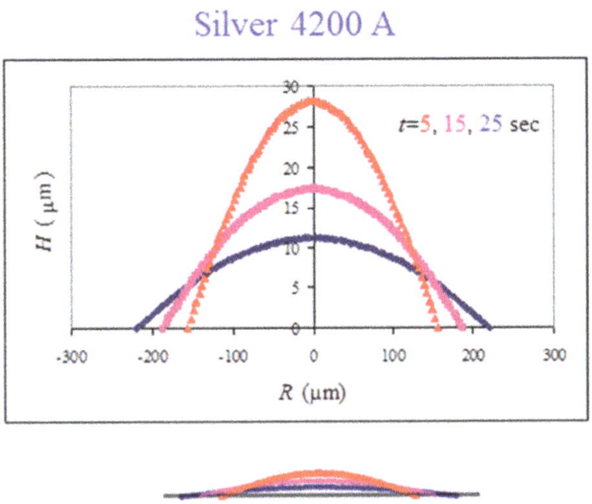

Figure 4. Side-view reconstruction of the droplet shape in the spreading dynamics of mercury droplet on silver thin film. The bottom sketch is the real scale.

3. The Spreading Process

Using the reconstructed side-view shape profiles, one can follow the details of the spreading process of the reactive metal droplet on the metal-on-glass [30,31]. This was found to consist of several consecutive *sub*-steps (Figure 5) in the "bulk spreading" regime prior to the "kinetic roughening" regime, with a fast flow region in between.

After placing the mercury droplet on the thin silver layer (a), the droplet starts to spread and react (or dissolve) with the film (b). This "bulk spreading regime" is mainly controlled by chemical reaction and diffusion, both on top and inside the silver film. After crossing the entire film downwards, the mercury material touches the underlying glass substrate (c). This is a critical event, as the film-glass interface enters the game as a dominant player. At this very moment, a new spreading regime, which we call the "fast-flow regime," occurs abruptly. A new and thin front (500 Å in height), which optically looks like a halo, detaches from the bulk (d) and flows ahead throughout the entire film (e) with a much higher velocity (about two orders of magnitude faster). This is since there is effectively no energy barrier at the Hg-glass interface, and the fast flow at this interface can easily lower the entire system's energy. The fast-flow dynamics give rise to the formation of a reaction band on top of the mercury film surface (e), advancing with the average velocity of the bulk propagation regime, and containing various chemical compositions and intermetallic phases [30,31]. This reaction band is where the "kinetic roughening" regime takes place. This will be discussed in detail in the next section.

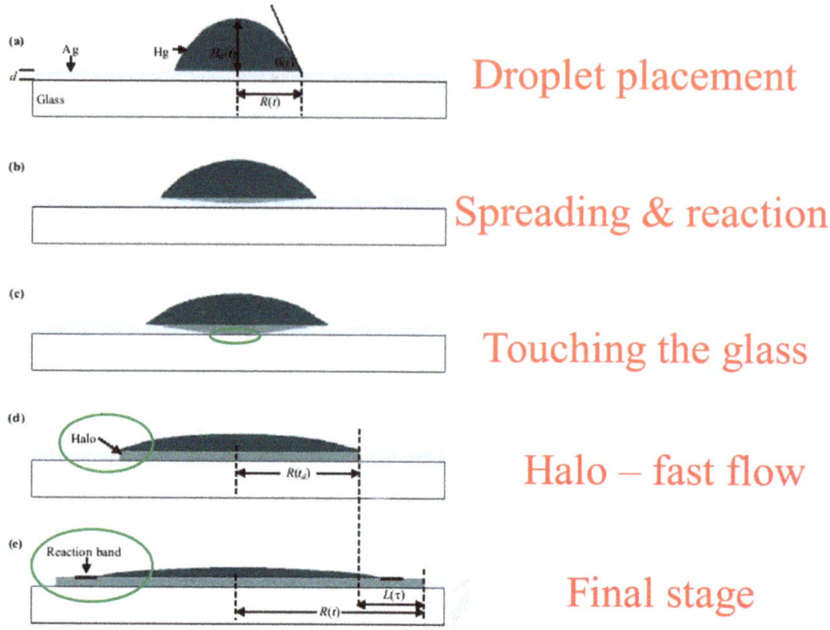

Figure 5. Schematic sequence of mercury droplet spreading on thin silver films at room temperature. (**a**) Droplet placement on the film. (**b**) Droplet bulk spreading and reacting with the underlying substrate. (**c**) Mercury touching the glass. (**d**) Start of the fast-flow regime, halo propagation. (**e**) Formation of a reaction band.

3.1. Bulk Spreading

In Figure 6, we show our empirical results for the parameters of the bulk spreading, the radius $R(t)$ and the angle $\theta(t)$, for the data of Figure 4. Initially, the droplet radius grows linearly in time, $R(t) \sim t$, with a constant velocity, in this case 2.1 m/s, and the angle of contact $\theta(t)$ decreases continuously. At about $t = 15$ s, $\theta(t)$ demonstrates a sudden step in its decreasing trend. This is exactly the time when the mercury crosses the entire silver layer for the first time and touches the underlying glass (Figure 5c). We define this characteristic time as t_d, where d is the thickness of the Ag layer. Hence, the step in the behavior of $\theta(t)$ at t_d is strongly *thickness dependent*. On the other hand, the velocity of the bulk propagation was found to be approximately the same for all Ag thicknesses around 4200 Å (from 2000 up to 6000 Å), having the average value of 2.5 ± 0.4 m/s. It is reasonable to assume that this velocity is *material dependent*, depending on the reaction of the mercury droplet with the silver surface. This surface velocity is much higher (about two orders of magnitude) than the mercury-silver reaction rate *inside* the silver film, which is attenuated by diffusion [30]. The constant velocity remains approximately the same also after t_d, with a very slight variation at t_d, the time of the fast flow burst. The halo velocity at this regime is also constant, but its specific value is expectedly to be thickness dependent. It is also important to point out that t_d is also the time where the kinetic roughening of the reaction band starts to show up (Figure 5e).

The bottom line is that the underlying glass plays a significant role in the transition of the system behavior from bulk spreading to kinetic roughening. Hence, the spreading of metal liquid on metal-*on-glass*, which is one of the features of our system, is quite different from the spreading of metal on metal. The glass is more than just a platform.

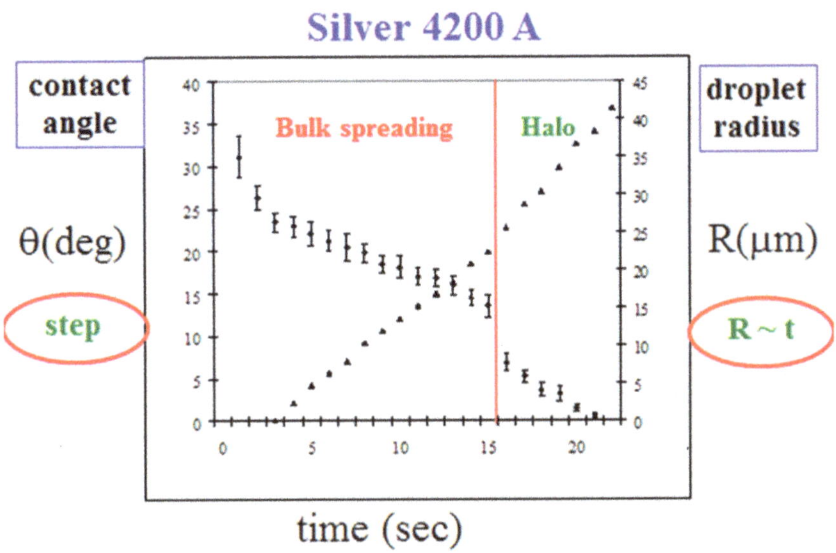

Figure 6. The dynamics of the droplet radius (right) and contact angle (left) as a function of time, in the bulk spreading regime.

3.2. Kinetic Roughening

In nonequilibrium surface and interface growth processes, kinetic roughening is the statistical physics framework to follow the stochastic dynamics in the processes, which induce scaling laws for the advancing interfaces with respect to time and length scales [37–47]. It is applicable to a large variety of systems, such as tumor growth [48,49], bacterial and colonies [50–52], biological systems [53], reaction-diffusion fronts [54], urban growth [55], urban skyline [56], forest fires [57], slow combustion of paper [58], drying wet paper [59], paper wetting [60], magnetic flux fronts [61], viscous fingering [62] and more [63,64]. Some recent related papers deal with kinetic roughening in thin film spreading [65,66], spreading fronts of wetting droplets [67,68], and contact line dynamics [69,70].

In Figure 7, we illustrate the top-view of an advancing interface, described by a function $h(x,t)$, for the horizontal "height" of the interface at position x at time t. The kinetic roughening properties of an interface of length L_0 are defined based on the fluctuations or the second moment of $h(x,t)$. This width function, $W(L,t)$, is defined as

$$W^2(L,t) = < h(x,t)^2 > - < h(x,t) >^2 \tag{1}$$

where the average is over all x values between 0 and the minimal length scale in the system (e.g., lattice unit) up to half of the system size $L_0/2$. Family-Vicsek [44] assumed that the width behavior obeys a scaling relation from which follows a dichotomic behavior,

$$W(L,t) \sim \begin{cases} t^\beta, & t \ll t_0 \\ L^\alpha, & t \gg t_0 \end{cases} \tag{2}$$

with α and β being the roughness and growth exponents, respectively, and t_0 is given in terms of the system size L_0 as

$$t_0 \sim {L_0}^{\alpha/\beta}. \tag{3}$$

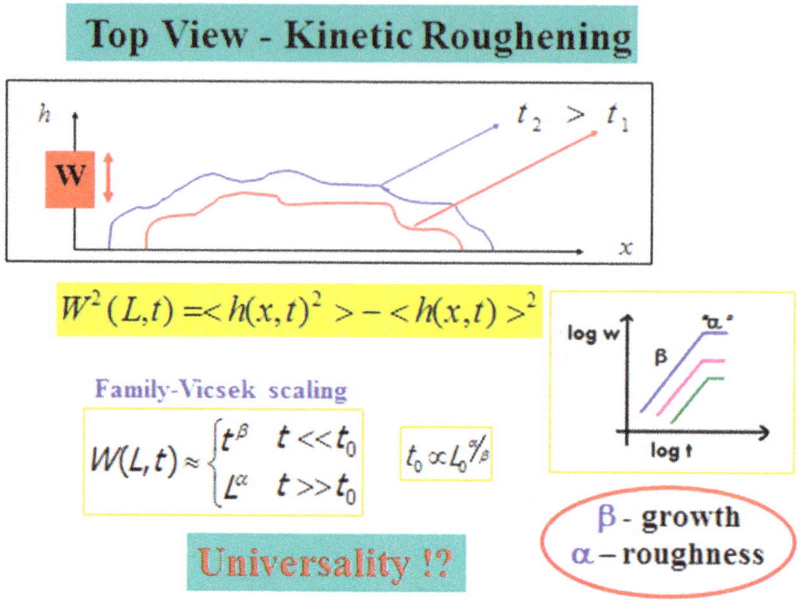

Figure 7. Kinetic roughening—The general scheme, Family-Vicsek scaling, and the growth and roughness scaling exponents.

At the beginning of the process, the interface width increases as t^β . This continues up to time t_0, when the interface width saturates. At this saturation regime, the width behaves as L^α, and one can extract the interface roughness defined by the roughness exponent α, which is related to the interface's autocorrelation. The growth exponent β describes the temporal dynamics of the interface, while the roughness exponent α, which is measured after the interface width reaches saturation, describes the geometric shape of its final stage. Based on these two exponents, one can classify the system into a certain *universality class* [40–43] that may shed light on the fundamental mechanisms in the process.

4. Scaling Exponents

4.1. The Roughness and Growth Exponents

In Figure 8, we show a typical set of kinetic roughening scaling exponents, obtained for a 2000 Å thick silver film evaporated on top of glass. The results are $\beta = 0.46 \pm 0.02$ for the growth exponent and $\alpha = 0.76$ for the roughness exponent. However, the experiments reveal another interesting characteristic of the roughness exponent. Plotting this exponent as a function of distance (according to Equation (2)), one obtains a crossover behavior from $\alpha = 0.76$ at short distances to around $\alpha = 0.47$ at longer distances. This calls for a deeper look into the meaning of the roughness exponent, which reflects the inner correlations of the advancing interface [28,29].

When $\alpha > 0.5$, there is a strong correlation between nearby interface points that tend to progress in a similar manner, with $\alpha \to 1$ for a smooth interface. $\alpha = 0.5$ describes a lack of correlation (random walk), where every point on the interface moves randomly and independently of the surroundings. $\alpha < 0.5$ describes anticorrelations between the points. Hence, the crossover behavior of α in Figure 8 from 0.76 to 0.47, represents a change of the correlation along the interface, from a relatively strong correlation at short distances to a complete loss of correlation at long distances. This is quite intuitive. The crossover occurs at some typical distance, which is termed the "correlation length" of the interface.

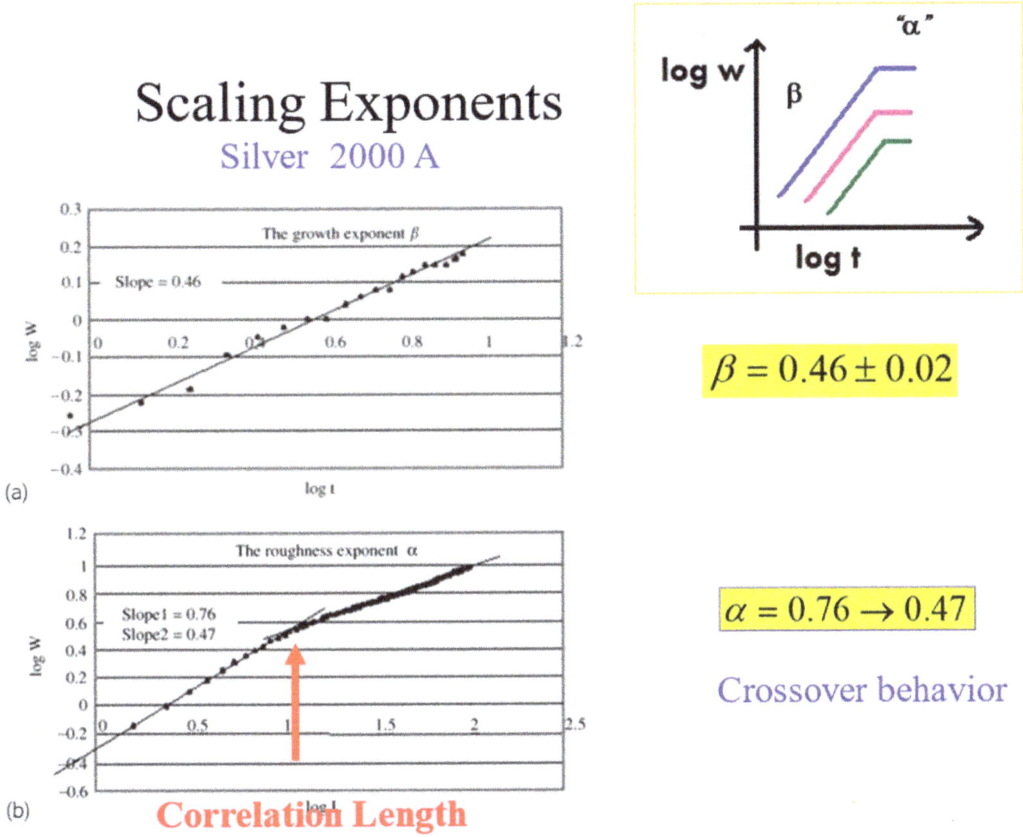

Figure 8. The scaling exponents α and β and the crossover behavior at the correlation length.

This correlation length can also be extracted from temporal width fluctuations of the growing interface in its growth regime. The fluctuating behavior results from competing mechanisms in the growth process, normal growth and the surface tension. This is discussed in detail in [28,29].

Regarding the classification of the system into a specific universality class, a common practice in statistical physics, a possibly relevant universality class is $\alpha + \alpha/\beta = 2$, valid for isotropic systems [40,41]. In Figure 9, we show some more examples of sets of exponents, obtained for a variety of thin films in various thicknesses at room temperature, which show that this is not the case. While the roughness exponent α seems to be always around 0.8 (at distances shorter than the correlation length) for all cases, the growth exponent β is significantly different in each system.

This can be attributed to the very different *substrate* roughness (prior to the mercury droplet deposition) resulting from different thicknesses and preparation conditions. It is also reflected in the different correlation lengths for each case, as shown next to the table. It is assumed that these different parameters affect the advancing interface growth process (hence the different growth exponent β), while the roughness exponent α at the saturation regime of the process is more universal.

Summary of Results –

Room Temperature

Material	Thickness	α	β
Silver	2000A	0.76 ± 0.03	0.46 ± 0.02
	0.1mm	0.82 ± 0.04	0.60 ± 0.02
Gold	1500A	0.85 ± 0.03	0.76 ± 0.03
	3000A	0.86 ± 0.04	1.00 ± 0.04

NOISE – Substrate Roughness

Correlation length

2μm

8μm

7μm

14μm

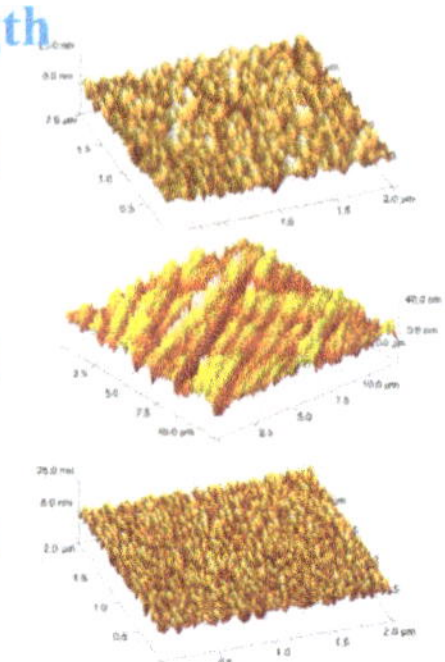

Universality Class ??

$\alpha + \alpha/\beta = 2$?

Figure 9. Scaling exponents at room temperature, for various materials and thicknesses. The corresponding correlation lengths and preparation techniques are shown on the right.

4.2. The Persistence Exponent

The microscopic dynamics of the interface can also be described by another measure, the *persistence* probability, defined as the probability that a *stochastic variable* will never cross some *reference level* within a time interval t [47]. In our system, the stochastic variable is the location of the interface point(s) along the interface line, and the reference level is the horizontal height $h(x,t)$ at the point location x at some initial time t_0 (see Figure 10). It is then averaged over all points in a computational method which is described in detail in Ref. [34]. This measure considers local microscopic fluctuations and is expected to reflect an additional statistical physics view on the dynamics of the interface, complementing the more traditional growth exponent.

The persistence probability was shown to obey a power law decay with a persistence exponent θ [47]

$$P(t) \sim t^{-\theta} \quad (4)$$

In linear systems, there is a simple relation between the two temporal measures, the growth and the persistence exponents [47]

$$\theta = 1 - \beta \quad (5)$$

Hence, in linear systems one can either measure the persistence exponent directly or indirectly using the growth exponent β.

What happens in a *nonlinear* system like reactive-wetting or droplet spreading? In [34], we report the direct persistence exponent results of the mercury droplet (150 μm) spreading on silver (4000 Å)-on-glass at room temperature. We were able to identify several distinct kinetic time regimes in this process. In the first one, while the interface (i.e., the contact line) is moving but its width is not yet growing, the persistence exponent is $\theta = 0.55 \pm 0.05$, which is typical for a random, noisy behavior. In the second regime, there is an effective growth of the interface width with a growth exponent $\theta = 0.67 \pm 0.06$, followed by a final regime of saturation, according to the Family-Vicsek description of interface growth. The persistence exponent in this regime is $\theta = 0.37 \pm 0.05$, which indicates that the hyper-scaling relation $\theta = 1 - \beta$ seems to hold even for this nonlinear experimental system. The roughness exponent α, calculated at the end of the process according to Equation (2), was found to be 0.83 ± 0.008 [32], in agreement with previous studies of this system [24,28,29].

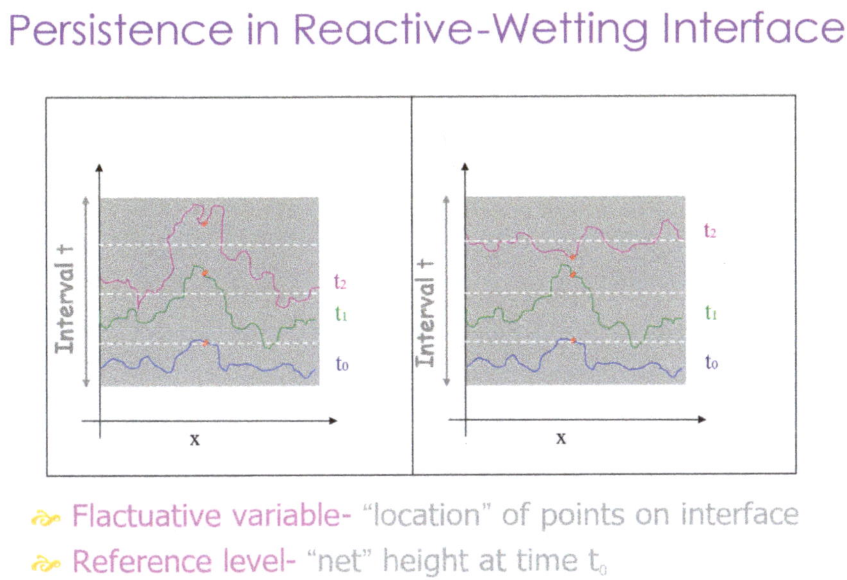

Figure 10. Persistence in reactive-wetting interfaces. The white dotted lines represent three consecutive times. The red dot is persistent on the left frame but not on the right one.

5. The Big Picture

Using this insight, we can now look at the "big picture" of the spreading process in our system. In Figure 11, we summarize the three sub-regimes in the kinetic roughening process of the interface: noise (light green), non-linear growth (purple), and surface tension relaxation (light blue). The upper curve shows the width of the interface $W(t)$ as a function of time, starting with a region of noisy constant width followed by regions of growth (with the growth exponent β) and then saturation (where the roughness exponent α is measured) as in the inset of Figures 7 and 8 above. The lower curve is the independent calculation of the complementary persistence curve, showing the random walk exponent -0.5 for the initial noisy region. The detailed calculation of the persistence exponent in the other two regions requires a shift of the initial time, as is explained in detail in Ref. [34].

The snapshots presented on the right side of the figure are examples of the time evolution of the interface at each of the three sub-regions. One can see the significant increase in the width in the purple growth region, compared with the constant (in time) width in both the preceding noise region (light green) and the succeeding surface tension relaxation/saturation region (light blue).

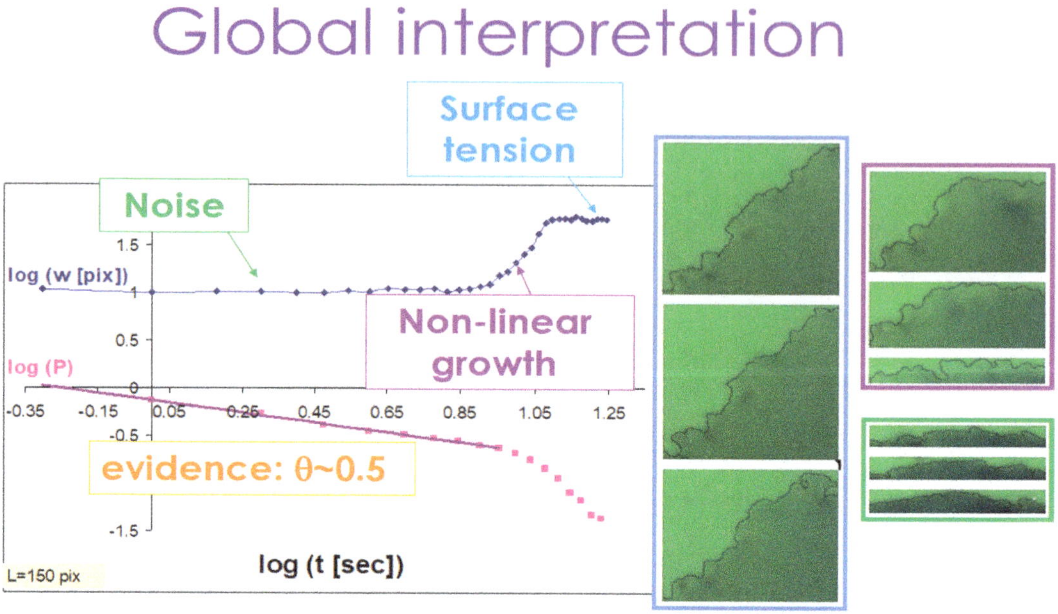

Figure 11. Noise, non-linear growth, and surface tension relaxation in the global kinetic roughening process.

6. Universality—High Temperatures

Most of the spreading results in *high*-temperature reactive-wetting systems were obtained using a *side* view of the spreading droplet (see, e.g., [22]). One of the exceptions is a series of *top*-view experiments performed by Singler's group at Binghamton [17,18,33]. They performed experiments of Sn droplets spreading on a 1 μm coating of Au on Cu of thickness 12 μm, deposited on glass, at several high temperatures. For detailed information on the experimental setup, see Yin et al. [17].

The kinetic roughening patterns at high temperatures are shown in Figure 12. They are quite similar to the room temperature patterns presented above. Analyzing the data in the same methodology and plotting the interface width W as a function of time, one obtains the three typical regimes of the interface dynamics [33]. It should be mentioned that in the high temperature system, there is an additional early and extremely short time regime in which the liquid spreads with no discernible morphological or chemical change of the interface. This very early regime is undetectable in the room temperature reactive-wetting experiment [33,34].

The persistence concept allows one to better understand and clearly define these regimes. The dimensionless comparison of the three regions in the room temperature Hg–Ag system and the high temperature Sn–Au/Cu system shown in Figure 13 speaks for itself.

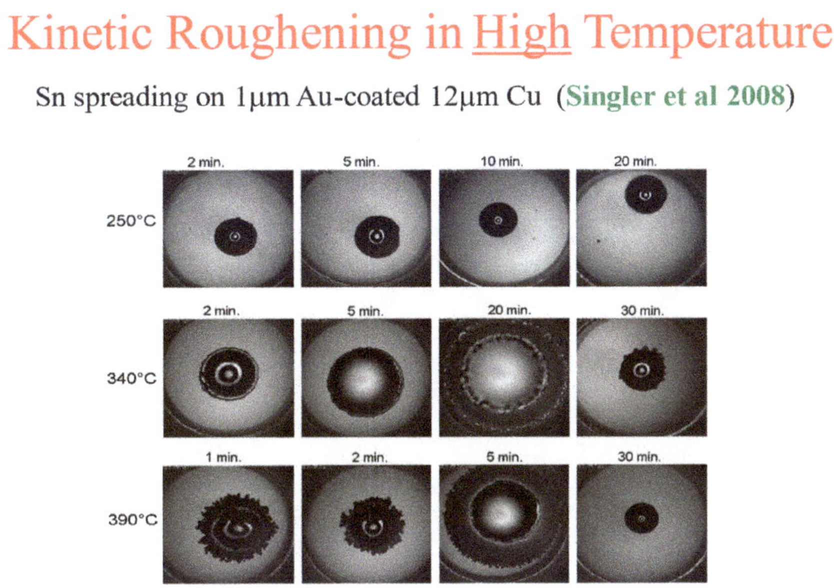

Figure 12. Kinetic roughening at high temperature [17].

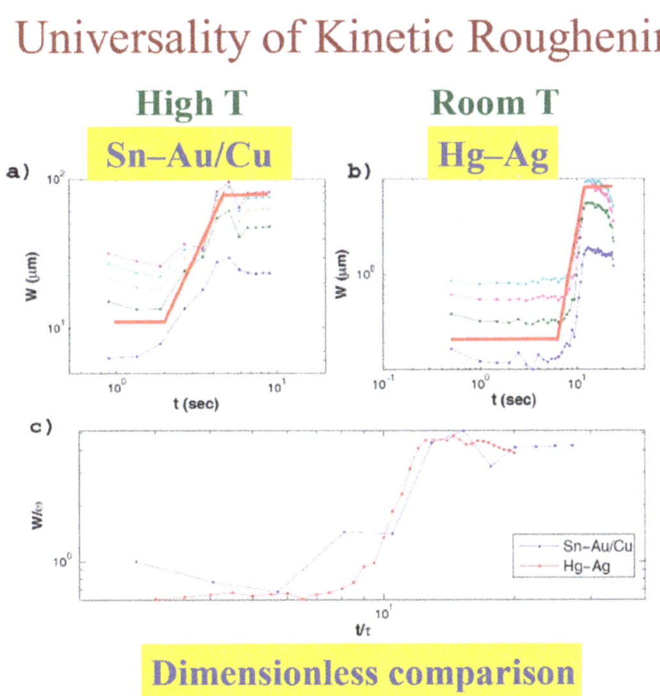

Figure 13. Universality of kinetic roughening: The interface width W as a function of time in high temperature (**a**) vs. room temperature (**b**). A dimensionless comparison (**c**).

This comparison allows one to compare quantitatively the similarity between reactive wetting experiments of different materials in various thicknesses, as reflected by the kinetic roughening exponents α and β (Figure 14). One can see that the growth exponent of the system under study, Sn–Au/Cu, is very close to the growth exponent of Hg on thick Ag (second and fourth lines). The value of the roughness exponent α is usually around 0.8 with variation that can be caused by other effects, like overhangs [32]. Therefore, it seems that the Sn–Au/Cu belongs to the same universality class as the Hg–Ag system. According to the value of the exponents, we conjecture that this could be the QKPZ universality class whose exponents are $\beta = 3/5$ and $\alpha = 3/4$ [40]. This conjecture should be confirmed by additional experiments. The Hg–Au system belongs to a different universality class; however, it obeys the same hyper-scaling relation, as discussed in detail in [25].

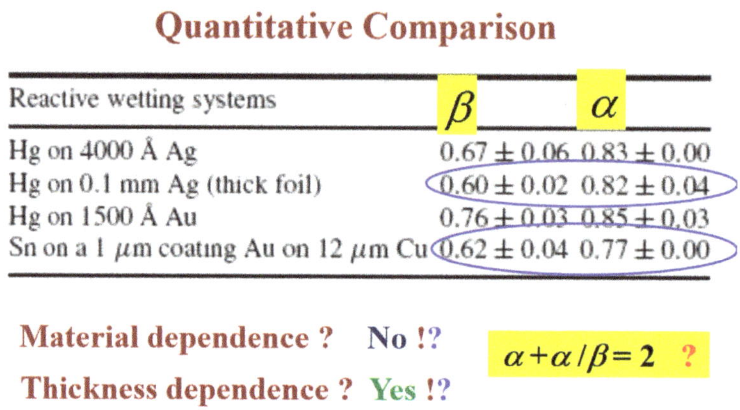

Figure 14. Room temperature (first three lines) vs. high temperature (last line): A quantitative comparison of the scaling exponents [33].

7. Kinetic Roughening, the QKPZ Equation, and the Ising Model

The natural question to be asked is whether there exists a well-defined model to describe the interface dynamics of the reactive-wetting system. In [32] we investigated two candidates for this task, the QKPZ equation and the Ising chain model at zero temperature. The QKPZ is an explicit equation for describing advancing interfaces, while a given configuration of the spins in an Ising ferromagnetic 1d chain can be converted into a landscape, i.e., can be thought of as an interface. The QKPZ equation is the quenched disorder version of the well-known KPZ (Kardar-Parisi-Zhang) equation [45]. It was studied in the context of its kinetic roughening exponents, but not for its persistence. The persistence exponent of the Ising model has been calculated, but the kinetic roughening exponents of an Ising chain were not. For more details, see Ref. [32].

In [32], we calculated for each model its persistence, growth, and roughness exponents, and compared them with the available experimental results at room temperature. The results are given in Table 1. We found that neither of these two models gives a complete description of the dynamics of the experimental reactive-wetting system, but each one of them describes quantitatively some aspects of the interface growth process.

Table 1. The persistence, growth and roughness exponents for reactive-wetting experiments, QKPZ and Ising simulations.

	Reactive-Wetting Experiment	QKPZ Simulations	Ising Simulations
Persistence θ	0.37 ± 0.05	0.76 ± 0.006	0.38 ± 0.0017
Growth β	0.67 ± 0.06	0.61 ± 0.02	0.29 ± 0.017
Roughness α	0.83 ± 0.008	0.72 ± 0.02	0.92 ± 0.012

We suggest that the behavior of our reactive-wetting system is dichotomic. While kinetic roughening has to do with the collective behavior of the advancing interface (allowing e.g., for estimation of the lateral correlation length [28,29]), the microscopic persistence measure is much more sensitive [47]. In the macroscopic scale, reflected by the growth exponent, the behavior of the system is QKPZ-like, but in the microscopic scale, reflected by the persistence exponent, the system shows Ising-like behavior. This might be supported by the observation that local surface tension resembles the neighboring interactions in the Ising chain, which are dominant for relatively small scales. In larger scales, however, the non-linear growth of the interface is more profound, due to the chemical interaction with the substrate. This growth is better described by the QKPZ equation. These conjectures should be further substantiated.

The interpretation of the results for the roughness exponent α [32] is based on the observation that in the final stages of the spreading process, the interface propagation is not necessarily perpendicular only. This results in an overhangs geometry of the advancing interface, which was shown to affect the value of the roughness exponent.

8. Effect of Temperature on Kinetic Roughening Exponents

So far, we have described spreading experiments around room temperature. A reasonable question is what is the effect of slight temperature variations *around* this temperature. To study experimentally such a possible effect, we have used a heating stage which allowed us to change the temperature in a controlled manner in the range $-15\ ^{\circ}\mathrm{C} < T < 25\ ^{\circ}\mathrm{C}$. Higher temperatures are more difficult to handle with respect to the mercury vapors. The spreading of a mercury droplet (~150 μm) on a silver substrate (4000 Å) at these temperatures was monitored using the optical microscope as in the room temperature experiments. The full details can be found in [35].

First, we studied the temperature effect on the wetting dynamics (droplet radius and velocity). We found that for all studied temperatures, the spreading radius $R(t)$ grows linearly with time, $R(t) \sim t$, with the specific velocity value (the proportionality constant) depending on the specific temperature. We also studied the temperature effect on the kinetic roughening properties of the advancing interface (growth and roughness exponents). Our results [35] are that the growth exponent β increases with temperature while the roughness exponent α is relatively constant for all studied temperatures, with a value around 0.8. In the literature, there are a variety of dependencies of kinetic roughening exponents on temperature [35]. Our results show that it depends on the specific system.

In sum, the temperature does play a role in the behavior of reactive-wetting systems, not only in very high temperatures, but also around room temperature.

9. Summary

The spreading of a droplet on metal-on-glass is highly non-linear and very sensitive to minor fluctuations. It's extremely difficult to construct a comprehensive model that fully describes it. Alternatively, one can look at the top-view data of the dynamics and geometry of the horizontal morphology of the advancing triple line. A statistical physics analysis of the common kinetic roughening properties of similar systems can yield universal insights (Figure 15), allowing for a better understanding of the dynamical aspects of the process.

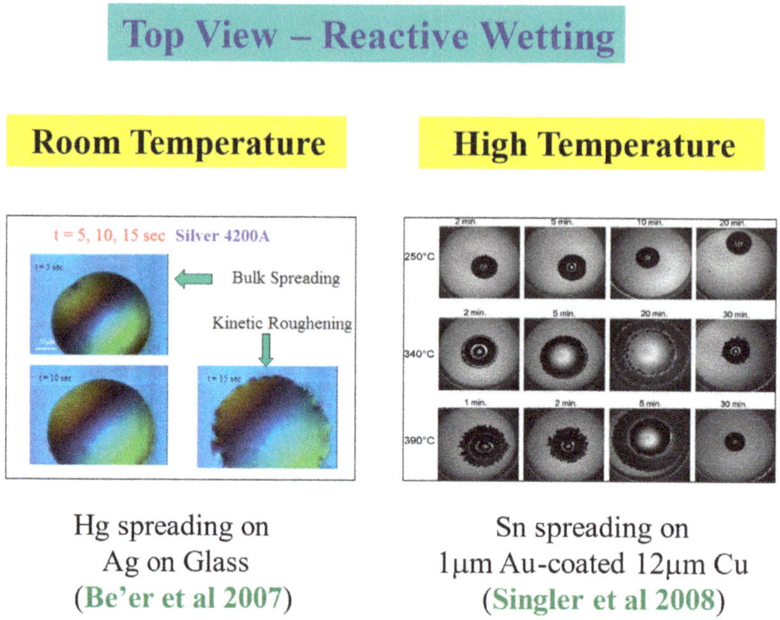

Figure 15. Room temperature [30] vs. high temperature [17]: A qualitative comparison.

Funding: This review paper received no external funding.

Acknowledgments: This paper is a written summary of numerous talks presented on this subject in international conferences and workshops. I am indebted to my former students Avraham Be'er, Inbal Hecht, Yael Efraim, Meital Harel, and late Tal Grynpas-Ya'akov for their immense contribution to this project throughout the years. I am highly thankful to my colleagues Yossi Lereah, Aviad Frydman, and of course to Timothy Singler and his group at Binghamton University.

Conflicts of Interest: The authors declare no conflict of interest.

References

1. de Gennes, P.G. Wetting: Statics and Dynamics. *Rev. Mod. Phys.* **1985**, *57*, 827. [CrossRef]
2. Leger, L.; Joanny, J.F. Liquid Spreading. *Rep. Prog. Phys.* **1992**, *55*, 431. [CrossRef]
3. Oron, A.; Davis, S.H.; Bankoff, S.G. Long-Scale Evolution of Thin Liquid Films. *Rev. Mod. Phys.* **1997**, *69*, 931. [CrossRef]
4. de Gennes, P.G. The Dynamics of Reactive Wetting on Solid Surfaces. *Physica A* **1998**, *249*, 196. [CrossRef]
5. Bonn, D.; Eggers, J.; Indekeu, J.; Meunier, J.; Rolley, E. Wetting and Spreading. *Rev. Mod. Phys.* **2009**, *81*, 739. [CrossRef]
6. Tadmor, R. Approaches in Wetting Phenomena. *Soft Matter* **2011**, *7*, 1577. [CrossRef]
7. Karim, A.M. A Review of Physics of Moving Contact Line Dynamics Models and its Applications in Interfacial Science. *J. Appl. Phys.* **2022**, *132*, 080701. [CrossRef]
8. Karim, A.M.; Suszynski, W.J. Physics of Dynamic Contact Line: Hydrodynamics Theory versus Molecular Kinetic Theory. *Fluids* **2022**, *7*, 318. [CrossRef]
9. Tanner, L.H. The Spreading of Silicone Oil Drops on Horizontal Surfaces. *J. Phys. D* **1979**, *12*, 1473. [CrossRef]
10. Lopez, J.; Miller, C.A.; Ruckenstein, E. Spreading Kinetics of Liquid Drops on Solids. *J. Colloid Interface Sci.* **1976**, *56*, 460. [CrossRef]
11. Blake, T.D. *Dynamic Contact Angles and Wetting Kinetics in Wettability*; Berg, J.C., Ed.; Marcel Dekker: New York, NY, USA, 1993.
12. Davidovitch, B.; Moro, E.; Stone, H.A. Spreading of Viscous Fluid Drops on a Solid Substrate Assisted by Thermal Fluctuations. *Phys. Rev. Lett.* **2005**, *95*, 244505. [CrossRef] [PubMed]
13. Landry, K.; Eustathopoulos, N. Dynamics of Wetting in Reactive Metal/Ceramic Systems: Linear Spreading. *Acta Mater.* **1996**, *44*, 3923. [CrossRef]
14. Eustathopoulos, N. Dynamics of Wetting in Reactive Metal/Ceramic Systems. *Acta Mater.* **1998**, *46*, 2319. [CrossRef]

15. Eustathopoulos, N.; Nicholas, M.G.; Drevet, B. *Wettability at High Temperatures*; Pergamon: New York, NY, USA, 1999.
16. Kumar, G.; Prabhu, K.N. Review of Non-Reactive and Reactive Wetting of Liquids on Surfaces. *Adv. Colloid Interface Sci.* **2007**, *133*, 61. [CrossRef]
17. Yin, L.; Chauhan, A.; Singler, T.J. Reactive Wetting in Metal/Metal systems: Dissolutive versus Compound-Forming Systems. *Mater. Sci. Eng. A* **2008**, *495*, 80. [CrossRef]
18. Singler, T.J.; Su, S.; Yin, L.; Murray, B.T. Modeling and Experiments in Dissolutive Wetting: A Review. *J. Mater. Sci.* **2012**, *47*, 8261. [CrossRef]
19. Bormashenko, E. *Wetting of Real Surfaces*; de Gruyter: Berlin, Germany, 2013.
20. Bormashenko, E. Apparent contact angles for reactive wetting of smooth, rough, and heterogeneous surfaces calculated from the variational principles. *J. Colloid Interface Sci.* **2019**, *537*, 597. [CrossRef]
21. Lin, Q.; Xie, K.; Sui, R.; Mu, D.; Cao, R.; Chang, J.; Qiu, F. Kinetic analysis of wetting and spreading at high temperatures: A review. *Adv. Colloid Interface Sci.* **2022**, *305*, 102698. [CrossRef] [PubMed]
22. Sobczak, N.; Agathopoulos, S. (Eds.) *HTC 2022: Book of Abstracts of the 10th International Conference on High Temperature Capillarity, Krakow, Poland, 12–16 September 2022*; Polish Foundrymen's Association: Kraków, Poland, 2022; ISBN 978-83-963247-1-9.
23. Be'er, A.; Lereah, Y.; Taitelbaum, H. The Dynamics and Geometry of Solid-Liquid Reaction Interface. *Physica A* **2000**, *285*, 156. [CrossRef]
24. Be'er, A.; Lereah, Y.; Hecht, I.; Taitelbaum, H. The Roughness and Growth of a Silver-Mercury Reaction Interface. *Physica A* **2001**, *302*, 297. [CrossRef]
25. Be'er, A.; Lereah, Y.; Frydman, A.; Taitelbaum, H. Spreading of a Mercury Droplet on Thin Gold Films. *Physica A* **2002**, *314*, 325. [CrossRef]
26. Be'er, A.; Lereah, Y. Time-resolved, three-dimensional quantitative microscopy of a droplet spreading on solid substrates. *J. Microsc.* **2002**, *208*, 148. [CrossRef]
27. Hecht, I.; Taitelbaum, H. Roughness and Growth in a Continuous Fluid Invasion Model. *Phys. Rev. E* **2004**, *70*, 046307. [CrossRef]
28. Be'er, A.; Hecht, I.; Taitelbaum, H. Interface Roughening Dynamics: Temporal Width Fluctuations and the Correlation Length. *Phys. Rev. E* **2005**, *72*, 031606. [CrossRef]
29. Hecht, I.; Be'er, A.; Taitelbaum, H. Single Interface Growth: Fluctuations and the Correlation Length. *Fluct. Noise Lett.* **2005**, *5*, L319. [CrossRef]
30. Be'er, A.; Lereah, Y.; Frydman, A.; Taitelbaum, H. Spreading of Mercury Droplets on Thin Silver Films at Room Temperature. *Phys. Rev. E* **2007**, *75*, 051601. [CrossRef] [PubMed]
31. Be'er, A.; Lereah, Y.; Taitelbaum, H. Reactive-Wetting of Hg-Ag System at Room Temperature. *Mater. Sci. Eng. A* **2008**, *495*, 102. [CrossRef]
32. Efraim, Y.; Taitelbaum, H. Can Ising model and/or QKPZ equation properly describe reactive-wetting interface dynamics? *Cent. Eur. J. Phys.* **2009**, *7*, 503. [CrossRef]
33. Yin, L.; Murray, B.T.; Su, S.; Sun, Y.; Efraim, Y.; Taitelbaum, H.; Singler, T.J. Reactive Wetting in Metal-Metal Systems. *J. Phys. Condens. Matter* **2009**, *21*, 464130. [CrossRef]
34. Efraim, Y.; Taitelbaum, H. Persistence in Reactive-Wetting Interfaces. *Phys. Rev. E* **2011**, *84*, 050602. [CrossRef]
35. Harel, M.; Taitelbaum, H. Effect of Temperature on the Dynamics and Geometry of Reacting-Wetting Interfaces around Room Temperature. *Phys. Rev. E* **2017**, *96*, 062801. [CrossRef] [PubMed]
36. Barlow, M.; Planting, P.J. The Wetting of Metal Surfaces by Liquid Mercury. *Z. Met.* **1969**, *60*, 10. [CrossRef]
37. Vicsek, T. *Fractal Growth Phenomena*, 2nd ed.; World Scientific: Singapore, 1992.
38. Bunde, A.; Havlin, S. *Fractals in Science*, 2nd ed.; Springer: Berlin/Heidelberg, Germany, 1994.
39. Bunde, A.; Havlin, S. *Fractals and Disordered Systems*, 2nd ed.; Springer: Berlin/Heidelberg, Germany, 1996.
40. Barabasi, A.L.; Stanley, H.E. *Fractal Concepts in Surface Growth*; Cambridge University Press: Cambridge, UK, 1995.
41. Meakin, P. *Fractals, Scaling, and Growth Far from Equilibrium*; Cambridge University Press: Cambridge, UK, 1998.
42. Halpin-Healy, T.; Zhang, Y.-C. Kinetic Roughening Phenomena, Stochastic Growth, Directed Polymers and all that. Aspects of Multidisciplinary Statistical Mechanics. *Phys. Rep.* **1995**, *254*, 215. [CrossRef]
43. Alava, M.; Dube, M.; Rost, M. Imbibition in Disordered Media. *Adv. Phys.* **2004**, *53*, 83. [CrossRef]
44. Family, F.; Vicsek, T. Scaling of the Active Zone in the Eden Process on Percolation Networks and the Ballistic Deposition Model. *J. Phys. A* **1985**, *18*, L75. [CrossRef]
45. Kardar, M.; Parisi, G.; Zhang, Y.-C. Dynamic Scaling of Growing Interfaces. *Phys. Rev. Lett.* **1986**, *56*, 889. [CrossRef]
46. Hentschel, H.G.E.; Family, F. Scaling in Open Dissipative Systems. *Phys. Rev. Lett.* **1991**, *66*, 1982. [CrossRef]
47. Majumdar, S.N. Persistence in Nonequilibrium Systems. *Curr. Sci.* **1999**, *77*, 370. [CrossRef]
48. Brú, A.; Pastor, J.M.; Fernaud, I.; Brú, I.; Melle, S.; Berenguer, C. Super-Rough Dynamics on Tumor Growth. *Phys. Rev. Lett.* **1998**, *81*, 4008. [CrossRef]
49. Khain, E.; Sander, L.M. Dynamics and Pattern Formation in Invasive Tumor Growth. *Phys. Rev. Lett.* **2006**, *96*, 188103. [CrossRef]

50. Ben-Jacob, E.; Schochet, O.; Tenenbaum, A.; Cohen, I.; Czirok, A.; Vicsek, T. Generic Modelling of Cooperative Growth Patterns in Bacterial Colonies. *Nature* **1994**, *368*, 46. [CrossRef] [PubMed]
51. Huergo, M.A.C.; Pasquale, M.A.; González, P.H.; Bolzán, A.E.; Arvia, A.J. Growth Dynamics of Cancer Cell Colonies and their Comparison with Noncancerous Cells. *Phys. Rev. E* **2012**, *85*, 011918. [CrossRef] [PubMed]
52. Santalla, S.N.; Rodríguez-Laguna, J.; Abad, J.P.; Marín, I.; Espinosa, M.M.; Muñoz-García, J.; Vázquez, L.; Cuerno, R. Nonuniversality of Front Fluctuations for Compact Colonies of Nonmotile Bacteria. *Phys. Rev. E* **2018**, *98*, 012407. [CrossRef] [PubMed]
53. Santalla, S.N.; Ferreira, S.C. Eden Model with Nonlocal Growth Rules and Kinetic Roughening in Biological Systems. *Phys. Rev. E* **2018**, *98*, 022405. [CrossRef]
54. Barreales, B.G.; Meléndez, J.J.; Cuerno, R.; Ruiz-Lorenzo, J.J. Kardar–Parisi–Zhang universality class for the critical dynamics of reaction–diffusion fronts. *J. Stat. Mech. Theor. Exp.* **2020**, *2020*, 023203. [CrossRef]
55. Makse, H.A.; Havlin, S.; Stanley, H.E. Modelling urban growth patterns. *Nature* **1995**, *377*, 608. [CrossRef]
56. Najem, S.; Krayem, A.; Ala-Nissila, T.; Grant, M. Kinetic Roughening of the Urban Skyline. *Phys. Rev. E* **2020**, *101*, 050301. [CrossRef]
57. Maunuksela, J.; Myllys, M.; Timonen, J.; Alava, M.J.; Ala-Nissila, T. Kardar–Parisi–Zhang Scaling in Kinetic Roughening of Fire Fronts. *Physica A* **1999**, *266*, 372–376. [CrossRef]
58. Maunuksela, J.; Myllys, M.; Kähkönen, O.-P.; Timonen, J.; Provatas, N.; Alava, M.J.; Ala-Nissila, T. Kinetic Roughening in Slow Combustion of Paper. *Phys. Rev. Lett.* **1997**, *79*, 1515. [CrossRef]
59. Balankin, A.S.; Morales, D.; Susarrey, O.; Samayoa, D.; Trinidad, J.M.; Marquez, J.; García, R. Self-Similar Roughening of Drying Wet paper. *Phys. Rev. E* **2006**, *73*, 065105. [CrossRef]
60. Balankin, A.S.; Paredes, R.G.; Susarrey, O.; Morales, D.; Vacio, F.C. Kinetic Roughening and Pinning of Two Coupled Interfaces in Disordered Media. *Phys. Rev. Lett.* **2006**, *96*, 056101. [CrossRef] [PubMed]
61. Barness, D.; Efraim, Y.; Taitelbaum, H.; Sinvani, M.; Shaulov, A.; Yeshurun, Y. Kinetic Roughening of Magnetic Flux Fronts in $Bi_2Sr_2CaCu_2O_{8+\delta}$ Crystals with Columnar Defects. *Phys. Rev. B* **2012**, *85*, 174516. [CrossRef]
62. Hasegawa, K.; Kishimoto, Y. Fingering Instability of Binary Droplets on Oil Pool. *Fluids* **2023**, *8*, 138. [CrossRef]
63. Santalla, S.; Rodríguez-Laguna, J.; Cuerno, R. Circular Kardar-Parisi-Zhang Equation as an Inflating, Self-Avoiding Ring Polymer. *Phys. Rev. E* **2014**, *89*, 010401. [CrossRef]
64. Barreales, B.G.; Meléndez, J.J.; Cuerno, R.; Ruiz-Lorenzo, J.J. Large-Scale Kinetic Roughening Behavior of Coffee-Ring Fronts. *Phys. Rev. E* **2022**, *106*, 044801. [CrossRef]
65. Harel, M.; Taitelbaum, H. Non-Universal Dynamic Exponents for Thin-Film Spreading. *Europhys. Lett.* **2018**, *122*, 26002. [CrossRef]
66. Harel, M.; Taitelbaum, H. Non-Monotonic Dynamics of Thin Film Spreading. *Eur. Phys. J. E* **2021**, *44*, 69. [CrossRef] [PubMed]
67. Marcos, J.M.; Rodríguez-López, P.; Meléndez, J.J.; Cuerno, R.; Ruiz-Lorenzo, J.J. Spreading Fronts of Wetting Liquid Droplets: Microscopic Simulations and Universal Fluctuations. *Phys. Rev. E* **2022**, *105*, 054801. [CrossRef]
68. Marcos, J.M.; Meléndez, J.J.; Cuerno, R.; Ruiz-Lorenzo, J.J. Microscopic Fluctuations in the Spreading Fronts of Circular Wetting Liquid Droplets. *Phys. Rev. E* **2025**, *111*, 045504. [CrossRef]
69. Iliev, S.; Pesheva, N.; Iliev, P. Dependence of the Contact Line Roughness Exponent on the Contact Angle on Substrates with Dilute Mesa Defects: Numerical Study. *Eur. Phys. J. E* **2022**, *45*, 66. [CrossRef]
70. Iliev, S.; Pesheva, N.; Iliev, P. Roughness Exponents of the Liquid/Vapor/Solid Contact Line on Surfaces with Dilute Random Gaussian Defects: Numerical Study. *Eur. Phys. J. E* **2025**, *48*, 32. [CrossRef] [PubMed]

fluids

MDPI

Article

Molecular Dynamics of Nanodroplet Coalescence in Quasi-Saturated Vapor

Dmitry Beloborodov [1] and Aleksey Vishnyakov [2,*]

1 Skolkovo Institute of Science and Technology, 121205 Moscow, Russia
2 Thermal Petrophysics Inc., 3 Nobel St., 143026 Moscow, Russia
* Correspondence: easygear3428@gmail.com

Abstract: The dynamics of coalescence of small Lennard–Jones droplets as a function of droplet size and temperature is explored with molecular simulations. Droplet sizes vary from several hundred to several thousand molecules, and three different temperatures are explored. As the droplets establish contact, a liquid-like bridge between them forms and grows, ultimately leading to a complete coalescence. The dynamics of the bridge growth are consistent with the "collective molecular jumps" mechanism reported in the literature rather than with the continuous interpretation of the coalescence process in terms of capillary and viscous forces. The effective coalescence time shows a linear growth with the droplet sizes. The influence of the larger droplet size is weaker but non-negligible. Surprisingly, practically no dependence of the coalescence time on the temperature is observed. Comparison of the coalescence times with the droplet lifespan in a suspension shows that for reasonably dense suspensions and small droplet sizes, the coalescence time becomes significant and should be accounted for in the theoretical models of aggregation.

Keywords: droplet coalescence; suspension; molecular dynamics

1. Introduction

Droplet collision is an integral part of many natural and industrial processes, such as cloud formation [1,2], suspension aggregation [3,4], inkjet printing [5,6], etc. The major mechanisms of droplet aggregation in vapor are (1) classical Ostwald ripening that occurs via evaporation of fluid from smaller droplets and condensation at the surfaces of larger droplets (because the equilibrium pressure over small droplets is higher); this process is described by the Lifshitz–Slyozov theory [7] and its various modifications [8]; (2) the coalescence of droplets upon a physical contact induced by the capillary forces. The higher the equilibrium vapor pressure over the droplets, the faster the Ostwald ripening. The contact coalescence is usually controlled by the diffusion of droplets in the atmosphere of vapor (unless the droplets experience long-range forces, e.g., electrostatic repulsion between charged droplets). The overall kinetics of the contact-driven aggregation in droplet suspensions is routinely described by the Smoluchowski equation [9], which (in the classical version) considers the relaxation of the coalescing droplets into one larger droplet as a very fast process in comparison with the droplet motion in the vapor atmosphere.

The dynamics of droplet coalescence upon contact have been examined in the literature, with most attention paid to collision-driven coalescence. Droplet collision could result in "shattering," "stretching separation," or "coalescence" [10–16]. Shattering means that two droplets split into several smaller ones due to the mechanical impact; in the stretching separation, they temporarily unite only to separate shortly afterward. The published statistical analysis includes the dependence of the outcome probability on the relative velocity, surface tension, angle of impact, and other parameters. The process of droplet coalescence was mostly explored for pendant or sessile droplets, with the attention primarily focused on the progression of the liquid bridge, which forms as the droplets come into contact. Negative curvature of the bridge region causes the flow of the molecules from the bulk of the droplets towards the meniscus, leading to the gradual expansion of the bridge. The evolution of the meniscus in time was explored experimentally [17,18] and via various hydrodynamics simulations, including Lattice–Boltzmann [19] and Smoothed Particle Hydrodynamics [20]. Quantification of the bridge growth rate allows analysis of the internal flow, i.e., the interplay between the capillary and viscous forces. First estimations under the assumption of Stokes flow were made by Frenkel [21] and then expanded by Hopper [22], who provided an analytical solution for the coalescence of two viscous liquid cylinders. Later, Eggers et al. [23] considered the coalescence of spherical droplets in 3D and argued that the Stokes regime was only valid for Reynolds numbers up to $\mathrm{Re} \approx 1$, after which the flow transitioned to a regime dominated by inertial forces. If described in terms of the reduced bridge radius $R = r_\mathrm{b}/R_0$, where r_b is the bridge radius and R_0 is the initial droplet radius (parent droplets of equal sizes are assumed), then for the initial viscous regime $R \sim t/\tau_v \ln(t/\tau_v)$ for $R < 0.03$, where $\tau_v = \eta R_0/\gamma$. For larger R, the flow transitions to the inertial regime, and the scaling becomes $R \sim (t/\tau_i)^{1/2}$ with the time scale $\tau_i = (\rho R_0^3/\gamma)^{1/2}$. Here, t is the time passed since the droplets came in contact, η is the dynamic viscosity, γ is the surface tension, and ρ is the mass density. Although the scaling for the inertial regime was confirmed by both experiment [18,24] and simulations [25], the logarithmic correction to the viscous scaling caused some discussions. For example, in an experimental study by Aarts et al. [18], the viscous regime was found to be well described by a linear scaling $R \sim t/\tau_v$. This inconsistency was resolved by Paulsen et al. [26], who proposed the existence of the inertia-limited viscous (ILV) asymptotic regime preceding the Hopper–Stokes regime. They argued that the viscous regime could not start until the surface tension forces were strong enough to initiate the translation of the droplets' center of mass. Until then, the coalescence process is defined by viscous, inertial, and surface tension forces together. Recently, Xia et al. [27] proposed a universality theory that combines the coalescence regimes described above, with ILV and inertial regimes being its two asymptotic approximations. Hence, we will refer to it as the "hydrodynamic regime" below.

Molecular simulations can help bypass some of the restrictions faced by empirical models and provide more physical details, as they do not require any assumptions about the nature of the system beyond the molecular interaction potential and can capture processes on scales that are out of reach of the current experimental methods. Recently, Perumanath et al. [28] peeked into the initial stages of the coalescence process with MD simulations. While from the hydrodynamics standpoint, the moment of contact creates a singularity due to infinite surface curvature, MD allows inspecting this moment more carefully. The authors proposed that the contact between the two droplets is initiated by thermal fluctuations on their surfaces. As the droplets approach each other, the surface molecules in an area around the line of approach randomly start "jumping" across the gap due to thermal motion and proximity to the opposing surface. These "collective molecular jumps" constitute a new coalescence regime, which the authors call the "thermal regime," preceding the hydrodynamic regime described above. This regime holds until the bridge radius grows to a certain threshold value l_T proportional to the width of the contact area on the surface, i.e., the area from which spontaneous molecular jumps can occur. MD is also used in the studies of more complicated cases of droplet collision, e.g., the collision of conducting droplets in an electric field [29]. In this case, droplets deform into a conical shape as they approach each other due to the electrostatic force, and the outcome of the collision depends on the cone angle and the critical electrical field strength. Nanoparticles covered by an adsorbed liquid-like film also show a bridge formation and growth that results in an effective attraction between them, although they cannot coalesce [30]. Another example is the recent study of the coalescence/sintering bifurcation phenomenon for nanoparticles [31]. With the help of isothermal MD simulations, the authors discovered that the temperature of the transition from sintering to coalescence for Au nanoparticles is lower than the nanoparticle melting point. In the vicinity of that critical temperature, the child particle can have either a crystalline structure or a liquid-like structure at the same temperature.

In this research note, we explore the characteristic time of the coalescence of two nanodroplets once contact between them has been established. The coalescence time is determined as a function of the sizes of the parent droplets and the temperature. We examine the meniscus evolution and compare the coalescence timescale with the characteristic lifetime of a droplet in a dense suspension to check whether the finite coalescence time can influence the aggregation dynamics. The study aims at shedding light on the coalescence mechanism at spatial and temporal scales unavailable for experimental studies. In particular:

- we explore the dependence of the characteristic coalescence time on the temperature and sizes of both droplets, which has not been explored in the experimental, simulation, and theoretical studies, as the literature review shows
- by examining the bridge growth and coalescence time, we explore in detail the recently discovered "thermal scaling regime" of the contact coalescence, which cannot be experimentally observed for the small temporal and spatial scales
- the characteristic coalescence times are for the first time compared with the characteristic lifetime of a droplet before it contacts another one in a suspension during the aggregation process; this comparison shows in what cases the finite coalescence times should be taken into account in the theoretical models and anticipates the studies of multiparticle collisions in the aggregation kinetics.

2. The Model and Simulation Details

The droplets are composed of Lennard–Jones (LJ) molecules. The LJ model is widely accepted for fluids with van der Waals interactions and quantitatively describes noble gases, nitrogen, and methane [32,33]. When working with the LJ potential, it is convenient to use the so-called reduced units, in which distances, masses, and energies are multiples of the parameters of the LJ potential σ (the effective molecular diameter), m (the molecular mass), and ϵ (the depth of the LJ potential, that is the minimum energy of interaction between two LJ particles), respectively. The density ρ is the number of LJ particles per volume equal to the cube of the effective diameter σ^3, thus $\rho\sigma^3 = 1$ in a simple cubic lattice, where the neighboring particles contact each other. All the values below are given in reduced units if not stated otherwise. The potential is cut and shifted at $r_c = 2.5\,\sigma$. Critical properties for such system are $kT_c/\epsilon = 0.937$ and $\rho_c\sigma^3 = 0.320$ [34,35]. Three temperatures were considered: $kT/\epsilon = 0.65$, 0.70, and 0.75. The corresponding saturated vapor densities, pressures, and viscosities are shown in Table 1. The rectangular simulation box has dimensions of (113.18 σ, 56.59 σ, 56.59 σ), (94.36 σ, 47.18 σ, 47.18 σ) and (80.08 σ, 40.04 σ, 40.04 σ) for the three temperatures, correspondingly. Periodic boundary conditions are applied to ensure constant volume and the number of LJ particles. Droplets are generated as spheres filled with LJ atoms arranged in face-centered cubic lattices and placed at a distance of 2.0 σ between their surfaces. This arrangement ensures that coalescence does not start right away, but rather the system equilibrates for a while, and droplets come in contact naturally by drifting towards each other due to the dispersion attraction. The moment of contact was determined using a clustering algorithm adopted from [36]. Initial radii of droplets span from 6.0 to 12.0 σ and vary by 0.5 σ steps. In the initial configuration, the droplets are aligned along the x-axis, and the left droplet is smaller or equal in size to the right one. In every particular simulation, the configuration is characterized by three parameters (T, r_l, r_r), where T is the system temperature, r_l is the radius of the smaller (left) droplet and r_r is the radius of the larger (right) droplet, e.g. (0.65, 7.0, 10.0) corresponds to a configuration of temperature 0.65 and initial droplet radii of 7.0 and 10.0 σ. Initial velocities of the molecules are randomly generated according to the temperature. One hundred simulations for each parameter set are performed to collect statistics. The necessity to conduct a large number of simulations to collect reasonable statistics prevents simulations of larger droplets due to computational expenses.

Table 1. Simulation parameters.

T	ρ_v	η_v	p_{sat}	ρ_l	η_l
0.65	0.0113	0.0241	0.0067	0.762	1.86
0.70	0.0195	0.0333	0.0119	0.729	1.50
0.75	0.0319	0.0490	0.0195	0.693	1.27

It should be noted that the vapor pressure (as well as the equilibrium density) depends on the droplet size, which means that the capillary coalescence and Ostwald ripening proceed concurrently. In the initial configuration, the two parts of the box are filled with LJ vapor in equilibrium with the droplet of particular radii (r_l, r_m). That is, the density is different in the two halves of the box, and after the start, the vapor starts mixing, which leads to evaporation from the surface of the smaller droplet and condensation at the surface of the larger one. The resulting droplet is larger than the parent droplets, and, therefore, the saturated vapor has a lower density, which has to be taken into account in the calculation of the coalescence time.

The simulations are carried out with LAMMPS [37,38], and the temperature is maintained with the Nose thermostat [39]. The simulation is gradually sped up: it is started by performing 30,000 timesteps of size 0.0001 in reduced units, then 30,000 timesteps of 0.001, and finally 45,000 timesteps of 0.01. For trajectories, the coordinates were stored every 1000 steps for the first two stages and every 250 steps for the last stage.

3. Results and Discussion

Figure 1 illustrates the coalescence process. The droplets attracted to each other by the van der Waals forces eventually establish contact. A bridge forms and grows around the point of contact until the parent droplets finally merge into a larger one. To follow the process, we apply a clusterization algorithm that assigns any two atoms located closer than 1.3σ to each other to the same cluster. The frame when the algorithm detected only one large cluster is identified as the moment when the contact is established and, therefore, as the starting moment of the coalescence process ($t = 0$). Correspondingly, the radius of gyration of the cluster is assumed as the main order parameter that describes the coalescence process. The characteristic coalescence time is estimated from its evolution, as described below. At the same time, we follow the dynamics of the liquid bridge growth characterized by the density profile built in two dimensions: tangential, i.e., along the line connecting the centers of mass of the merging droplets (identified as the axis with respect to which the cluster had the lowest radius of inertia at any given moment) and the radial one. The number density profiles are built in this coordinate system with the droplet domain divided into cylindrical slices of equal width of about $2\sigma/3$ along the tangential direction. From the profile and the liquid density, the bridge radius is calculated. Figure 2 shows the dependence of the evolution of the average bridge radius on the time after the contact for select sets of parameters. At the starting moment of coalescence $r_b \approx 1\sigma$, because the bridge consists of one molecule. The bridge radius grows approximately linearly with time, and then saturation is observed as the bridge radius becomes comparable to the droplet radius. The steep linear growth is consistent with the "collective molecular jumps" mechanism observed first by Perumanath et al. [28]. Heinen et al. [40] also observed a linear bridge growth in the initial stages of their MD simulations of droplet coalescence. The linear thermal regime then seemingly shifts to the less steep ILV regime and comes to a plateau shortly after. The dynamics of other scaling regimes are difficult to determine because the curve flattens too quickly. At higher temperatures, the thermal regime has a longer duration. However, the plateau value depends only on the radii of the coalescing droplets.

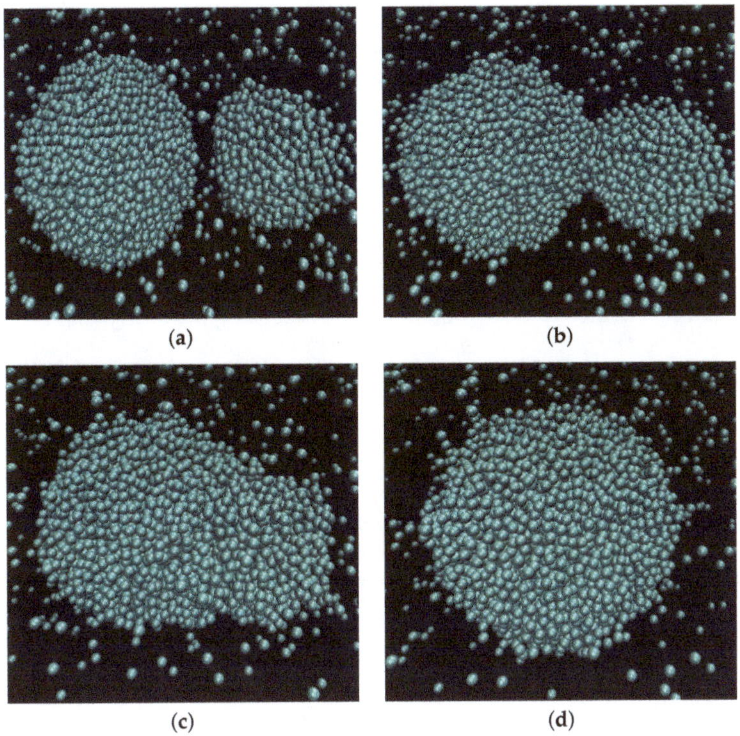

Figure 1. Snapshots of molecular configurations at different stages of coalescence process. (**a**) Before coalescence; (**b**) Beginning of coalescence; (**c**) In the process of relaxation; (**d**) Coalescence finished.

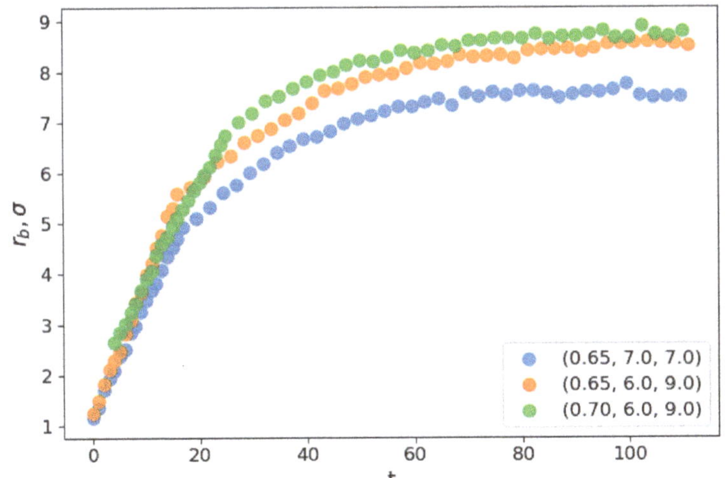

Figure 2. Mean radius of the liquid bridge as a function of reduced time for three different system configurations.

The bridge formation is accompanied by the reduction of the gyration radius of the newly forming droplet. A set of typical examples is shown in Figure 3a. The vdW attraction causes particle motion to the bridge zone resulting in a rapid decrease of the gyration radius until the parent droplets finally merge into a new one. It should be noted that along with the capillary-driven coalescence, we concurrently observed the ripening mechanism: the LJ molecules evaporated from the areas of high curvature and condensed at the areas of lower curvature ($r_l < r_r$). Since the radius of the resulting droplet is higher, and the curvature is lower than those of the parent droplets, the resulting droplet actually somewhat grows in size after the shape relaxation has finished due to vapor condensation on its surface (Figure 3a).

The calculation of the coalescence time is demonstrated in Figure 3b. The $R_g(t)$ dependence is augmented with an interval of constant radius $R_u(t) = R_g(t = 0)$, effectively assuming that before the contact $R_g(t)$ is the same as at the moment of the contact. Then we assume that during the condensation rebound, the droplet size linearly increases with time R_g, and the condensation rebound process starts after time t_m that corresponds to the maximum slope of $R_g(t)$ dependence. These assumptions create the "upper" and "lower" baselines. $R_u(t)$ is constant, while $R_l(t)$ is a piece-linear function, constant at $t \leq t_m$ and fitting the condensation rebound regime at $t > t_m$ (Figure 3b). The $R_g(t)$ dependence is fitted using the hyperbolic tangent as:

$$R_g^*(t) = R_u \tanh(-D(t + t_T)) + R_l(t)(1 - \tanh(D(t + t_T))), \quad (1)$$

where t_T and D are the adjustable parameters, which characterize the "turning point" of the coalescence and the half-duration of the relaxation process. The moment t_c that corresponds to the minimum value of R_g^* was assumed as the end of the coalescence and, therefore, as the characteristic coalescence time.

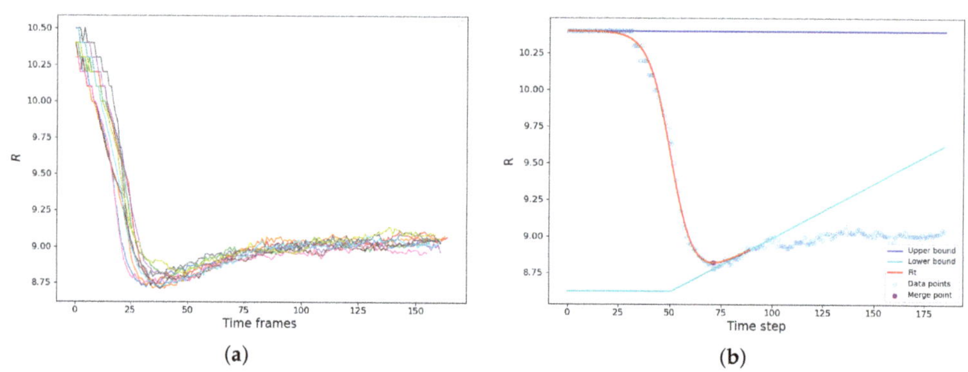

Figure 3. (**a**) Gyration radius of the composite droplet as a function of time for 10 simulations of configuration (0.65, 7.0, 10.0). The sharp exponential drop corresponds to coalescence, followed by fluctuations around the "equilibrium" value. (**b**) Demonstration of the coalescence time approximation procedure.

With this procedure, the dependence of $t_c(r_l, r_r, T)$ is obtained. Figure 4 shows how coalescence time depends on three major configuration parameters. The mean values of time obtained for each set of conditions are plotted. It can be seen that the coalescence time scales linearly with both initial droplet sizes. Interestingly enough, the temperature does not seem to accelerate the coalescence process in the mean but rather increases the standard deviation from the fitted plane. We trained linear regression models for every temperature separately and found the scores to be approximately equal 0.85 for $T = 0.65kT/\epsilon$, 0.69 for $T = 0.70kT/\epsilon$ and 0.77 for $T = 0.75kT/\epsilon$. The fairly unrestricted nature of our simulation setup is partly to blame since the 2.0 σ of initial separation between the outer droplet surfaces does not guarantee an immediate coalescence after equilibration, so by the time the droplets eventually come into contact, their sizes may have changed somewhat from the initial values. Since the droplet radii features have the same span of values, we can judge their importance based on the fitted weight coefficients. The smaller droplet has the larger weight between the two, making it the main parameter. The weight of the larger droplet is about 20 times smaller for all temperatures. To determine the influence of the temperature, we trained the linear regression model on the entire dataset. Considering the scatter of target values at higher temperatures and the span of considered temperature values, the corresponding weight is insignificant. The addition of Lasso regularization with constant $\alpha = 0.15$ sets the temperature weight to zero and gives an overall better fit for temperatures $T = 0.65kT/\epsilon$ and $T = 0.75kT/\epsilon$. It should be noted that the larger droplet's coefficient does not vanish in this model: the role of the size of the larger droplet remains significant. The above Lasso model predicts zero coalescence time for two droplets of size 1 σ (that is, containing only one molecule), which is the correct asymptotic limit since merging of two single LJ molecules into a dimer "droplet" needs no internal restructuring. It was reported that an increase in temperature difference between the parent droplets accelerates the mixing between their particles [41]. We did not attempt to validate those claims during the current work, but the temperature difference between the coalescing drops would lead to non-uniformity of the particle diffusion, which, in turn, could affect the coalescence time.

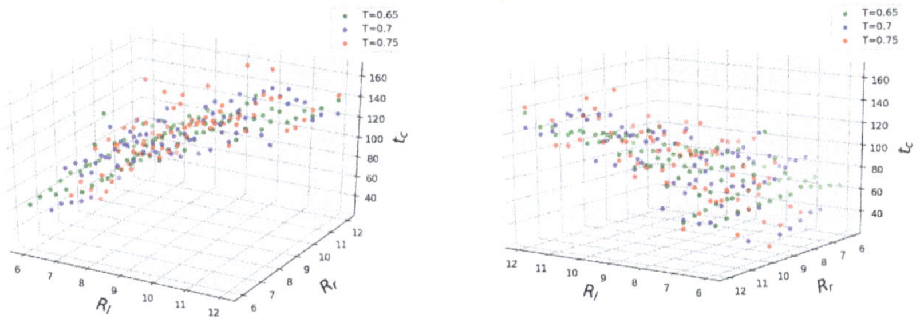

Figure 4. Coalescence time plotted as a function of initial droplet radii ($r_l \leq r_r$). Different colors represent different system temperatures.

It is interesting to compare the coalescence times with the characteristic lifetimes of the droplet in a droplet suspension in vapor. To estimate the characteristic lifetimes, we conducted Langevin dynamics (LD) simulations. Each simulation modeled the Brownian motion of a particle whose mass and size corresponded to a liquid droplet at certain conditions in the medium with characteristics (density and viscosity) corresponding to the equilibrium vapor. The trajectory of the Brownian motion is recorded; after the simulation, we calculate the volume of the exclusion sphere (the exclusion sphere radius equals the particle diameter) covered during its motion per unit of time. The product of the volume by the concentration is the inverse average lifetime of the droplet with respect to the contact-driven coalescence if we assume that each contact leads to the merging of the two droplets (which is reasonable since the droplet velocities are very low). This estimation is fully similar to that of the free path of a molecule in a gas, but in the Brownian, rather than in the ballistic, regime of motion.

Here we consider only equally sized droplets. The larger the droplet sizes, the slower their motion is due to the increased mass, but the lower the viscosity of the vapor. Obviously, the droplet lifetime with the current assumptions decreases linearly with the volume fraction occupied by the liquid phase. LD simulations are carried out for the spherical particles with the mass and size corresponding to droplets containing 500, 750, 1000, 2500, 5000, and 10,000 molecules with the time step of 5×10^{-5}. Reduced temperatures are once again 0.65, 0.70, and 0.75 kT/ϵ. The coalescence times reported above only depend on the droplet size and grow proportional to r^3.

Figure 5 shows the ratio of the droplet lifetime to the coalescence time. The plot spans the values of liquid phase volume fractions from 0.01 to 0.1 with a step of 0.01. The ratio decreases exponentially with the increase in liquid volume fraction, and for smaller droplets with $N \approx 1000$ reaches the value of 10 at 0.07–0.08 volume fraction, meaning the characteristic times become quite comparable at that point. Hence, for denser suspensions, coalescence time probably should be taken into account in aggregation theories. Temperature also has an influence on the ratio; although weaker than the dependence on the droplet size, it is still considerable for smaller droplets.

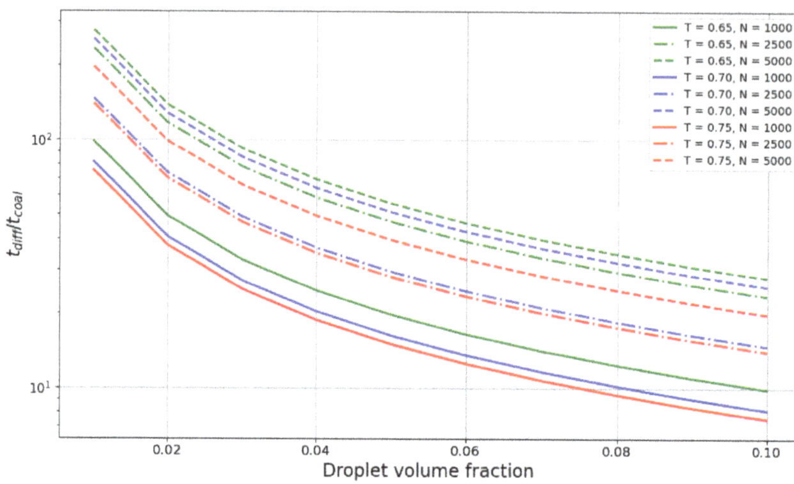

Figure 5. Dependence of the droplet lifetime to coalescence time ratio on the volume fraction of the liquid phase for the reduced temperatures of 0.65 (green line), 0.70 (blue line) and 0.75 (red line) kT/ϵ. Linestyle represents the droplet size in particle number N: N = 750 – dotted line, N = 1000 – solid line, N = 2500 – dash-dotted line, N = 5000 – dashed line.

4. Conclusions

We have performed molecular dynamics simulations of LJ nanodroplet contact-driven coalescence (that is, merging of two droplets into a larger one) in quasi-equilibrium vapor. The droplet sizes ranged from 6 to 12 molecular diameters σ, and the reduced temperature kT/ϵ ranged from 0.65 to 0.75. Upon establishing contact, the parent droplets form a bridge, with the radius growing linearly with time. This scaling regime is consistent with the "collective molecular jumps" mechanism described recently [28] and is dominant for the considered length scales, with its duration increasing with temperature. Currently, this regime cannot be captured by experimental studies. Thus, for small droplets, the classical hydrodynamic regimes characterized by the power law (inertial) or the logarithmically corrected linear (viscous) bridge diameter dependence on time are not observed. The characteristic duration of the coalescence t_c is determined from the analysis of the gyration radius devolution in time. t_c shows a linear dependence on the initial radii of the smaller and larger droplets R_l, R_r with R_l as the main parameter with the largest weight. This is expected, because if $R_l << R_r$, t_c should depend on R_l only. Obviously, a twofold difference in the spatial size (eight-fold mass difference) is not sufficient to break the linear dependence. It is interesting that the temperature does not show a significant influence on t_c in the mean, despite the far greater molecular mobilities in the liquid phase. Removing the temperature feature from the model altogether even produces a better fit to the simulation results.

In practical conditions, the droplets are not spherical, and their motion is not purely diffusive; they often collide at considerable velocities. Since Perumanath et al. [28], in the original article mentioning the "thermal scaling regime," considered both cylindrical and spherical droplets and obtained qualitatively similar results, we would expect that the linear scaling of the characteristic time observed in this work would be valid for droplets whose shape is substantially non-spherical. Droplet velocities (especially for smaller droplets because that case corresponds to higher Reynolds numbers) should qualitatively affect the coalescence time. We may speculate that the linear scaling will hold when the droplet velocities are low (that is if the time within which the droplet travels the distance equal to its diameter is larger than the droplet coalescence time) and will break at higher velocities. However, the Reynolds number threshold for transitions between coalescence/stretching separation/shattering for nanodroplets is a potentially fruitful topic for future investigations.

Another outcome of this work is the comparison of the coalescence time t_c to the "free diffusion time" t_d in dense suspensions. This comparison also sheds light on the importance of multiparticle collisions in liquid droplet suspensions because if the coalescence time is negligible, the number of multiparticle contacts is negligible as well. The t_d/t_c ratio appears to show a strong dependence on size (as droplet diffusion slows down faster than coalescence as the droplet grows) and a much weaker dependence on temperature. The results show that in very dense suspensions (droplet volume fraction above 0.1) and small sizes ($N \approx 1000$), the coalescence time is undoubtedly comparable with the droplet lifetime and should be taken into account. In less extreme cases, the coalescence can be considered a fast process, although, of course, it depends on the required precision of the model.

Author Contributions: Conceptualization, A.V.; methodology, A.V.; validation, D.B. and A.V.; formal analysis, D.B.; investigation, D.B.; writing, D.B. and A.V. All authors have read and agreed to the published version of the manuscript.

Funding: This research received no external funding.

Data Availability Statement: Not applicable.

Acknowledgments: We thank Nicholas V. Brilliantov (Skolkovo Institute of Science and Technology) for fruitful discussions. The calculations were performed at Zhores' computational facilities [42].

Conflicts of Interest: The authors declare no conflict of interest.

Abbreviations

The following abbreviations are used in this manuscript:

LJ	Lennard-Jones
MD	Molecular Dynamics

References

1. Illingworth, A.J. The Formation of Rain in Convective Clouds. *Nature* **1988**, *336*, 754–756. [CrossRef]
2. Xue, Y.; Wang, L.P.; Grabowski, W.W. Growth of Cloud Droplets by Turbulent Collision–Coalescence. *J. Atm. Sc.* **2008**, *65*, 331–356. [CrossRef]
3. Dreher, T.M.; Glass, J.; O'Connor, A.J.; Stevens, G.W. Effect of rheology on coalescence rates and emulsion stability. *AIChE J.* **1999**, *45*, 1182–1190. [CrossRef]
4. Schröder, A.; Sprakel, J.; Schroën, K.; Spaen, J.N.; Berton-Carabin, C.C. Coalescence stability of Pickering emulsions produced with lipid particles: A microfluidic study. *J. Food Eng.* **2018**, *234*, 63–72. [CrossRef]
5. Bartolo, D.; Josserand, C.; Bonn, D. Singular Jets and Bubbles in Drop Impact. *Phys. Rev. Lett.* **2006**, *96*, 124501. [CrossRef]
6. Eggers, J.; Villermaux, E. Physics of liquid jets. *Rep. Prog. Phys.* **2008**, *71*, 036601. [CrossRef]
7. Lifshitz, I.M.; Slyozov, V.V. The Kinetics of Precipitation From Supersaturated Solid Solutions. *J. Phys. Chem. Solids* **1961**, *19*, 35–50. [CrossRef]
8. Niethammer, B.; Otto, F.; Velázquez, J.J.L. On the Effect of Correlations, Fluctuations, and Collisions in Ostwald Ripening. In *Analysis, Modeling and Simulation of Multiscale Problems*; Springer: Berlin/Heidelberg, Germany, 2006; pp. 501–530.
9. Smoluchowski, M. Mathematical Theory of the Kinetics of the Coagulation of Colloidal Solutions. *Z. Phys. Chem.* **1917**, *29*, 129.
10. Adam, J.R.; Lindblad, N.R.; Hendricks, C.D. The Collision, Coalescence, and Disruption of Water Droplets. *J. Appl. Phys.* **1968**, *39*, 5173–5180. [CrossRef]
11. Ashgriz, N.; Poo, J.Y. Coalescence and separation in binary collisions of liquid drops. *J. Fluid Mech.* **1990**, *221*, 183–204. [CrossRef]
12. Pan, Y.; Suga, K. Numerical simulation of binary liquid droplet collision. *Phys. Fluids* **2005**, *17*, 082105. [CrossRef]
13. Ming, L.; Marković, N.; Svanberg, M.; Pettersson, J.B.C. Collision Dynamics of Large Argon Clusters. *J. Phys. Chem. A* **1997**, *101*, 4011–4018. [CrossRef]

14. Kalweit, M.; Drikakis, D. Collision Dynamics of Nanoscale Lennard-Jones Clusters. *Phys. Rev. B* **2006**, *74*, 235415. [CrossRef]
15. Demidovich, A.V.; Kralinova, S.S.; Tkachenko, P.P.; Shlegel, N.E.; Volkov, R.S. Interaction of Liquid Droplets in Gas and Vapor Flows. *Energies* **2019**, *12*, 4256. [CrossRef]
16. Dolganov, P.V.; Zverev, A.S.; Baklanova, K.D.; Dolganov, V.K. Dynamics of capillary coalescence and breakup: Quasi-two-dimensional nematic and isotropic droplets. *Phys. Rev. E* **2021**, *104*, 014702. [CrossRef]
17. Qian, J.; Law, C.K. Regimes of coalescence and separation in droplet collision. *J. Fluid Mech.* **1997**, *331*, 59–80. [CrossRef]
18. Aarts, D.G.A.L.; Lekkerkerker, H.N.W.; Guo, H.; Wegdam, G.H.; Bonn, D. Hydrodynamics of Droplet Coalescence. *Phys. Rev. Lett.* **2005**, *95*, 164503. [CrossRef]
19. Baroudi, L.; Nagel, S.R.; Morris, J.F.; Lee, T. Dynamics of viscous coalescing droplets in a saturated vapor phase. *Phys. Fluids* **2015**, *27*, 121702. [CrossRef]
20. Acevedo-Malavé, A. Hydrodynamics coalescence collision of three liquid drops in 3D with smoothed particle hydrodynamics. *AIP Adv.* **2012**, *2*, 042106. [CrossRef]
21. Frenkel, J.J. Viscous Flow of Crystalline Bodies under the Action of Surface Tension. *J. Phys.* **1945**, *9*, 385.
22. Hopper, R.W. Plane Stokes flow driven by capillarity on a free surface. *J. Fluid Mech.* **1990**, *213*, 349–375. [CrossRef]
23. Eggers, J.; Lister, J.R.; Stone, H.A. Coalescence of liquid drops. *J. Fluid Mech.* **1999**, *401*, 293–310. [CrossRef]
24. Case, S.C. Coalescence of low-viscosity fluids in air. *Phys. Rev. E* **2009**, *79*, 026307. [CrossRef]
25. Eiswirth, R.T.; Bart, H.J.; Ganguli, A.A.; Kenig, E.Y. Experimental and numerical investigation of binary coalescence: Liquid bridge building and internal flow fields. *Phys. Fluids* **2012**, *24*, 062108. [CrossRef]
26. Paulsen, J.D.; Burton, J.C.; Nagel, S.R.; Appathurai, S.; Harris, M.T.; Basaran, O.A. The inexorable resistance of inertia determines the initial regime of drop coalescence. *Proc. Natl. Acad. Sci. USA* **2012**, *109*, 6857–6861. [CrossRef]
27. Xia, X.; He, C.; Zhang, P. Universality in the viscous-to-inertial coalescence of liquid droplets. *Proc. Natl. Acad. Sci. USA* **2019**, *116*, 23467–23472. [CrossRef]
28. Perumanath, S.; Borg, M.K.; Chubynsky, M.V.; Sprittles, J.E.; Reese, J.M. Droplet Coalescence is Initiated by Thermal Motion. *Phys. Rev. Lett.* **2019**, *122*, 104501. [CrossRef]
29. Wang, B.B.; Wang, X.D.; Yan, W.M.; Wang, T.H. Molecular Dynamics Simulations on Coalescence and Non-coalescence of Conducting Droplets. *Langmuir* **2015**, *31*, 7457–7462. [CrossRef]
30. Vishnyakov, A.; Shen, Y.; Tomassone, M.S. Interactions of silica nanoparticles in supercritical carbon dioxide. *J. Chem. Phis.* **2008**, *129*, 174704. [CrossRef]
31. Samsonov, V.M.; Puytov, V.V.; Talyzin, I.V. Bifurcation phenomenon in molecular dynamics model of coalescence/sintering on the nanoscale. *J. Phys. Conf. Ser.* **2021**, *2052*, 012037. [CrossRef]
32. Ravikovitch, P.I.; Vishnyakov, A.; Neimark, A.V. Density functional theories and molecular simulations of adsorption and phase transitions in nanopores. *Phys. Rev. E* **2001**, *64*, 011602. [CrossRef]
33. Vrabec, J.; Kedia, G.K.; Fuchs, G.; Hasse, H. Comprehensive study of the vapour–liquid coexistence of the truncated and shifted Lennard–Jones fluid including planar and spherical interface properties. *Mol. Phys.* **2006**, *104*, 1509–1527. [CrossRef]
34. Johnson, J.K.; Zollweg, J.A.; Gubbins, K.E. The Lennard-Jones equation of state revisited. *Mol. Phys.* **1993**, *78*, 591–618. [CrossRef]
35. Errington, J.R.; Debenedetti, P.G.; Torquato, S. Quantification of order in the Lennard-Jones system. *J. Chem. Phys.* **2003**, *118*, 2256–2263. [CrossRef]
36. Vishnyakov, A.; Lee, M.T.; Neimark, A.V. Prediction of the critical micelle concentration of nonionic surfactants by dissipative particle dynamics simulations. *J. Phys. Chem. Lett.* **2013**, *4*, 797–802. [CrossRef]
37. Plimpton, S. Fast Parallel Algorithms for Short-Range Molecular Dynamics. *J. Comput. Phys.* **1995**, *117*, 1–19. [CrossRef]
38. Thompson, A.P.; Aktulga, H.M.; Berger, R.; Bolintineanu, D.S.; Brown, W.M.; Crozier, P.S.; in't Veld, P.J.; Kohlmeyer, A.; Moore, S.G.; Nguyen, T.D.; et al. LAMMPS—A flexible simulation tool for particle-based materials modeling at the atomic, meso, and continuum scales. *Comp. Phys. Comm.* **2022**, *271*, 108171. [CrossRef]
39. Nose, S. A unified formulation of the constant temperature molecular dynamics methods. *J. Chem. Phys.* **1984**, *81*, 511–519. [CrossRef]
40. Heinen, M.; Hoffmann, M.; Diewald, F.; Seckler, S.; Langenbach, K.; Vrabec, J. Droplet coalescence by molecular dynamics and phase-field modeling. *Phys. Fluids* **2022**, *34*, 042006. [CrossRef]
41. Li, T.; Li, M.; Zhang, L.; Yan, M.; Li, H. Molecular Dynamics Study of the Temperature-Dependent Coalescence of Liquid Nanodrops: Implications for Microfluidics. *ACS Appl. Nano Mater.* **2019**, *2*, 7978–7988. [CrossRef]
42. Zacharov, I.; Arslanov, R.; Gunin, M.; Stefonishin, D.; Bykov, A.; Pavlov, S.; Panarin, O.; Maliutin, A.; Rykovanov, S.; Fedorov, M. "Zhores" — Petaflops supercomputer for data-driven modeling, machine learning and artificial intelligence installed in Skolkovo Institute of Science and Technology. *Open Eng.* **2019**, *9*, 512–520. [CrossRef]

fluids

MDPI

Article

Influence of the Deposition Rate and Substrate Temperature on the Morphology of Thermally Evaporated Ionic Liquids

Rita M. Carvalho [1], Cândida Neto [2], Luís M. N. B. F. Santos [1], Margarida Bastos [1] and José C. S. Costa [1,*]

[1] CIQUP, Institute of Molecular Sciences (IMS), Department of Chemistry and Biochemistry, Faculty of Science, University of Porto, Rua do Campo Alegre s/n, P4169-007 Porto, Portugal
[2] Faculty of Science, University of Porto, Rua do Campo Alegre s/n, P4169-007 Porto, Portugal
* Correspondence: jose.costa@fc.up.pt

Abstract: The wetting behavior of ionic liquids (ILs) on the mesoscopic scale considerably impacts a wide range of scientific fields and technologies. Particularly under vacuum conditions, these materials exhibit unique characteristics. This work explores the effect of the deposition rate and substrate temperature on the nucleation, droplet formation, and droplet spreading of ILs films obtained by thermal evaporation. Four ILs were studied, encompassing an alkylimidazolium cation ($C_nC_1$im) and either bis(trifluoromethylsulfonyl)imide (NTf_2) or the triflate (OTf) as the anion. Each IL sample was simultaneously deposited on surfaces of indium tin oxide (ITO) and silver (Ag). The mass flow rate was reproducibly controlled using a Knudsen cell as an evaporation source, and the film morphology (micro- and nanodroplets) was evaluated by scanning electron microscopy (SEM). The wettability of the substrates by the ILs was notably affected by changes in mass flow rate and substrate temperature. Specifically, the results indicated that an increase in the deposition rate and/or substrate temperature intensified the droplet coalescence of [C_2C_1im][NTf_2] and [C_2C_1im][OTf] on ITO surfaces. Conversely, a smaller impact was observed on the Ag surface due to the strong adhesion between the ILs and the metallic film. Furthermore, modifying the deposition parameters resulted in a noticeable differentiation in the droplet morphology obtained for [C_8C_1im][NTf_2] and [C_8C_1im][OTf]. Nevertheless, droplets from long-chain ILs deposited on ITO surfaces showed intensified coalescence, regardless of the deposition rate or substrate temperature.

Keywords: vapor deposition; thin films; micro- and nanodroplets; nucleation and growth; coalescence; mass flow rate; surface and interfacial tension; Ag; ITO-coated glass; SEM

1. Introduction

Ionic liquids (ILs) are a family of materials typically formed by a large organic cation and an organic/inorganic anion. Most of the studied ILs are liquids at room temperature (RTILs) [1–6]. Their low melting points derive from the asymmetry of the cation-anion pairs and the dispersion of charges, which altogether have an impact on the supramolecular structure, stabilizing the ionic liquid (IL), mainly by entropic factors [5–9]. ILs have received special attention because of their applications in different areas, such as electrochemistry (in the electrodeposition of metals), as gas sensors and biosensors, as additives (lubricants and detergents), or as "designer solvents" [4,10–15]. In some of these applications, the ILs have to interact with solid surfaces. Hence, it is important to study the behavior of the IL at interfaces in order to design better applications based on solid grounds [9,12,16–19]. ILs can be divided into different groups based on the identity of the cation and the anion: alkylammonium, dialkylimidazolium, phosphonium, and N-alkylpyridinium-based ILs are widely used as cations. Tetrafluoroborate (BF_4), hexafluorophosphate (PF_6), bis-trifluoromethanesulfonimide (NTf_2), and trifluoromethanesulfonate (OTf) are some typical ILs' anions. The conjugation of different cation-anion pairs creates ILs with different physicochemical properties. Nevertheless, most ILs share some common traits such as high viscosity, ionic conductivity, very low vapor pressure, high thermal and chemical stability, and an unusual wetting behavior [1–3,11,12,14,15,20–25]. Structurally, ILs have a network made of polar (where the electrostatic energy is concentrated) and non-polar (leading to nanostructuration at a molecular level) domains. Both polar and non-polar domains influence the bulk and surface properties of the ILs. The main interactions between the cation and anion are electrostatic, together with other weaker intermolecular interactions such as hydrogen bonds or van der Waals forces [1,2,4,11,26–29]. The segregation of the cation in distinct domains is highly dependent on the alkyl side chain length: for shorter alkyl chains, the non-polar domains are less relevant to the structural organization, whereas for larger alkyl chains, both domains have a significant impact on the structural organization of the ILs [1–3,9,23,30–33].

This work characterizes the production of ILs' micro- and nanodroplets (precursor of thin films of ILs) by vacuum thermal evaporation—one of the physical vapor deposition (PVD) methods [9,34–37]. The thin IL films are the result of the deposition of multiple monolayers (ML), and their morphology is highly dependent on the nanostructuration of the ILs, the methodology used to deposit the material, and the substrate characteristics [9,35,38,39]. The process of formation of IL films by PVD methods starts with the arrival of the ion pairs at the surface, with specific kinetic energy and subsequent adsorption in energetically stable places, creating clusters. This first step is called nucleation and is strongly influenced by the minimum free area of nucleation (MFAN), which in turn is dependent on the interactions between the IL and the substrate [9,35,40–43]. The higher the substrate-IL affinity, the lower MFAN. The progress of deposition leads to an increase in the number of clusters/nanodroplets formed, which tend to merge into larger droplets. These coalescence processes can occur in two ways (details are presented in the scheme of Figure S1 as Supplementary Material, SM): the coalescence of native clusters (first order of coalescence) or the coalescence of clusters previously coalesced (second order of coalescence). The processes of nucleation and growth are influenced by the wettability of the substrate to the ILs. Spherical IL droplets are formed when surfaces solvophobic to the ILs are used [9,34,35,40–43]. The shape of the droplets can be analyzed to evaluate the affinity of the ILs to a solid surface and, consequently, the wetting behavior of the film. In a macroscopic and ideal scenario (an inert and smooth substrate), Young's equation can be used to evaluate the contact angle, which is correlated with the different interfacial tensions of the system, according to Equation (1):

$$\cos(\theta_c) = \frac{\gamma(s-v) - \gamma(s-l)}{\gamma(l-v)} \quad (1)$$

where θ_c is the contact angle, $\gamma(s-v)$ is the solid-vapor interfacial tension, $\gamma(s-l)$ is the solid-liquid interfacial tension, and $\gamma(l-v)$ is the liquid-vapor interfacial tension. This contact angle evaluation must be carried out with some care for nanoscale droplets. We are far from the ideal situation because of the surface roughness of the substrates, which affects the surface energy and, consequently, the wetting behavior of ILs at the nanoscale [9,34,35,44–50]. The morphological characterization of IL droplets and a prediction of their contact angles can be experimentally performed by high-precision techniques such as scanning electron microscopy (SEM) and atomic force microscopy (AFM) [9,34–37,43,46]. Microdroplets of ILs with contact angles between 15 and 20 degrees have been reported [9]. The influence of different substrates in the nucleation and growth of different ILs has been studied by different groups. The adsorption and molecular arrangement of the first ML of ILs on metal surfaces were proven to be very dependent on the chemical structure of the IL, the interfacial tension IL-substrate, and the surface roughness [39,51–54]. Adventitious carbon contamination in the surfaces might also play a relevant role in the nucleation process [48,55–57]. Some studies on the mesoscopic scale report IL droplet formation on a wide range of solid substrates [35,47,54,58–62]. For large thicknesses, there is no particular time-dependent growth behavior of IL films after their deposition [9,46,63]. On the other hand, some post-deposition treatments can change the film morphology. For instance, the droplet coalescence is induced by external interferences, such as plasma surface treatments [34,36,43,64,65].

In our recent research, we examined the droplet spreading of various classes of ILs that were thermally evaporated on different substrates using both a constant deposition rate and substrate temperature [9,46]. As such, no in-depth study has been conducted on the effects of varying deposition parameters. Therefore, in this work, we studied the influence of the deposition rate and substrate temperature on the nucleation and growth of different ILs deposited by the PVD method on two common solid surfaces: indium tin oxide (ITO) and silver (Ag). The samples analyzed were alkylimidazolium-based ILs with different cation alkyl chain lengths (C_2C_1im and C_8C_1im) and two different anions (NTf_2 and OTf). The molecular structure of the cation-anion pairs ([C_8C_1im][NTf_2], [C_8C_1im][OTf], [C_2C_1im][NTf_2], and [C_2C_1im][OTf]) studied is presented in Figure 1. The deposition of films was performed by a customized procedure of PVD using Knudsen effusion cells as the evaporation sources [9,35,38,46]. Different mass flow rates were obtained by changing the temperature of the evaporation source and/or the orifice of the Knudsen effusion cell. Moreover, different temperatures of the substrate were employed, and for each IL, the deposition occurred simultaneously on the surfaces of ITO and Ag. For all experiments, the ILs were obtained with a constant thickness of 40 ML to investigate the morphology, shape, and droplet size distribution for the same amount of IL deposited at different experimental conditions. The morphology of the samples was analyzed by scanning electron microscopy (SEM), which provided detailed information on the shape and size of the micro- and nanodroplets formed. A tendency for 3D growth (island growth) was observed for all the examined films. For the same amount of IL, a very similar dependence of the droplet size distribution on the deposition rate was found for [C_2C_1im][NTf_2] and [C_2C_1im][OTf]. Intrinsically, the droplets formed by PVD of these ILs have a lower tendency to coalesce by second order mechanisms when compared to the long-chain imidazolium-based ILs, as reported for ITO/glass surfaces in recent works [9,34–36,46]. However, their first order coalescence mechanisms were intensified when higher mass flow rates were used. For [C_8C_1im][NTf_2] and [C_8C_1im][OTf], there was no significant effect of the deposition rate on the morphology of the droplets. Hence, even at very low mass flow rates, the second order coalescence processes of the droplets were found to be self-induced by the inherent wetting behavior of these ILs. The experimental results further showed that the substrate temperature had a less significant impact on the nucleation and growth of all studied ILs deposited on the Ag surfaces. The higher adhesion of the IL droplets on this surface avoids their aggregation, even with the increase in the substrate temperature.

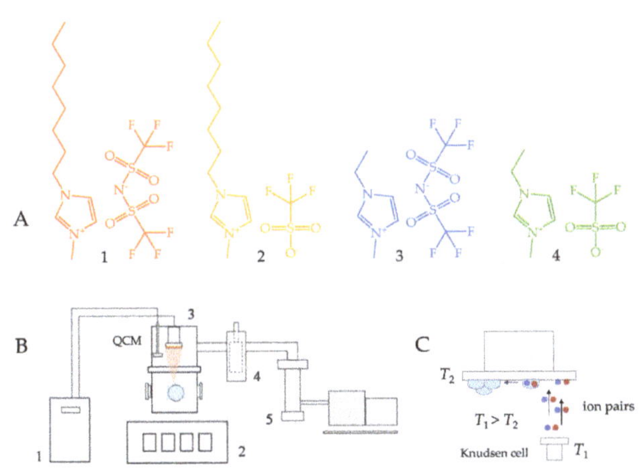

Figure 1. **(A)**—Molecular structures of the studied ionic liquids: (1) 1-octyl-3-methylimidazolium bis(trifluoromethyl)sulfonyl]imide, [C_8C_1im][NTf_2]; (2) 1-octyl-3-methylimidazolium triflate, [C_8C_1im][OTf];

(3) 1-ethyl-3-methylimidazolium bis(trifluoromethyl)sulfonyl]imide, [C_2C_1im][NTf_2]; (4) 1-ethyl-3-methylimidazolium triflate, [C_2C_1im][OTf]. (**B**)—Schematic representation of the thermal evaporation method: (1) cooling system; (2) instrumentation box; (3) vacuum chamber; (4) N_2 (l) metallic trap; (5) vacuum pumping system). (**C**)—Schematic representation of the effusion process of ionic liquids from a Knudsen cell (T_1 and T_2 denote the effusion temperature and the substrate temperature, respectively).

2. Materials and Methods

2.1. Samples

The ILs studied in this work are constituted by an alkylimidazolium cation ([$C_nC_1$im]) and by the triflate (OTf) or the bis[(trifluoromethyl)sulfonyl]imide (NTf_2) anions: 1-ethyl-3-methylimidazolium bis[(trifluoromethyl)sulfonyl]imide, ([C_2C_1im][NTf_2]); 1-octly-3-methylimidazolium bis[(trifluoromethyl)sulfonyl]imide, ([C_8C_1im][NTf_2]); 1-ethyl-3-methylimidazolium triflate, ([C_2C_1im][OTf]); and 1-octly-3-methylimidazolium triflate, ([C_8C_1im][OTf]). The IL samples were purchased from IoLiTec with a state purity higher than 99%. Prior to the deposition of IL films, the possible volatile contaminations were removed by thermal evaporation under reduced pressure ($p < 0.1$ Pa) at $T = 423$ K.

2.2. Substrates

Each sample was deposited on surfaces of indium tin oxide (ITO)/glass and silver (Ag)/ITO/glass by using substrate supports developed in our laboratory. The ITO/glass substrates are constituted by a glass substrate (dimensions of 10 mm × 10 mm × 1.1 mm) coated with a thin film of ITO (≈180 nm of thickness). The substrates were commercialized by the Präzisions Glas & Optik GmbH supplier. Before any deposition, the substrates were cleaned by typical procedures using high-purity ethanol in an ultrasonic bath. The substrates were further dried with pure argon and stored in a desiccator. The Ag/ITO/glass surfaces were prepared by sputter deposition of Ag films on the ITO/glass substrates. A Cressington 108 Auto Sputter Coated instrument was used for sputtering the metallic film. An Ag target (purity > 99.9%) was employed, and a plasma of argon ions was used to bombard the cathode and release the Ag atoms of the target. High-purity (>99.995%) argon was used as the sputter gas, and the sputtering process was performed through a current discharge of 40 mA. The Ag films were obtained with a thickness of ≈100 nm. After the deposition of the metallic film, the substrates were handled under an inert atmosphere to minimize contaminations (e.g., carbon contamination).

2.3. Thin-Film Deposition

The evaporation process of the ILs and consequent formation of films was carried out by a customized system of PVD developed in our laboratory [35,38]. The experimental procedure consists of a vacuum thermal evaporation methodology based on the Knudsen effusion technique. Details of the deposition method are presented in Figures S2–S4 (SM). The process occurs in a low-pressure environment ($p < 10^{-4}$ Pa), facilitating the evaporation of the ion pairs and subsequent condensation on the substrates. The Knudsen effusion cells (evaporation sources) allow good control of the mass flow rate effused. These cells are divided into three main parts: the cell lid, the cell body, and a disk with an orifice. The mass flow rate on the substrate surface and the film thickness are monitored by a quartz crystal microbalance (QCM, Inficon model STM-2). Equation (2) shows the relation between the mass flow rate on the substrate surface ($\Phi_{\text{substrate}}$), the mass flow rate from the Knudsen cell orifice ($\Phi_{\text{Knudsen cell}}$), and a geometric factor (g). $\Phi_{\text{Knudsen cell}}$ is obtained by the derivation of the Knudsen equation, where p is the equilibrium vapor pressure, ω_o is the transmission probability factor, M is the molar mass of the effused vapor, T is the evaporation temperature, m is the mass of the compound, and A_o is the area of Knudsen cell orifice [9,38,43,46].

$$\Phi_{\text{substrate}} = g \cdot \Phi_{\text{Knudsen cell}} = g \cdot \frac{p \cdot \omega_o \cdot \sqrt{M}}{\sqrt{2\pi RT}} = g \cdot \frac{m}{A_o \cdot t} \tag{2}$$

The effusion time was monitored in each experiment, in order to obtain the intended film thickness (40 ML). To study the influence of the evaporation temperature ($T_{\text{evap.}}$) and the resulting mass flow rate in the morphology of the IL droplets and their size distribution, the ILs were evaporated at different temperature intervals: $T_{\text{evap.}}$ = (473 ± 5) K; emphT$_{\text{evap.}}$ = (503 ± 10) K; $T_{\text{evap.}}$ = (533 ± 5) K. For the first study, the substrate temperature ($T_{\text{subst.}}$) was kept constant, at $T_{\text{subst.}}$ = (283.2 ± 0.2) K. $T_{\text{subst.}}$ was controlled by regulating the temperature of a refrigerated liquid bath (Huber Minichiller). The circulating bath was used to refrigerate a copper tube where the substrate's support system (Figure S4) is attached. For each $T_{\text{evap.}}$, the mass flow rate was also changed by using two different orifice diameters in the Knudsen effusion cells: 2 or 3 mm. Additional experiments were performed by removing the disk containing the orifice, using only the cell body (diameter of 10 mm), to maximize the mass flow rate. To study the influence of $T_{\text{subst.}}$ in the nucleation and growth mechanisms of ILs (second study), each sample was evaporated under similar experimental conditions ($T_{\text{evap.}}$ = (533 ± 5) K, but only with an orifice diameter of 2 mm), but three different $T_{\text{subst.}}$ were used: $T_{\text{subst.}}$ = (273.2 ± 0.2) K; $T_{\text{subst.}}$ = (283.2 ± 0.2) K; $T_{\text{subs.}}$ = (298.2 ± 0.2) K. Additional experiments were performed without thermal regulation of the substrate temperature. All details concerning the experimental setup (evaporation temperature, vapor pressure, deposition rate, and time of deposition) used for the deposition of films are presented in Tables S1 and S2 as SM. For the deposition of IL films, ITO/glass and Ag/ITO/glass substrates were rapidly introduced into the vacuum chamber that was degassed at $\approx 10^{-4}$ Pa. Each IL film was obtained with a thickness of 40 ML. An estimation of the height (h) of 1 ML can be performed according to Equation (3), where M is the molar mass, ρ is the density of the IL, and N_A is the Avogadro's constant. For the studied ILs, [C_2C_1im][NTf_2], [C_8C_1im][NTf_2], [C_2C_1im][OTf], and [C_8C_1im][OTf], 1 ML corresponds to a height of 0.75 nm, 0.84 nm, 0.68 nm, and 0.78 nm, respectively [9,36,39,43,46].

$$h = \sqrt[3]{\frac{M}{(N_A \times \rho)}} \tag{3}$$

2.4. Morphological Characterization

The morphological characterization of the films was performed by a compact scanning electron microscope (SEM, Hitachi, FlexSEM 1000). With this technique, it was possible to obtain detailed micrographs of the IL microdroplets (deposited on ITO and Ag surfaces) at different magnifications (800×, 1600×, and 4000×) by using a backscattered electron detector (BSE). SEM imaging of samples deposited on ITO and Ag surfaces was obtained by employing accelerating voltages of 3 kV and 5 kV, respectively. The detector used a working distance between 6 and 9 mm. Afterwards, a deep analysis of the micrographs was made by the ImageJ software (version 1.52a) [66], in order to study the shape, surface coverage, and size distribution of the microdroplets formed.

3. Results

3.1. The Effect of the Mass Flow Rate on the Morphology of ILs Deposited on ITO and Ag Surfaces

The change in the evaporation temperature and diameter of the Knudsen cell orifice were used to investigate the effect of different mass flow rates on the morphology of IL films deposited on both ITO and Ag surfaces. The wetting behavior (in a mesoscopic scale) of the ILs on both surfaces was inferred based on the analysis and processing of the SEM micrographs. A detailed morphology of the substrates is presented in Figure S5 of the SM. The topographies of ITO/glass and Ag/ITO/glass surfaces look similar, suggesting that the roughness of both substrates does not differ significantly. The IL films were obtained at different deposition rates by changing the evaporation temperature ($T_{evap.}$) (from 473 to 533 K), the diameter of the Knudsen cell orifice (2, 3 mm), or using only the cell body. The overall results obtained for each IL are presented in Figures S6–S21 (SM). Increasing the evaporation temperature as well as the use of larger Knudsen cell orifices lead to an increase in the mass flow rate (for the same substrate temperature of T = 283 K). The increase in the evaporation temperature on other hand also leads to an increase in the gas flow temperature, contributing to a higher ion pair initial mobility at the surface. As representative results, Figures 2–5 present the morphology of the micro- and nanodroplets and the respective size distribution for $[C_2C_1im][NTf_2]$, $[C_2C_1im][OTf]$, $[C_8C_1im][NTf_2]$, and $[C_8C_1im][OTf]$, respectively. These results were obtained using a constant diameter of the cell orifice (3 mm) but at different evaporation temperatures (473, 503, and 533 K). The combination of these two variables allowed us to obtain IL films (40 ML) deposited at different deposition rates (Φ): $[C_2C_1im][NTf_2]$ (Φ from 0.08 to 1.5 Å/s); $[C_2C_1im][OTf]$ (Φ from 0.02 to 0.51 Å/s); $[C_8C_1im][NTf_2]$ (Φ from 0.08 to 1.5 Å/s); $[C_8C_1im][OTf]$ (Φ from 0.04 to 0.73 Å/s) (Figures 2–5 and Table S1). Considering the same evaporation temperature, the deposition rate was found to be lower for the OTf-based ILs. This is due to their lower volatility in comparison to the congeners NTf_2-based ILs [67]. A global analysis of the SEM images reveals that, for the same mass flow rate, the size of the IL droplets is always lower on the surface of Ag. This is clearly observed by the comparison between images A, C, and E (SEM images on the left—ILs deposited on ITO) with images B, D, and F (SEM images on the right—ILs deposited on Ag) in Figures 2–5. For the same IL, the higher number of droplets formed on Ag surfaces is explained by the lower MFAN of IL films on the metallic surface, as we recently reported [9,46]. Another obvious outcome derived from the morphological analysis and clear by eye is the larger size of the droplets obtained for $[C_8C_1im][NTf_2]$ and $[C_8C_1im][OTf]$. There is a noticeable differentiation between the wetting behavior of short- and long-chain imidazolium-based ILs. More intense coalescence mechanisms are observed for $[C_8C_1im][NTf_2]$ and $[C_8C_1im][OTf]$, which emphasizes the higher wettability of both ITO and Ag substrates to long-chain alkylimidazolium cations [9,46]. After further processing of the micrographs using ImageJ software (version 1.52a), the total number of droplets formed was counted, their size was measured, and the surface coverage was evaluated. The resulting droplet size distribution was then used to create histograms (graphs 1 to 6 of Figures 2–5) that illustrate how different deposition rates on the substrate surface affect the size distribution of the formed droplets. The deposition rate had no significant effect on the surface coverage of the IL microdroplets (40 ML) on both ITO and Ag surfaces. The films obtained for $[C_2C_1im][NTf_2]$ were found to have surface coverages of (37 $\pm$ 2)% and (35 $\pm$ 1)% on ITO and Ag surfaces, respectively, whereas values of (47 $\pm$ 2)% and (40 $\pm$ 2)% were obtained for $[C_2C_1im][OTf]$. An increase in the mass flow rate led to similar changes for $[C_2C_1im][NTf_2]$ (Figure 2) and $[C_2C_1im][OTf]$ (Figure 3): larger droplets formed on ITO as the flow rate increases (comparing graphs 1, 3, and 5 in both Figures), whereas the size distribution of droplets deposited on Ag shows no difference for the two smaller deposition rates and small difference at the highest one (comparing graphs 2, 4, and 6). Translating these conclusions into numbers, we can see that, for $[C_2C_1im][NTf_2]$ (Figure 2), the most frequent sizes (modal diameter, MD—the diameter value that corresponds to the highest frequency or peak in the distribution) of the droplets obtained on ITO/glass surfaces as we change the deposition rate are: between 1.1 and 1.3 μm for Φ of 0.12 Å/s (graph 1); between 1.4 and 1.8 μm, Φ of 0.54 Å/s (graph 3); and between 1.5 and 2.1 μm, Φ of 1.5 Å/s (graph 5). Under the same experimental conditions, only a small shift is observed in the MD of droplets deposited on Ag: MD changes from 0.7–0.8 μm (graph 2) to 0.7–0.9 μm (graph 4) and 1.1–1.2 μm (graph 6). An overall conclusion indicates that the increase in mass flow rate leads to more intense first order coalescence mechanisms of $[C_2C_1im][NTf_2]$ droplets formed on ITO/glass. This result might be related to the increase in the ion pair mobility at the surface by increasing T_{evap}. Very similar behavior is observed for the congener $[C_2C_1im][OTf]$ (graphs 1, 3, and 5 of Figure 3). ITO is a somewhat solvophobic surface to the ILs and, as a consequence, the increase in Φ promotes the aggregation of the small droplets. On the Ag surfaces, a negligible effect of Φ on the MD is seen—MD values between 0.6 and 0.8 µm are observed for Φ of 0.06, 0.18, and 0.51 Å/s (graphs 2, 4, and 6). The small influence of the deposition rate on the wetting behavior of ILs on Ag surfaces on the mesoscopic scale is probably related to the higher affinity of the imidazolium cations to the metallic surface. Hence, due to good adhesion, there is a lower tendency for the coalescence of the nanodroplets even when large deposition rates are applied. For 40 ML, a very reduced number of microdroplets with a diameter above 3 µm was obtained for the C_2C_1im samples. A very distinct behavior was obtained when long-chain ILs ([C_8C_1im][NTf_2] and [C_8C_1im][OTf]) were deposited on ITO. For the same number of ML, 40 ML, MDs above 6 µm were often obtained (graphs 1 and 3 of Figure 4, graphs 1, 3, and 5 of Figure 5). Curiously, the behavior of [C_8C_1im][NTf_2] on the Ag surface (graphs 2, 4, and 6 of Figure 4) is very similar to that observed for [C_2C_1im][NTf_2] and [C_2C_1im][OTf] on the same substrate—there is a shift in the MD when increasing Φ from 0.11 to 1.5 Å/s. [C_8C_1im][OTf] formed larger microdroplets (graphs 2, 4, and 6 of Figure 5), but the coalescence mechanisms were not as intense as observed for the ITO surface. On ITO, at odds, there is no visible dependence between Φ and MD since the coalescence of the droplets occurred even at the lowest Φ. The coalescence processes may be self-induced by the inherent wetting behavior of [C_8C_1im][NTf_2] and [C_8C_1im][OTf]. In comparison to C_2C_1im, the IL film obtained from PVD on ITO surfaces for the larger C_8C_1im is heterogeneous, as the microdroplets depict an irregular shape. This morphology arises from the typical second order coalescence mechanisms observed for the long-chain imidazolium-based ILs [9,46].

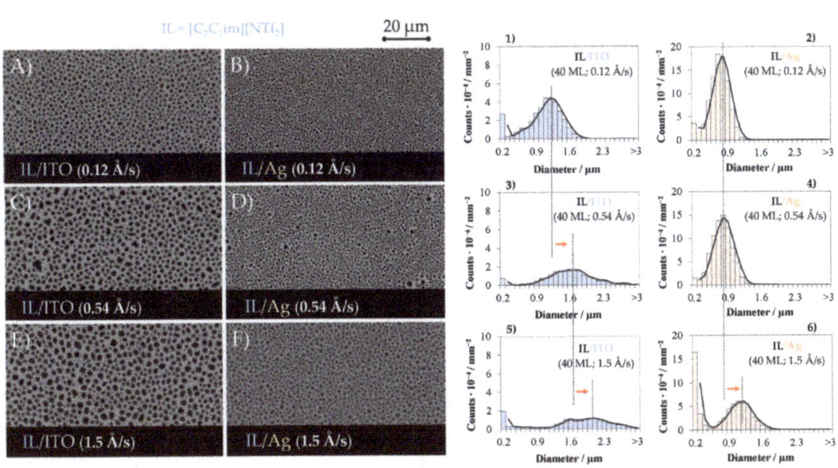

Figure 2. Morphology of the thin films (micro- and nanodroplets) of [C_2C_1im][NTf_2] (**A–F**) and the respective size distribution of the droplets (histograms (**1–6**)). The IL films (40 ML) were deposited (by PVD) simultaneously on surfaces of ITO (images (**A**,**C**,**E**)) and Ag (images (**B**,**D**,**F**)). The IL film surfaces were obtained through different mass flow rates in the deposition process: 0.12, 0.54, and 1.5 Å/s (evaporation temperatures of 473, 503, and 533 K, respectively, and using a Knudsen cell orifice with 3 mm of thickness). The red arrows indicate the successive shift of the modal diameter with an increase in the deposition rate. The temperature of the substrates was kept at T = 283 K. Surface coverages of 35.9, 33.8, 36.1, 34.6, 39.8, and 36.1% were derived by image processing of (**A–F**), respectively. Top views (**A–F**) were acquired through scanning electron microscopy by using a backscattered electron detector.

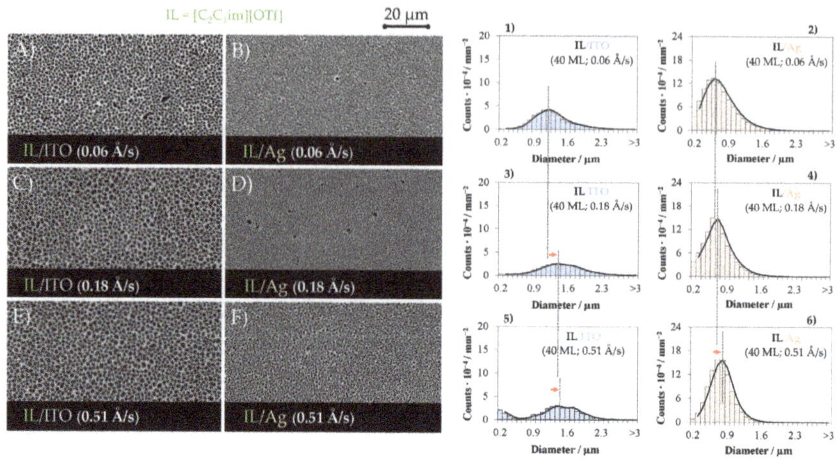

Figure 3. Morphology of the thin films (micro- and nanodroplets) of $[C_2C_1im][OTf]$ (**A–F**) and the respective size distribution of the droplets (histograms (**1–6**)). The IL films (40 ML) were deposited (by PVD) simultaneously on surfaces of ITO (images (**A,C,E**)) and Ag (images (**B,D,F**)). The IL film surfaces were obtained through different mass flow rates in the deposition process: 0.06, 0.18, and 0.51 Å/s (evaporation temperatures of 473, 503, and 533 K, respectively, and using a Knudsen cell orifice with 3 mm of thickness). The red arrows indicate the successive shift of the modal diameter with an increase in the deposition rate. The temperature of the substrates was kept at T = 283 K. Surface coverages of 46.2, 41.9, 48.6, 38.1, 44.8, and 40.6% were derived by image processing of (**A–F**), respectively. Top views (**A–F**) were acquired through scanning electron microscopy by using a backscattered electron detector.

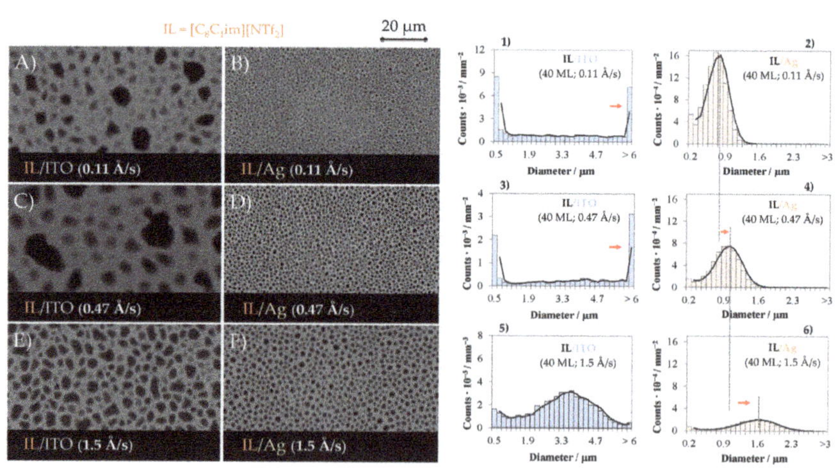

Figure 4. Morphology of the thin films (micro- and nanodroplets) of $[C_8C_1im][NTf_2]$ (**A–F**) and the respective size distribution of the droplets (histograms (**1–6**)). The IL films (40 ML) were deposited (by PVD) simultaneously on surfaces of ITO (images (**A,C,E**)) and Ag (images (**B,D,F**)). The IL film surfaces were obtained through different mass flow rates in the deposition process: 0.11, 0.47, and 1.5 Å/s (evaporation temperatures of 473, 503, and 533 K, respectively, and using a Knudsen cell orifice with 3 mm of thickness). The red arrows indicate the successive shift of the modal diameter with an increase in the deposition rate. The temperature of the substrates was kept at T = 283 K. The image processing was not able to derive with accuracy the surface coverage of the films (it is estimated to be >40%). Top views (**A–F**) were acquired through scanning electron microscopy by using a backscattered electron detector.

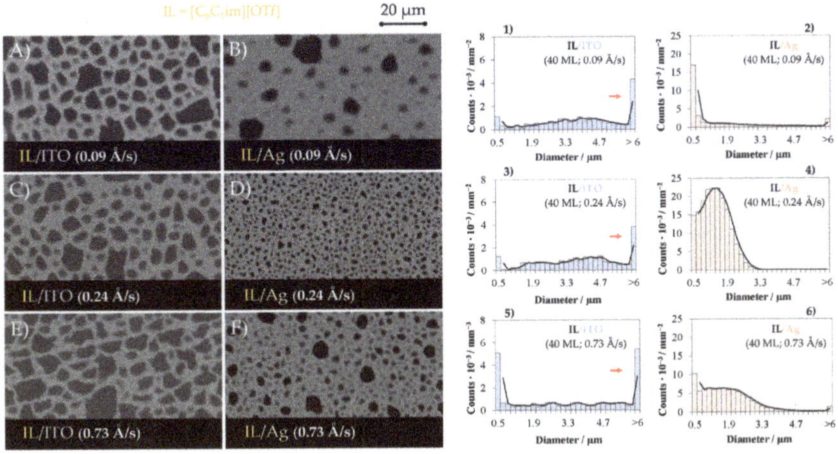

Figure 5. Morphology of the thin films (micro- and nanodroplets) of [C_8C_1im][OTf] (**A–F**) and the respective size distribution of the droplets (histograms (**1–6**)). The IL films (40 ML) were deposited (by PVD) simultaneously on surfaces of ITO (images (**A**,**C**,**E**)) and Ag (images (**B**,**D**,**F**)). The IL film surfaces were obtained through different mass flow rates in the deposition process: 0.09, 0.24, and 0.73 Å/s (evaporation temperatures of 473, 503, and 533 K, respectively, and using a Knudsen cell orifice with 3 mm of thickness). The red arrows indicate the successive shift of the modal diameter with an increase in the deposition rate. The temperature of the substrates was kept at T = 283 K. The image processing was not able to derive with accuracy the surface coverage of the films (it is estimated to be >40%). Top views (**A–F**) were acquired through scanning electron microscopy by using a backscattered electron detector.

3.2. Effect of the Substrate Temperature on the Morphology of ILs Deposited on ITO or Ag Surfaces

The influence of the substrate temperature ($T_{subst.}$) on the morphology and size distribution of the IL droplets was examined using a constant deposition rate, with the deposition carried out simultaneously on the two surfaces (ITO and Ag), kept at the same substrate temperature ($T_{subst.}$), which was changed in the different experiments. Detailed results obtained for each IL are presented in Figures S22–S29 and Table S2 (SM). Figure 6 shows the SEM micrographs and the respective droplets' size distribution for the [C_2C_1im][NTf_2] sample deposited on surfaces at 273 K (images A and B, graphs 1 and 2), 283 K (images C and D, graphs 3 and 4), 298 K (images E and F, graphs 5 and 6), and surfaces without thermal regulation of the substrate temperature (with the refrigerated circulating bath turned off) (images G and H, graphs 7 and 8). For [C_2C_1im][NTf_2], the SEM images show very similar morphologies and identical distribution of the droplets for the films obtained on surfaces kept at different temperatures (273, 283, and 298 K). Further, for all experiments, the droplets are formed with a larger size on the ITO surface (images A, C, E, and G). On ITO surfaces, the MD of the droplets is between 1.3 and 1.4 μm (graphs 1, 3, and 5) whereas on the Ag surface, it was between 0.6 and 0.7 μm (graphs 2, 4, 6). Similar results were obtained without controlling $T_{subst.}$, where $T_{subst.}$ could have been much higher than the highest temperature used before, 298 K, as the substrates were not refrigerated, and hence their temperature depends on the vacuum chamber walls, which can be affected by the temperature of the evaporation ovens.

Overall, we observed that the increase in the substrate temperature had a small effect on the morphology and size distribution of [C_2C_1im][NTf_2] droplets, especially when deposited on the Ag metallic surface. The results presented further highlight the stability of the IL droplets and the reproducibility of the PVD method for the deposition of IL, as independent experiments led to the formation of droplets with similar morphology and size. As regarding [C_2C_1im][OTf], the effect of $T_{subst.}$ on the MD of the droplets deposited on Ag was also minor (Figures S24 and S25 of the SM), with the MD between 0.6 and 0.7 μm, indicating an identical morphological behavior for $[C_2C_1im][NTf_2]$ and $[C_2C_1im][OTf]$. On ITO, the results show a clear increase in the MD when $T_{subst.}$ was increased. The MD values obtained were 0.8–1.0 μm, 1.0–1.1 μm, and 1.2–1.4 μm, for substrate temperatures of 273, 283, and 298 K, respectively. For $[C_8C_1im][NTf_2]$ (Figures S26 and S27), the droplets obtained on substrates at 273 K had a different appearance compared to those obtained at 283 K and 298 K for both the ITO and Ag surfaces. Further morphological differentiation was noted for $[C_8C_1im][OTf]$ (Figures S27 and S28). As the substrate temperature progressively increased from 273 K to 298 K, a clear increase in droplet size was observed on the Ag surfaces. The droplets of $[C_8C_1im][OTf]$ on ITO were consistently found to be very large in all experiments due to their tendency to coalesce, independent of the mass flow rate or substrate temperature used. In general, the droplet size was consistently larger for both C_2C_1im and C_8C_1im samples when the ILs were deposited on ITO/glass surfaces.

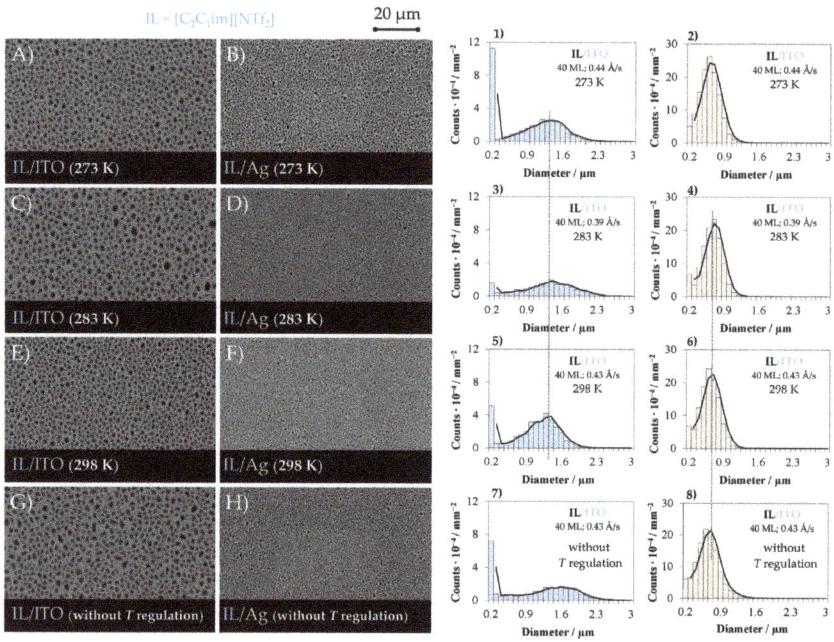

Figure 6. Morphology of the thin films (micro- and nanodroplets) of $[C_2C_1im][NTf_2]$ (**A–H**) and the respective size distribution of the droplets (histograms (**1–8**)). The IL films (40 ML) were deposited (by PVD) simultaneously on surfaces of ITO (images (**A,C,E,G**)) and Ag (images (**B,D,F,H**)). The IL film surfaces were obtained on substrates kept at different temperatures: T = 273 K (**A,B**); T = 283 K (**C,D**); T = 298 K (**E,F**); without thermal regulation of the substrate temperature (**G,H**). In all experiments, the deposition process was carried out by using a similar mass flow rate (≈0.4 Å/s). Top views (**A–H**) were acquired through scanning electron microscopy by using a backscattered electron detector.

4. Discussion

Reproducible morphological changes were obtained for the four studied ILs, with the molten salts like (shorter alkyl size) ILs $[C_2C_1im][NTf_2]$ and $[C_2C_1im][OTf]$ forming small droplets whose characteristics are particularly sensitive to changes in the mass flow rate, and the long-chain ILs ($[C_8C_1im][NTf_2]$ and $[C_8C_1im][OTf]$) showing an inherent tendency to form higher-size/coalesced droplets.

To better visualize these results, Figure 7 shows the graphical representations of the total number of IL microdroplets obtained (on both ITO and Ag surfaces) as a function of the deposition rate for $[C_2C_1im][NTf_2]$ (graph A), $[C_2C_1im][OTf]$ (graph B), $[C_8C_1im][NTf_2]$ (graph C), and $[C_8C_1im][OTf]$ (graph D). A glance at the vertical scale of the graphs clearly indicates that a higher number of droplets was formed for the shorter alkyl size imidazolium-based ILs (C_2C_1im, graphs A and B). This arises from the less intense coalescence mechanisms of these ILs on the studied surfaces. The better wettability of long-chain imidazolium-based ILs led to the formation of larger droplets for $[C_8C_1im][NTf_2]$ and $[C_8C_1im][OTf]$, and consequently, the total number of microdroplets formed was lower (graphs C and D). Figure 7 also evidences the higher number of droplets formed on Ag surfaces, in good agreement with previous works [9,46].

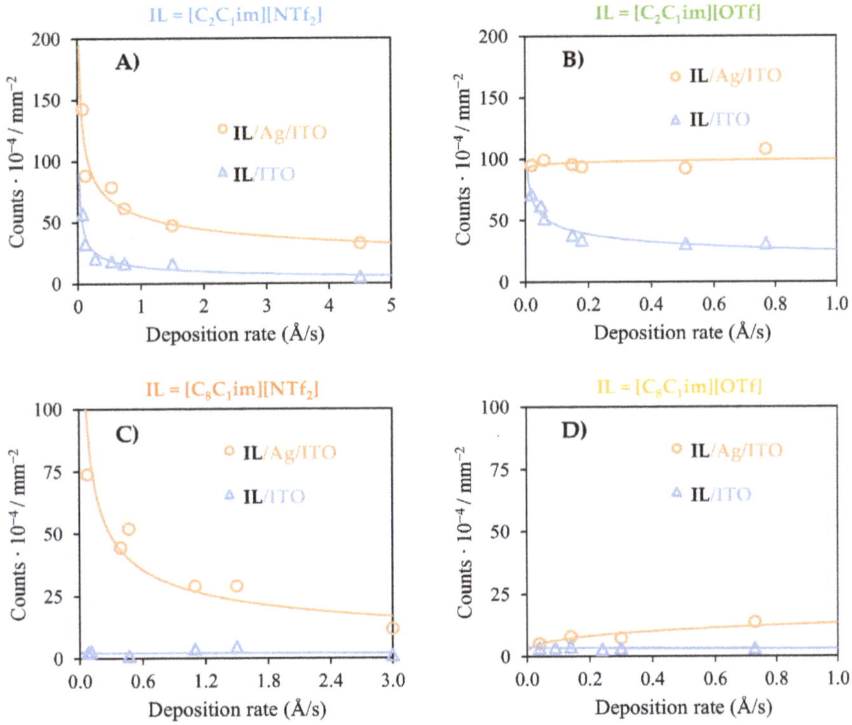

Figure 7. Graphical representations of the number of IL droplets obtained (all microdroplets with a minimum area of 0.02 μm^2) as a function of the deposition rate for $[C_2C_1im][NTf_2]$ (**A**); $[C_2C_1im][OTf]$ (**B**); $[C_8C_1im][NTf_2]$ (**C**); and $[C_8C_1im][OTf]$ (**D**). These results were derived from image processing of the SEM micrographs relative to the simultaneous deposition of each IL film (40 ML) on surfaces of ITO and Ag/ITO.

A detailed observation of curve shapes shows that similar profiles (although with different magnitudes) were obtained for the number of microdroplets as a function of the deposition rate for $[C_2C_1im][NTf_2]$/Ag and $[C_2C_1im][NTf_2]$/ITO (graph A). At lower deposition rates, the number of droplets formed is maximum (but higher on Ag surfaces) and the successive increase in Φ leads to an exponential decay of the total droplet count until reaching a plateau. This suggests that the coalescence mechanisms increase progressively when applying higher deposition rates until reaching a limit where they level off. A similar profile was obtained for $[C_2C_1im][OTf]$/ITO (graph B). A peculiar result is observed for $[C_2C_1im][OTf]$/Ag as the results indicate no influence of Φ on the droplet counts. This observation agrees with the size distribution of the droplets presented above in Figure 3 (graphs 2, 4, and 6)—only a small effect of Φ in the MD, profile of droplet size distribution, as well as on the total droplet count is observed. We interpret this as resulting from the small submicrodroplets of $[C_2C_1im][OTf]$ being very stable and well adsorbed to the Ag surface. For this IL, both MFAN and coalescence mechanisms are intrinsically reduced on the Ag surface and the coalescence of the droplets (first order mechanisms) is not intensified by increasing Φ. For the longer ILs, we observe that the intrinsic tendency of the C_8C_1im droplets to coalesce when deposited on ITO/glass leads to the profiles presented in graphs C and D. The wetting behavior of $[C_8C_1im][NTf_2]$ and $[C_8C_1im][OTf]$ on ITO leads to the formation of very large droplets (Figures 4 and 5), independently of the experimental conditions of PVD. This behavior arises from the typical second order coalescence mechanisms of the long-chain ILs [9,46]. Similar behavior is observed for $[C_8C_1im][OTf]$ deposited on the Ag surface (graph D). Curiously, $[C_8C_1im][NTf_2]$ deposited on Ag shows a very distinct behavior, as the dependence of the droplet counts with the deposition rate is closer to observed for the short-chain ILs. This result agrees with the size distribution of the droplets presented in Figure 4 (graphs 2, 4, and 6). Clearly, droplets of $[C_8C_1im][NTf_2]$ deposited on Ag (images B, D, and F of Figure 4) are smaller in comparison to the congener $[C_8C_1im][OTf]$ deposited on the same surface (images B, D, and F of Figure 5). Overall, these results indicate that, independently of the deposition rate, the second order coalescence processes of $[C_8C_1im][OTf]$ on Ag are clearly higher when compared to what we observed for $[C_8C_1im][NTf_2]$.

For all the studied ILs, the wetting behavior did not reflect an excellent film-substrate affinity since a continuous film was not obtained, even for the long-chain ILs. Despite the C_8C_1im-based ILs having a lower film-vapor interfacial tension [24], the magnitude of the substrate-film interfacial tension might be even high, as the obtained 3D film growth mode (island growth) reveals. Although longer chain ILs have a greater droplet spreading, the contact angle on both ITO and Ag surfaces is still considerable ($\approx$20 degrees) [9]. Other authors have also investigated the island growth of ILs on the mesoscopic scale; employing vacuum deposition methods, similar film morphologies (the formation of ionic liquid microdroplets) were reported [35,68–70]. Further, there are many reports highlighting the initial formation of a 2D wetting layer followed by droplet formation (3D growth) on a wide range of solid surfaces [55,56,59,71]. Good wettability and spreading behavior have been observed on gold [9,39,46,63,64,72,73] or organic surfaces [43,70] since the ion pairs can interact with the surface by the formation of image dipoles (with gold) or through dispersive interactions (with organic materials).

Herein, the experimental variables studied were found to have a visible influence on the droplet formation and droplet spreading of the small ILs (C_2C_1im). Our results led us to raise the hypothesis that an increase in the deposition rate could have an impact on the kinetics of the droplet formation and coalescence mechanisms, leading to higher first order coalescence in the short-chain ILs, characteristic of high surface mobility. On the other hand, the prevailing second order coalescence mechanisms known for long-chain ILs (C_8C_1im) on solvophobic surfaces could explain the formation of large droplets, regardless of the magnitude of the deposition rate. For a detailed comparison of the influence of the anion for the smaller ILs (C_2C_1im), Figure 8 represents the dependence of the MD of the droplets on the deposition rate (Φ, graph A) and the substrate temperature ($T_{subst.}$, graph B). Remarkably, the profiles observed for both ILs in each surface are very similar as regarding to the effect of Φ (A); the curves obtained are almost coincident, and superior MD values are always observed for samples deposited on ITO, as a result of more preponderant first order coalescence mechanisms. For both ILs deposited on ITO, a small increase in Φ (from 0.05 to $\approx$0.3 Å/s) leads to a large increment in the MD, whereas for higher Φ values, the MD increases more slowly. The similar profiles observed for both ILs might suggest the less significant role of the anion on the wetting behavior of the ILs on ITO.

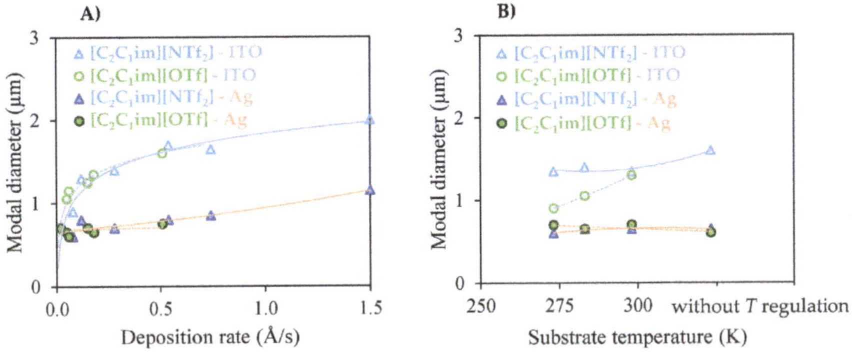

Figure 8. Graphical representations of the modal diameter of the IL droplets as a function of the deposition rate (**A**) and the substrate temperature (**B**) for $[C_2C_1im][NTf_2]$ and $[C_2C_1im][OTf]$. These results were derived from the analysis of the SEM micrographs relative to the simultaneous deposition of each IL film (40 ML) on surfaces of ITO and Ag/ITO.

On Ag surfaces, a small (close to linear) increase in the MD occurs for all range of Φ values. A similar increment in the MD as Φ increases above 0.3 Å/s can be derived from these representations. For both C_2C_1im samples, the droplets' MD on ITO is (0.8 ± 0.1) μm higher than the value obtained for the Ag surface. Considering the effect of the substrate temperature (for the same deposition rate, graph B), the MD of the microdroplets was, as expected, lower on Ag surfaces. According to the results, the substrate temperature influences the MD of droplets deposited on ITO, especially for $[C_2C_1im][OTf]$. The formation of larger droplets is a consequence of the increased surface mobility caused by the higher substrate temperature [74]. On the other hand, the substrate temperature did not influence the deposition process of the ILs on Ag, and a constant MD of (0.6 ± 0.1) μm was observed for both $[C_2C_1im][NTf_2]$ and $[C_2C_1im][OTf]$. These observations emphasize the higher affinity/stability of the droplets on the Ag/ITO surface.

5. Conclusions

The impact of deposition rate and substrate temperature on the morphology of ionic liquid (IL) films deposited by PVD on ITO and Ag/ITO surfaces was investigated for four imidazolium-based ionic liquids (ILs): $[C_2C_1im][NTf_2]$, $[C_2C_1im][OTf]$, $[C_8C_1im][NTf_2]$, and $[C_8C_1im][OTf]$. The film growth was characterized by the formation of three-dimensional droplets/islands (3D growth). The increase in deposition rate and a small change in substrate temperature were found to have a minor effect on the wettability of Ag towards ILs with shorter alkyl chains (C_2C_1im), indicating good adsorption and stability of IL droplets on the metallic surface. However, on the ITO substrate, a higher degree of droplet coalescence was observed due to the higher surface mobility of the IL on the substrate. It is noteworthy that the increase in deposition rate had a very similar effect on the formation of films for both $[C_2C_1im][NTf_2]$ and $[C_2C_1im][OTf]$, with the modal diameter of the droplets obtained being identical for each deposition rate. The droplet morphology of the long-chain ILs (C_8C_1im) was found to be influenced by changes in the deposition rate and/or substrate temperature. Nevertheless, very large droplets were consistently observed on the ITO surface, regardless of the deposition parameters.

In summary, the experimental data presented in this study provides reliable information on the influence of deposition parameters on the nucleation and growth of IL films. Our findings demonstrate that changes in the mass flow rate and substrate temperature have a greater impact on the coalescence stage and morphology of IL droplets deposited on ITO surfaces than those observed on Ag surfaces.

Supplementary Materials: The following supporting information can be downloaded at: https://www.mdpi.com/article/10.3390/fluids8030105/s1, Figure S1: Schematic illustration of the typical mechanisms of nucleation and growth of ionic liquid films; Figure S2: Schematic representation of the vacuum thermal evaporation methodology; Figure S3: Schematic representation of the ovens of the ThinFilmVD apparatus and an image of an individual oven; Figure S4: Schematic representation and images of the substrates support system; Figure S5: Morphology of the substrates (ITO/glass and Ag/ITO/glass); Figures S6–S17: Morphology of $[C_2C_1im][NTf_2]$ (S6), $[C_2C_1im][OTf]$ (S9), $[C_8C_1im][NTf_2]$ (S12), and $[C_2C_1im][OTf]$ (S15) films on ITO and Ag using Knudsen effusion cells with different orifice diameters and employing different evaporation temperatures and the respective size distribution of the droplets (S7, S8, S10, S11, S13, S14, S16, and S17); Figures S18–S21: Morphology of micro-/nanodroplets of $[C_2C_1im][NTf_2]$ (S18), $[C_2C_1im][OTf]$ (S19), $[C_8C_1im][NTf_2]$ (S20), and $[C_2C_1im][OTf]$ (S21) fabricated onto ITO/glass and Ag/ITO/glass using Knudsen effusion cells with different orifice diameters; Figures S22–S29: Morphology of micro-/nanodroplets of $[C_2C_1im][NTf_2]$ (S22), $[C_2C_1im][OTf]$ (S24), $[C_8C_1im][NTf_2]$ (S26), and $[C_2C_1im][OTf]$ (S28) fabricated onto ITO/glass and Ag/ITO/glass employing different substrate temperatures and the respective size distribution (S23, S25, S27, and S29); Table S1: Experimental conditions related to the study of the effect of the mass flow rate on the morphology of the ionic liquid films deposited on ITO/glass and Ag/ITO/glass surfaces; Table S2: Experimental conditions related to the study of the effect of the substrate temperature on the morphology of the ionic liquid films deposited on ITO/glass and Ag/ITO/glass surfaces.

Author Contributions: Conceptualization, J.C.S.C.; methodology, R.M.C., C.N. and J.C.S.C.; investigation, R.M.C. and J.C.S.C.; writing—original draft preparation—R.M.C.; writing—review and editing, M.B., L.M.N.B.F.S. and J.C.S.C.; supervision, M.B. and J.C.S.C.; project administration, M.B. and J.C.S.C.; funding acquisition, L.M.N.B.F.S. All authors have read and agreed to the published version of the manuscript.

Funding: The Portuguese Foundation for Science and Technology (FCT) for financial support to CIQUP, Faculty of Science, University of Porto (Project UIDB/00081/2020), and IMS-Institute of Molecular Sciences (LA/P/0056/2020), Faculty of Science, University of Porto.

Institutional Review Board Statement: Not applicable.

Informed Consent Statement: Not applicable.

Data Availability Statement: Not applicable.

Acknowledgments: The authors thank the Portuguese Foundation for Science and Technology (FCT) for financial support to CIQUP, Faculty of Science, University of Porto (Project UIDB/00081/2020), and IMS-Institute of Molecular Sciences (LA/P/0056/2020), Faculty of Science, University of Porto.

Conflicts of Interest: The authors declare no conflict of interest. The funders had no role in the design of the study; in the collection, analyses, or interpretation of data; in the writing of the manuscript; or in the decision to publish the results.

References

1. Hayes, R.; Warr, G.G.; Atkin, R. Structure and Nanostructure in Ionic Liquids. *Chem. Rev.* **2015**, *115*, 6357–6426. [PubMed]
2. Canongia Lopes, J.N.A.; Pádua, A.A.H. Nanostructural Organization in Ionic Liquids. *J. Phys. Chem. B* **2006**, *110*, 3330–3335. [CrossRef]
3. Del Pópolo, M.G.; Voth, G.A. On the Structure and Dynamics of Ionic Liquids. *J. Phys. Chem. B* **2004**, *108*, 1744–1752. [CrossRef]
4. Ghandi, K. A Review of Ionic Liquids, Their Limits and Applications. *Green Sustain. Chem.* **2014**, *04*, 44–53. [CrossRef]
5. Lei, Z.; Chen, B.; Koo, Y.-M.; MacFarlane, D.R. Introduction: Ionic Liquids. *Chem. Rev.* **2017**, *117*, 6633–6635. [CrossRef]
6. Ren, F.; Wang, J.; Xie, F.; Zan, K.; Wang, S.; Wang, S. Applications of Ionic Liquids in Starch Chemistry: A Review. *Green Chem.* **2020**, *22*, 2162–2183. [CrossRef]
7. Esperança, J.M.S.S.; Canongia Lopes, J.N.; Tariq, M.; Santos, L.M.N.B.F.; Magee, J.W.; Rebelo, L.P.N. Volatility of Aprotic Ionic Liquids—A Review. *J. Chem. Eng. Data* **2009**, *55*, 3–12. [CrossRef]
8. Rocha, M.A.A.; Coutinho, J.A.P.; Santos, L.M.N.B.F. Vapor Pressures of 1,3-Dialkylimidazolium Bis(Trifluoromethylsulfonyl)Imide Ionic Liquids with Long Alkyl Chains. *J. Chem. Phys.* **2014**, *141*, 134502. [CrossRef]
9. Costa, J.C.S.; Alves, A.; Bastos, M.; Santos, L.M.N.B.F. The Impact of the Cation Alkyl Chain Length on the Wettability of Alkylimidazolium-Based Ionic Liquids at the Nanoscale. *Phys. Chem. Chem. Phys.* **2022**, *24*, 13343–13355. [CrossRef]

10. Shahiduzzaman; Muslih, E.Y.; Hasan, A.K.M.; Wang, L.; Fukaya, S.; Nakano, M.; Karakawa, M.; Takahashi, K.; Akhtaruzzaman; Nunzi, J.-M.; et al. The Benefits of Ionic Liquids for the Fabrication of Efficient and Stable Perovskite Photovoltaics. *Chem. Eng. J.* **2021**, *411*, 128461. [CrossRef]
11. Pei, Y.; Zhang, Y.; Ma, J.; Fan, M.; Zhang, S.; Wang, J. Ionic Liquids for Advanced Materials. *Mater. Today Nano* **2022**, *17*, 100159. [CrossRef]
12. Kaur, G.; Kumar, H.; Singla, M. Diverse Applications of Ionic Liquids: A Comprehensive Review. *J. Mol. Liq.* **2022**, *351*, 118556. [CrossRef]
13. Sun, P.; Armstrong, D.W. Ionic Liquids in Analytical Chemistry. *Anal. Chim. Acta* **2010**, *661*, 1–16. [CrossRef]
14. Zhang, S.; Sun, N.; He, X.; Lu, X.; Zhang, X. Physical Properties of Ionic Liquids: Database and Evaluation. *J. Phys. Chem. Ref. Data* **2006**, *35*, 1475–1517. [CrossRef]
15. Plechkova, N.V.; Seddon, K.R. Applications of Ionic Liquids in the Chemical Industry. *Chem. Soc. Rev.* **2008**, *37*, 123–150. [CrossRef] [PubMed]
16. Borghi, F.; Podestà, A. Ionic Liquids under Nanoscale Confinement. *Adv. Phys. X* **2020**, *5*, 1736949. [CrossRef]
17. Calandra, P.; Szerb, E.I.; Lombardo, D.; Algieri, V.; De Nino, A.; Maiuolo, L. A Presentation of Ionic Liquids as Lubricants: Some Critical Comments. *Appl. Sci.* **2021**, *11*, 5677. [CrossRef]
18. Yoo, E.; Kim, J.; Hosono, E.; Zhou, H.; Kudo, T.; Honma, I. Large Reversible Li Storage of Graphene Nanosheet Families for Use in Rechargeable Lithium Ion Batteries. *Nano Lett.* **2008**, *8*, 2277–2282. [CrossRef]
19. Erto, A.; Silvestre-Albero, A.; Silvestre-Albero, J.; Rodríguez-Reinoso, F.; Balsamo, M.; Lancia, A.; Montagnaro, F. Carbon-Supported Ionic Liquids as Innovative Adsorbents for CO_2 Separation from Synthetic Flue-Gas. *J. Colloid Interface Sci.* **2015**, *448*, 41–50. [CrossRef]
20. Fredlake, C.P.; Crosthwaite, J.M.; Hert, D.G.; Aki, S.N.V.K.; Brennecke, J.F. Thermophysical Properties of Imidazolium-Based Ionic Liquids. *J. Chem. Eng. Data* **2004**, *49*, 954–964. [CrossRef]
21. Wishart, J.F. Energy Applications of Ionic Liquids. *Energ. Environ. Sci.* **2009**, *2*, 956. [CrossRef]
22. Hallett, J.P.; Welton, T. Room-Temperature Ionic Liquids: Solvents for Synthesis and Catalysis. 2. *Chem. Rev.* **2011**, *111*, 3508–3576. [CrossRef] [PubMed]
23. Rocha, M.A.A.; Lima, C.F.R.A.C.; Gomes, L.R.; Schröder, B.; Coutinho, J.A.P.; Marrucho, I.M.; Esperança, J.M.S.S.; Rebelo, L.P.N.; Shimizu, K.; Lopes, J.N.C.; et al. High-Accuracy Vapor Pressure Data of the Extended $[C_nC_1Im][Ntf_2]$ Ionic Liquid Series: Trend Changes and Structural Shifts. *J. Phys. Chem. B* **2011**, *115*, 10919–10926. [CrossRef] [PubMed]
24. Almeida, H.F.D.; Freire, M.G.; Fernandes, A.M.; Lopes-da-Silva, J.A.; Morgado, P.; Shimizu, K.; Filipe, E.J.M.; Canongia Lopes, J.N.; Santos, L.M.N.B.F.; Coutinho, J.A.P. Cation Alkyl Side Chain Length and Symmetry Effects on the Surface Tension of Ionic Liquids. *Langmuir* **2014**, *30*, 6408–6418. [CrossRef]
25. Ebrahimi, M.; Moosavi, F. The Effects of Temperature, Alkyl Chain Length, and Anion Type on Thermophysical Properties of the Imidazolium Based Amino Acid Ionic Liquids. *J. Mol. Liq.* **2018**, *250*, 121–130. [CrossRef]
26. Katsyuba, S.A.; Vener, M.V.; Zvereva, E.E.; Fei, Z.; Scopelliti, R.; Laurenczy, G.; Yan, N.; Paunescu, E.; Dyson, P.J. How Strong Is Hydrogen Bonding in Ionic Liquids? Combined X-ray Crystallographic, Infrared/Raman Spectroscopic, and Density Functional Theory Study. *J. Phys. Chem. B* **2013**, *117*, 9094–9105. [CrossRef]
27. Kurnia, K.A.; Lima, F.; Cláudio, A.F.M.; Coutinho, J.A.P.; Freire, M.G. Hydrogen-Bond Acidity of Ionic Liquids: An Extended Scale. *Phys. Chem. Chem. Phys.* **2015**, *17*, 18980–18990. [CrossRef]
28. Bica, K.; Deetlefs, M.; Schröder, C.; Seddon, K.R. Polarisabilities of Alkylimidazolium Ionic Liquids. *Phys. Chem. Chem. Phys.* **2013**, *15*, 2703. [CrossRef]
29. Fadeeva, Y.A.; Gruzdev, M.S.; Kudryakova, N.O.; Shmukler, L.E.; Safonova, L.P. Physico-Chemical Characterization of Alkyl-Imidazolium Protic Ionic Liquids. *J. Mol. Liq.* **2020**, *297*, 111305. [CrossRef]
30. Sanchora, P.; Pandey, D.K.; Kagdada, H.L.; Materny, A.; Singh, D.K. Impact of Alkyl Chain Length and Water on the Structure and Properties of 1-Alkyl-3-Methylimidazolium Chloride Ionic Liquids. *Phys. Chem. Chem. Phys.* **2020**, *22*, 17687–17704. [CrossRef]
31. Rocha, M.A.A.; Neves, C.M.S.S.; Freire, M.G.; Russina, O.; Triolo, A.; Coutinho, J.A.P.; Santos, L.M.N.B.F. Alkylimidazolium Based Ionic Liquids: Impact of Cation Symmetry on Their Nanoscale Structural Organization. *J. Phys. Chem. B* **2013**, *117*, 10889–10897. [CrossRef] [PubMed]
32. Xue, L.; Gurung, E.; Tamas, G.; Koh, Y.P.; Shadeck, M.; Simon, S.L.; Maroncelli, M.; Quitevis, E.L. Effect of Alkyl Chain Branching on Physicochemical Properties of Imidazolium-Based Ionic Liquids. *J. Chem. Eng. Data* **2016**, *61*, 1078–1091. [CrossRef]
33. Perkin, S.; Crowhurst, L.; Niedermeyer, H.; Welton, T.; Smith, A.M.; Gosvami, N.N. Self-Assembly in the Electrical Double Layer of Ionic Liquids. *Chem. Commun.* **2011**, *47*, 6572. [CrossRef] [PubMed]
34. Costa, J.C.S.; Mendes, A.; Santos, L.M.N.B.F. Morphology of Imidazolium-Based Ionic Liquids as Deposited by Vapor Deposition: Micro-/Nanodroplets and Thin Films. *ChemPhysChem* **2016**, *17*, 2123–2127. [CrossRef] [PubMed]
35. Costa, J.C.S.; Coelho, A.F.S.M.G.; Mendes, A.; Santos, L.M.N.B.F. Nucleation and Growth of Microdroplets of Ionic Liquids Deposited by Physical Vapor Method onto Different Surfaces. *Appl. Surf. Sci.* **2018**, *428*, 242–249. [CrossRef]
36. Campos, R.M.; Alves, A.C.P.M.; Lima, M.A.L.; Farinha, A.F.M.; Cardoso, J.P.S.; Mendes, A.; Costa, J.C.S.; Santos, L.M.N.B.F. Morphology, Structure, and Dynamics of Pentacene Thin Films and Their Nanocomposites with $[C_2C_1Im][NTf_2]$ and $[C_2C_1Im][OTF]$ Ionic Liquids. *ChemPhysChem* **2020**, *21*, 1814–1825. [CrossRef] [PubMed]

37. Cremer, T.; Killian, M.; Gottfried, J.M.; Paape, N.; Wasserscheid, P.; Maier, F.; Steinrück, H.-P. Physical Vapor Deposition of [EMIM][Tf_2N]: A New Approach to the Modification of Surface Properties with Ultrathin Ionic Liquid Films. *ChemPhysChem* **2008**, *9*, 2185–2190. [CrossRef]
38. Costa, J.C.S.; Rocha, R.M.; Vaz, I.C.M.; Torres, M.C.; Mendes, A.; Santos, L.M.N.B.F. Description and Test of a New Multilayer Thin Film Vapor Deposition Apparatus for Organic Semiconductor Materials. *J. Chem. Eng. Data* **2015**, *60*, 3776–3791. [CrossRef]
39. Rietzler, F.; May, B.; Steinrück, H.-P.; Maier, F. Switching Adsorption and Growth Behavior of Ultrathin [C_2C_1Im][OTf] Films on Au(111) by Pd Deposition. *Phys. Chem. Chem. Phys.* **2016**, *18*, 25143–25150. [CrossRef]
40. Richey, N.E.; de Paula, C.; Bent, S.F. Understanding Chemical and Physical Mechanisms in Atomic Layer Deposition. *J. Chem. Phys.* **2020**, *152*, 040902. [CrossRef]
41. Venables, J.A.; Spiller, G.D.T.; Hanbucken, M. Nucleation and Growth of Thin Films. *Rep. Prog. Phys.* **1984**, *47*, 399–459. [CrossRef]
42. Ratsch, C.; Venables, J.A. Nucleation Theory and the Early Stages of Thin Film Growth. *J. Vac. Sci. Technol. A* **2003**, *21*, S96–S109. [CrossRef]
43. Costa, J.C.S.; Campos, R.M.; Castro, A.C.M.; Farinha, A.F.M.; Oliveira, G.N.P.; Araújo, J.P.; Santos, L.M.N.B.F. The Effect of Ionic Liquids on the Nucleation and Growth of Perylene Films Obtained by Vapor Deposition. *CrystEngComm* **2023**, *25*, 913–924. [CrossRef]
44. Young, T., III. An Essay on the Cohesion of Fluids. *Philos. Trans. Roy. Soc. Lond.* **1805**, *95*, 65–87.
45. ADAM, N.K. Use of the Term "Young's Equation" for Contact Angles. *Nature* **1957**, *180*, 809–810. [CrossRef]
46. Teixeira, M.S.M.; Santos, L.M.N.B.F.; Costa, J.C.S. Nucleation, Coalescence, and Thin-Film Growth of Triflate-Based Ionic Liquids on ITO, Ag, and Au Surfaces. *Colloids Interfaces* **2022**, *6*, 46. [CrossRef]
47. Tariq, M.; Freire, M.G.; Saramago, B.; Coutinho, J.A.P.; Lopes, J.N.C.; Rebelo, L.P.N. Surface Tension of Ionic Liquids and Ionic Liquid Solutions. *Chem. Soc. Rev.* **2012**, *41*, 829–868. [CrossRef]
48. Dragoni, D.; Manini, N.; Ballone, P. Interfacial Layering of a Room-Temperature Ionic Liquid Thin Film on Mica: A Computational Investigation. *ChemPhysChem* **2012**, *13*, 1772–1780. [CrossRef]
49. Weijs, J.H.; Marchand, A.; Andreotti, B.; Lohse, D.; Snoeijer, J.H. Origin of Line Tension for a Lennard-Jones Nanodroplet. *Phys. Fluid.* **2011**, *23*, 022001. [CrossRef]
50. Zhang, J.; Wang, P.; Borg, M.K.; Reese, J.M.; Wen, D. A Critical Assessment of the Line Tension Determined by the Modified Young's Equation. *Phys. Fluid.* **2018**, *30*, 082003. [CrossRef]
51. Lexow, M.; Talwar, T.; Heller, B.S.J.; May, B.; Bhuin, R.G.; Maier, F.; Steinrück, H.-P. Time-Dependent Changes in the Growth of Ultrathin Ionic Liquid Films on Ag(111). *Phys. Chem. Chem. Phys.* **2018**, *20*, 12929–12938. [CrossRef] [PubMed]
52. Lexow, M.; Heller, B.S.J.; Maier, F.; Steinrück, H.-P. Anion Exchange at the Liquid/Solid Interface of Ultrathin Ionic Liquid Films on Ag(111). *ChemPhysChem* **2018**, *19*, 2978–2984. [CrossRef] [PubMed]
53. Lexow, M.; Heller, B.S.J.; Partl, G.; Bhuin, R.G.; Maier, F.; Steinrück, H.-P. Cation Exchange at the Interfaces of Ultrathin Films of Fluorous Ionic Liquids on Ag(111). *Langmuir* **2018**, *35*, 398–405. [CrossRef] [PubMed]
54. Liu, Y.; Zhang, Y.; Wu, G.; Hu, J. Coexistence of Liquid and Solid Phases of Bmim-PF_6 Ionic Liquid on Mica Surfaces at Room Temperature. *J. Am. Chem. Soc.* **2006**, *128*, 7456–7457. [CrossRef]
55. Lexow, M.; Maier, F.; Steinrück, H.-P. Ultrathin Ionic Liquid Films on Metal Surfaces: Adsorption, Growth, Stability and Exchange Phenomena. *Adv. Phys. X* **2020**, *5*, 1761266. [CrossRef]
56. Delcheva, I.; Ralston, J.; Beattie, D.A.; Krasowska, M. Static and Dynamic Wetting Behaviour of Ionic Liquids. *Adv. Colloid Interface Sci.* **2015**, *222*, 162–171. [CrossRef]
57. Deyko, A.; Cremer, T.; Rietzler, F.; Perkin, S.; Crowhurst, L.; Welton, T.; Steinrück, H.-P.; Maier, F. Interfacial Behavior of Thin Ionic Liquid Films on Mica. *J. Phys. Chem. C* **2013**, *117*, 5101–5111. [CrossRef]
58. Gong, X.; Li, L. Understanding the Wettability of Nanometer-Thick Room Temperature Ionic Liquids (RTILs) on Solid Surfaces. *Chin. Chem. Lett.* **2017**, *28*, 2045–2052. [CrossRef]
59. Beattie, D.A.; Espinosa-Marzal, R.M.; Ho, T.T.M.; Popescu, M.N.; Ralston, J.; Richard, C.J.E.; Sellapperumage, P.M.F.; Krasowska, M. Molecularly-Thin Precursor Films of Imidazolium-Based Ionic Liquids on Mica. *J. Phys. Chem. C* **2013**, *117*, 23676–23684. [CrossRef]
60. Borghi, F.; Mirigliano, M.; Lenardi, C.; Milani, P.; Podestà, A. Nanostructure Determines the Wettability of Gold Surfaces by Ionic Liquid Ultrathin Films. *Front. Chem.* **2021**, *9*, 619432. [CrossRef]
61. Li, S.; Han, K.S.; Feng, G.; Hagaman, E.W.; Vlcek, L.; Cummings, P.T. Dynamic and Structural Properties of Room-Temperature Ionic Liquids near Silica and Carbon Surfaces. *Langmuir* **2013**, *29*, 9744–9749. [CrossRef] [PubMed]
62. Pensado, A.S.; Malberg, F.; Gomes, M.F.C.; Pádua, A.A.H.; Fernández, J.; Kirchner, B. Interactions and Structure of Ionic Liquids on Graphene and Carbon Nanotubes Surfaces. *RSC Adv.* **2014**, *4*, 18017–18024. [CrossRef]
63. Meusel, M.; Gezmis, A.; Jaekel, S.; Lexow, M.; Bayer, A.; Maier, F.; Steinrück, H.-P. Time- and Temperature-Dependent Growth Behavior of Ionic Liquids on Au(111) Studied by Atomic Force Microscopy in Ultrahigh Vacuum. *J. Phys. Chem. C* **2021**, *125*, 20439–20449. [CrossRef]
64. Cremer, T.; Stark, M.; Deyko, A.; Steinrück, H.-P. Liquid/Solid Interface of Ultrathin Ionic Liquid Films: [C_1C_1Im][Tf_2N] and [C_8C_1Im][Tf_2N] on Au(111). *Langmuir* **2011**, *27*, 3662–3671. [CrossRef] [PubMed]
65. Rola, K.P.; Zajac, A.; Szpecht, A.; Kowal, D.; Cybinska, J.; Smiglak, M.; Komorowska, K. Interaction of electron beam with ionic liquids and its application for micropatterning. *Eur. Polym. J.* **2021**, *156*, 110615. [CrossRef]

66. Schneider, C.A.; Rasband, W.S.; Eliceiri, K.W. NIH Image to ImageJ: 25 Years of Image Analysis. *Nat. Methods* **2012**, *9*, 671–675. [CrossRef]
67. Zaitsau, D.H.; Topp, A.; Siegesmund, A.; Papcke, A.; Kockerling, M.; Verevkin, S.P. In the Footsteps of August Michaelis: Syntheses and Thermodynamics of Extremely Low-Volatile Ionic Liquids. *ChemistryOpen* **2021**, *10*, 243–247. [CrossRef]
68. Maruyama, S.; Takeyama, Y.; Taniguchi, H.; Fukumoto, H.; Itoh, M.; Kumigashira, H.; Oshima, M.; Yamamoto, T.; Matsumoto, Y. Molecular Beam Deposition of Nanoscale Ionic Liquids in Ultrahigh Vacuum. *ACS Nano* **2010**, *4*, 5946–5952. [CrossRef]
69. Matsumoto, Y.; Maruyama, S. Laser Deposition of Nano-Ionic Liquids and their Process Applications in a Vacuum. *Ionic Liquid Devices* **2017**, 136–167. [CrossRef]
70. Takeyama, Y.; Mantoku, S.; Maruyama, S.; Matsumoto, Y. Growth Behaviours of Pentacene Films Confined in Engineered Shapes of Ionic-Liquid in Vacuum. *CrystEngComm* **2014**, *16*, 684–689. [CrossRef]
71. Delcheva, I.; Beattie, D.A.; Ralston, J.; Krasowska, M. Dynamic Wetting of Imidazolium-Based Ionic Liquids on Gold and Glass. *Phys. Chem. Chem. Phys.* **2018**, *20*, 2084–2093. [CrossRef] [PubMed]
72. Hessey, S.G.; Jones, R.G. On the Evaporation, Bonding, and Adsorbate Capture of an Ionic Liquid on Au(111). *Chem. Sci.* **2013**, *4*, 2519–2529. [CrossRef]
73. Rietzler, F.; Piermaier, M.; Deyko, A.; Steinruck, H.-P.; Maier, F. Electrospray Ionization Deposition of Ultrathin Ionic Liquid Films: [C_8C_1Im]Cl and [C_8C_1Im][Tf_2N] on Au(111). *Langmuir* **2014**, *30*, 1063–1071. [CrossRef] [PubMed]
74. Chopra, K.L.; Randlett, M.R. Influence of Deposition Parameters on the Coalescence Stage of Growth of Metal Films. *J. Appl. Phys.* **1968**, *39*, 1874–1881. [CrossRef]

fluids

MDPI

Article

Thin Film Evaporation Modeling of the Liquid Microlayer Region in a Dewetting Water Bubble

Ermiyas Lakew [1], Amirhosein Sarchami [1], Giovanni Giustini [2], Hyungdae Kim [3] and Kishan Bellur [1,*]

[1] Department of Mechanical and Materials Engineering, University of Cincinnati, P.O. Box 210072, Cincinnati, OH 45221, USA

[2] Department of Mechanical, Aerospace & Civil Engineering, School of Engineering, University of Manchester, Manchester M13 9PL, UK

[3] Department of Nuclear Engineering, Kyung Hee University, Yongin 17104, Republic of Korea

* Correspondence: bellurkn@ucmail.uc.edu

Abstract: Understanding the mechanism of bubble growth is crucial to modeling boiling heat transfer and enabling the development of technological applications, such as energy systems and thermal management processes, which rely on boiling to achieve the high heat fluxes required for their operation. This paper presents analyses of the evaporation of "microlayers", i.e., ultra-thin layers of liquid present beneath steam bubbles growing at the heated surface in the atmospheric pressure nucleate of boiling water. Evaporation of the microlayer is believed to be a major contributor to the phase change heat transfer, but its evolution, spatio-temporal stability, and impact on macroscale bubble dynamics are still poorly understood. Mass, momentum, and energy transfer in the microlayer are modeled with a lubrication theory approach that accounts for capillary and intermolecular forces and interfacial mass transfer. The model is embodied in a third-order nonlinear film evolution equation, which is solved numerically. Variable wall-temperature boundary conditions are applied at the solid–liquid interface to account for conjugate heat transfer due to evaporative heat loss at the liquid–vapor interface. Predictions obtained with the current approach compare favorably with experimental measurements of microlayer evaporation. By comparing film profiles at a sequence of times into the ebullition cycle of a single bubble, likely values of evaporative heat transfer coefficients were inferred and found to fall within the range of previously reported estimates. The result suggests that the coefficients may not be a constant, as previously assumed, but instead something that varies with time during the ebullition cycle.

Keywords: evaporation; thin film; contact line; microlayer; boiling

1. Introduction

The phase change flux from a thin film is several orders of magnitude greater than the bulk liquid [1]. Hence, researchers have recently increased the use of thin films in high-performance phase change heat transfer devices such as spray coolers, heat pipes, capillary pumped loops, and grooved evaporators [2]. Even in a nucleate boiling scenario, a thin film is momentarily formed underneath a bubble just prior to departure and is generally referred to as the "microlayer" (Figure 1). The superheated thin film forms a conduction path between the solid–liquid interface (heated wall) and the liquid–vapor interface, where evaporation takes place. This results in thermal resistance and is critical to overall heat transfer and fluid flow. The evaporation from this superheated liquid film causes the bubble to grow while diminishing film thickness at the same time. As the bubble grows in size, a net upward force develops due to the density gradient and vapor recoil. The upward forces finally overcome surface tension when the bubble becomes large enough to detach itself from the heated surface. As seen in Figure 1, the liquid region can be delineated into three main parts based on the dominant pressure component. In the bulk region, the surface/capillary force is generally dominant and influences the macroscale bubble shape. However, in the nanoscale adsorbed (also referred to as a 'precursor') film region, intermolecular forces have a dominating effect and are characterized by a disjoining pressure [3]. Thus, the combined action of the disjoining pressure and capillary pressure causes the liquid to flow from the bulk region to the evaporating thin film region [4]. Due to a coupled interplay of both thermal resistances and dominant pressure components, the "microlayer" is essentially a transition region connecting the adsorbed film and bulk liquid and is the focus of the study.

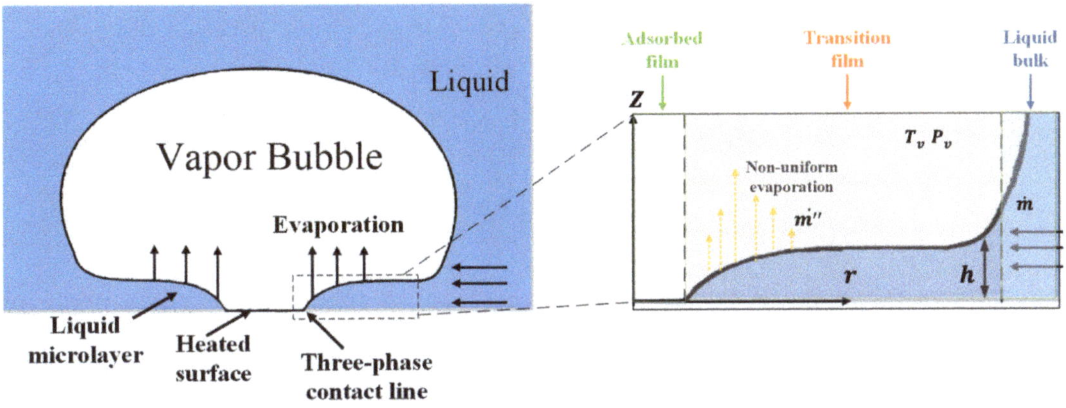

Figure 1. Schematic of an evaporating extended meniscus in the "microlayer" region.

Although a lot of experimental and theoretical work has been conducted to show the existence of the microlayer and its contribution to phase change heat transfer, its evolution, spatio-temporal stability, and impact on macroscale bubble dynamics are still relatively unknown. Therefore, for a better understanding of thin film evaporation and mass transfer phenomena, more numerical and experimental studies are required. A few studies [5–9] that worked on the thin film phenomenon are summarized in Table 1. Most prior investigations have used a Cartesian coordinate system, often resulting in one radius of curvature. Importantly, most studies had a constant wall temperature boundary condition and assumed a value of unity for the accommodation coefficient. To the best of the authors' knowledge, variable wall temperature boundary conditions and the influence of accommodation coefficients have not been studied.

Table 1. Summary of prior and current approaches of modeling evaporation at the thin film.

Study	Coordinate System	Radii of Curvature	Wall Boundary Condition	Start of Integration	Accommodation Coefficient
Akkuş and Dursunkaya, 2016 [6]	Cartesian	1	Constant	Bulk	1
Wang et al., 2009 [8]	Cartesian	1	Constant	Adsorbed	1
Ahmed and Pandey, 2019 [5]	Cartesian	1	Constant	Adsorbed	1
Dwivedi et al., 2020 [7]	Cartesian	1	Constant	Adsorbed	1
Wee et al., 2005 [10]	Cylindrical	2	Constant	Adsorbed	1
Zheng et al., 2016 [9]	Cartesian	1	Constant	Adsorbed	1
Current study	Cylindrical	2	Variable	Bulk	Estimated

This study aims to build upon previously conducted research on traditional thin film evaporation for an enhanced understanding of the microlayer evolution and elucidate new connections between mathematical modeling and experimental data. A numerical analysis is used throughout this study to solve the 3rd order steady nonlinear ordinary differential equation (described in Section 3.2) which governs the thin film/transition film region. Our results are then compared with experimental data to gain insights into the fundamental physics of the problem. In this study, the cylindrical coordinate system with two radii of curvature is used, allowing for three-dimensional curvature estimation. Additionally, for the first time, a variable wall temperature and accommodation coefficient are accounted for by using experimental data for different time steps to comprehensively evaluate the evaporation flux in the microscale thin film. Contrary to several previous models, the solution begins at the thicker region of the film and proceeds in the direction of reducing film thickness. This alleviates the need for guessed or "tuned" boundary conditions at the adsorbed film since they are difficult to measure [11].

The present study has four main objectives:

- Develop a new thin film solution methodology that alleviates guessed boundary conditions by starting the solution at the thicker region of the film and proceeding in the direction of reducing film thickness.
- Predict the evaporation mass flux in the microlayer.
- Estimate the accommodation coefficient and explore its dependence.
- Investigate the effect of simplifying assumptions, such as slip velocity and constant/variable wall temperature.

2. Background

2.1. The Bulk Region

The bulk liquid area is the largest part of the bubble [12]. In this region, intermolecular forces are negligible and the region is dominated by capillary pressure (P_c). The region also depends on the curvature and surface tension of the thin film [13]:

$$P_c = \sigma K \tag{1}$$

where K is the curvature and σ is the surface tension.

2.2. The Adsorbed Region

In this region, no evaporation occurs because the liquid molecules are subject to strong attractive forces from the wall molecules, resulting in a nanometer-scale thickness. Many studies [14–16] assume that the adsorbed region has a uniform, constant film thickness for simplicity. Nevertheless, Wayner [17] indicated that the thickness of the adsorbed film varies with temperature, and that adopting a constant thickness would significantly undermine the total heat transfer. Furthermore, it was proven that the intermolecular interactions between the fluid and the wall significantly alter the film profile and phase change flux [5]. This intermolecular force between the wall and fluid in the nano-scale thin film is characterized as a disjoining pressure, which is applied in thin film modeling to simulate the impact of nanoscale interactions [10]. In the literature, several different approaches are proposed to model disjoining pressure for water [18,19]. Many studies used Equation (2) to estimate the disjoining pressure in the thin film based on the non-polar model [1,14–16]:

$$P_d = \frac{A}{h^3} \tag{2}$$

where A represents the dispersion (also called Hamaker) constant and h denotes the film thickness. It should be noted that Equation (2) is strictly applicable to non-polar molecules, yet it is widely used for water due to its simplicity [20,21]. The results are in good agreement with experimental data and the model performs well when accounting for temperature dependent thermo-physical properties [22]. Nonetheless, water is a polar liquid; hence, polarity might affect the thermal properties of the thin film, especially in the adsorbed region. Therefore, Holm and Goplen [23] presented a logarithmic model (Equation (3)) for water as a polar liquid:

$$P_d = \rho \overline{R} T_v \ln\left(a h^b\right) \tag{3}$$

where P_d, ρ, $\overline{R}$, T_v are the disjoining pressure, density, universal gas constant, temperature of vapor, respectively. Additionally, a and b are empirical coefficients.

2.3. The Thin/Transition Film Region

The transition film area, which lies between the adsorbed film and the bulk meniscus, has characteristics of both the bulk meniscus and adsorbed region. As a result, the contributions of disjoining pressure and capillary pressure have equal importance in this region for both fluid flow and heat transfer [10,22]. A switch in dominating pressure forces, from disjoining to capillary, occurs as the thin film approaches the bulk region, which means that the disjoining pressure increases when the film thickness decreases, and in contrast, the capillary pressure increases when the film thickness increases [15]. Moving from the adsorbed to the bulk regions, the thickness of the liquid film increases from nanometers to the order of micrometers, which forces the molecules to 'loosen up' from the intermolecular forces and evaporate [5].

2.4. Augmented Young–Laplace

In order to analyze the thermodynamic effects of the interfacial tension across the thin film, the augmented Young–Laplace equation is utilized, which is similar to the Young–Laplace equation, with the addition of disjoining pressure. The Augmented Young–Laplace equation is essentially a multi-scale force balance at the liquid–vapor interface, including disjoining and capillary effects [1,12,24]:

$$P_v = P_l + P_c + P_d \tag{4}$$

where P_v, P_l, P_c, and P_d are vapor, liquid, capillary, and disjoining pressures, respectively.

2.5. Kinetic Theory of Phase Change

Classic kinetic theory is a statistical description of the behavior of gases based on the velocities of the component molecules. In the case of mass transport across the liquid–vapor phase, kinetic theory has provided the basis for modeling phase change [25,26]. In general, under equilibrium conditions, the velocity distribution of the vapor molecules follows a Maxwell–Boltzmann distribution, and the vapor close to the liquid–vapor interface can be taken as an ideal gas [27]. Nonetheless, phase change is a dynamic process; hence, the liquid–vapor system undergoes simultaneous condensation and evaporation. Therefore, a net phase change flux is expressed as a sum of condensation and evaporation fluxes [25,26]. The first mass flux relationship was developed by Hertz [28]; however, it was found to be inconsistent with many experimental studies since it only represented a theoretical maximum rate. Knudsen [29] introduced condensation and evaporation coefficients to account for the derivation from the theoretical maximum rate, as shown in Equation (5):

$$\dot{m}'' = \sqrt{\frac{m}{2\pi K_b}}\left(\alpha_e \frac{P_l}{\sqrt{T_i}} - \alpha_c \frac{P_v}{\sqrt{T_v}}\right) \tag{5}$$

where $\dot{m}''$ is the evaporation mass flux, m is the molecule mass, K_b is the Boltzmann's constant, P_l is the liquid pressure at the interface, P_v is the vapor pressure at the interface, T_i is the temperature of the liquid at the interface, T_v is the vapor temperature, α_e is the evaporation coefficient, and α_c is the condensation coefficient.

Marek and Straub [30] defined the evaporation coefficient as the ratio of the number of molecules transferred to the vapor phase to the number of molecules emitted from the liquid phase and the condensation coefficient as the ratio of the number of molecules absorbed by the liquid phase to the number of molecules impinging on the liquid phase. By this definition, α_e and α_c must have a value between 0 and 1. Although assuming that α_e and α_c are equal may not strictly be true and it is common to assume equality and call it an "accommodation coefficient" [5–10]. Accordingly, Schrage [31] developed Equation (6) using a drift velocity concept, and this is currently the most popular kinetic approach used to model phase change at a planar surface:

$$\dot{m}'' = \frac{2\alpha}{2-\alpha}\left(\frac{m}{2\pi K_b}\right)^{\frac{1}{2}}\left(\frac{P_l}{\sqrt{T_i}} - \frac{P_v}{\sqrt{T_v}}\right) \quad (6)$$

where α is the accommodation coefficient and can be defined as the fraction of molecules that undergo a phase change. Wayner [16,24] adapted the planar kinetic model for a curved interface by including disjoining pressure and curvature contributions:

$$\dot{m}'' = \frac{2\alpha}{2-\alpha}\left(\frac{M}{2\pi\overline{R}T_i}\right)^{\frac{1}{2}}\left[\frac{P_v M h_{fg}}{\overline{R}T_v T_i}(T_i - T_v) - \frac{V_l P_v}{\overline{R}T_i}(P_d + P_c)\right] \quad (7)$$

where $\overline{R}$ is the universal gas constant, M is the molar mass, h_{fg} is the enthalpy of vaporization, and V_l is the molar volume. In Equations (6) and (7), the accommodation coefficients are a necessary parameter. However, there is a significant discrepancy in the reported values of the accommodation coefficient. Depending on the researcher or experimental technique employed, the coefficients for water vary by almost three orders of magnitude [11,30]. Many researchers assume the accommodation coefficient as unity even though this is simply the theoretical maximum. In reality, the value is much lower, depending on the fluid temperature, pressure, property, wall/container material, and geometry [26,30,32]. The discrepancy in the reported coefficients is yet to be resolved and estimating coefficients in microlayer evaporation is one of the objectives of the study.

3. Methodology

3.1. Experimental Data

In this investigation, the experimental data measured by Jung and Kim [33,34] were utilized to develop appropriate boundary conditions and to validate the numerical results. The experiment consisted of the optical characterization of the microlayer using interferometry and infrared analysis to obtain time and space-resolved experimental data of boiling heat transfer and fluid flow near a wall (Figure 2a). DI water was used as the working fluid in a cylindrical chamber with a 50 mm base diameter at atmospheric pressure. A 700 nm thick ITO (Indium Tin Oxide) with a surface of 8 mm × 15 mm was used as a thin film heater. The ITO film was placed on top of a 10 mm thick calcium fluoride (CaF_2) plate.

A high-speed infrared camera was employed for the measurements of the temperature distribution on the boiling surface at the center of the substrate. A high-coherency He-Ne laser and a high-speed camera were utilized to measure the total reflection and laser interferometry simultaneously. The laser light was expanded and directed to the boiling surface and the camera by using a beam expander, two prisms, and mirrors. An interference filter was placed in front of the high-speed camera. An example of the interferometric pattern is shown in Figure 2b. The fringe patterns spacing between the white and dark rings is characterized by Equation (8):

$$d_{m+1} - d_m = \frac{\lambda}{2n_l cos\theta_l} \quad (8)$$

where d_m is the thickness of the microlayer, λ is the wavelength, n_l is the index of refraction, and θ_l is the angle of refraction into liquid from the substrate. The experimental study reported spatio-temporal variations of the film thickness and wall/evaporation heat fluxes. Using Equation (8) the film thickness profiles were calculated and are illustrated in Figure 3a as a function of time. The surface temperature was measured using an IR camera pointed at the bottom of the ITO surface. The CaF_2 plate used in the experiment is transparent to infrared light and the ITO being opaque in the infrared range makes the infrared camera reading an accurate representation of the actual temperature on top of the ITO surface [33]. Using measured variable wall temperature and microlayer thickness (Figure 3a,b), wall and evaporation heat fluxes were calculated based on Equations (9) and (10). Figure 3c shows the computed heat flux values at a time = 1.9 ms. Data at other time steps are not shown here for brevity. As seen in Figure 3c, the wall and evaporation heat fluxes are in good agreement with each other.

$$q''_{wall} = -k\frac{\partial T(r,\ t)}{\partial z} \tag{9}$$

$$q''_{evap} = \rho_f h_{fg}\frac{\partial h(r,t)}{\partial t} \tag{10}$$

(a) (b)

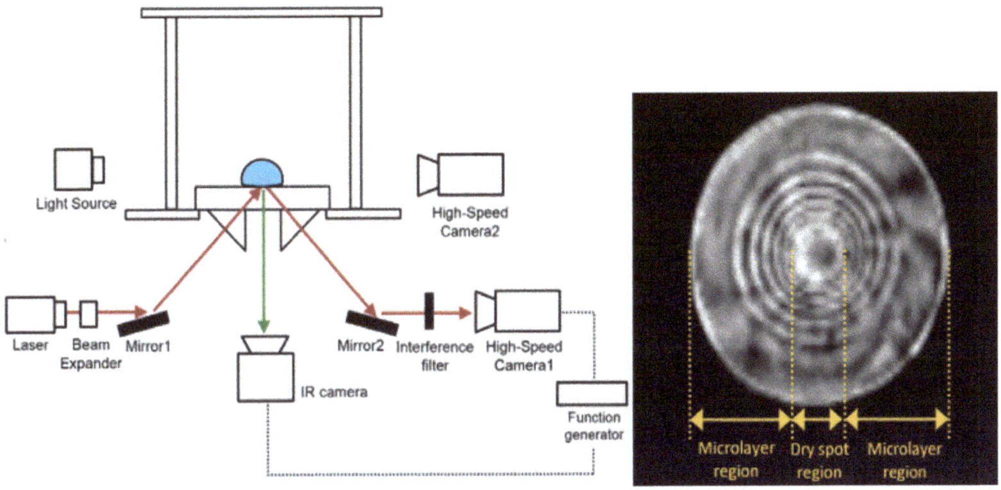

Figure 2. (**a**) Schematic diagram of the experimental setup; (**b**) Example image of interferometric pattern from bubble total reflection [33,34].

(a) (b)

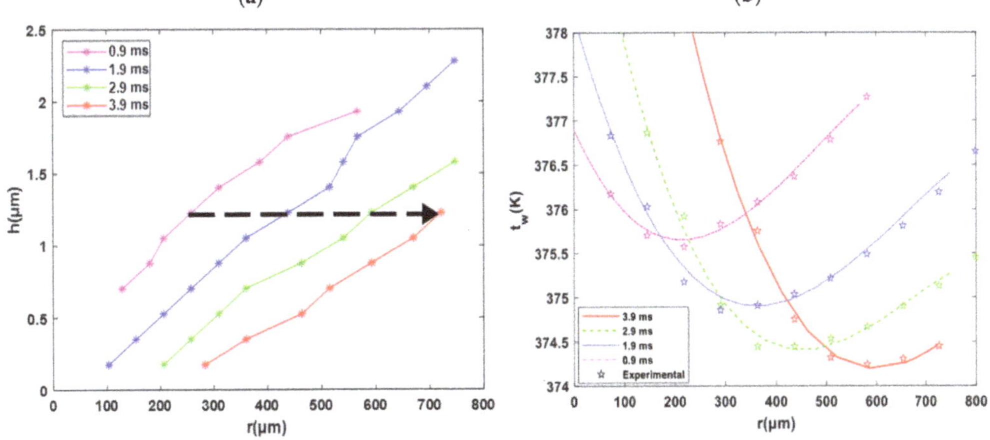

Figure 3. *Cont.*

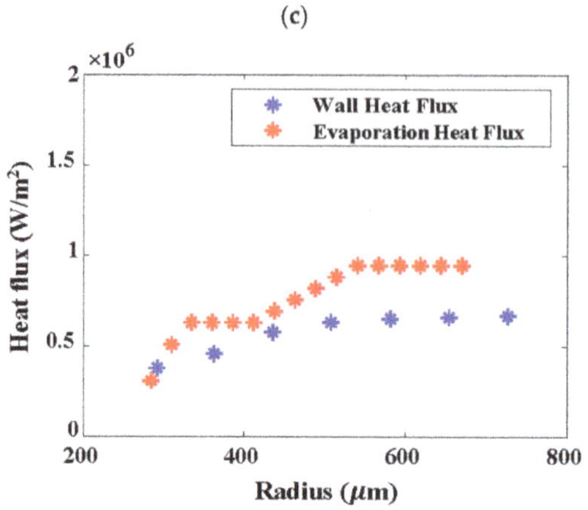

Figure 3. Radius evolutions of microlayer film thickness (**a**), variable wall temperature (**b**), and calculations of heat flux using two different methods (**c**). The black arrow in Figure 3a shows the temporal displacement for a constant film thickness. For brevity, heat flux data at only 3.9 ms is shown here. Please refer to [33–35] for additional details.

The film profiles (Figure 3a) were used to obtain the necessary boundary conditions for the model described in the proceeding section. The heat flux data was used to calculate mass flux and net evaporation rates. This allows for a direct comparison to the evaporation rate predicted by the model, enabling the estimation of the accommodation coefficient.

3.2. Mathematical Model

In the present study, different regions of the microlayer are delineated based on the dominant pressure component. The transition film region modeled in this study is illustrated in Figure 1. To numerically model the 3D structure of the microlayer, a cylindrical coordinate system is employed. The origin of the coordinate system is placed at the surface of the heated wall at the center of the bubble. The z coordinate defines the film thickness and r coordinate is along the length of the film. The following assumptions are made:

- The working fluid is a pure single component liquid with no disolved gas.
- Fluid flow is assumed to be uni-directional, along the wall in the negative r direction, and there is no liquid flow in the z direction.
- The vapor temperature is assumed to be constant.
- The wall is smooth and chemically neutral.

3.2.1. Force Balance

The mechanical force balance in the thin film can be modeled by using the augmented Young–Laplace equation (Equation (4)) [36]. Furthermore, the surface tension and curvature that are used to calculate the capillary pressure (Equation (1)) are defined below. The surface tension is modeled as a temperature-dependent quantity:

$$\sigma = \sigma_1 + \gamma\, T_i \tag{11}$$

where γ is the slope of the interfacial temperature and surface tension. To calculate the total curvature in the bubble geometry, the curvature is represented by Equation (12) [37]:

$$K = \frac{h''}{(1+h'^2)^{\frac{3}{2}}} + \frac{h'}{r(1+h'^2)^{\frac{1}{2}}} \tag{12}$$

where h', h'', and r are the first and second derivatives of the film thickness, and radius of the bubble, respectively.

3.2.2. Thin-Film Evolution

An expression for the film thickness profile is computed by substituting Equations (1), (2), (11) and (12) into Equation (4), assuming P_v is constant and differentiating it with respect to r. Thus, a 3rd order steady nonlinear ordinary differential equation is achieved:

$$h''' = \frac{3h''^2 h'}{(1+h'^2)} + \frac{h'' h'^2}{r} - \frac{h''(1+h'^2)}{r} - \frac{\gamma}{\sigma}\left(h'' + \frac{h'(1+h'^2)}{r}\right)\frac{dT_i}{dr} + \frac{1}{\sigma}\left(1+h'^2\right)^{\frac{3}{2}}\left(-\frac{dP_d}{dr} - \frac{dP_l}{dr}\right) \quad (13)$$

where $\frac{dT_i}{dr}$ is the temperature gradient, $\frac{dP_l}{dr}$ is the liquid pressure gradient, and $\frac{dP_d}{dr}$ is the disjoining pressure gradient. In Equation (13), all terms are presented in terms of film thickness. Solving Equation (13) will result in a film thickness profile from which all other quantities (mass flux, curvature, disjoining pressure, etc.) can be computed. However, the temperature, liquid pressure and disjoining pressure gradients must first be evaluated. These are calculated using the conservation equations described in the proceeding subsections.

3.2.3. Energy Conservation

Conduction heat transfer in the thin film region as heat is applied through the wall and is represented by Equation (14) [38,39]:

$$\frac{1}{r}\frac{d}{dr}\left(k\,r\frac{dT}{dr}\right) + \frac{1}{r^2}\frac{d}{dy}\left(k\frac{dT}{dy}\right) + \frac{d}{dz}\left(k\frac{dT}{dz}\right) + q = \rho C_P\frac{dT}{dt} \quad (14)$$

where k, ρ, and C_P are the thermal conductivity, density, and specific heat, respectively. For thin microscale films, it can be assumed that only heat is transferred through the liquid in the direction perpendicular to the substrate (z direction) [12]. Hence, Equation (14) is simplified to Equation (15) to obtain the interfacial temperature:

$$\frac{d}{dz}\left(k\frac{dT_i}{dz}\right) = 0 \quad (15)$$

The interfacial temperature can be calculated by applying boundary conditions:
At the wall when $z = 0$;

$$t = t_W \quad (16)$$

Additionally, at the liquid–vapor interface when $z = h$;

$$-K_l\frac{dT}{dz} = m'' h_{fg} \quad (17)$$

Using a specified variable temperature boundary condition at the wall (Figure 3b) and mass flux calculated from the kinetic model (Equation (7)), Equation (18) is obtained:

$$\frac{dT_i}{dr} = \left(\frac{2T_v}{h^3}\right)\left[\frac{h\prime\left(k\,h^2\left(T_i^{\frac{5}{2}} - T_w T_i^{\frac{3}{2}}\right) - 3AV_l\eta\right) + V_l\eta\sigma h^4 k\prime}{k\,T_v\left(5T_i^{\frac{3}{2}} - 3T_w T_i^{\frac{1}{2}}\right) + 2h\eta\left(Mh_{fg} - V_l T_v\gamma k\right)}\right] \quad (18)$$

where η is defined as:

$$\eta = \left(\frac{\left(\frac{2\alpha}{2-\alpha}\right)^2 M}{2\pi\overline{R}}\right)^{\frac{1}{2}}\left(\frac{P_v h_{fg}}{\overline{R}}\right) \quad (19)$$

From Equation (18), the interfacial temperature gradient can be estimated and used for calculating the film thickness profile. However, the pressure gradients are still required for solving Equation (13) and are discussed in the proceeding subsection.

3.2.4. Momentum Conservation

In this study, the cylindrical coordinate is used, assuming the fluid flows only in the r direction due to the small liquid thin film thickness. Hence, using the lubrication approximation, the complete form of the Navier–Stokes equation is reduced to Equation (20) [10]:

$$\frac{dP_l}{dr} = \mu \frac{d^2u}{dz^2} \tag{20}$$

To obtain a velocity profile, desirable boundary conditions are applied to Equation (20). At the wall when $z = 0$, a slip velocity (u_s) is applied:

$$u = u_s \tag{21}$$

At the liquid–vapor interface when $z = h$, a free surface boundary condition is applied:

$$\mu \frac{du}{dz} = \frac{d\sigma}{dr} \tag{22}$$

The free surface boundary accounts for thermocapillarity [11]. Using these conditions, the velocity profile can be derived:

$$u = \frac{1}{\mu}\left[\frac{dP_l}{dr}\left(\frac{z^2}{2} - hz\right) + \frac{d\sigma}{dr}z\right] + u_s \tag{23}$$

Initially, the dewetting velocity was calculated by tracking the radius displacement in a constant film thickness which is indicated by a black arrow, with respect to time, as depicted in Figure 3a. The calculation is then repeated for different film thicknesses and time steps, and the values are within 9%. Since the variation is small, the calculated dewetting velocity is considered equivalent to the slip velocity (u_s) used in Equation (23).

3.2.5. Pressure Gradients

An explicit expression for the liquid pressure gradient is obtained by applying the velocity profile (Equation (23)) into a continuity equation ($\dot{m} = \rho u \overline{A}$):

$$\frac{dP_l}{dr} = \frac{3}{2h}\frac{d\sigma}{dr} + \frac{3V\rho u_s}{h^2} - \frac{3V}{2h^3 r\pi}\dot{m} \tag{24}$$

where u_s = 0.165 m/s is a slip velocity obtained from experiments. The mass flow rate can be calculated by differentiating Equation (4) to evaluate $\frac{dP_l}{dr}$ and by using Equation (24):

$$\dot{m} = -\frac{2\pi r h^3}{3V}\left(-\frac{dP_c}{dr} - \frac{dP_d}{dr}\right) + \frac{\pi r h^2}{V}\frac{d\sigma}{dr} + 2\rho\pi r h u_s \tag{25}$$

Additionally, the disjoining pressure gradient is obtained by differentiating Equation (2):

$$\frac{dP_d}{dr} = \frac{-3Ah'}{h^4} \tag{26}$$

Hence, all temperature and pressure gradients can be derived using the governing equations. The only unknowns are film thickness and its derivatives. Experimentally measured film profiles (Figure 3a) are used as boundary conditions in the bulk region to solve Equation (13) as described below.

3.3. Numerical Model

The thin film evolution equation (Equation (13)) is solved numerically, based on cylindrical coordinates, using the ODE 45 method in MATLAB. In this work, the solution begins at the thicker end of the transition region and proceeds in the direction of reducing film thickness. Wee et al. [10] indicated that the solution of the thin film evolution is very sensitive to the physical boundary conditions at the adsorbed region, and these are often difficult to measure experimentally. Hence, they ended up being guesses or tuned for numerical stability [13,40]. Thus, the solution of the thin film evolution equation (Equation (13)) requires the initial film thickness (h) and its derivatives (h', and h'') at the adsorbed film as boundary conditions.

As illustrated in Figure 4, the initial interfacial temperature is estimated from the experimental data and used to calculate the local evaporative mass flux using Equation (7). Then, the calculated mass flux is used to calculate the next interfacial temperature based on an updated film thickness by solving Equation (13). Integrating the solution from the thicker region and marching down to the thinner region will minimize the uncertainties to the boundary conditions at the adsorbed region [11]. Based on the depicted solution setup (Figure 4), the accommodation coefficient, dispersion constant, estimated interfacial temperature, and vapor temperature are necessary inputs. Initial conditions including r, u_s, h, h', and h'' are extracted from the experimental data (Figure 3a). The wall temperature derived from experiments is used as a thermal boundary condition (Figure 3b). Using an estimated value of interfacial temperature (T_{i_n}), Equations (1), (2), (7), (13), (18) and (22)–(24) are solved in a coupled manner using a variable step 4th order Runge Kutta solver (ODE 45) in MATLAB to obtain a new interfacial temperature (T_{i_n+1}), evaporative mass flux, pressure gradients, and thin film thickness, as shown in Figure 4. The computed T_{i_n+1}, $\dot{m}''_{i\_n}$, r_{i_n+1}, h_{i_n+1}, h'_{i_n+1}, and h''_{i_n+1} are utilized as inputs in the next step, and the process is continued until an adsorbed film with a slope of zero is obtained.

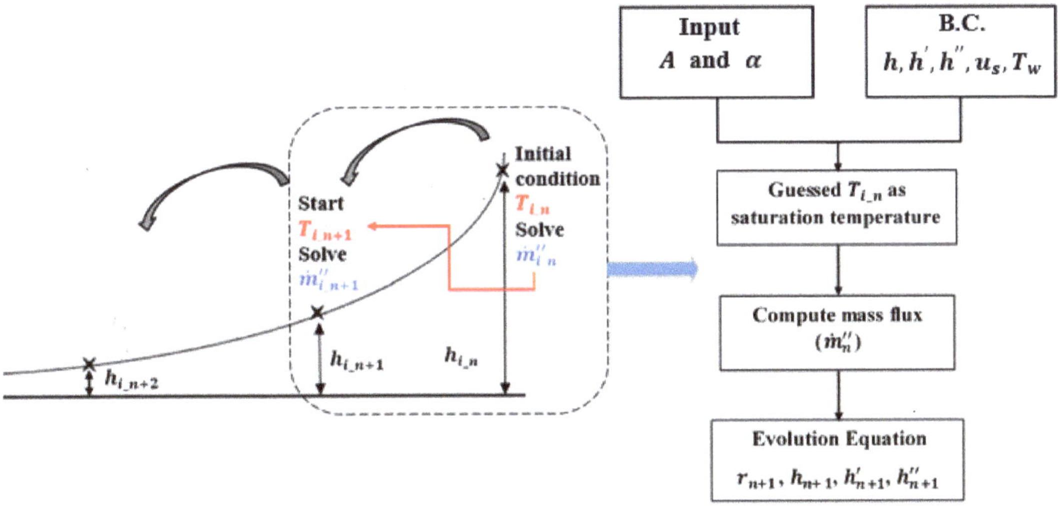

Figure 4. Solution setup used to solve the interfacial temperature and evaporative mass flux in the thin film.

4. Results and Discussion

The thin film region of the evaporating meniscus is modeled numerically using MATLAB to calculate the film thickness in different regions, interfacial temperature, and evaporative mass flux. In this section, the results of the numerical model are provided for different conditions, and comprehensive evaluations are carried out.

4.1. Baseline Result

In this part of the study, the baseline results are provided to serve as typical results from the model. Then, by changing parameters such as wall temperature, accommodation coefficient, disjoining pressure models, and dispersion constant, their influences on the evaporative thin film are carefully investigated. The baseline result uses an experimentally measured wall temperature, a non-polar disjoining pressure model with a dispersion constant (A) of 2.87×10^{-21} J, an accommodation coefficient (α) value of 0.01, and an experimentally obtained slip velocity of 0.165 m/s. The local surface tension and vapor tension are obtained from the NIST (National Institute of Standards and Technology) thermo-physical property database.

First, to validate the accuracy of the microlayer model for the thin film, the film thickness predicted numerically is compared with the experimental result [33,34]. As shown in Figure 5, good agreement can be seen between the numerically predicted film thickness and the experimental data validating the baseline version of the model.

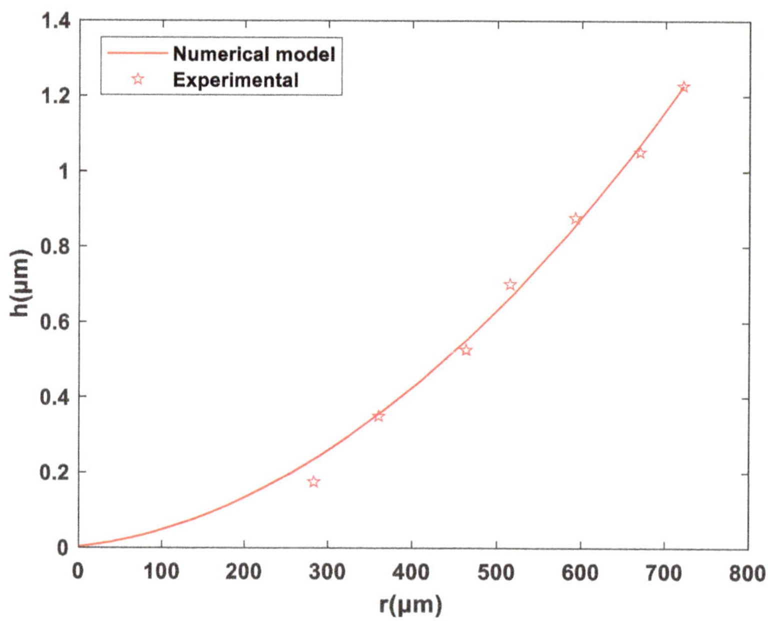

Figure 5. Comparison of the numerical model and experimental results based on film thickness at a time step of 3.9 ms [33,34].

The mass flow rate, interfacial temperature, and evaporation mass flux are reported in Figure 6. Disjoining pressure (P_D) is dominant in the adsorbed region due to van der Waals force at the nanoscale [41], resulting in an abrupt decrease in the evaporation flux ($\dot{m}''$). Figure 6b indicates the amount of liquid flow from the bulk liquid into the thin film region per unit of time. The model predicts a higher mass flow rate at the thicker region and a smaller mass flow rate close to zero at the thinnest region. As seen in Figure 6c, the interfacial temperature shows a trend equivalent to the wall temperature at the thinnest film region. Additionally, as discussed before, the van der Waals force is dominant when the film thickness is small. This resulted in a sharp reduction in evaporation mass flux, as indicated in Figure 6d.

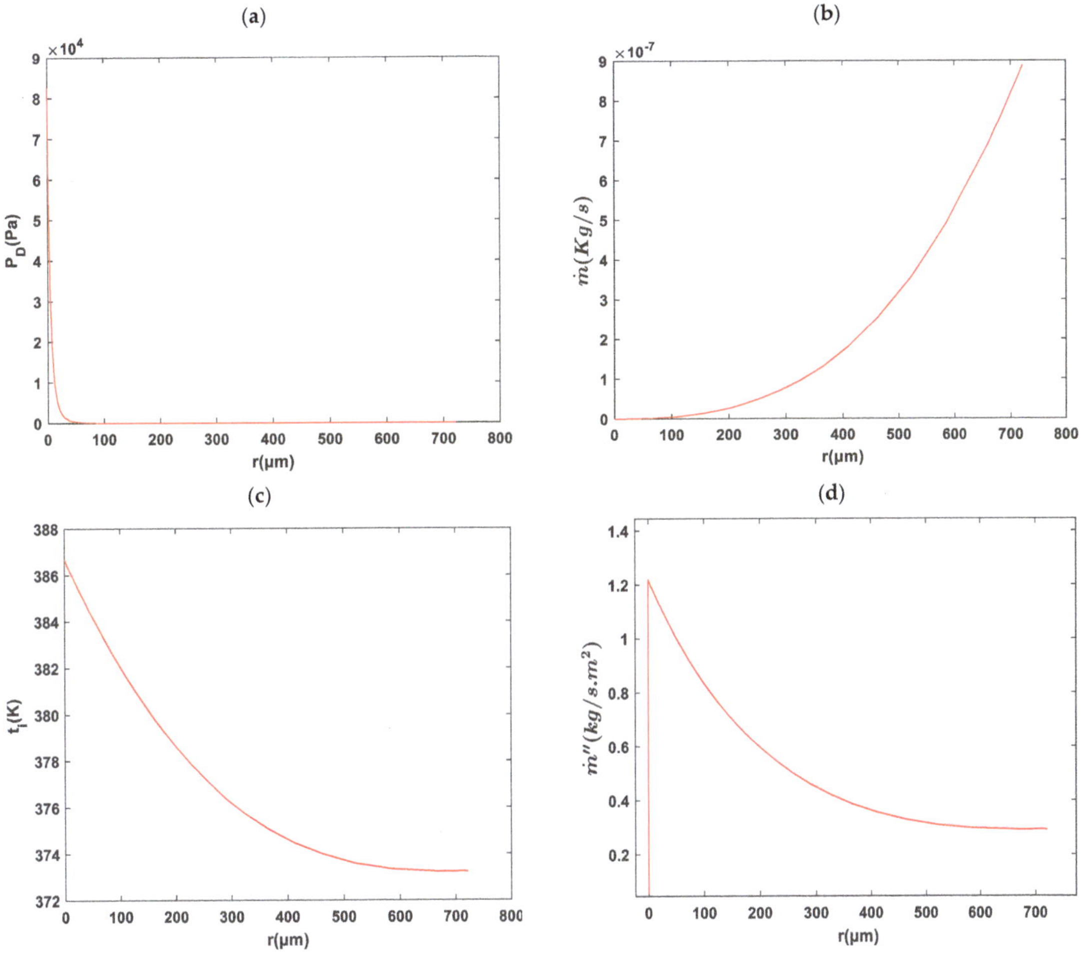

Figure 6. Radius evolutions of (**a**) disjoining pressure; (**b**) mass flow rate; (**c**) interfacial temperature; and (**d**) evaporation mass flux for baseline parameters at a time step of 3.9 ms.

4.2. Effect of Wall Temperature

In this section, Figure 7 depicts the influence of the constant vs. variable wall temperature on interfacial temperature, evaporation mass flux, and film thickness. In contrast to the previous studies (Table 1), this study applied a variable wall temperature at the solid–liquid interface instead of a constant wall temperature boundary condition. When the wall temperature was held constant (377.15 K) to the variable wall temperature, a smaller interfacial temperature and mass flux variation were identifiable, compared to the case with variable wall temperatures (Figure 7a,b). The peak magnitude of the evaporative mass flux of the variable case is more than double when compared with the constant wall temperature case. At the thinnest region, the interfacial temperature becomes equal to the applied constant wall temperature showing the same trend as the interfacial temperature with the applied variable wall temperature. Nevertheless, the constant wall temperature model exhibits limited variation in interfacial temperature. As indicated in Figure 7c, even though there are notable variations in interfacial temperature and mass flux, applying variable wall temperature provides no significant change in film thickness. The difference in film thickness is in the order of nanometers.

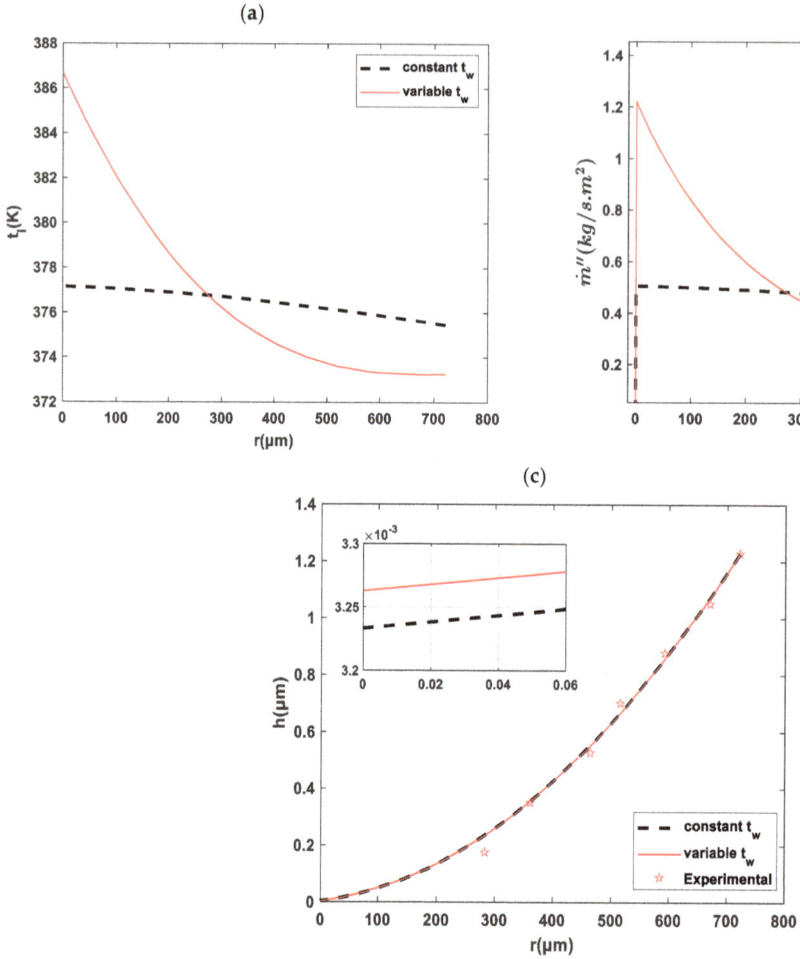

Figure 7. Radius evolutions of (**a**) interfacial temperature; (**b**) evaporative mass flux; and (**c**) film thickness for different conditions at a time step of 3.9 ms.

4.3. *Effect of Slip Velocity*

To study the influence of the slip velocity, a slip velocity of 0.165 m/s is considered, as obtained from the experimental data, and applied to the model to compare with a no-slip velocity condition. As shown in Figure 8a, when a slip velocity is applied to the model it leads to a higher liquid mass flow rate under the bubble. Furthermore, Figure 8b shows that, when a slip velocity was applied, the thickness of the film indicated no significant change. Hence, it is important to account for slip conditions while assembling properties of the thin film region because mass flow can be under-predicted if slip is ignored.

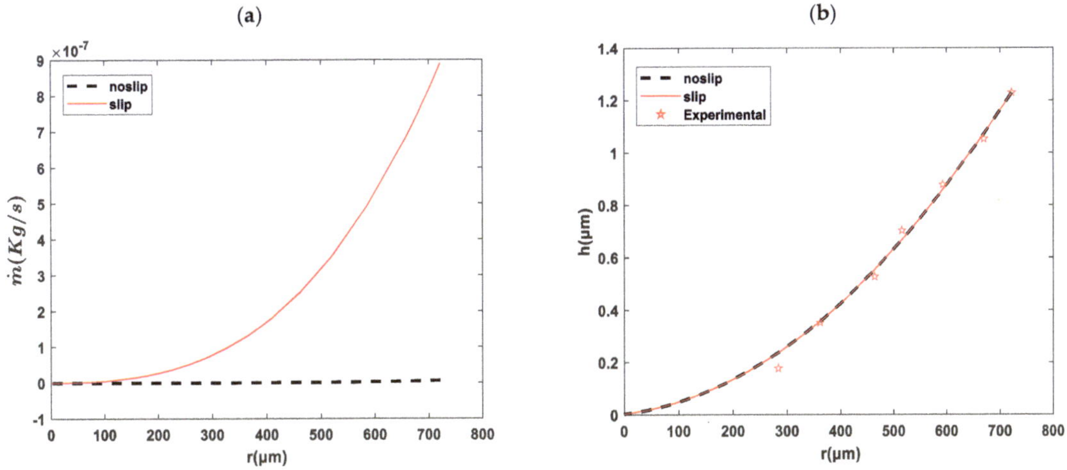

Figure 8. Radius evolutions of (**a**) evaporative mass flux; and (**b**) film thickness for no slip and slip conditions at a time step of 3.9 ms.

4.4. Effect of Disjoining Pressure

In this section, different disjoining pressure models such as negative/positive logarithmic models, and a non-polar model are used, based on Equations (2) and (3), to evaluate the influence of disjoining pressure in the adsorbed region. As seen in Figure 9a–d, the effects of various disjoining pressure models on the value of disjoining pressure, mass flow rate, film thickness, and evaporative mass flux are illustrated, respectively. The behavior of the logarithmic model differs from that of the non-polar model. For smaller film thicknesses, the logarithmic model predicts a noticeably larger amount of disjoining pressure, and for radius above 1.43×10^{-4} m, it predicts a negative value (Figure 9a). Some studies also used a positive logarithmic equation to fit in their model [22]; however, the positive logarithmic model predicts a negative disjoining pressure for smaller film thickness which contradicts physical values. Therefore, the disjoining pressure is overestimated by the logarithmic models, and their results do not agree with physical values. Based on Figure 9b, although the mass flow rate increased at the thicker region using the positive logarithmic model, it decreased considerably when the model changed to the negative logarithmic model. This indicates that there is a liquid flow back in the direction of the positive r direction. Eventually, Figure 9c,d show that different disjoining pressure models do not have an impact on the film thickness and mass flux. Accordingly, the non-polar model is utilized for the following section because it is an acceptable model to accurately calculate the disjoining pressure.

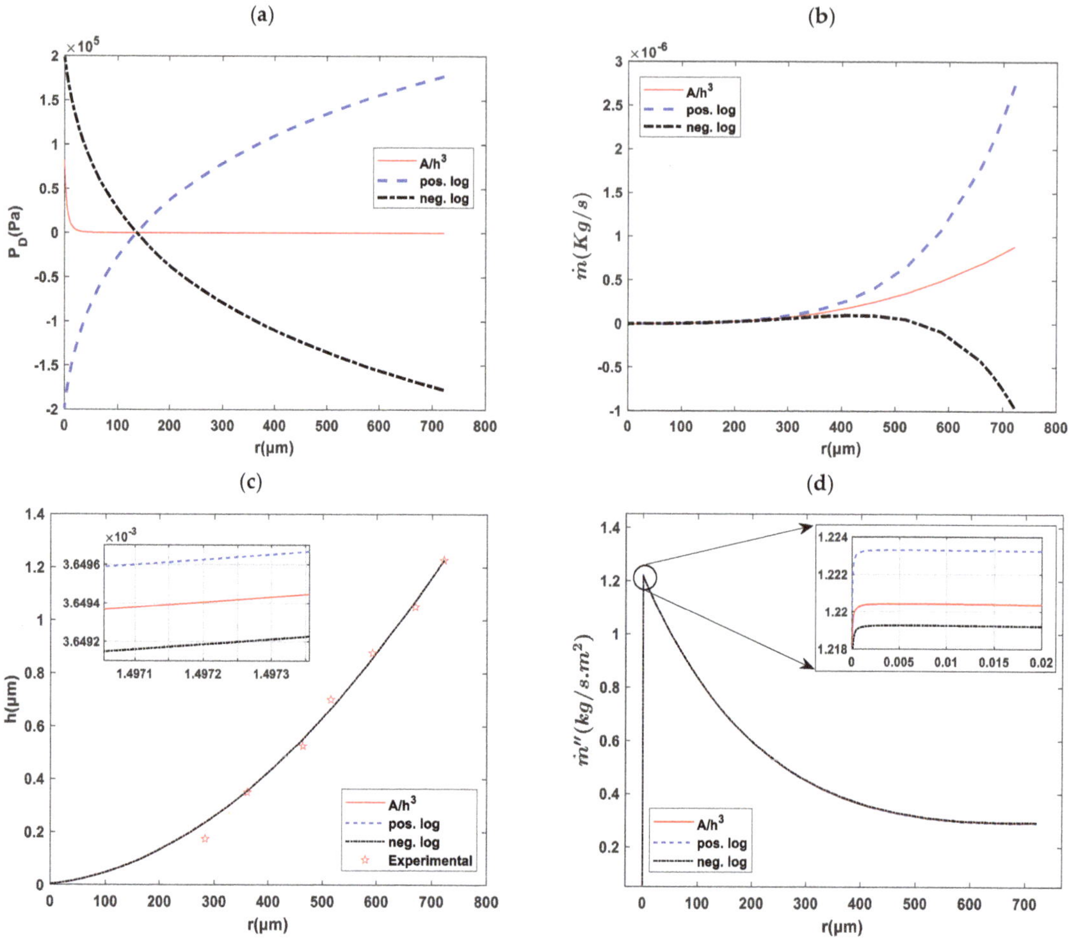

Figure 9. Radius evolutions of (**a**) disjoining pressure; (**b**) mass flow rate; (**c**) film thickness; and (**d**) evaporative mass flux for different disjoining pressure models at a time step of 3.9 ms.

4.5. Effect of Dispersion Constant

The effect of the dispersion constant on the non-polar disjoining pressure model is studied to observe its influence on the thin film results. For this study, the dispersion constant was set for a range of values at -2.8×10^{-21} J, 0.8×10^{-21} J, and 2.8×10^{-21} J to evaluate its effect on evaporation. As can be seen in Figure 10, increasing the dispersion constant results in higher disjoining pressure, while a negative dispersion constant results in negative disjoining pressure. In terms of evaporation, when the dispersion constant alters the peak evaporation mass flux slightly, but the difference is not significant. A negative dispersion constant outputs a slightly higher peak in evaporation mass flux, as shown in Figure 10b.

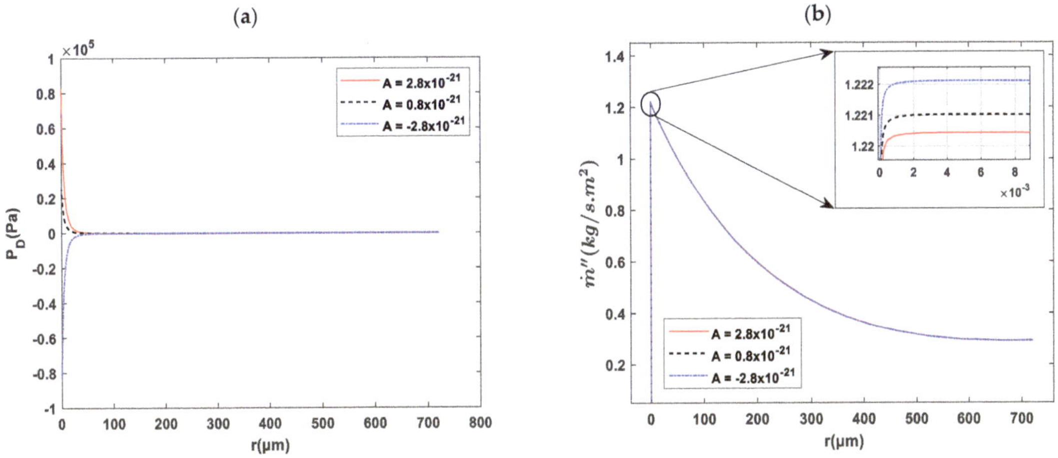

Figure 10. Radius evolutions of (**a**) disjoining pressure; and (**b**) evaporative mass flux for different dispersion constants at a time step of 3.9 ms.

4.6. Effect of Accommodation Coefficient

This section explores the sensitivity of results by changing the value of the accommodation coefficient. Many studies used unity for the accommodation coefficient, which indicates a perfect evaporation capacity. This means that, for every liquid molecule released, none are rebounded and reabsorbed [42]. Using unity for the accommodation coefficient might be unrealistic, and it is very difficult to have extreme vapor purity in an experimental setup. It should be noted that purity is one of many factors that can reduce the accommodation coefficient from its theoretical upper limit. However, varying the accommodation coefficient results in very minimal film thickness change, as shown in Figure 11a. The variations of interfacial temperature and evaporative mass flux are indicated in Figure 11b,c. With a decreasing accommodation coefficient, the peak magnitude of the evaporative mass flux decreases, and the corresponding interfacial temperature indicates a higher temperature at the thickest region. The peak of evaporative mass flux is highly dependent on the coefficient value used. These results indicate that using different accommodation coefficients has a significant impact on the thin film's evaporative properties.

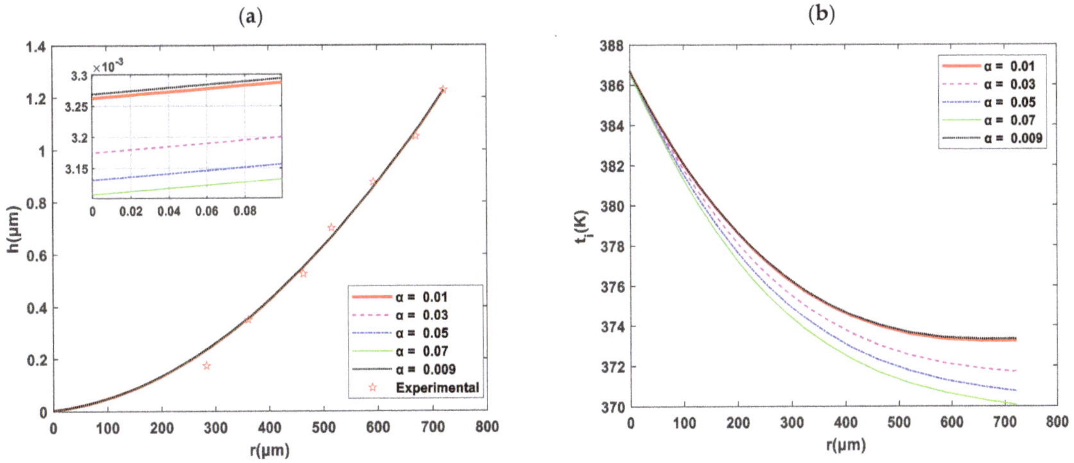

Figure 11. *Cont.*

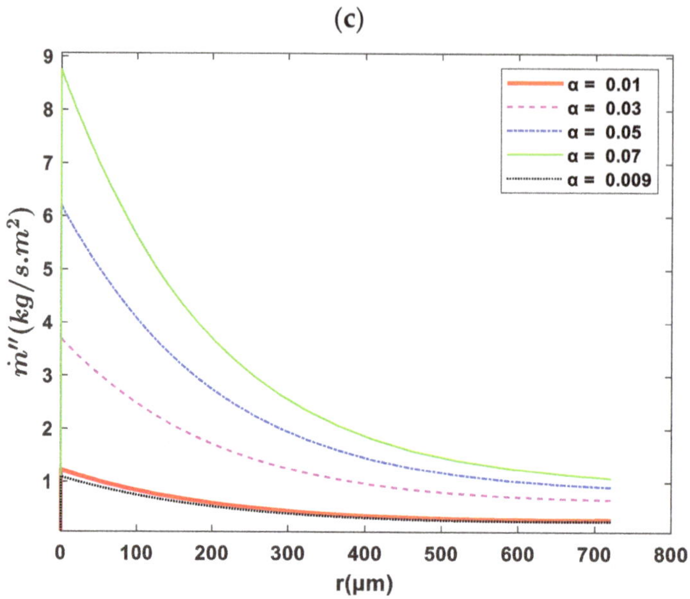

Figure 11. Radius evolutions of (**a**) film thickness; (**b**) interfacial temperature; and (**c**) evaporative mass flux for different accommodation coefficients at a time step of 3.9 ms.

4.7. Effect of Time Step

All the results discussed in the previous sections are illustrated at a snapshot in time (time step 3.9 ms). However, experimental data was obtained at 1 ms intervals and comparisons can be made at each time step. This section expands the previous discussion for multiple time steps. The numerical model was run for different time steps at 0.9, 1.9, 2.9, and 3.9 ms with variable wall temperature, applied slip velocity, and an accommodation coefficient of 0.01. Initially, the variations of the wall temperature for different time steps are depicted and compared with the experimental results, which can be seen in Figure 3b. Wall temperature from the experimental data shows that initially, the wall temperature is highest in the thick region of the film. However, as time progresses, the wall temperature peak shifts to the thinner region of the film. When this is accounted for, the thin film profile from the model is in good agreement with the experimental results for all time steps. Changes in the film thickness, mass flow rate, interfacial temperature, and evaporative mass flux, are depicted in Figure 12a–d, respectively. As can be seen in Figure 12a, the radius of the bubble clearly increased at each region of the film by increasing the time step. Additionally, Figure 12b shows that, at a particular point of the bubble radius, a higher mass flow rate is observed for smaller time steps in the thicker region and obtained a near-zero flow rate at the thinnest region. The interfacial temperature shows an increase in peak magnitude as time increases (Figure 12c). Similarly, the mass flux peak at the thinnest region increases with time (Figure 12d).

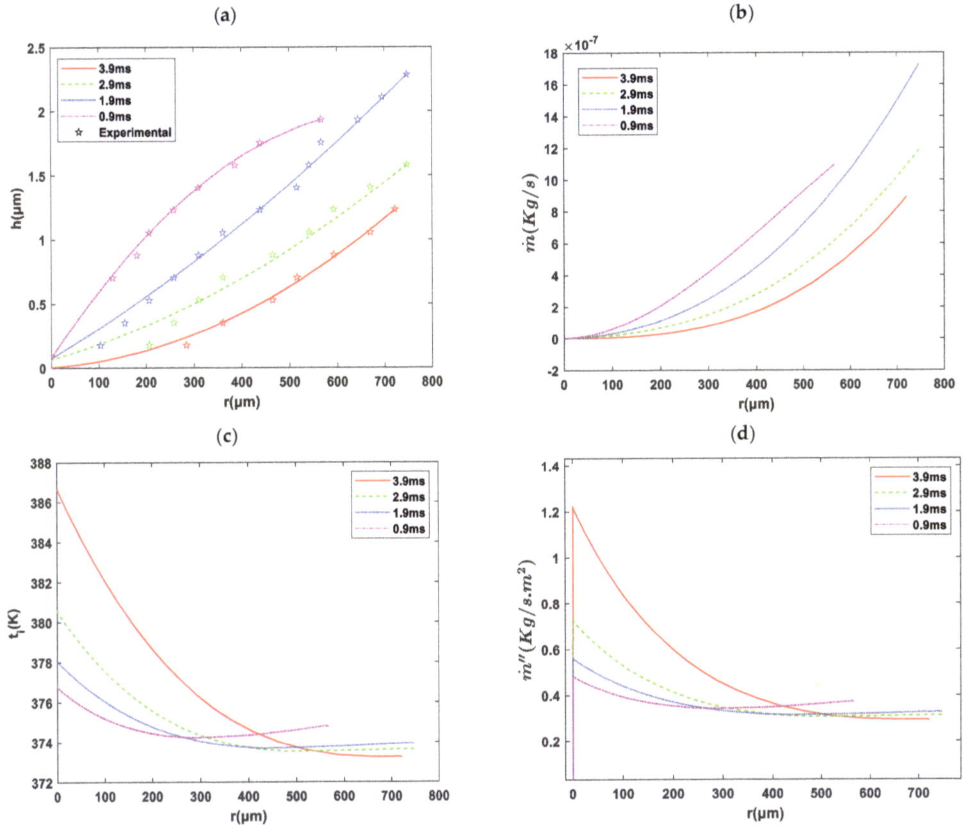

Figure 12. Radius evolutions of (**a**) film thickness; (**b**) mass flow rate; (**c**) interfacial temperature; and (**d**) evaporative mass flux for different time steps.

4.8. Net Evaporation Rate of the Microlayer

The local mass flux distribution from the numerical model (Figure 11c) is integrated over the entire three-dimensional surface area of the bubble (based on Figure 11a) to obtain the total evaporation rate from the microlayer region. Mass fluxes are calculated from the experimentally measured wall, and evaporation heat fluxes across the microlayer using Equation (27) are also calculated based on Figure 3c:

$$\dot{m}'' = \frac{q''}{h_{fg}} \quad (27)$$

The mass flux obtained from both the model (Figure 12d) and the experiments (Figure 3c) could be integrated over the film profile (Figure 12a) to obtain an overall evaporation rate. The accommodation coefficient can then be varied to match the evaporation rates from both the model and the experiment. Figure 13 shows the comparison between the net evaporation rate from the model and the experimental wall heat flux derived evaporation rate at various time steps. Figure 13a shows that the time step of 0.9 ms, in conjunction with an accommodation coefficient value of 0.0198, results in a match between numerical and experimental results of the net evaporation rates. However, the same coefficient results in a systemic underprediction of the modeled rate at higher time steps, as seen in Figure 13a. Similar trends are seen at 0.9 ms, 1.9 ms, 2.9, and 3.9 ms, where accommodation coefficients of 0.019, 0.015, 0.0186, and 0.00893 match the experimentally derived wall heat flux data. These results suggest that the accommodation coefficient may not be a constant and is likely time dependent. The rates derived from the evaporation heat flux (Figure 3c) were also compared to the model. Similarly, as seen in Figure 14, the comparison between the experimental evaporation heat flux and the net evaporation rate from the model derived evaporation rate at various time steps, which represented different matching values of the accommodation coefficient. Figure 14a–d indicate that the accommodation coefficients of the experimental data (evaporation heat flux) and numerical model are matched at 0.0214, 0.0124, 0.011, and 0.005 at time steps of 0.9, 1.9, 2.9, and 3.9, respectively. Based on the results in Figures 13 and 14, the matching accommodation coefficients decrease with time, suggesting that the accommodation coefficient may not be a constant. This observation was also recently made by Tecchio [43], where "interfacial resistance" increased with time as the microlayer was formed. An increase in the interfacial resistance is analogous to a decrease in the value of the accommodation coefficient, as both effects reduce the overall evaporation flux. The accumulation of impurities at the interface between the liquid and vapor as the bubble grows could explain this reduction [43] and is the likely reason for the systemic decrease in the accommodation coefficient observed here.

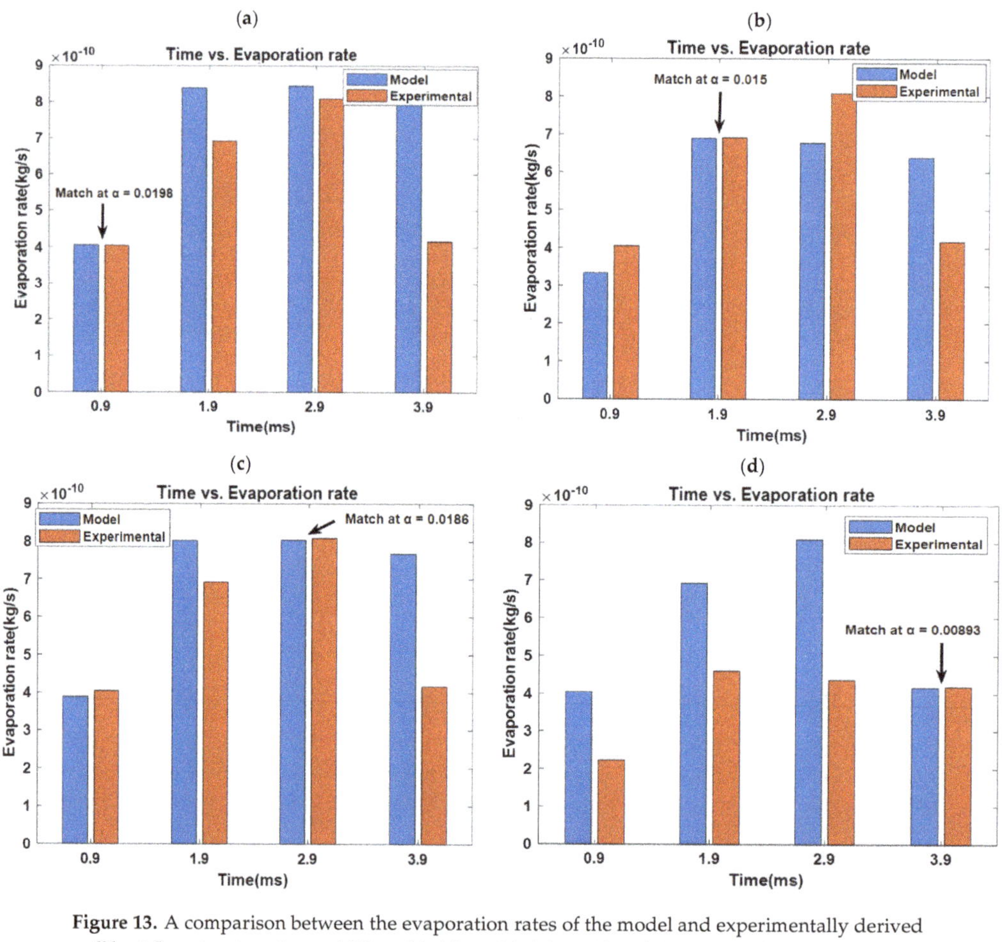

Figure 13. A comparison between the evaporation rates of the model and experimentally derived wall heat flux at various times of 0.9 ms (**a**), 1.9 ms (**b**), 2.9 ms (**c**), and 3.9 ms (**d**). The accommodation coefficient could be estimated through iterative matching. The results suggest that the matched coefficient reduces with time.

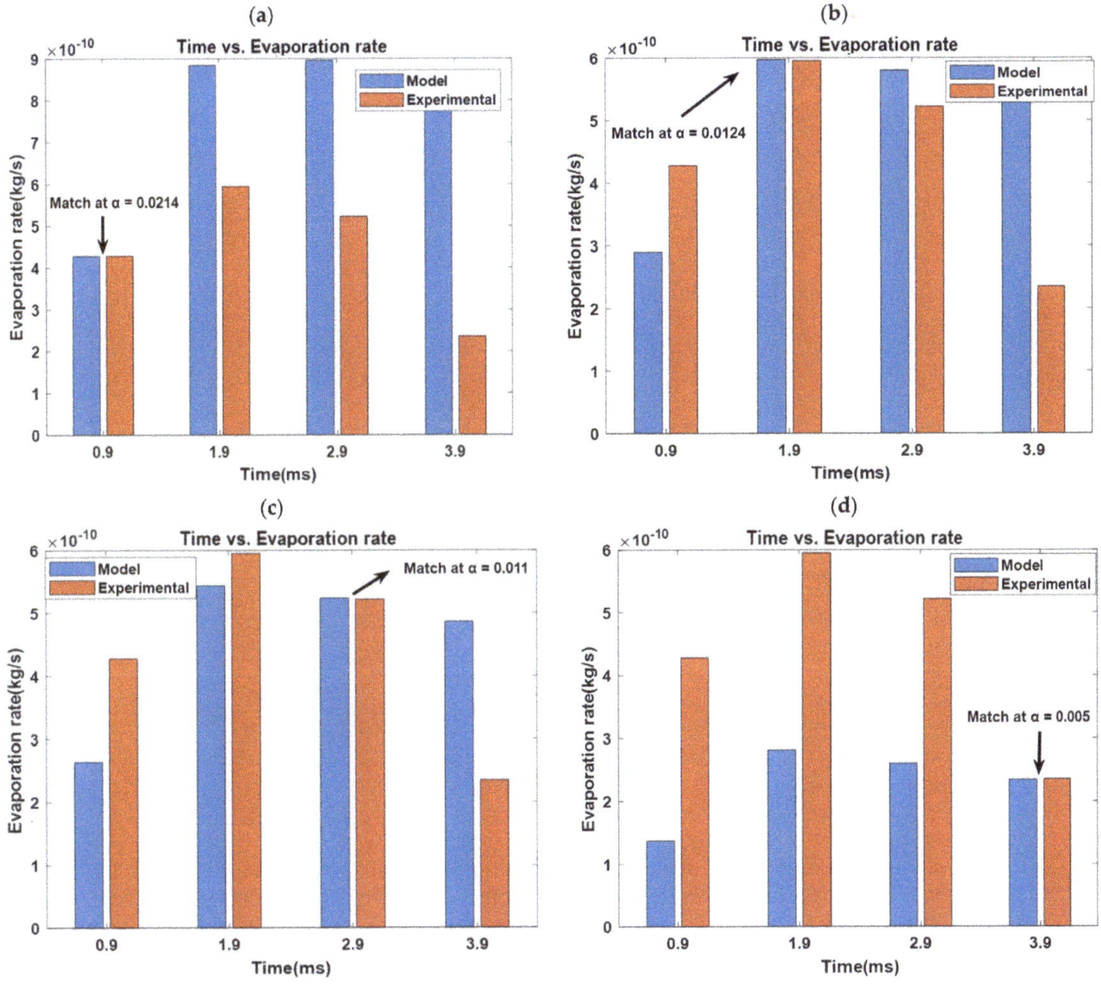

Figure 14. A comparison between evaporation rates between the model and experimentally derived evaporation heat flux at various times of 0.9 ms (**a**), 1.9 ms (**b**), 2.9 ms (**c**), and 3.9 ms (**d**). The accommodation coefficient could be estimated through iterative matching. The results suggest that the matched coefficient reduces with time.

Giustini et al. [35], and Bures and Sato [44] also estimated the accommodation coefficient from the microlayer to be 0.03 and 0.034, respectively. These values are higher than the ones reported here, and the discrepancy is because we allow for both a variable wall and interfacial temperature while Giustini et al. [35] used a constant wall temperature and Bures and Sato [44] assumed isothermal interface temperature. This suggests that assuming an isothermal condition, either at the wall or at the interface, has the potential to overestimate the coefficient in the microlayer. Finally, the fact that all of the values here are much less than unity suggests that only a fraction of impinging vapor molecules is absorbed by a liquid surface and that there is a high possibility of molecular reflection at the liquid–vapor interface.

5. Conclusions

Understanding the microlayer is critical to advancement in nucleate boiling since it is a major contributor to the phase change heat transfer. However, thin film evolution, spatio-temporal stability, and the impact on macroscale bubble dynamics are still poorly understood. Using a lubrication theory approach that accounts for mass, momentum, and energy balances, a film evolution equation is developed. The interfacial mass transfer is modeled using kinetic theory where the accommodation coefficient is a necessary input. While most prior models begin the solution at the nanoscale adsorbed film, the current study reverses the integration path. Contrary to most prior models, the current formulation uses a variable wall temperature boundary rather than a constant wall temperature boundary. Experimentally determined film thickness derivatives are used as boundary conditions at the thicker end of the film, and the model is solved till an absorbed film is obtained. The model is validated by comparing the film thickness with experimental results. The phase change flux was integrated over the film to obtain a net rate. By comparing the net phase change flux from the model with the independently measured experimental values, the accommodation coefficient was estimated for various time steps. Key findings are summarized below:

- Starting with experimentally measurable initial conditions in the bulk region removes the need to guess or tune boundary conditions at the adsorbed film, as performed in prior studies. Consequently, the film thickness and derivatives in the adsorbed film become outputs of the model rather than guessed inputs.
- A constant wall temperature boundary suppresses the evaporation non-uniformity and shifts the mass flux profiles toward the thicker film. On the other hand, a variable wall temperature allows for a sharper peak in evaporation flux in the thin film vicinity.
- Applying the slip velocity condition in the model led to a higher mass flow rate, allowing more bulk liquid flow under the bubble compared to a no slip boundary.
- The value of the accommodation coefficient affects the peak magnitude of evaporative mass flux significantly. By comparing the experimentally measured spatiotemporally resolved heat flux measurements with the results from the model, the accommodation coefficient could be estimated. The results show that the accommodation coefficient most likely reduces with time as the microlayer evolves. It is likely that this systemic variation in the accommodation coefficient is due to either the accumulation of impurities [43] or an increase in surface interface temperature with time [45]. Future investigations on spatio-temporal variation in accommodation coefficient are recommended.

Author Contributions: Conceptualization, G.G. and H.K.; formal analysis, K.B. and E.L.; investigation, K.B., E.L. and G.G.; methodology, K.B.; supervision, K.B., G.G. and H.K.; visualization, E.L. and A.S.; writing—original draft, E.L. and A.S.; writing—review and editing, K.B., A.S., G.G. and H.K. All authors have read and agreed to the published version of the manuscript.

Funding: This research was partly funded by the United Kingdom Engineering and Physical Sciences Research Council (EPSRC) through grant EP/T027061/2 and the APC was funded by the University of Cincinnati through the Department of Mechanical and Materials Engineering.

Data Availability Statement: The data presented in this study are available on request from the corresponding author.

Acknowledgments: We acknowledge support from the University of Cincinnati in form of aGraduate Incentive Award for Ermiyas Lakew and a Graduate Assistant Scholarship for Amirhosein Sarchami.

Conflicts of Interest: The authors declare no conflict of interest.

Nomenclature

List of symbols			
A	Dispersion constant (J)	u_s	Velocity of thin film (m s^{-1})
C_p	Specific heat (J kg^{-1} K^{-1})	V_l	Molar volume (m^3 mol^{-1})
d_m	Microlayer thickness (m)	Greek symbols	
h	Film thickness (m)	α	Accommodation coefficient (-)
h', h'', h'''	Film thickness derivatives (-)	μ	Dynamic viscosity (kg m^{-1} s^{-1})
h_{fg}	Latent heat (J kg^{-1})	ρ	Mass density (kg m^{-3})
K	Surface curvature (m^{-1})	λ	Wavelength (m)
K_b	Boltzmann's constant (J s^{-1} m^{-2} K^{-3})	γ	Slope of surface tension against temperature (N m^{-1} K^{-1})
k	Thermal conductivity (W m^{-1} K^{-1})	σ	Surface tension (N m^{-1})
M	Molar mass (kg mol^{-1})	θ	Refraction angle (°)
$\dot{m}''$	Evaporation mass flux (kg s^{-1} m^{-2})	Subscripts	
n_l	Refractive index (-)	c	Capillary
P	Pressure (Pa)	d	Disjoining
q''	Heat flux ()	i	Interfacial
r	Radius (m)	l	Liquid
$\overline{R}$	Universal gas constant (J kg^{-1} K^{-1})	v	Vapor
T	Temperature (K)	w	Wall

References

1. Polansky, J. Numerical Model of an Evaporating Thin Film. Ph.D. Thesis, Carleton University, Ottawa, ON, Canada, 2011.
2. Sodtke, C.; Kern, J.; Schweizer, N.; Stephan, P. High Resolution Measurements of Wall Temperature Distribution underneath a Single Vapour Bubble under Low Gravity Conditions. *Int. J. Heat Mass Transf.* **2006**, *49*, 1100–1106. [CrossRef]
3. Hu, C.; Pei, Z.; Shi, L.; Tang, D.; Bai, M. Phase Transition Properties of Thin Liquid Films with Various Thickness on Different Wettability Surfaces. *Int. Commun. Heat Mass Transf.* **2022**, *135*, 106125. [CrossRef]
4. Matin, M.H.; Moghaddam, S. Thin Liquid Films Formation and Evaporation Mechanisms around Elongated Bubbles in Rectangular Cross-Section Microchannels. *Int. J. Heat Mass Transf.* **2020**, *163*, 120474. [CrossRef]
5. Ahmed, S.; Pandey, M. New Insights on Modeling of Evaporation Phenomena in Thin Films. *Phys. Fluids* **2019**, *31*, 092001. [CrossRef]
6. Akkuş, Y.; Dursunkaya, Z. A New Approach to Thin Film Evaporation Modeling. *Int. J. Heat Mass Transf.* **2016**, *101*, 742–748. [CrossRef]
7. Dwivedi, R.; Pati, S.; Singh, P.K. Combined Effects of Wall Slip and Nanofluid on Interfacial Transport from a Thin-Film Evaporating Meniscus in a Microfluidic Channel. *Microfluid. Nanofluid.* **2020**, *24*, 84. [CrossRef]
8. Wang, H.; Pan, Z.; Chen, Z. Thin-Liquid-Film Evaporation at Contact Line. *Front. Energy Power Eng. China* **2009**, *3*, 141–151. [CrossRef]
9. Zheng, Z.; Zhou, L.; Du, X.; Yang, Y. Numerical Investigation on Conjugate Heat Transfer of Evaporating Thin Film in a Sessile Droplet. *Int. J. Heat Mass Transf.* **2016**, *101*, 10–19. [CrossRef]
10. Wee, S.-K.; Kihm, K.D.; Hallinan, K.P. Effects of the Liquid Polarity and the Wall Slip on the Heat and Mass Transport Characteristics of the Micro-Scale Evaporating Transition Film. *Int. J. Heat Mass Transf.* **2005**, *48*, 265–278. [CrossRef]
11. Bellur, K.; Médici, E.F.; Choi, C.K.; Hermanson, J.C.; Allen, J.S. Multiscale Approach to Model Steady Meniscus Evaporation in a Wetting Fluid. *Phys. Rev. Fluids* **2020**, *5*, 024001. [CrossRef]
12. Ball, G. Numerical Analysis of the Heat Transfer Characteristics within an Evaporating Meniscus. Ph.D. Thesis, Carleton University, Ottawa, ON, Canada, 2012.
13. Kou, Z.-H.; Lv, H.-T.; Zeng, W.; Bai, M.-L.; Lv, J.-Z. Comparison of Different Analytical Models for Heat and Mass Transfer Characteristics of an Evaporating Meniscus in a Micro-Channel. *Int. Commun. Heat Mass Transf.* **2015**, *63*, 49–53. [CrossRef]
14. Park, K.; Lee, K.-S. Flow and Heat Transfer Characteristics of the Evaporating Extended Meniscus in a Micro-Capillary Channel. *Int. J. Heat Mass Transf.* **2003**, *46*, 4587–4594. [CrossRef]
15. Park, K.; Noh, K.-J.; Lee, K.-S. Transport Phenomena in the Thin-Film Region of a Micro-Channel. *Int. J. Heat Mass Transf.* **2003**, *46*, 2381–2388. [CrossRef]
16. Wayner, P.C., Jr.; Kao, Y.K.; LaCroix, L.V. The Interline Heat-Transfer Coefficient of an Evaporating Wetting Film. *Int. J. Heat Mass Transf.* **1976**, *19*, 487–492. [CrossRef]
17. Wayner, P.C., Jr. Intermolecular Forces in Phase-Change Heat Transfer: 1998 Kern Award Review. *AIChE J.* **1999**, *45*, 2055–2068. [CrossRef]
18. Chatterjee, A.; Plawsky, J.L.; Wayner, P.C., Jr. Disjoining Pressure and Capillarity in the Constrained Vapor Bubble Heat Transfer System. *Adv. Colloid Interface Sci.* **2011**, *168*, 40–49. [CrossRef]

19. Khurshid, I.; Al-Shalabi, E.W. New Insights into Modeling Disjoining Pressure and Wettability Alteration by Engineered Water: Surface Complexation Based Rock Composition Study. *J. Pet. Sci. Eng.* **2022**, *208*, 109584. [CrossRef]
20. Do, K.H.; Kim, S.J.; Garimella, S.V. A Mathematical Model for Analyzing the Thermal Characteristics of a Flat Micro Heat Pipe with a Grooved Wick. *Int. J. Heat Mass Transf.* **2008**, *51*, 4637–4650. [CrossRef]
21. Zhao, J.-J.; Duan, Y.-Y.; Wang, X.-D.; Wang, B.-X. Effects of Superheat and Temperature-Dependent Thermophysical Properties on Evaporating Thin Liquid Films in Microchannels. *Int. J. Heat Mass Transf.* **2011**, *54*, 1259–1267. [CrossRef]
22. Azarkish, H.; Behzadmehr, A.; Frechette, L.G.; Sheikholeslami, T.F.; Sarvari, S.M.H. A Modified Disjoining Pressure Model for Thin Film Evaporation of Water. In Proceedings of the ASME International Mechanical Engineering Congress and Exposition, San Diego, CA, USA, 15–21 November 2013; American Society of Mechanical Engineers: New York, NY, USA; Volume 56321, p. V07BT08A007.
23. Holm, F.W.; Goplen, S.P. Heat Transfer in the Meniscus Thin-Film Transition Region. *J. Heat Transf.* **1979**, *101*, 543–547. [CrossRef]
24. Wayner, P.C., Jr. The Effect of Interfacial Mass Transport on Flow in Thin Liquid Films. *Colloids Surf.* **1991**, *52*, 71–84. [CrossRef]
25. Akkus, Y.; Gurer, A.T.; Bellur, K. Drifting Mass Accommodation Coefficients: In Situ Measurements from a Steady State Molecular Dynamics Setup. *Nanoscale Microscale Thermophys. Eng.* **2021**, *25*, 25–45. [CrossRef]
26. Bellur, K.S. A New Technique to Determine Accommodation Coefficients of Cryogenic Propellants. Ph.D. Thesis, Michigan Technological University, Houghton, MI, USA, 2018.
27. Carey, V.P. *Liquid-Vapor Phase-Change Phenomena: An Introduction to the Thermophysics of Vaporization and Condensation Processes in Heat Transfer Equipment*; CRC Press: Boca Raton, FL, USA, 2020.
28. Hertz, H. Ueber die Verdunstung der Flüssigkeiten, Insbesondere des Quecksilbers, im Luftleeren Raume. *Ann. Phys.* **1882**, *253*, 177–193. [CrossRef]
29. Knudsen, M. *The Kinetic Theory of Gases: Some Modern Aspects*; Methuen & Company Limited: London, UK, 1934.
30. Marek, R.; Straub, J. Analysis of the Evaporation Coefficient and the Condensation Coefficient of Water. *Int. J. Heat Mass Transf.* **2001**, *44*, 39–53. [CrossRef]
31. Schrage, R.W. A Theoretical Study of Interphase Mass Transfer. In *A Theoretical Study of Interphase Mass Transfer*; Columbia University Press: New York, NY, USA, 1953.
32. Bellur, K.; Médici, E.F.; Kulshreshtha, M.; Konduru, V.; Tyrewala, D.; Tamilarasan, A.; McQuillen, J.; Leão, J.B.; Hussey, D.S.; Jacobson, D.L. A New Experiment for Investigating Evaporation and Condensation of Cryogenic Propellants. *Cryogenics* **2016**, *74*, 131–137. [CrossRef]
33. Jung, S.; Kim, H. An Experimental Method to Simultaneously Measure the Dynamics and Heat Transfer Associated with a Single Bubble during Nucleate Boiling on a Horizontal Surface. *Int. J. Heat Mass Transf.* **2014**, *73*, 365–375. [CrossRef]
34. Jung, S.; Kim, H. An Experimental Study on Heat Transfer Mechanisms in the Microlayer Using Integrated Total Reflection, Laser Interferometry and Infrared Thermometry Technique. *Heat Transf. Eng.* **2015**, *36*, 1002–1012. [CrossRef]
35. Giustini, G.; Jung, S.; Kim, H.; Ardron, K.H.; Walker, S.P. Microlayer Evaporation during Steam Bubble Growth. *Int. J. Therm. Sci.* **2019**, *137*, 45–54. [CrossRef]
36. DasGupta, S.; Schonberg, J.A.; Wayner, P.C., Jr. Investigation of an Evaporating Extended Meniscus Based on the Augmented Young–Laplace Equation. *J. Heat Transf.* **1993**, *115*, 201–208. [CrossRef]
37. Adamson, A.W.; Gast, A.P. *Physical Chemistry of Surfaces*; Interscience Publishers: New York, NY, USA, 1967; Volume 150.
38. Li, J.; Yang, Z.; Duan, Y. Numerical Simulation of Single Bubble Growth and Heat Transfer Considering Multi-Parameter Influ-ence during Nucleate Pool Boiling of Water. *AIP Adv.* **2021**, *11*, 125207. [CrossRef]
39. Liu, Y.; Xing, Y.; Yang, C.; Li, C.; Xue, C. Simulation of Heat Transfer in the Progress of Precision Glass Molding with a Finite Element Method for Chalcogenide Glass. *Appl. Opt.* **2019**, *58*, 7311–7318. [CrossRef] [PubMed]
40. Du, S.-Y.; Zhao, Y.-H. Numerical Study of Conjugated Heat Transfer in Evaporating Thin-Films near the Contact Line. *Int. J. Heat Mass Transf.* **2012**, *55*, 61–68. [CrossRef]
41. Emelyanenko, K.A.; Emelyanenko, A.M.; Boinovich, L.B. Disjoining Pressure Analysis of the Lubricant Nanofilm Stability of Liquid-Infused Surface upon Lubricant Depletion. *J. Colloid Interface Sci.* **2022**, *618*, 121–128. [CrossRef]
42. Kaya, T.; Ball, G.; Polansky, J. Investigation of Particular Features of the Numerical Solution of an Evaporating Thin Film in a Channel. *Front. Heat Mass Transf.* **2013**, *4*, 013002. [CrossRef]
43. Tecchio, C. Experimental Study of Boiling: Characterization of Near-Wall Phenomena and Bubble Dynamics. Ph.D. Thesis, Université Paris-Saclay, Paris, France, 2022.
44. Bureš, L.; Sato, Y. Comprehensive Simulations of Boiling with a Resolved Microlayer: Validation and Sensitivity Study. *J. Fluid Mech.* **2022**, *933*, A54. [CrossRef]
45. Nagayama, G.; Takematsu, M.; Mizuguchi, H.; Tsuruta, T. Molecular Dynamics Study on Condensation/Evaporation Coefficients of Chain Molecules at Liquid–Vapor Interface. *J. Chem. Phys.* **2015**, *143*, 014706. [CrossRef]

fluids

Article

Fingering Instability of Binary Droplets on Oil Pool

Koji Hasegawa * and Yuya Kishimoto

Department of Mechanical Engineering, Kogakuin University, Tokyo 163-8677, Japan
* Correspondence: kojihasegawa@cc.kogakuin.ac.jp

Abstract: The interfacial instability of a complex fluid in a multiphase flow system is ubiquitous in both nature and industry. We experimentally investigated the spreading and interfacial instability dynamics of a binary droplet (a water and 2-propanol (IPA) mixture) on an immiscible (sunflower oil) pool. For droplets of 40 wt% IPA solution on sunflower oil, fingering instability occurred at the spreading liquid front. To reveal the interfacial characteristics of the spreading and fingering processes, we analyzed the interplay among the speed, diameter, and number of fingers on the spreading front. Based on our observations, the finger length, wavelength between the fingers, head length, and neck length were quantified. Our experimental results clearly demonstrate that fingering instability can be driven by the capillary effect for a liquid–liquid system as well as the Plateau–Rayleigh instability. We hope that our results will inspire further experimental and numerical investigations to provide deeper insights into the interfacial dynamics of multicomponent droplets in a liquid pool.

Keywords: fingering instability; binary droplets; droplet spreading

1. Introduction

The wetting, spreading of droplets, and the accompanying fingering instability have attracted considerable attention as natural phenomena and those that occur in various industries [1–9]. Understanding the interfacial dynamics is crucial for exploring the impingement of droplets on solid and liquid substrates [10–13]. To better understand spreading characteristics, the evaporation-induced Marangoni effect has been extensively investigated over the past few decades [14–35]. For solid substrates, Mouat et al. [20] investigated contact line dynamics and fingering patterns in volatile liquid mixtures on a silicon wafer. They demonstrated two distinct regimes of interfacial instability: surface wetting and solutal Marangoni forces. The spontaneous spreading of a thin volatile film along the surface of a liquid layer with a higher surface tension has been studied experimentally and theoretically by Troian et al. [21–24]. In the case of a deep liquid layer, although previous studies of the spreading dynamics of non-volatile immiscible thin films on a deep liquid layer have shown that the spreading front advances in time as $t^{3/4}$ as predicted by laminar boundary layer theory, they found that the leading edge of volatile immiscible spreading films also advances with time as a power law, t^{α}, where $\alpha \sim 1/2$ [22]. This indicates that the differences in liquid vapor pressure or spreading coefficient can affect the velocity of the spreading front and suggests the presence of a universal scaling law. A recent work by Keiser et al. [25] revealed the instability that was observed when a two-component drop of water and volatile alcohol is deposited on a bath of sunflower oil. The drop temporarily spreads and spontaneously breaks up into thousands of tiny droplets owing to the Marangoni effect, which is induced by the evaporation of the IPA component. To understand the spreading dynamics caused by the concentration gradient of the IPA–water interface, Kim et al. [26] analyzed the spreading owing to the solutal Marangoni effect in fully miscible liquids by employing a scaling analysis. Even for non-Newtonian fluids, Ma et al. [29] discussed fingering instability during Marangoni spreading on a deep layer of the polymer. For a surfactant-coated droplet, Matar and Troian [35] investigated the linear stability of a coupled set of equations describing the Marangoni spreading of a surfactant monolayer on a thin liquid film. They found that the inclusion of additional forces acting on the surfactant-coated spreading film could form regions of unfavorable mobility gradients known to produce finger instabilities in other fluid flows.

Although several researchers have explored the dynamics of the spreading and fingering instabilities associated with the Marangoni effect, more quantitative and predictive insights into the interfacial instability during the spreading and self-atomization processes are needed. A better understanding of the spreading and fingering instability characteristics will enable the development of a promising atomization technology for a liquid–liquid interface that is easy to operate and can produce thousands of microdroplets within a short time. This study aimed to identify the spreading and fingering instability characteristics of an evaporating liquid pool. We visualized and quantified the dynamic behavior of IPA solution droplets on an oil pool. During the spreading process of the released droplet, the fingering instability in the circumferential and radial directions was characterized on the spreading front via the capillary and critical instability lengths. These experimental findings may also be relevant to other fascinating Marangoni-driven dynamics, such as the swimming of droplets [36] and particles [37], and other interfacial phenomena [38–42].

2. Materials and Methods

Figure 1 represents a schematic of the experimental setup in the present study. The experimental setup consisted of an oil pool in a Petri dish, a pipette for an injection of droplets, and a digital camera for a visualization. The droplets were placed manually on the oil pool. The droplet behavior was captured using a digital video camera (Nikon, D5600) backlit with an LED light (WTH-C2020, Nissin Electronics Co., Ltd., Hachioji, Japan). To enhance the contrast, the droplets were dyed blue using food coloring. The obtained images were processed using the ImageJ software (1.52v) to estimate the spreading and fingering instability of the droplets. In this study, a 40 wt% 2-propanol (IPA)/water mixture and sunflower oil were used as the droplet and pools, respectively. In a previous study [33], we investigated the speed dynamics of 20, 40, 60, and 80 wt% IPA solutions in sunflower oil. For the 20 wt% IPA solution, the released droplet floated on sunflower oil (no obvious change occurred even after 20 s of contact). For higher concentrations of IPA solution (i.e., 60 and 80 wt%), although spontaneous spreading and atomization were observed on the sunflower oil, the finger length was too short, and the atomized daughter droplet size was too small to observe and quantify the entire process. To better understand the spreading and fingering instability dynamics, we focused on the 40 wt% IPA solution. The depth of the oil pool was fixed at 10 mm. The inner diameter of each Petri dish was 85.7 mm. The droplets released on the oil pool are ≈2.0 mm in diameter. The ambient temperature and relative humidity were 25 °C and 50%, respectively. The droplet motion was recorded at 60 fps with a spatial resolution of ≈85 μm/pixel.

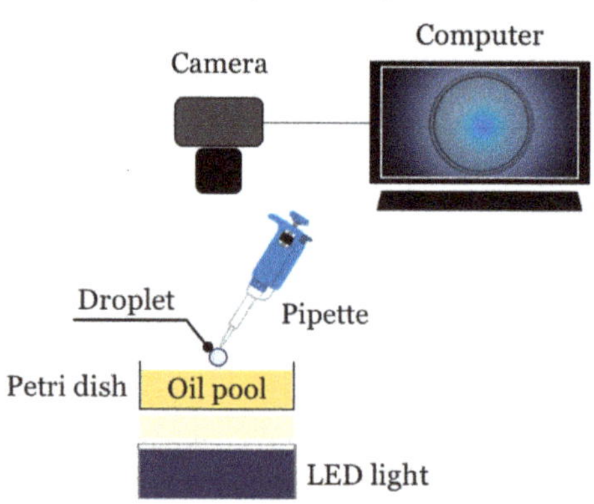

Figure 1. Schematic of the experimental setup.

Table 1 lists the thermophysical properties of the tested liquids. The density and kinematic viscosity of sunflower oil were measured. All surface tensions listed in Table 1 are the surface tensions between each liquid and air.

Table 1. Thermophysical properties of the test fluids.

Test Sample	Density ρ (kg/m^3)	Kinematic Viscosity ν (mm^2/s)	Surface Tension σ (mN/m)
Water [1]	997	0.89	72
IPA [1]	781	2.61	21
Sunflower oil	916	58	32 [2]

[1] Ref. [43]. [2] Ref. [25].

3. Results and Discussion

3.1. Droplet Spreading

Figure 2 shows the effect of 40 wt% IPA solution on sunflower oil. After contact with oil (time t = 0 s; Figure 2a), the released droplets started wetting and spreading within 1 s (Figure 2b). Fingering instability and atomization were observed after 6 s (Figure 2c). The spreading front reaches its maximum diameter after 12 s (Figure 2d). Subsequently, the spreading front gradually shrunk (Figure 2e) and finally vanished after 26 s (Figure 2f). During this process, the released droplet forms a wetting film that exhibits fingering at the spreading front. Furthermore, it is reasonable to assume that brightness corresponds to the IPA concentration distribution in the spreading film, allowing us to predict the concentration. The initial stage of droplet spreading (≈6 s) is darker blue (higher IPA concentration) near the center of the spreading film, which gradually changes to a lighter shade of blue (lower IPA concentration) in the later stage (≈26 s) with the evaporation of the IPA component. Based on these results, we quantified the brightness value to analyze the spreading behavior.

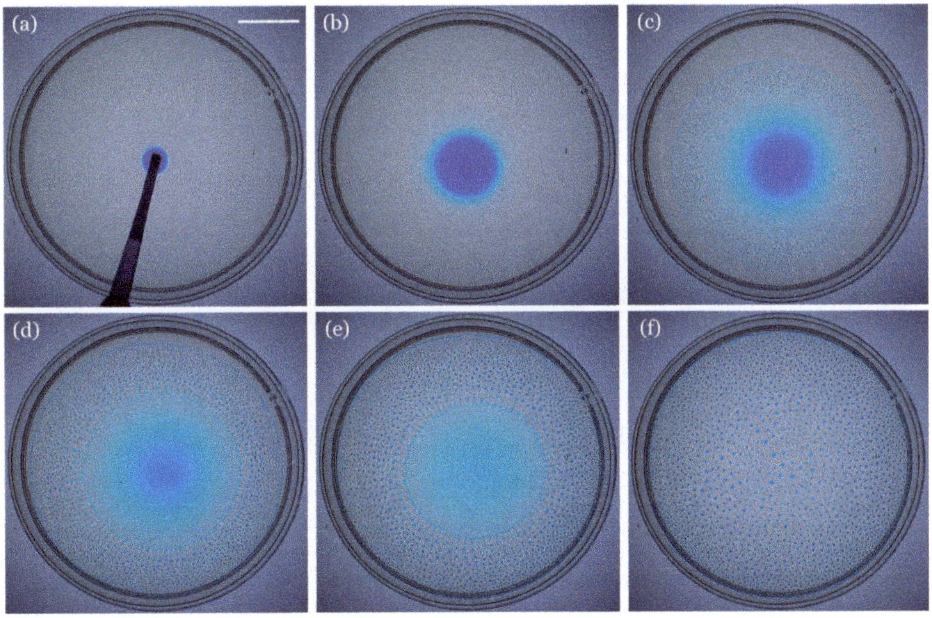

Figure 2. Snapshots of dyed 40 wt% IPA solutions on the sunflower oil for (**a**) t = 0 s, (**b**) 1 s, (**c**) 6 s, (**d**) 12 s, (**e**) 18 s, and (**f**) 26 s, respectively. t = 0 s indicates the time at which the droplet contacted the oil pool. Scale bar in (**a**): 20 mm.

Figure 3 represents the brightness information of the spreading films. The data for the spreading times of 1, 6, 12, and 18 s correspond to the snapshots in Figure 2. The brightness changed with time, as shown in Figure 3a, allowing us to quantify the position of the spreading front on the oil pool, which reached 12 mm at 1 s, 20 mm at 6 s, and 22 mm at 12 s, and shrank subsequently. The brightness fluctuation beyond the spreading front is due to the pinch-off of the droplets owing to the fingering instability. Figure 3b represents the brightness decay at the center of the spreading film (R = 0 mm in Figure 3a). The linear decrease in the brightness value was consistent with the IPA components in the spreading binary liquid film, which evaporated linearly with time. Consequently, the evaporation-induced Marangoni force can drive the spontaneous spreading and shrinking processes [29]. According to a previous analysis [25], it can be assumed that the spreading coefficient S ($=\sigma_{oa} - \sigma_{sa} - \sigma_{so}$) [44] becomes positive for the 40 wt% IPA solution and increases with the increasing IPA concentration, forcing the droplets to stretch and spread to the outer edge in the early stage. The parameters σ_{oa}, σ_{sa}, and σ_{so} represent the oil–air, solution–air, and solution–oil interfacial tensions, respectively. For lower concentrations of IPA solution (<35 wt%), the spreading coefficient was negative, and the droplet floated on the pool [29]. Notably, an appropriate amount of water (≥35 wt% IPA solution) plays a critical role in the spontaneous spreading dynamics. In the present case, the surface tensions of sunflower oil and 40 wt% IPA solution in air are 32 and 25 mN/m, respectively, and the interfacial tension is approximately 3 mN/m [25]. Using these figures, the initial spreading coefficient in the present case can be 32 − (25 + 3) = 4 mN/m. This indicates that the 40 wt% IPA solution spontaneously spreads over the sunflower oil pool initially.

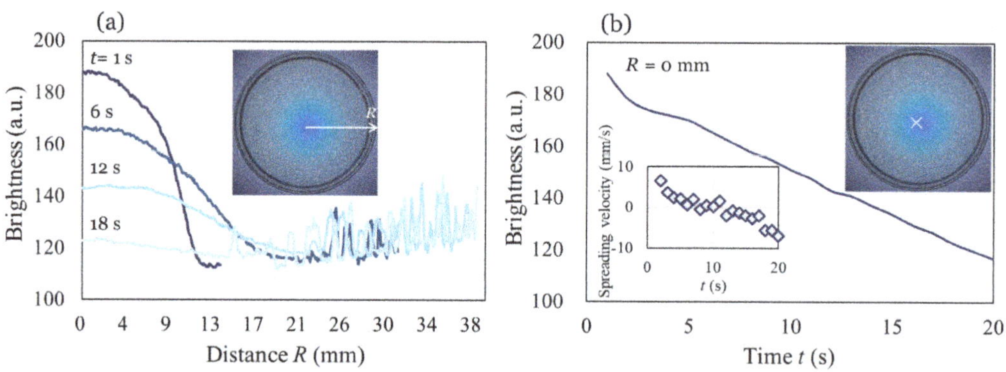

Figure 3. Brightness information obtained during the spreading of 40 wt% IPA solutions on sunflower oil. (**a**) Brightness profiles at 1, 6, 12, and 18 s. Inset: coordinates of the distance R. (**b**) Time evolution of the brightness at R = 0 mm (center of spreading film marked as ×). To eliminate the effects of the injection process, the brightness data from 1 s onwards are plotted. Inset: the film-spreading velocity.

Spontaneous spreading determines the contact line speed. The contact line speed affected the evaporation rate of the IPA component by increasing the surface area of the 40 wt% IPA solution in air on sunflower oil. Therefore, the evaporation of the IPA component can be enhanced using faster contact line speeds for a larger spreading coefficient (higher IPA concentration), resulting in a stronger Marangoni effect on the spreading film. To explain this quantitatively, we estimated the Marangoni velocity of the spreading film for a 40 wt% IPA solution. We assumed competition between Marangoni stress and viscous stress on the spreading film. The Marangoni and viscous stresses on the film can be described by $\frac{\Delta\sigma}{R_{max}}$ and $\mu_p \frac{V_m}{H}$, respectively. By balancing the tangential stresses at the interface, we derived the Marangoni velocity as shown in Equation (1).

$$V_m \sim \frac{\Delta\sigma H}{\mu_p R_{max}} \tag{1}$$

where $\Delta\sigma$ is the surface tension deference between the front and center of the spreading film due to the evaporation, H is the oil-pool depth, μ_p is the viscosity of the sunflower oil, and R_{max} is the maximum spreading radius. By substituting $\Delta\sigma \sim O(1)$ mN/m, $H = 10$ mm, $\mu_p = 63$ mPa·s, and $R_{max} = 45$ mm into Equation (1), we obtain a Marangoni velocity of $O(10^{-3})$ m/s. This estimate is in quantitative agreement with the film-spreading velocity shown in the inset of Figure 3b.

3.2. Fingering Instability on Spreading Front

During droplet spreading on the oil pool, fingering instability was generated on the spreading front, as shown in Figure 2c–e. Figure 4 represents the fingering dynamics as a function of the characteristic length. A typical snapshot of the fingering instability of one-quarter of a spreading film is shown in Figure 4a. The number of fingers increased and decreased with the droplet spreading and shrinking, respectively, as shown in Figure 4b. For one-quarter of the spreading area, when the spreading film was fully developed for $5\ \mathrm{s} < t < 14$ s, over 20 fingers were observed at the spreading front. Notably, capillary-driven fingering instability is strongly related to the evaporation of the IPA component. To reveal these fingering characteristics, we quantified the four characteristic lengths depicted in the inset of Figure 4a: finger length, λ_r; wavelength, λ_θ; head length, D; and neck length, d. Figure 4c plots the time evolution of these characteristic lengths. For λ_θ and d, the experimental data was almost constant during the spreading process, while λ_r and D increased with time. This indicates that the instability mechanism must be distinguished between the circumferential and radial directions.

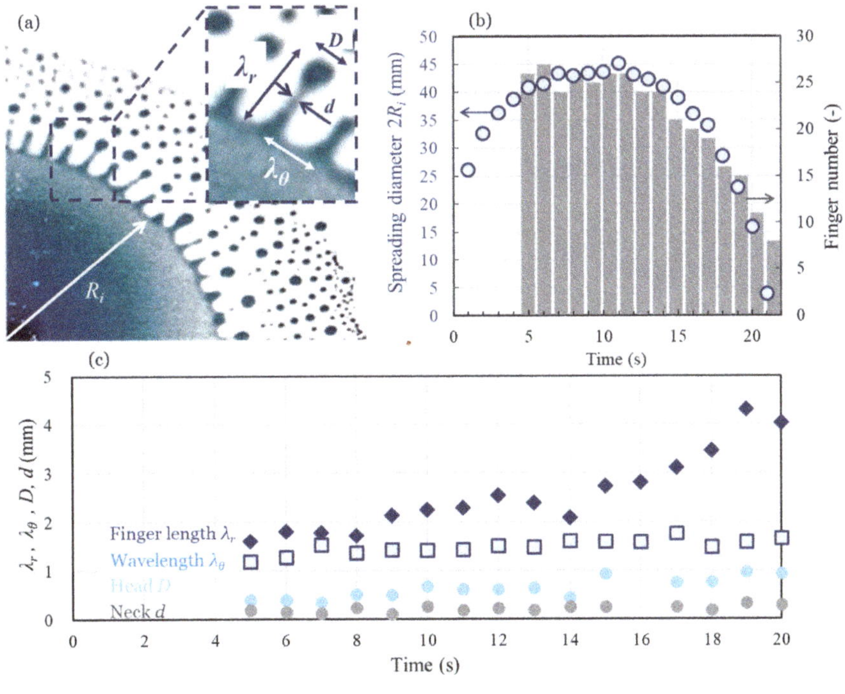

Figure 4. Fingering characteristics on the spreading front. (**a**) Snapshot of the fingering instability. R_i represents the radius from the spreading center to the front. Four characteristic lengths are defined in the close-up view. (**b**) Relationship between the spreading diameter and finger number on the spreading front. (**c**) Time evolution of the finger length, wavelength, head length, and neck length.

For the circumferential direction, the capillary length of the liquid–liquid (IPA solution–sunflower oil) interface, λ_c can be calculated using Equation (2) [1]:

$$\lambda_c \sim \sqrt{\frac{\sigma_{so}}{(\rho_o - \rho_{IPA})g}} \tag{2}$$

where σ_{so} is the interfacial tension between the IPA solution and sunflower oil, ρ_o is the density of sunflower oil, ρ_{IPA} is the density of IPA, and g is the gravitational acceleration. Figure 5a shows the time evolution of the wavelength between the fingers in the circumferential direction. The wavelength was normalized to the capillary length, as described in Equation (2). For $\sigma_{so} \sim 3$ mN/m [25], $\rho_o = 916$ kg/m^3, $\rho_s = 781$ kg/m^3, and $g = 9.81$ m/s^2, the predicted wavelength of $\approx$1.5 mm agrees well with the experimental data. This model allows us to conclude that a higher concentration of IPA leads to lower interfacial tension and wavelength [29]. In Figure 5, although we characterized the time evolution of the wavelength between the fingers in the circumferential direction by the capillary length for the liquid–liquid interface, the experimental data showed a slight increase with time. This may be due to the effect of evaporation on the interfacial tension caused by a change in the IPA concentration in the spreading film. To verify the effect of interfacial tension distribution by evaporation, future work is expected to measure the IPA concentration distribution during film spreading.

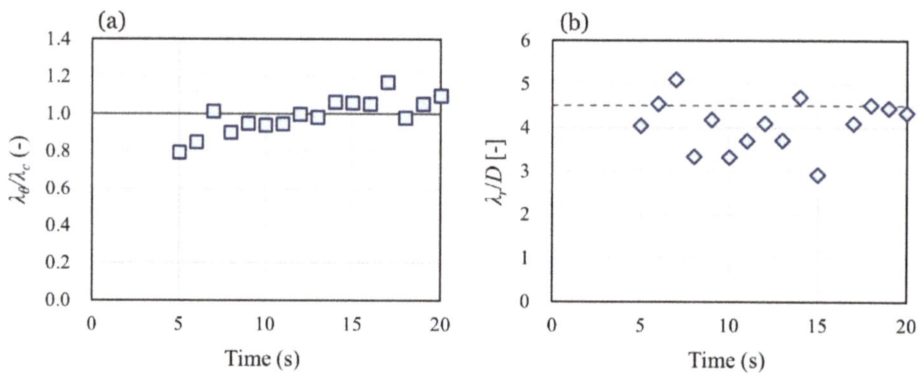

Figure 5. Time evolution of characteristic finger length in the (**a**) circumferential direction and (**b**) radial direction. The solid line in (**a**) and dashed line in (**b**) indicate the capillary length of the liquid–liquid (IPA solution–sunflower oil) interface predicted using Equation (2) and the most unstable wavelength predicted by the Plateau–Rayleigh instability in Equation (4), respectively.

In the radial direction, the pinch-off process is reminiscent of the classical Plateau–Rayleigh instability of liquid cylinders [45]. Based on this assumption, the experimental results were compared with those of the theoretical model. The key expression that describes the development of the Rayleigh–Plateau instability is the dispersion relation between the growth rate $\omega$and wave number k (=2π/λ, where λ is the wavelength on liquid cylinder interface). For axisymmetric capillary waves on a liquid cylinder with radius R_0:

$$\omega^2 = \frac{\sigma}{\rho R_0^3} \frac{I_1(kR_0)}{I_0(kR_0)} kR_0 \left(1 - (kR_0)^2\right), \tag{3}$$

where σ is the surface tension of the liquid cylinder and I_0 and I_1 are the modified Bessel functions of the first kind. From this relationship, if the right-hand side of Equation (3) becomes positive ($kR_0 < 1$), the perturbation grows exponentially and eventually forces the system to become unstable. The fastest growing mode occurred for $kR_0 = 0.697$, i.e., when the wavelength of the disturbance $\lambda_{crit} \sim 9.02R_0$. By substituting $R_0 = D/2$ into

$\lambda_{crit}/R_0 \sim 9.02$, we calculated the critical wavelength of the Plateau–Rayleigh instability, λ_{crit}, according to Equation (4) [7,45].

$$\frac{\lambda_{crit}}{D} \sim 4.51, \tag{4}$$

Figure 5b shows the time evolution of the finger length in the circumferential direction. The finger length was normalized to the critical wavelength predicted using Equation (4). Our experimental results roughly agree with theoretical predictions. A reason for the slight difference between the experimental data and theoretical values is the interplay between the fingers. Although the Plateau–Rayleigh instability assumes only one liquid cylinder, multiple fingers exist, and neighboring fingers may dynamically interact in the present case. Another possible reason for this is the effect of the flow field on the spreading film. For an evaporation-driven flow near the spreading front, the flow fluctuation generated pressure fluctuations and enhanced the interfacial deformation for finger formation.

One fingering instability application resulted in tiny and myriad daughter droplets after the pinch-off. Figure 6a shows the generation frequency of daughter droplets owing to fingering instability. In this study, we analyzed five representative fingers. According to our results, the majority of the daughter droplets measured less than 1 mm, with frequencies in the range of 1–4 Hz. By counting the droplets produced from five representative fingers, Figure 6b shows the probability density function (PDF) histogram for the daughter droplet diameter. For all 131 data points associated with the five representative fingers, the average diameter and standard deviation of the daughter droplets were 0.7 and 0.2 mm, respectively. According to Figure 4b, nearly one-hundred fingers occupied the entire spreading front during the spreading phenomenon; therefore, the total number of daughter droplets can exceed 2000 for the entire spreading circle within tens of seconds [29].

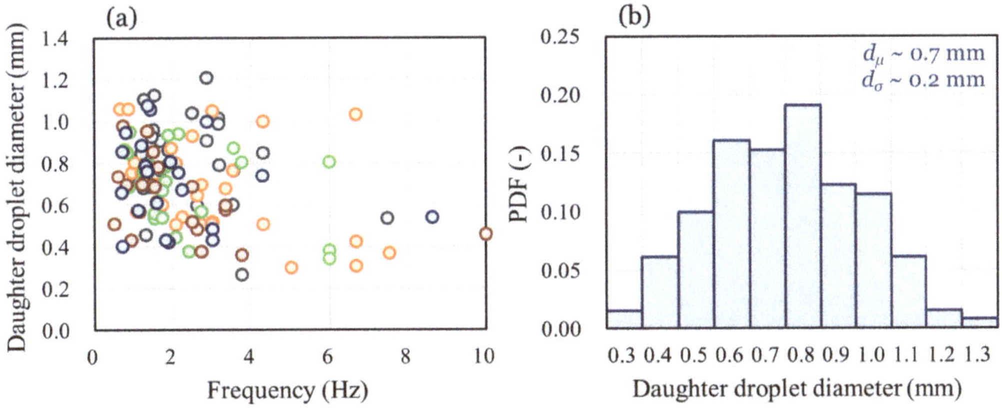

Figure 6. Atomization characteristics of the fingering instability. (**a**) Generation frequency of the daughter droplets for five representative fingers during the spreading process; each finger can be distinguished by the plot color. (**b**) PDF histogram for experimental data in (**a**). The number of samples was 131.

4. Conclusions

We experimentally investigated the spontaneous spreading and fingering instability of binary droplets on sunflower oil. The droplets released from a 40 wt% IPA solution exhibited spontaneous spreading and fingering instability and underwent a pinch-off process. The spreading behavior of the 40 wt% IPA solution was quantitatively studied. The linear decrease in the brightness of the spreading film could be due to evaporation of the IPA component in the spreading binary liquid film. This evaporation-induced Marangoni force drives the spreading and shrinking processes by temporarily changing the surface tension of the spreading film. We identified the interplay between the spreading diameter and number of fingers for fingering instability on the spreading front. To elucidate the instability mechanism, the characteristic circumferential and radial lengths of the spreading front are discussed. In the circumferential direction, the capillary length for the IPA solution–sunflower oil interface can reasonably characterized the wavelength between the fingers. In the radial direction, the pinch-off process is described by the classical Plateau–Rayleigh instability of the liquid cylinders. Our experimental results agreed with the theoretical results. The final feature of the fingering instability phenomenon is the pinch-off dynamics. Owing to fingering instability, a myriad of tiny droplets was produced. Most daughter droplets measured less than 1 mm and had frequencies in the range 1–4 Hz. Thus, the generated daughter droplets were distributed across hundreds of micrometers. For the entire spreading film, this spontaneous pinch-off process enables the generation of myriad daughter droplets using a simple sequence.

Although we attempted to characterize the fingering dynamics on the spreading front of the binary droplets on the oil pool as described above, it was difficult to quantify the flow field induced by the Marangoni effect. The flow field in a spreading micrometer-thick film on an oil pool can play an important role in the inception of fingering instability; however, this is beyond the scope of this study. To achieve this, the flow fields within the thin film during the spreading and evaporation of the IPA components must be visualized. This investigation represents a step toward understanding and predicting the fingering instability dynamics of spreading binary droplets on an oil pool to develop a promising atomization technology.

Author Contributions: Conceptualization, K.H.; methodology, Y.K.; formal analysis, Y.K. and K.H.; resources, K.H.; writing—original draft preparation, K.H.; writing—review and editing, K.H.; supervision, K.H. All authors have read and agreed to the published version of the manuscript.

Funding: This work was financially supported by the Research Institute for Science and Technology of Kogakuin University.

Institutional Review Board Statement: Not applicable.

Informed Consent Statement: Not applicable.

Data Availability Statement: The data presented in this study are available upon request from the corresponding author.

Conflicts of Interest: The authors declare no conflict of interest.

References

1. De Gennes, P.G.; Brochard-Wyart, F.; Quéré, D. *Capillarity and Wetting Phenomena: Drops, Bubbles, Pearls, Waves*; Springer: New York, NY, USA, 2004.
2. Troian, S.M.; Wu, X.L.; Safran, S.A. Fingering instability in thin wetting films. *Phys. Rev. Lett.* **1989**, *62*, 1496. [CrossRef] [PubMed]
3. Cazabat, A.M.; Heslot, F.; Troian, S.M.; Carles, P. Fingering instability of thin spreading films driven by temperature gradients. *Nature* **1990**, *346*, 824–826. [CrossRef]
4. Neitzel, G.P.; Dell'Aversana, P. Noncoalescence and nonwetting behavior of liquids. *Annu. Rev. Fluid. Mech.* **2002**, *34*, 267–289. [CrossRef]
5. Yarin, A.L. Drop impact dynamics: Splashing, spreading, receding, bouncing. *Annu. Rev. Fluid. Mech.* **2006**, *38*, 159–192. [CrossRef]
6. Venerus, D.C.; Nieto Simavilla, D. Tears of wine: New insights on an old phenomenon. *Sci. Rep.* **2015**, *5*, 16162. [CrossRef] [PubMed]
7. Nikolov, A.; Wasan, D.; Lee, J. Tears of wine: The dance of the droplets. *Adv. Colloid Interface Sci.* **2018**, *256*, 94–100. [CrossRef] [PubMed]
8. Lohse, D.; Zhang, X. Physicochemical hydrodynamics of droplets out of equilibrium. *Nat. Rev. Phys.* **2020**, *2*, 426–443. [CrossRef]
9. Wang, Z.; Orejon, D.; Takata, Y.; Sefiane, K. Wetting and evaporation of multicomponent droplets. *Phys. Rep.* **2022**, *960*, 1–37. [CrossRef]
10. Cira, N.J.; Benusiglio, A.; Prakash, M. Vapour-mediated sensing and motility in two-component droplets. *Nature* **2015**, *519*, 446–450. [CrossRef]

11. Sanatkaran, N.; Kulichikhin, V.G.; Malkin, A.Y.; Foudazi, R. Spreading of oil-in-water emulsions on water surface. *Langmuir* **2018**, *34*, 10974–10983. [CrossRef]
12. Wodlei, F.; Sebilleau, J.; Magnaudet, J.; Pimienta, V. Marangoni-driven flower-like patterning of an evaporating drop spreading on a liquid substrate. *Nat. Commun.* **2018**, *9*, 820. [CrossRef] [PubMed]
13. Motaghian, M.; Shirsavar, R.; Erfanifam, M.; Sabouhi, M.; Van Der Linden, E.; Stone, H.A.; Bonn, D.; Habibi, M. Rapid spreading of a droplet on a thin soap film. *Langmuir* **2019**, *35*, 14855–14860. [CrossRef] [PubMed]
14. Fanton, X.; Cazabat, A.M. Spreading and instabilities induced by a solutal Marangoni effect. *Langmuir* **1998**, *14*, 2554–2561. [CrossRef]
15. Deodhar, S.; Thampi, S.P.; Basavaraj, M.G. Drops spreading on fluid surfaces: Transition from Laplace to Marangoni regime. *Phys. Rev. Fluids* **2021**, *6*, L112001. [CrossRef]
16. Jia, F.; Wang, T.; Peng, X.; Sun, K. Three stages of Marangoni-driven film spreading for miscible fluids. *Phys. Fluids* **2022**, *34*, 121705. [CrossRef]
17. Motaghian, M.; van der Linden, E.; Habibi, M. Surfactant-surfactant interactions govern unusual Marangoni spreading on a soap film. *Colloids Surf. A Physicochem. Eng. Asp.* **2022**, *653*, 129747. [CrossRef]
18. Baumgartner, D.A.; Shiri, S.; Sinha, S.; Karpitschka, S.; Cira, N.J. Marangoni spreading and contracting three-component droplets on completely wetting surfaces. *Proc. Natl. Acad. Sci. USA* **2022**, *119*, e2120432119. [CrossRef]
19. Liu, D.; Tran, T. Vapor-induced attraction of floating droplets. *J. Phys. Chem. Lett.* **2018**, *9*, 4771–4775. [CrossRef]
20. Mouat, A.P.; Wood, C.E.; Pye, J.E.; Burton, J.C. Tuning contact line dynamics and deposition patterns in volatile liquid mixtures. *Phys. Rev. Lett.* **2020**, *124*, 064502. [CrossRef]
21. Troian, S.M.; Herbolzheimer, E.; Safran, S.A.; Joanny, J.F. Fingering instabilities of driven spreading films. *Europhys. Lett.* **1989**, *10*, 25–30. [CrossRef]
22. Dussaud, A.D.; Troian, S.M. Dynamics of spontaneous spreading with evaporation on a deep fluid layer. *Phys. Fluids* **1998**, *10*, 23–38. [CrossRef]
23. Darhuber, A.A.; Troian, S.M. Marangoni driven structures in thin film flows. *Phys. Fluids* **2003**, *15*, S9. [CrossRef]
24. Dussaud, A.D.; Matar, O.K.; Troian, S.M. Spreading characteristics of an insoluble surfactant film on a thin liquid layer: Comparison between theory and experiment. *J. Fluid Mech.* **2005**, *544*, 23–51. [CrossRef]
25. Keiser, L.; Bense, H.; Colinet, P.; Bico, J.; Reyssat, E. Marangoni bursting: Evaporation-induced emulsification of binary mixtures on a liquid layer. *Phys. Rev. Lett.* **2017**, *118*, 074504. [CrossRef] [PubMed]
26. Kim, H.; Muller, K.; Shardt, O.; Afkhami, S.; Stone, H.A. Solutal Marangoni flows of miscible liquids drive transport without surface contamination. *Nat. Phys.* **2017**, *13*, 1105–1110. [CrossRef]
27. Pinto, R.L.; Le Roux, S.; Cantat, I.; Saint-Jalmes, A. Enhanced interfacial deformation in a Marangoni flow: A measure of the dynamical surface tension. *Phys. Rev. Fluids* **2018**, *3*, 024003. [CrossRef]
28. Kim, S.; Kim, J.; Kim, H.Y. Dewetting of liquid film via vapour-mediated Marangoni effect. *J. Fluid Mech.* **2019**, *872*, 100–114. [CrossRef]
29. Ma, X.; Zhong, M.; He, Y.; Liu, Z.; Li, Z. Fingering instability in Marangoni spreading on a deep layer of polymer solution. *Phys. Fluids* **2020**, *32*, 112112. [CrossRef]
30. Néel, B.; Villermaux, E. The spontaneous puncture of thick liquid films. *J. Fluid Mech.* **2018**, *838*, 192–221. [CrossRef]
31. Aljedaani, A.B.; Wang, C.; Jetly, A.; Thoroddsen, S.T. Experiments on the breakup of drop-impact crowns by Marangoni holes. *J. Fluid Mech.* **2018**, *844*, 162–186. [CrossRef]
32. Villermaux, E. Fragmentation versus cohesion. *J. Fluid Mech.* **2020**, *898*, P1. [CrossRef]
33. Hasegawa, K.; Manzaki, Y. Marangoni fireworks: Atomization dynamics of binary droplets on an oil pool. *Phys. Fluids* **2021**, *33*, 034124. [CrossRef]
34. Zhao, W.; Ma, H.; Ji, W.; Li, W.; Wang, J.; Yuan, Q.; Wang, Y.; Lan, D. Marangoni-driven instability patterns of an N-hexadecane drop triggered by assistant solvent. *Phys. Fluids* **2021**, *33*, 024104. [CrossRef]
35. Matar, O.K.; Troian, S.M. Linear stability analysis of an insoluble surfactant monolayer spreading on a thin liquid film. *Phys. Fluids* **1997**, *9*, 3645–3657. [CrossRef]
36. Schmitt, M.; Stark, H. Marangoni flow at droplet interfaces: Three-dimensional solution and applications. *Phys. Fluids* **2016**, *28*, 012106. [CrossRef]
37. Pototsky, A.; Thiele, U.; Stark, H. Mode instabilities and dynamic patterns in a colony of self-propelled surfactant particles covering a thin liquid layer. *Eur. Phys. J. E* **2016**, *39*, 51. [CrossRef]
38. Sergievskaya, I.; Ermakov, S.; Lazareva, T.; Guo, J. Damping of surface waves due to crude oil/oil emulsion films on water. *Mar. Pollut. Bull.* **2019**, *146*, 206–214. [CrossRef]
39. Ermakov, S.A.; Sergievskaya, I.A.; Gushchin, L.A. Damping of gravity-capillary waves in the presence of oil slicks according to data from laboratory and numerical experiments. *Izv. Atmos. Ocean. Phys.* **2012**, *48*, 565–572. [CrossRef]
40. Rajan, G.K. Dissipation of interfacial Marangoni waves and their resonance with capillary-gravity waves. *Int. J. Eng. Sci.* **2020**, *154*, 103340. [CrossRef]
41. Rajan, G.K. Solutions of a comprehensive dispersion relation for waves at the elastic interface of two viscous fluids. *Eur. J. Mech. B Fluids* **2021**, *89*, 241–258. [CrossRef]

42. Stetten, A.Z.; Moraca, G.; Corcoran, T.E.; Tristram-Nagle, S.; Garoff, S.; Przybycien, T.M.; Tilton, R.D. Enabling Marangoni flow at air-liquid interfaces through deposition of aerosolized lipid dispersions. *J. Colloid Interface Sci.* **2016**, *484*, 270–278. [CrossRef] [PubMed]
43. Japan Society of Mechanical Engineers (JSME). *JSME Data Book: Thermophysical Properties of Fluids*; JSME: Tokyo, Japan, 1983.
44. Rosen, M.J. *Surfactants and Interfacial Phenomena*; John Wiley & Sons: Hoboken, NJ, USA, 2004.
45. Mesyats, G.A.; Zubarev, N.M. The Rayleigh–Plateau instability and jet formation during the extrusion of liquid metal from craters in a vacuum arc cathode spot. *J. Appl. Phys.* **2015**, *117*, 043302. [CrossRef]

fluids

MDPI

Article

Fine Flow Structure at the Miscible Fluids Contact Domain Boundary in the Impact Mode of Free-Falling Drop Coalescence

Yuli D. Chashechkin * and Andrey Yu. Ilinykh

Laboratory of Fluid Mechanics, Ishlinsky Institute for Problems in Mechanics RAS, 119526 Moscow, Russia; ilynykh@ipmnet.ru

* Correspondence: yulidc@gmail.com; Tel.: +7-(495)-434-0192

Abstract: Registration of the flow pattern and the matter distribution of a free falling liquid drop in a target fluid at rest in the impact mode of coalescence when the kinetic energy (KEn) of the drop exceeds its available surface potential energy (ASPe) was carried out by photo and video recording. We studied the evolution of the fine flow structure at the initial stage of the cavity formation. To carry out color registration, the observation field was illuminated by several matrix LED and fiber-optic sources of constant light. The planning of experiments and interpretation of the results were based on the properties of the complete solutions of the fundamental equations of a fluid mechanics system, including the transfer and conversion of energy processes. Complete solutions of the system of equations describe large-scale flow components that are waves or vortices as well as thin jets (ligaments, filaments, fibers, trickles). In experiments, the jets are accelerated by the converted available surface potential energy (ASPe) when the free surfaces of merging fluids were eliminated. The experiments were performed with the coalescence of water, solutions of alizarin ink, potassium permanganate, and copper sulfate or iron sulfate drops in deep water. In all cases, at the initial contact, the drop begins to lose its continuity and breaks up into a thin veil and jets, the velocity of which exceeds the drop contact velocity. Small droplets, the size of which grows with time, are thrown into the air from spikes at the jet tops. On the surface of the liquid, the fine jets leave colored traces that form linear and reticular structures. Part of the jets penetrating through the bottom and wall of the cavity forms an intermediate covering layer. The jets forming the inside layer are separated by interfaces of the target fluid. The processes of molecular diffusion equalize the density differences and form an intermediate layer with sharp boundaries in the target fluid. All noted structural features of the flow are also visualized when a fresh water drop isothermally spreads in the same tap water. Molecular diffusion processes gradually smooth out the fast-changing boundary of merging fluids, which at the initial stage has a complex and irregular shape. Similar flow patterns were observed in all performed experiments; however, the geometric features of the flow depend on the individual thermodynamic and kinetic parameters of the contacting fluids.

Keywords: drop; impact; experiment; fine structure; substance transfer

1. Introduction

The registration of stably reproducible components of flows in the pattern of free-falling drop coalescence with a target fluid at rest had a great impact on the development of a number of sciences such as hydrodynamics, physics in general, mathematics, biology and many others. At first, these components were ring vortices [1] and multi-tiered vortex systems [2]. With the development of scientific photography, they were supplemented by more short-lived components such as cavity, crown, sprays, and splash [3]. Publications [1–3] quickly became widespread among physicists and scientists of other specialties. The establishment of the physical nature of a number of biological patterns [4], based on sketches of the cascade of vortices formation [2], had a great influence on the development of the theory of evolution and biology in general.

With time, scientific results in the study of droplet flows were being implemented in solving applied problems of ocean acoustics [5,6], creating new technologies in oil, bio, chemical, pharmaceutical, metallurgical and other industries. At the same time, the experimental technique was developing: that is, more and more powerful controlled light sources were being created [7]. The range of illuminating wavelengths was expanding (X-ray, optical and infrared waves are now used in experiments [8,9]). Photo, film and video recording techniques were improved as well: the shooting speed of modern video cameras exceeds several million shots per second [10,11].

The number of experimental and theoretical studies began to increase rapidly, and currently, the number of annual publications on the topic of "drop impact" is approaching a thousand [12,13]. The application of high-resolution instruments allowed studying in detail the evolution of the flow components identified in the first observations [3]: in particular, the change in the shape of the cavity and crown [14,15]. The patterns of the capillary waves on their surfaces [16] was traced as well. Important features in the droplet impact flows caused by the merging of both single droplets and synchronously contacting with the target fluid pairs and even groups of drops were recognized [17].

The hypothesis of "passive admixture", which is often used in fluid mechanics in general [18], was applied interpreting observations of the impact flows of a free-falling drop as well. It allows using images of the redistribution of substance density, which is an independent physical quantity and is described by the additional equation in the complete system of fundamental equations [19–21], as an indicator of the field of a flow velocity structure. The hypothesis was used to identify compact vortices at various phases of the flow created by a fallen drop in a fluid at rest [1,2,22].

At a low contact velocity of the drop, the shape and internal structure of the area of colored liquid in the cuvette resembles a "mushroom-shaped" dusty cloud of a ground-based nuclear explosion [23–25]. In this range of parameters, the drop inflows directly into the liquid, and the cavity is formed with a delay of about 10 ms [26].

Gradually, the number of experimental, analytical, and, with the development of hard and soft technologies, numerical studies of the shape of the cavern, crown, flying spray and capillary waves began to grow rapidly. The range of parameters, in which new regimes of the flows under study were discovered, has also expanded. The conditions for the rebound of single and multiple sequentially falling small drops with a diameter $D < 1.2$ mm were experimentally determined [27]. The momentum and kinetic energy of the drop were chosen as the basic parameters. In experiments [27], single small drops did not bounce off an undisturbed surface.

Side-view photographs of the developing cavern and the surrounding crown, which is converted into a gas bubble in some flow regimes, graphs of the cavity depth as a function of time and the maximum depth as a function of kinetic energy, and estimates of the total energy components are given in [28,29]. Recording the process of a drop (colored with a thymol blue solution) merging with water allowed distinguishing new features of the flow geometry in addition to the found dependences of the maximum cavity size on the diameter and contact velocity of a drop [30]. In particular, the difference between the irregular shape of the boundary of the colored liquid region and the classical smooth cavity shell was shown. Observations of the dynamics of the process of cavity formation in a deep liquid were continued in [31,32].

The technique developed in [33] for calculating the merging of liquids, taking into account the influence of surface tension, was incorporated into a numerical model of cavity formation to describe the shape of detached gas bubbles [34]. Cavity contour calculations were also independently performed in [35]. High-frequency oscillations of detached gas cavities play an important role in the processes of generating sound packets [6,36,37], which were recorded in both air [38] and aquatic environments [39].

In the impact coalescence mode, the deepening cavity is surrounded by a crown rising above the surface of the fluid that is produced by a growing hollow cylindrical jet with a jagged outer edge [3,15]. Sequences of small droplets—sprays flying from the tops of thin spikes at the tops of the teeth—were visualized in [40]. At the initial stage of the merger, thin layers of liquid—ejecta and lamellae—fly out from the contact of the liquids domain [41]. The expanding crown generates a group of annular capillary waves, whose patterns of propagation over relatively large distances are satisfactorily described by linear theory [42].

The synchronicity of the generation of a thin splash flying off from the cavity bottom and a submerging jet with a vortex head in the center of the cavern was noted in experiments [43]. The evolution of the flow as the drop velocity increases was traced in [44]. Calculations of cavity geometry, regime map, and comparison results with observations of pigmented droplet merging were performed in [45,46].

The high-speed video cameras of a new generation [10] allowed tracing the fine details of the flow pattern during the initial contact of a drop with a stationary target fluid. The fast evolution of the air gap shape between the merging drop and the deforming surface of the liquid, the process of formation of isolated gas bubbles, and their sequences was traced. Illumination of the liquid coalescence domain was carried out by electromagnetic waves in the optical [47,48] and X-ray ranges [49]. Depending on the size, shape, and dynamic state of the falling drop, a "necklace" of gas bubbles [47], a single central bubble [36,46], or a complex system of bubbles, the parameters of which depend on the size, speed, and shape of the drop, can be formed in the merging domain [9]. The generation of an initial (impact) high-frequency sound packet is associated with oscillations of small gas bubbles [37,39].

Estimates of the influence of viscosity on the formation of a cavern, as well as the possibility of using experimental results in describing meteorite craters on icy planets, are contained in [50].

A number of studies have studied the influence of the depth of the target fluid on the general structure and dynamics of impact flows. The formation of a cavity in a deep fluid was traced in [32]. Comparisons of the observed pattern of distribution of a tinted drop with calculations of the shape of the free surface allowed identifying different modes of formation of a "thin" and "thick" splash and determining the critical conditions for the restructuring of the flow depending on the values of the Weber and Ohnesorge numbers [8,51]. Numerical simulations have also been used to study the spreading behavior of a composite droplet [52]. A review article [53] is devoted to the analysis of the processes of a drop spreading in a thin layer of liquid and the determination of heat transfer parameters.

Calculations of the axisymmetric flow pattern, taking into account the contribution of the ASPE transformation processes during the elimination of the free surface during the merging of liquids and the formation of a new one with the flow evolution, were carried out in [54,55], following [28].

In most experiments, the flow pattern was studied when a drop falls normally onto a smooth surface of a liquid at rest. However, a number of works present the results of experimental studies of the influence of the shape of the perturbed surface, the type and dynamics of the flow on the pattern of merging of a freely falling drop. Photographs of the merging of a drop ejected along an inclined trajectory with a horizontal surface of a quiescent fluid and a classification of flow regimes are presented in [56]. The pictures of the collision of a drop with a diameter of $2.5 < D < 4.1$ mm, flying in a wind tunnel along an inclined trajectory at a speed of $U = 6.7$ m/s, with a liquid in a cuvette at its bottom, was recorded in the side and frontal projections [57]. The distortion of the cavity shape and the elongation of the spikes on its edge were traced.

The dependences of the shape of the cavity, the inclination of the splash and the direction of the preferential emission of the sprays on the position of the point of primary contact on the contour of the liquid surface deformed by a solitary wave are given in [58]. Calculations of the shape of the wave surface of a liquid by a deformed falling drop were performed in [59]. The influence of the composition of the gaseous medium on the formation of a cavity and a cloud of spray from a drop falling into a thin layer of liquid at the end of a rotating disk was studied in [60,61].

Systematic experiments have allowed identifying two modes of merging of a freely falling drop with a target fluid: intrusive, when the liquid of the drop initially smoothly inflows into the thickness of the target fluid, and impact, in which a cavity begins to form from the moment of initial contact. The condition for restructuring the flow pattern is determined by the ratio of the components of the total energy of the falling drop: kinetic energy KEn and ASPE $R_{En} = En_k/En_\sigma$. Here, $En_k = MU^2/2$ is the kinetic energy, and $En_\sigma = \sigma_d^a S_\sigma$ is the ASPE of a droplet with a diameter D, surface area S_σ, and mass M falling at a speed U. Here, the coefficient of surface tension at the interface of the liquid droplet with the air environment is σ_d^a. The intrusive mode is observed at $R_{En} < 1$ [62].

In the impact mode, when the drop kinetic energy exceeds ASPE and $R_{En} > 1$, thin jets are formed in the domain where a drop merges with a thin layer of liquid on a solid substrate [63]. Thin jets were also observed in a deep fluid. The colored fibrous traces of fine jets form linear and reticular meshes the surface of the cavity and crown [64].

The formation of thin jets into the free-falling droplet spreading pattern is explained by the intrinsic features of the inhomogeneous distribution of internal energy in fluids with a free surface [65] and the properties of complete solutions of the fundamental equations system [19–21] for inhomogeneous fluids.

In today's description, one of the types of thermodynamic potentials, namely, free enthalpy—the Gibbs potential, is chosen as the main physical quantity of a fluid medium (liquid, gas, plasma) [66]. Coefficients of the Gibbs potential differential form $dG = sdT + VdP + S_\sigma d\sigma + \mu_i dS_i$ determine the traditional thermodynamic physical quantities of an inhomogeneous droplet fluid with a free surface area S_σ. Here, V is the specific volume, $P(t,\ x_1,\ x_2,\ x_3)$ is pressure, $T(t,\ x_1,\ x_2,\ x_3)$ is temperature, $S_i(x_1,\ x_2,\ x_3)$ is the concentration of dissolved substances or suspended particles, μ_i is chemical potential and $\rho(t,\ x_1,\ x_2,\ x_3) = 1/V$ is density. All of the mentioned quantities have a clear physical meaning and are available for observation with an error estimation.

In modern fluid mechanics, some parameters of the medium, such as density and pressure, are considered to be quantities of a mixed, namely, mechanical and thermodynamic nature. Empirical relations between thermodynamic quantities constitute equations of state [66–68], among which the main one is the dependence of density $\rho = \rho(P,\ T,\ S_i)$ on pressure P, temperature T, and the concentration of dissolved substances and suspended particles S_i. The thermodynamic state of the medium is characterized by the spatial distributions of Gibbs potential and its derivatives, which include the density, temperature, salinity and pressure, which change over time.

Along with thermodynamic quantities, the medium is also characterized by kinetic coefficients that determine the molecular transfer of momentum (dynamic μ and kinematic viscosity $\nu = \mu/\rho$), heat (thermal or temperature conductivity coefficient κ_T) and substance (diffusion coefficient κ_S) as well as the transfer parameters of other physical quantities (velocity of sound, index of refraction and others). In weakly dissipative media, which include a large number of fluids and gases found in natural and industrial conditions (in particular, water, aqueous solutions of salts or air), kinetic coefficients have small values.

Complete solutions of the system of fundamental equations [19–21] for weakly dissipative media containing small coefficients at higher derivatives are searched by the theory of singular perturbations [69]. The calculations have shown that the complete solutions of the system of fluid mechanics equations contain functions of two classes—regular in a small parameter and singularly perturbed [70]. In periodic flows, regular solutions describe waves, and singular solutions describe the accompanying components that are fibers and highly gradient interfaces forming the fine structure of the medium. The complete dispersion relations for surface waves and ligaments in a viscous stratified liquid are given in [71]. The analysis of the intrinsic scales of singular solutions complementing inertial, internal, acoustic and hybrid waves in the fluid thickness is carried out in [72].

In the experiment, the fine structure of droplet flows on the surface and in the thickness of the target fluid was previously visualized in [64]. The fine components of flows (fibers or jets, trickles) are formed as a result of the internal or available potential surface energy conversion into other forms. They are described by singular solutions of the fundamental equations system [71,72].

The theory and methodology of studying fluid flows was created within the framework of the "continuous medium" concept, all the physical properties of which as well as parameter of flows were described by continuous functions. This approach does not agree with the actually observed discrete atomic–molecular and finer nuclear structure of matter. Atoms and molecules forming real liquids or gases are combined into various associates of physical and chemical nature. Various structure combinations are registered in the experiments such as complexes, clathrates, clusters, and voids with individual atoms, with physical and chemical bonds [73,74]. The observations and calculations show that the internal energy and one of the forms of its representation, namely the Gibbs potential, are distributed non-uniformly in the fluid.

The linear scales of the associates are in the $10^{-8} < \delta < 10^{-6}$ cm range, where the lifetime of single elements is $10^{-12} < \tau_e < 10$ s. Each of the elements of the fluid structure is characterized by its own internal energy, which accumulates during its formation and is converted into other forms during the rearrangement of the structure and the elimination of the elements' boundaries, which ensures the fluidity of liquids [66,70]. The available surface potential energy at the liquid boundary with air or vacuum (ASPE [65]) is the most important. It is concentrated in a near-surface layer with a thickness of $\delta_c \sim 10^{-6}$ cm in droplet and target fluids. The accumulation of internal energy occurs rather slowly with characteristic free surface formation in macroscopic flows for times of the order of $10^{-3} < \tau < 10^{-2}$ s and more.

By methods of optical and X-ray reflectometry and atomic force microscopy, it was found out that the density, the dielectric permittivity, and the dipole moment in the fluid thickness and in a structurally distinguished near-surface layer with a thickness of the size order of a molecular cluster ($\delta_\sigma \sim 10^{-6}$ cm) differ noticeably [75,76]. A more complex distribution of free enthalpy could be observed in a thin surface layer with a thickness of the order of molecular size $\delta_c \sim 10^{-8}$ cm, where supramolecular structures are expressed even more clearly. In a surface layer with a thickness $\delta_b \sim 10^{-8}$ cm, the self-ionization of matter can occur due to the anisotropy of atomic–molecular interactions.

When the free surfaces are eliminated during the coalescence of liquids, the ASPE quickly transforms into other forms (at a coalescence velocity of about 1 m/s per time $10^{-10} < \tau < 10^{-8}$ s). The processes of ASPE conversion form ligaments (fast thin jets [70]), colored traces of which are observed when a free-falling drop merged with a fluid at rest in the impact mode at $En_k > En_\sigma$ [62].

The inclusion of the Gibbs potential in the description of liquids and gases flows allows taking into account the action of four energy transfer mechanisms: with currents at the velocity of U, with the group velocity of various waves c_g^w, the dissipative–diffusion one with coefficients ν, κ_T, κ_S and the internal energy conversion effects. The release and accumulation of the internal energy processes includes, in particular, transformations of available potential surface energy En_σ of the liquid as a whole and its single structural components, which are complexes, hydrates, clathrates, clusters, voids and other combinations of atoms and molecules of physical and chemical nature [75,76].

Thin fast jets and longitudinal oscillations of the expanding circular boundary of the drop coalescence domain with a thin fluid layer on the surface of a slide were visualized in [11]. In the experiments, it was observed that the jets accelerated when they crossed the contact domain boundary of merging liquids in a cavity, which was formed in a deep liquid [77]. The evolution of the pattern of thin jets penetrating the bottom of a growing cavity at the initial stage of fluids coalescence was first traced in [78]. The purpose of this study is to visualize the flow fine structure in the vicinity of the contact domain boundary of a droplet coalescing with deep water. In the impact mode, a cavity in a deep pool begins to form at the initial contact of the droplet boundary with the free surface of the target fluid at rest if $En_k > En_\sigma$: that is, the drop KEn $En_k = MU^2/2$ exceeds the ASPE $En_\sigma = \sigma_d^a S_\sigma$.

2. Parametrization

The experimental methodology was developed taking into account the definition of a liquid as a fluid medium, the properties of which are characterized by the distributions of the Gibbs potential, its derivatives defining thermodynamic quantities, kinetic and other physical coefficients [66,67]. The flow is defined as the combined transfer of momentum, energy, and matter, which causes the corresponding changes in the physicochemical parameters of the medium [70]. The universal classification of flow components is based on the description of complete solutions of linearized and weakly nonlinear forms of the fundamental equations systems [70,71]. They are constructed by the theory of singular perturbations, taking into account the explicit compatibility condition. The classification includes large-scale components (waves, vortices, jets) and thin ligaments, the spatial structure of which is characterized by their intrinsic scales [72,79].

The Gibbs potentials of the droplet G_d, the air medium G_a and the target liquid G_t are included as the main dimensional parameters characterizing the flows under study, considering the complex and tunable internal medium structure, which affects the dynamics and energy of droplet flows. The indices indicate the belonging of the parameter medium. The media are also characterized by density $\rho_{d,a,t}$, kinematic $\nu_{d,a,t}$ and dynamic $\mu_{d,a,t} = (\rho\nu)_{d,a,t}$ viscosities; full σ_d^a, σ_t^a and normalized surface tension coefficients based on the density of fluids $\gamma_d^a = \sigma_d^a/\rho_d$, $\gamma_t^a = \sigma_t^a/\rho_t$; and the equivalent diameter D, surface area S_σ, volume V_d, mass $M = \rho V_d$, momentum $p_d = MU$ and velocity U of the droplet at the moment of initial contact with the target liquid.

The influencing parameters include extensive kinetic energy (KEn) $En_k = MU^2/2$ and available surface potential energy (ASPe) $En_\sigma = \sigma_d^a S_\sigma$, which are formed due to the anisotropy of atomic–molecular interactions in the drop shell. The ASPE is contained in a thin near-surface layer with a thickness of the order of the molecular cluster size $\delta_\sigma \sim 10^{-6}$ cm, and its density is $W_d^\sigma = En_\sigma/V_\sigma$. The extensive KEn density is $W_d^k = En_k/V_d$. The ratio of the drop energy components $R_{En} = En_d^k/En_d^\sigma$ can be both small and large, and the ratio of their densities $R_W = W_d^k/W_d^\sigma \sim \delta_\sigma/D$ is a small value under the conditions of these experiments.

The energy parameters of the drop, falling in a gravitational field with a free-fall acceleration g, also include potential energy $En_p = MgD$ in scale D. An additional parameter is the diffusion coefficient of the pigment, used for tinting the drop, in the target liquid κ_S.

From the system of equations and physically valid boundary conditions, it follows that the basic group of linear scales, which are determined by the physical properties of media, includes the capillary–gravitational ratio $\delta_g^\gamma = \sqrt{\gamma/g}$. It belongs to the dispersion equation of short surface waves [19] as well as to the dissipative–capillary scale $\delta_\gamma^\nu = \nu^2/\gamma$. The group of linear scales depending on the droplet velocity includes viscous velocity $\delta_U^\nu = \nu/U$, capillary $\delta_U^\gamma = \gamma/U^2$ and diffusion $\delta_U^{\kappa_S} = \kappa_S/U$ scales.

Accordingly, the first part of the time scales of the task comprises only the parameters of the medium—$\tau_\gamma^\nu = \nu^3/\gamma^2$, $\tau_g^\gamma = \sqrt[4]{\gamma/g^3}$, the second one contains the drop size—$\tau_\gamma^D = \sqrt{D^3/\gamma}$, $\tau_\gamma^{\nu D} = \nu D/\gamma$, the third part embraces its contact velocity, that is kinematic $\tau_g^U = U/g$, as well as dynamic parameters determined by the drop size $\tau_U^D = D/U$ and the thickness of the shell $\tau_U^\sigma = \delta_\sigma/U$. The duration of the transferring matter and surface energy processes of the merging drop is determined $\tau_U^D = D/U$ by the drop size and the thickness of the subsurface layer $\tau_U^\sigma = \delta_\sigma/U$.

The process of converting ASPE into other forms with the elimination of the free surface of the merging liquids proceeds in a short time of the order of $\tau_U^\sigma \sim \delta_\sigma/U \sim 10^{-8}$ s for typical conditions of the experiments with free-falling droplets. Rapid processes of ASPE transformation into other forms at the circular boundary of the fluid coalescence contribute to the formation of thin jets and the generation of capillary waves in the target fluid [11,42,78].

A large number of scales of the same dimension reflect the diversity and complexity of processes occurring in a wide range of scales—from supramolecular ones of the order of

$\delta_\sigma \sim 10^{-6}$ cm in the processes of release and accumulation of ASPE to the full size ones of the flow domain.

The relations of scales of the same dimension define a set of traditional dimensionless parameters, which includes the following numbers: Reynolds $\mathrm{Re} = D/\delta_U^\nu = UD/\nu$, Froude $\mathrm{Fr} = En_k/En_p = U^2/gD$, Bond $\mathrm{Bo} = D^2/(\delta_g^\gamma)^2 = gD^2/\gamma$, Ohnesorge $\mathrm{Oh} = \sqrt{\delta_\gamma^\nu/D} = \nu/\sqrt{\gamma D}$, Weber $\mathrm{We} = D/\delta_U^\gamma = DU^2/\gamma$, and Schmidt $\mathrm{Sc} = \nu/\kappa_S$. The relations of the energy components form additional dimensionless combinations $\mathrm{R}_{En}^{k,\sigma} = En_k/En_\sigma$ and $\mathrm{R}_W = W_d^k/W_d^\sigma \sim \delta_\sigma/D << 1$.

The intrinsic scales of the task determine the requirements for choosing the size of the observation area, the spatial–temporal resolution of the instruments, and the duration of recording the flow pattern. Dimensionless relations allow us to assess the relative contribution of processes of different nature to the overall flow pattern and compare the conditions of independent experiments.

3. Experimental Setup

The experiments were carried out on a modified stand for studying the fine structure of fast processes (FSPs), which is part of the Unique Research Facility "HPC IPMech RAS" [80]. The installation, which is presented in Figure 1, included a cuvette with a target fluid 1, time delay unit 2, fiber-optic light source with fiber light guides 3, a recording photo or video camera 4, matrix LED sources 5, monitor and a computer for controlling the experiment and for data collecting 6, drop fluid discharge valve 7, drop generating dispenser with normally cut capillary 8, and a photodetector 9 that marks the pass of a drop intersecting the light beam. White dashed lines are the vertical trajectory of the drop and the inclined line of sight of the optical system.

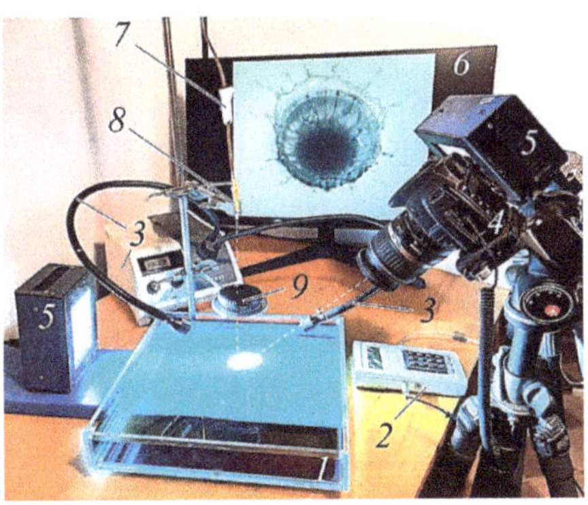

Figure 1. Experimental setup: 1—cuvette with a target fluid, 2—time delay unit, 3—fiber-optic light source with fiber light guides, 4—photo or video camera, 5—matrix LED sources, 6—monitor and a computer, 7—discharge valve, 8—dispenser, 9—photodetector.

The drop formed in a dispenser *8* with a replaceable capillary of a diameter $0.8 < d_o < 4$ mm with a flat cross-section broke off under the gravity and fell on the surface of partially degassed tap water in a target cuvette *1* of the size $10 \times 10 \times 7$ or $30 \times 30 \times 5$ cm^3. The pattern of coalescence of single water drops, aqueous solutions of alizarin ink or metal salts (potassium permanganate $KaMnO_4$, copper $CuSO_4$ and iron $FeSO_4$ sulfates) was studied. The target fluid, which was partially tinted with a drop pigment, was renewed after each experiment.

The registration system included an Optronics CR 300 × 2 video camera or a Canon EOS 350D camera, the sight line position of which was selected in favor of the greatest clarity of the boundaries of the recorded pattern components. Two directions of the sight line were mainly implemented: at an angle to the horizontal $\vartheta = 0°$ when registering the flow pattern in the vertical plane or $\vartheta = 65°–70°$ when observing the free surface (the distance from the lens to the center of the flows was from 12 to 40 cm, the pixel sizes in the experiments ranged from 10 to 50 microns). The shutter speed was chosen to be minimal for a given level of spatial resolution, the size of the recorded area and the required illumination.

The fluids merging domain was illuminated by two Optronis MultiLED matrix constant light sources with a luminous flux of 7700 lm and also two fiber-optic sources of the Schott KL2500LCD (the power was 250 W, the luminous flux was 1300 lm). Special attention was paid to the organization of lighting in order to exclude the shading of the examined components of the flow or the effects of total internal reflection.

Traditionally, the technique of "back-light" illumination with a bright continuous [81] or pulsed [82] light source with an arc, gas-discharge or stroboscopic lamp is used in the droplet flows observation experiments. Recently, LED [44], fiber-optic [62] and laser [83] sources have become widespread. The emitted light is directly projected on to the studied flow domain, and sometimes, it passes through a semitransparent screen (for example, tracing paper [47]) for more uniform illumination of the field of view. To increase the spatial resolution of the flow pattern components in the fluid thickness, a light laser knife and shadowgraphy were used as well [84].

In these experiments, combined lighting, including multi-element LED and fiber-optic sources, was used to ensure the contrast of the droplet residue image, the continuously changing cavity shape and the distribution of the droplet matter in the liquid. The location of the light sources was chosen based on the maximum contrast of the residue image at the bottom of a continuously growing cavity. The simultaneous rapid change in the location and shape of the flow components led to some defocusing of the image, which made it difficult to conduct further analysis. The stability of the vertical position of the contact line moving along the surface of a thin fluid layer on the glass slide contributed to the improved clarity of the video shots [11].

When the detached drop crossed the light ray, the signal of the photodetector launched the registration system with the selected time delay. The droplet velocity was estimated by its displacements in the last shots before the contact with the target liquid and the duration of the signal delay from the photodetector to start the flow pattern recording. The image was scaled according to photos of the marker photos with size from 1 to 10 mm. Photometry and data processing were carried out in the Matlab and Kompas envelope (CAD).

4. Main Results

In each series of experiments, the individual adjustment of several sources was used instead of the traditional "back-light" illumination for the flow pattern lighting. Simultaneous visualization of the shape of a rapidly deforming free surface during the formation of a cavity, a crown, a veil on its edge, as well as the registration of the drop substance position was organized. The lighting technique allowed us to carry out both color and black-and-white registration of the flow pattern. To control the universality of the proposed flow structural features classification, the spreading of uniformly colored drops of dilute solutions of alizarin ink, potassium permanganate, copper and iron sulfates, as well as tap water was studied. In all cases, the image geometry of both the moving free surface and the drop matter distribution in a target fluid was recorded. While the overall fine flow structure was preserved, the individual properties of small elements turned out to depend on the composition of the media.

4.1. Drop Spreading of Alizarin Ink Solution in Water

The evolution of the fine structure of the matter transfer pattern during the free-falling drop coalescence of a dilute 1:200 solution of blue alizarin ink is illustrated by the video shots shown in Figure 2. The line of sight in these experiments is inclined at an angle

$\vartheta = 65°$ to the horizon. The shooting speed is 4000 fps, the shutter speed is 1/5000 s, and the spatial resolution is 30 microns/pixel (μm/pix).

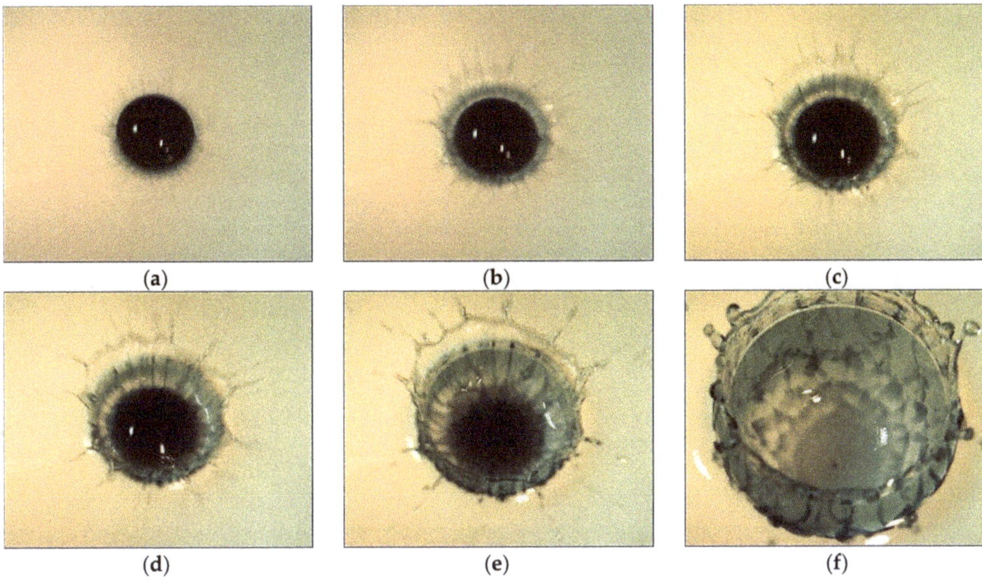

Figure 2. Evolution of the drop fluid distribution pattern (an aqueous ink solution diluted in 1:200 ratio) in water ($\rho_d = 1$ g/cm^3, $\sigma_d^a = 73$ g/s^2, $\nu = 0.01$ cm^2/s, $D = 4.3$ mm, $U = 3.1$ m/s, $En_\sigma = 4$ μJ, $En_k = 200$ μJ, Re = 13,300, Fr = 230, We = 570, Bo = 2.5, Oh = 0.0018, $R_{En} = E_k/E_\sigma = 48$, $R_W = 1.66 \cdot 10^{-3}$). (**a**)—$t = 0.2$ ms, (**b**)—$t = 0.45$ ms, (**c**)—$t = 0.7$ ms, (**d**)—$t = 1$ ms, (**e**)—$t = 1.7$ ms, (**f**)—$t = 6$ ms.

The primary contact of the drop is accompanied by the release of a veil (a gray strip $\Delta r_v = 0.45$ mm wide along the main diameter $d_{cr} = 5.6$ mm), which is pierced by thin spikes. A sequence of drops is thrown from the tops of the spikes (Figure 2, $t = 0.2$ ms). The deviation of the pattern from strict symmetry is due to the inclination of the line of sight and some asymmetry of the merging drop shape.

The lengths of separate jets, which are elongated traces of thrown droplets, is four times the width of the strip. In general, the trace pattern radially diverges, but some colored jets are located at an angle to the local radius vector. As it follows from the analysis of variations in the illumination distribution pattern along an arc $\Delta r_l = 0.2$ mm away from the contact line, the thickness of the droplet traces is $d_s \sim 50$ microns, the length reaches $l_s \sim 1.8$ mm, and the inclination angle from the direction of the local radius vector lies in the range $0 < \varphi < 35°$. The dispersion of the angular positions of the spray trajectories can be caused by variations in the angle between the shells of the veil edge in the vicinity of the spike from which the small sprays are thrown out.

The general asymmetry of the flow pattern is because of the difference in the shape of the incoming drop and the spherical one due to Rayleigh oscillations and capillary waves traveling along its surface [85]. The general gray background of the veil and darker lines, which are traces of the thrown liquid layer and single ligaments (thin jets that arise at the contact line of the drop with the target liquid), indicate the complexity of the emerging flow pattern. It expresses both two-dimensional continuous veil and fast spikes that are three-dimensional components.

The ratios of the sizes (length and diameter) of individual strokes, which are the extensions of spikes (blurred images of thrown droplets—their velocity is $u_s = l_s/\Delta t$, $l_s \sim 1.8 \div 2$ mm—length of droplet traces, $\Delta t = 0.25$ ms—exposition time, $u_s \sim 7 \div 8$ m/s), show that their velocity noticeably exceeds the drop velocity [86]. The rapid flow of liquid in the jets is provided by the additional ASPE released when the free surface of the merging liquids is eliminated. This energy is transformed into other forms, including the energy of mechanical motion.

During the drop coalescence, the flow pattern, in which one can distinguish the veil in the upper left part and the crown, becomes more complicated (Figure 2b, $t = 0.45$ ms). The visible edge of the tinted domain (the trace of the drop residue) loses its smoothness. Colored fibers come out of individual protrusions, which spread at the cavity bottom, go through the crown wall and the veil, and form spikes on its outer border. The continuity of the jet traces (linear elements of the flow geometry) indicates the structural stability of the developing flow. Jets are formed at the initial coalescence phase and persist for a long time.

The radial inhomogeneity of the image brightness allows us to confidently identify at $t = 0.45$ ms the individual flow components. They include the drop residue with a diameter of $d_r = 4.74$ mm and the cavity bottom. A dark circular line is the inner boundary of the crown wall; the circular boundary line is the crown outer edge, which is continuing by veil. The outer edge of the veil is tightening and thickening; individual fibers on its surface and their extensions are spikes coming out from the tops of the teeth. Strokes, which proceed the spikes, are traces of fast flying sprays.

The jet wakes (dark fibers) are adjacent to the protrusions at the visible contact boundary of the drop residue and the deepening cavity. The slope of the cavity bottom explains some clarity loss of the image of the crown moving boundary and the drop residue spreading along the cavern bottom. However, here, the extension of jet traces connecting the spikes on the veil edge with the boundary of the drop residue is well traced.

As the flow evolves, the diameter of the central spot slowly grows (the drop remnant spreads along the cavity bottom). At the same time, the adhesion of the colored fibers to the boundary of the merging liquids is more and more clearly expressed (Figure 2c, $t = 0.7$ ms).

With time, the diameter of the droplet residue (the central undisturbed spot) decreases somewhat due to the formation of new tiers at its edge of a discrete distribution pattern of the droplet matter (Figure 2d, 1 ms). The boundary between the drop residue and the cavity bottom becomes more and more indented and thickened. There are separate protrusions on it that are analogs of spikes on the veil edge in the air. Over time, the diameters of the protrusions grow, and the contrast decreases.

The elements of the fibrous reticular structure are traced inside the tinted central spot with a diameter of $d_r = 5.45$ mm at $t = 1.7$ ms. Gradually, the linear structure on the outer cavity boundary is replaced by a reticular one; at $t = 6$ ms, triangular cells appear on the crown walls (Figure 2f). Five tiers of cells are allocated inside the cavity.

The photometry results of the relative illumination $I(l_\varphi)$ in the flow pattern presented in Figure 2e at $t = 1.7$ ms and the dependence of the relative energy spectrum $S(\lambda)$ on the scale λ are shown in Figure 3. The illumination values $I(l_\varphi)$ are determined in a circular spot with a diameter $d_p = 30$ µm, which moves a distance l_φ along the arcs of circles with a radius $r_\varphi = 3.15$ and 2.6 mm.

In the middle of the cavity sidewall at $r_\varphi = 3.15$ mm, where linear structures in the distribution of the drop matter are distinguished, the peaks on the spectrum correspond to scales $\lambda = 0.42$, 0.47, 0.54, 0.63 and 0.8 mm. In the transition zone between the drop residue and the cavity bottom at $r_\varphi = 2.6$ mm, where reticular structures with triangular cells are traced, the peaks at scales $\lambda = 0.39$, 0.49, 0.6, and 0.71 mm are distinguished in the spectrum. The smallest scales characterize the thickness of single fibers.

As can be seen in Figure 2, and a number of subsequent ones, the color image of the drop coalescence expresses a number of structural components identified in [67,77], which were not previously studied systematically. For convenience of further reading, photographs of the flow are given in Figure 4 with designations of the components under discussion and an indication of their sizes.

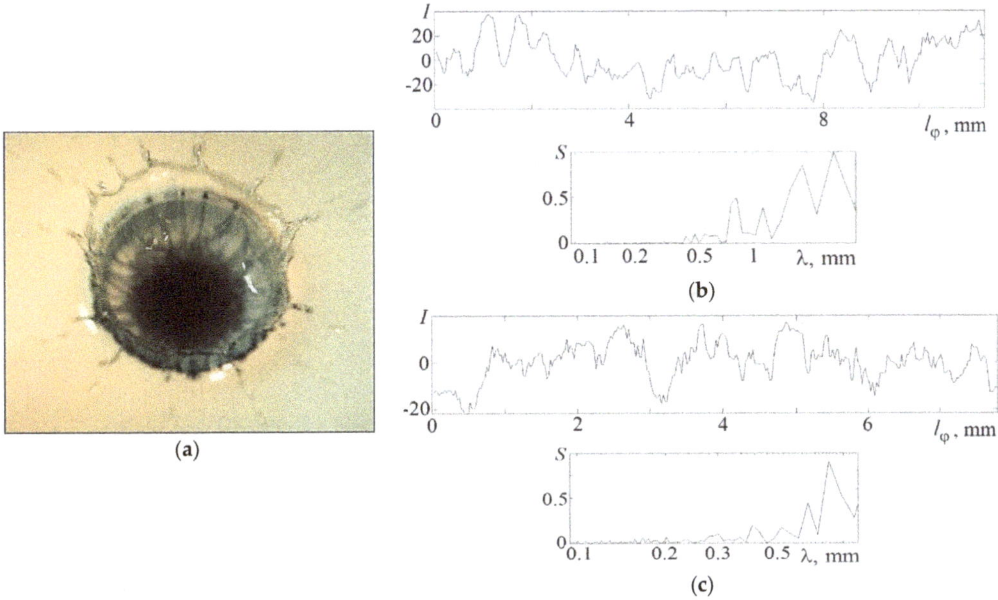

Figure 3. The fine flow structure: (**a**) shot at $t = 1.7$ ms; (**b**,**c**) relative illumination distributions $I(l_\varphi)$ along the $r_\varphi = 3.15$ and 2.6 mm circumference arcs and their spectra $S(\lambda)$.

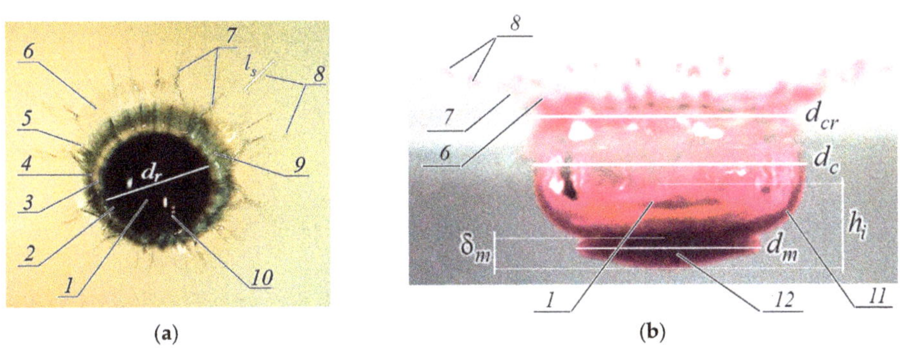

Figure 4. Structural components of the flow pattern in the impact mode of droplet coalescence: (**a**,**b**) frontal and side view (1 is the remainder of a coalescing drop with diameter d_r; 2 is the boundary of the fluid merging domain; 3 is the bottom of cavity; 4 is the area of the cavity to the crown transition; 5 is the outer boundary of the crown edge; 6 is a veil; 7 are spikes; 8 are traces of droplets (sprays); 9 are ligaments—fibrous traces of colored jets on the walls of the cavity and crown; 10 is the system of short capillary waves on the bottom of the submerging drop, 11 is densely colored layer at the cavity, 12 is intermediate layer. (**a**)—$t = 0.7$ ms. (**b**)—$t = 2.25$ ms.

In the frontal image of the drop spreading pattern in Figure 4a, the following structural components are indicated: the remainder of a coalescing drop *1* with diameter d_r at the bottom of the cavity; the boundary of the fluid merging domain of complex shape *2*; the bottom of cavity *3*; the area of transition of the cavity to the crown *4*; the outer boundary of the crown edge *5*; *6* is a veil; spikes are in *7*; the long stretched images of rapidly flying droplets (sprays) *8*; fibrous traces of colored jets on the walls of the cavity and crown—ligaments *9*; a system of short capillary waves on the bottom of the submerging drop *10*. Wakes of jets are continuous fibers colored by the substance of the drop, which are stretched from the boundary of the area where liquids merge to the edge of the crown and the veil. On the edge of the veil, continuations of the jets form spikes *7*, from the tops of which small, partially colored droplets *8* fly out sequentially. The lengths of several strokes (traces of small droplets) *I–III*, highlighted on the right side of the figure, are equal to $l_s = 1.77, 1.87,$ and 1.82 mm, respectively. Some spikes are turned into the flow center. The sprays flying from their tops form capillary waves on the back surface of the submerging drop *10*. Traces of the spray collision ЫI with the backside of the submerging drop were also visualized in [87].

In the lateral projection, only part of the flow structure components designed with the same numbers is presented. Sprays from different spikes and remnants of spike fly synchronously. Droplets on the tops on the edge of the crown with a diameter of d_{cr} are visible. Its outer boundary is marked by 5 in Figure 4a

Under synchronously flying sprays from different spikes and remnants of spikes, droplets on the tops are visible on the edge of the crown with a diameter of d_{cr} (its outer boundary is marked by *5* in Figure 4a). In the underwater part of the flow, there is a cavity with a diameter of d_c (corresponding to 4 in Figure 4a).

In the bottom of the cavity, an "intermediate layer" 12 with a diameter d_m and a height δ_m is adjacent. A dark color and a clear boundary structurally distinguish the "intermediate layer". The symbol h_i indicates the height of the densely colored layer in the lower part of flow 11, which includes the remnants of the drop at the cavity bottom and the adjacent intermediate layer.

The evolution of flow parameters is illustrated by the graphs presented in Figure 5. In the phase of active cavity growth, the outer diameter of the crown, represented by curve *1* in Figure 5a, grows monotonically and is approximated by the exponential function $d_{cr}(t) = 4.5t^{0.5} + 4$.

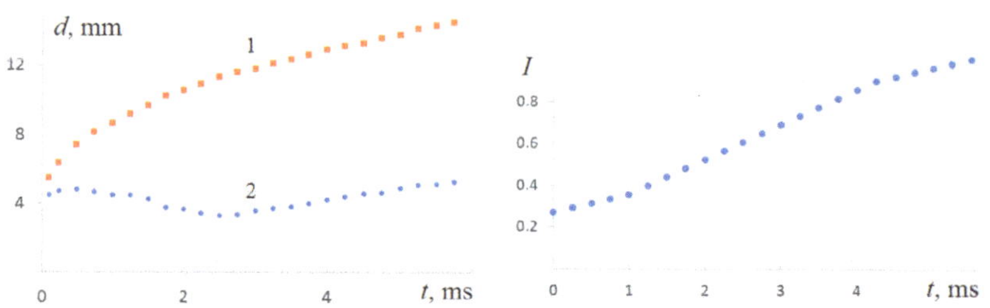

(a) **(b)**

Figure 5. Evolution of parameters of the ink pigmented drop coalescing with water: (**a**) curves *1*, *2* are diameters of the crown $d_{cr}(t)$ and the central colored spot $d_r(t)$; (**b**) changes in illumination in the center of the flow inside a circle with a diameter $d_p = 0.8$ mm.

The size of the central densely colored spot d_r changes non-monotonically with time. At first, it increases somewhat; then, in the phase of merging of the bottom part of the drop, it decreases somewhat and begins to increase again at $t > 2.5$ ms. The observed increase in the size of the spot is explained by the spreading of the previously formed intermediate layer, colored by the pigment of the drop, along the bottom of the growing cavity.

In this case, due to the dilution of the pigment and the spreading of the colored liquid, the illumination in the center of the cavity $I(t)$ gradually increases (Figure 5b). Three sections are highlighted in the dependence graph $I(t)$: slow growth at $t < 1$ ms, accelerated at $1.0 < t < 4$ ms and again slow at $t > 4$ ms.

The change in the rate of evolution of illumination is explained by the complexity of competing diffusion and hydrodynamic processes occurring on the cavity wall and in the intermediate layer under its bottom. With further evolution, the protrusions on the cavity sidewall are transformed into short jets, which, when the cavity collapses, form elongated colored loops in neutral and chemically reacting contacting media [88].

The pattern features of a drop spreading, colored with blue alizarin ink, in the fluid thickness are illustrated by frames from the video of the side view flow pattern shown in Figure 6. Drops begin to deform the free fluid surface, creating a cavity, from the moment of the initial contact. At the same time, the drop liquid leaks in thin jets through the cavity bottom and forms an intermediate layer in which the fibers of the colored liquid are separated by interfaces of the target fluid [78]. Over time, diffusion equalizes the difference in densities of the pigmented liquid of the drop and the target fluid, evens out the pigment distribution and forms an intermediate layer adjacent to the bottom of the cavity. The formed intermediate layer of liquid of its own density, separated from the receiving liquid by a high-gradient shell, has also been previously visualized in [84,89].

(a) (b) (c)

(d) (e) (f)

Figure 6. Side view of the flow pattern and the impurity distribution when a drop of ink solution is immersed in water (the experimental conditions are shown in Figure 2; the tag length is 5 mm). (**a**) $t = 0.75$ ms, (**b**) $t = 1$ ms, (**c**) $t = 1.75$ ms, (**d**) $t = 2.75$ ms, (**e**) $t = 5$ ms, (**f**) $t = 14.25$ ms.

In the first shot of the flow pattern in the lateral projection shown in Figure 6a, a dark strip with a flat bottom $d_c = 5.6$ mm in diameter, and $h_c = 0.56$ mm in height, adjacent to the free surface, visualizes the cavity partially covered with droplet liquid fibers.

Adjacent to the cavity from below is a growing intermediate layer with diameter $d_m = 3.5$ mm and height $h_m = 0.4$ mm, which is formed by fibers containing a drop matter that have passed through the cavity bottom. The fibers grow perpendicular to the cavity bottom at the initial stage of the coalescence process, when the size of the contact domain of liquids increases rapidly. In this case, the free surfaces of the contacting liquids are eliminated with the simultaneous conversion of ASPE into other forms, causing the formation of thin fast jets. The visualization of the intermediate layer fine structure at the later stages of droplet spreading was given in [84], and its thickness is measured in [89].

The fibrous structure of the flow pattern, which is presented along the entire boundary of the intermediate layer, is more clearly expressed on its lower edge in Figure 6b. In the upper part of the cavity with diameter $d_c = 6.12$ mm and depth $h_c = 0.96$ mm, an air layer of cavity with thickness $h_i = 0.5$ mm is visible. An intermediate layer with diameter $d_m = 3.9$ mm and thickness $h_m = 0.44$ mm adjoins the cavity from below.

As the cavity grows, the intermediate layer on its lower edge becomes thinner, and its outer border is leveled (Figure 6c). Here, the main part of the droplet substance is concentrated in the lower part of the cavity in a layer $h_i = 0.6$ mm high, including an advanced intermediate layer. Examination of the enlarged image shows that the lower part of the cavity is densely covered with colored liquid droplets, and in the upper part, single colored fibers are distinguished.

Furthermore, the intermediate layer becomes much thinner and covers the cavity surface with diameter $d_c = 8.7$ mm with a thin layer (up to $h_i = 0.3$ mm thick; see Figure 6d). The cavity height with an intermediate layer is $h_c = 3.42$ mm. Here, the intermediate layer is distributed more evenly along the cavity bottom; its thickness does not exceed $h_i = 0.4$ mm. In a densely colored domain, the droplet matter is collected in the lower part of the cavity in a $\Delta h_c = 0.9$ mm high layer. At the upper edge of the intermediate layer, the surface is tinted most densely. Crests and troughs $h_w \sim 2$ mm in size are traced on the border of the tinted layer.

After the crown reaches its maximum size (the height is $h_{cr} = 3.7$ mm, the diameter is $d_{cr} = 11.3$ mm) and begins to decline, the capillary waves propagate down from its upper edge. The shape of the phase surfaces of waves replicates the changing contour of the edge crown. As the cavity deepens further, the concentration of the drop pigment becomes more and more aligned in height (Figure 6e). Gradually, the pigment flows down, the side wall of the cavity become sufficiently transparent and vertical fibers are displayed on them (Figure 6f). At the same time, the lower edge of the intermediate layer begins to lose its smoothness again; protrusions and troughs appear in it. They are created by the flow of colored fluid along the fibers. The fluid accumulates at the reticular formation nodes, penetrates through the cavity surface, and forms small vortices that are gradually transformed into fiber loops [88]. At the same time, the crown edge is rounded, and capillary waves cover its entire surface.

The release of droplets from the tops of the spikes on the crown surface ceases over time; only their rounded remnants persist (Figure 6f, $t = 14.25$ ms). The inner surface of the cavity is covered with individual fibers and spots containing droplet matter. The intermediate layer on the outer surface of the cavity also loses its homogeneity; individual brightly colored domains appear in it.

The study of color flow patterns allows distinguishing between the processes of deformation of the liquid surface forming a cavity and a crown and the processes of droplet matter transfer. The fibers containing the droplet matter leak through the deformed boundary of the cavity and form a finely structured layer under the cavity bottom in the thickness of the target fluid. Gradually, diffusion aligns out the heterogeneity of the droplet pigment distribution, and a uniformly colored intermediate layer is formed under the cavity bottom. In the frontal projection, it corresponds to a more densely colored spot in the flow center. Over time, the thickness and size of the spot, as well as its shape, undergo significant changes due to the size and shape alteration of the cavity, the fluid flow in thin jets forming linear and reticular structures on the cavity inner surface and in filament loops under it.

4.2. Drop Spreading of Potassium Permanganate Solution in Water

The video shots of the drop coalescence of dilute potassium permanganate solution with water are presented in Figure 7 (the inclination of the line of sight is $\vartheta = 65°$ to the horizon, and the shooting speed is 4000 fps). When the chemical composition of the merging liquids changes, the general flow structure is preserved, but the parameters of the thin components change noticeably. The outer ring of the flow at the initial droplet contact consists of individual spikes $\delta_s \sim 0.1$ mm thick and $l_s < 2$ mm long, which are partially tinted with pigment droplets (Figure 7a, $t = 0.25$ ms).

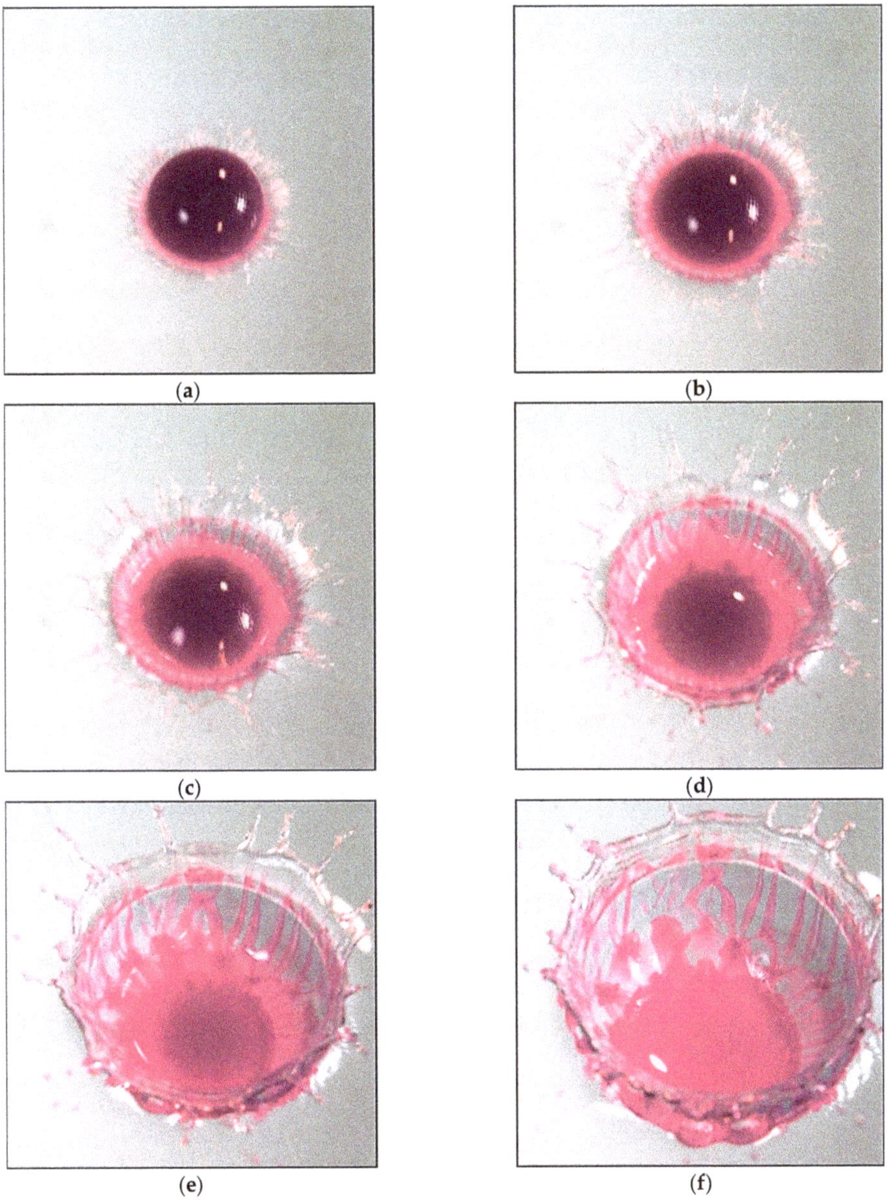

Figure 7. The flow pattern evolution on the water surface at the drop of aqueous potassium permanganate solution diluted in a 1:100 ratio coalescence ($\rho_d = 1$ g/cm^3, $\sigma_d^a = 73$ g/s^2, $\nu_d = 0.01$ cm^2/s, $D = 4.3$ mm, $U = 3.1$ m/s, $E_\sigma = 4.2$ μJ, $E_k = 200$ μJ, Re = 13,300, Fr = 230, We = 570, Bo = 2.5, Oh = 0.0018, $R_{En} = En_k/En_\sigma = 48$, $R_W = 1.66 \cdot 10^{-3}$). (**a**) $t = 0.25$ ms, (**b**) $t = 0.5$ ms, (**c**) $t = 0.75$ ms, (**d**) $t = 1.25$ ms, (**e**) $t = 2.25$ ms, (**f**) $t = 3.5$ ms.

The spikes are in contact with the edge of a thin, weakly colored light circular veil with a width $\delta_v = 0.57$ mm, in which thin brightly colored jets $\delta_l \sim 0.1$ mm are traced. They form a rather regular linear structure, which has been repeatedly observed in experiments with other substances [86]. Large splashes of light allow considering the veil surface as a continuous relatively smooth surface.

The veil is adjacent to a darker wide ring $\delta_{cr} = 0.2$ mm visualizing a growing crown. A brightly colored strip with a width $\delta_{ca} = 0.18$ mm is a growing cavity in contact with a drop residue $d_r = 4.5$ mm in diameter.

The examination of the enlarged image shows that colored fibers protrude from under the drop and continuously last in the cavity, crown, and veil. They form the core of a thin jet, which is a spike protruding from the crown edge. Thus, the flow of the spreading droplet has a predominantly radial direction.

The flow structure is preserved in the next shot at $t = 0.5$ ms. The dimensions of the above-mentioned structural components have grown: they are $\delta_s \approx 0.15$ mm and $0.5 < l_s < 2.5$ mm for the thickness and length of the spikes, $\delta_v = 0.74$ mm for the veil, $\delta_{cr} = 0.4$ mm for the crown and $\delta_{ca} = 0.36$ mm for the cavity. The diameter of the drop residue (a dark area with an uneven edge), the protrusions of which are adjacent to the colored fibers, is $d_r = 4.26$ m. The outer edge of the veil becomes more indented, the number of spikes increases (in the first frame you can count 17; in the second one, there are 19 teeth in the upper semicircle).

Over time, the clarity of the fiber pattern at the bottom of the cavity remains, which becomes more and more densely colored (Figure 7c). At the same time, the boundary between the cavity bottom and the drop residue with diameter $d_r = 4.33$ mm becomes increasingly indented. The preservation of the continuity of the colored fibers indicates the immutability of their angular position in the moving contact domain of the merging fluids.

As the crown width increases and the cavity deepens, the velocity of the colored jets decreases, the protrusions in the edge contour of the cavity deepen, and the spikes break up into sequences of individual droplets, the diameter of which grows over time (Figure 7d). A reticular pattern of individual fibers is visible under the colored drop liquid floating to the cavity bottom. The uneven boundary of the central spot shifts slightly to the flow center—$d_r = 4.6$ mm.

As the cavity deepens and the crown expands, the contrast of the fibers decreases, and their width increases (Figure 7e, $t = 2.25$ ms). A fracture in the angular position of the fibers indicates a sharp change in the inclination of the cavity walls, at the bottom of which the contours of the emerging reticular structure of the fibers are traced in a diffusely colored layer. In the drop residue, a densely colored core and a diffuse outer part are distinguished. The fibers move to the protrusions of its boundary. They are traced on the cavity walls, on remnants of the veil, spikes, and ejected sprays. The outer part of the cavity sidewall, visible in the lower part of the figure, loses its smoothness. Jets begin to bulge on it, gradually transforming into fibrous loops when the cavity collapses [88].

After a while, the contrast of the fibers is decreasing, and their width is increasing (Figure 7f). The color of the flow central domain becomes more uniform. The inhomogeneity of the outer boundary contour of the drop residue is preserved in the right part of the figure, where the fibers are visualized along the entire length. In general, the pattern of fiber distribution is being rebuilt, and its elements are being enlarged.

In the energy spectrum $S(\lambda)$ of the distribution of the relative illumination flow pattern $I(l_\varphi)$ along a circle with a radius $r_\varphi = 2.75$ mm in the upper half-ring shown in Figure 8 at $t = 1.7$ ms, the peaks at scales $\lambda = 0.2, 0.37, 0.44$, and 0.76 mm are distinguished. The smallest scales characterize the thickness of individual fibers.

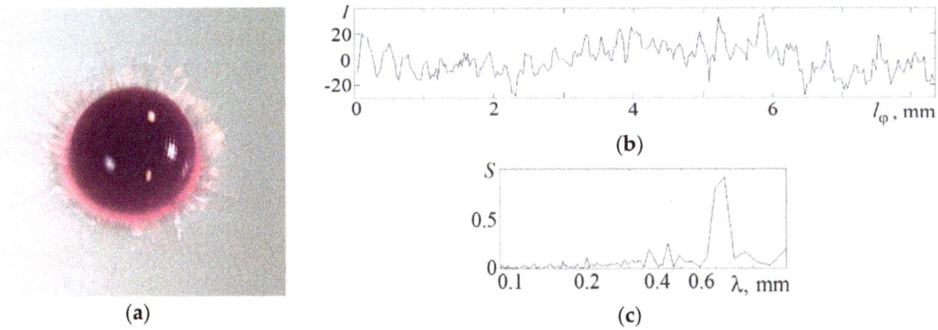

Figure 8. The fine structure of the flow: (**a**) the shot from Figure 6, $t = 1.7$ ms; (**b**,**c**) the distribution of relative illumination $I(l_\varphi)$ along the arc of the $r_\varphi = 2.75$ mm circle, in the middle of the side wall and in the transition zone between the drop residue and the cavity bottom, and their spectra $S(\lambda)$.

The peculiarities of the droplet matter transfer in the impact coalescence mode are explained by samples from the video film of the flow pattern in the vertical plane (Figure 9, side view image). At the initial stage of droplet coalescence, the flat bottom of the cavity is penetrated by thin fibers separated by interfaces of the target fluid. A fibrous layer of diameter $d_m = 3.44$ mm and height $\delta_m = 0.3$ mm with an uneven lower edge adjoins a flat brightly colored bottom of a growing cavity with width $d_c = 4.83$ mm and height $h_c = 0.52$ mm (Figure 9a). The cavity quickly deepens, and at $t = 0.5$ ms, a gas cavity with width $d_a = 3.3$ mm and height $h_a = 0.25$ mm begins to be visible through the colored wall. The total height of the cavity is $h_c = 0.55$ mm. The growth of the size of the fibrous layer adjacent to the cavity bottom slows down, and at $t = 0.5$ ms, its width is $d_m = 4$ mm and its height is $\delta_m = 0.42$ mm.

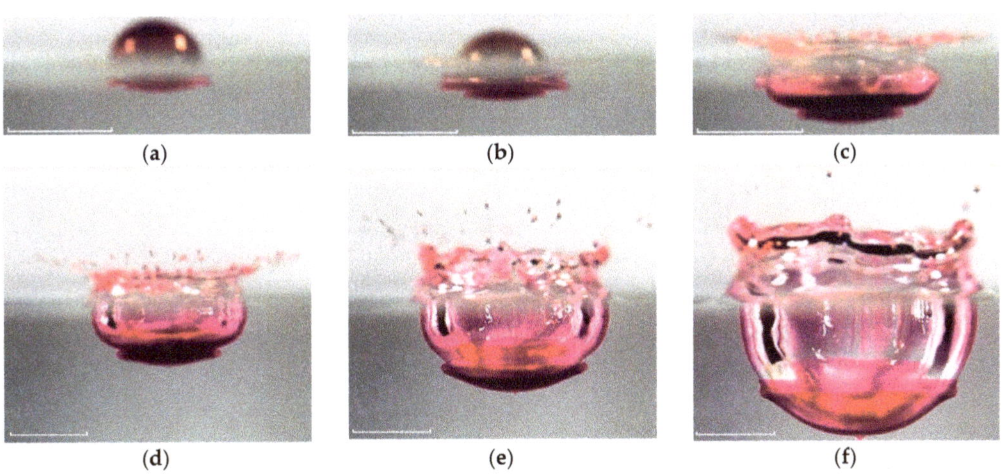

Figure 9. Side view of evolution of the impurity distribution pattern on the surface of the cavity and crown when a drop of an aqueous solution of potassium permanganate $KMnO_4$ is immersed in water (the parameters of the experiments are indicated in Figure 7, the tag length is 5 mm). (**a**) $t = 0.25$ ms, (**b**) $t = 0.5$ ms, (**c**) $t = 1.25$ ms, (**d**) $t = 2.25$ ms, (**e**) $t = 4$ ms, (**f**) $t = 8.25$ ms.

All the traditional components of the flow—a cavity, a growing crown with spikes and a spray cloud—are shown in Figure 9c. The diameter and overall height of the cavity with a brightly colored lower layer is $d_c = 8.2$ mm, and the height is $h_c = 1.3$ mm. The dimensions of the adjacent fibrous layer are $d_m = 5.24$ m and $\delta_m = 0.77$ mm at $t = 1.25$ ms in Figure 9c. The inhomogeneities of the pigment distribution begin to smooth out quickly by the processes of molecular diffusion in the intermediate layer under the cavity bottom ($d_c = 10.3$ mm and the height $h_c = 3.13$ mm in Figure 9d). Here on the side walls of the cavity, the bottom of which is gradually deformed from flat to convex, vertical fibers are viewed.

Further, the rapidly deepening cavity bottom "pushes apart" and throws the remnants of the intermediate layer, the boundaries of which can still be identified at $t = 4$ ms (Figure 9e). The crown begins to fall off, and the capillary waves appear at its edge.

The cavity takes a shape close to spherical at $t = 8.25$ ms (Figure 9f). A system of vertical fibers remains on its side surface. The crown walls are covered with capillary waves. An intermediate layer $\delta_m = 0.3$ mm thick and $h_m = 3.86$ mm high covers the entire cavity bottom with an even layer. The exception is its center, where a thin jet begins to grow.

The three-dimensional structure of the flow in the contact domain of the merging liquids includes thin flat jets running along the cavity bottom to the spikes on the crown edge, and even thinner fibers penetrating the cavity bottom, which form an intermediate layer. At the initial stage, the liquid interface alternately passes along the tops of the fibers with the drop pigment; then, it "jumps" to the tops of the layers of the target liquid, separating the colored fibers. That is, in some domains, it passes along the outer boundary of the fibrous intermediate layer, and in neighboring domains, it passes along the inner one. Qualitatively, we can assume that at the bottom of the growing cavity, the boundary of the coalescence region has a complex three-dimensional piecewise smooth shape.

A sample of shots from the video of the coalescence of a solution of potassium permanganate of higher density, as shown in Figure 10, indicates the scenario stability of the appearance of new components of the flow pattern and illustrates some subtle details of the flow pattern more clearly. The bright color of the spikes with splash droplets flying from their tops confirms the tendency of the rapid flow of the drop pigmented fluid. The velocity of the first splashes in these experiments, which is $u_s = 5.6$ m/s, noticeably exceeds the drop velocity $U = 3.5$ m/s (Figure 10a,b).

The processes of ASPE conversion in a thin layer of merging near-surface layers contact [70,78] provide the inflow of additional energy, which ensures the rapid flow in spikes and sprays. The length of the spikes does not exceed $l_s < 2.7$ mm; their thickness is $\delta_s \sim 0.1$ mm. A colored disk $\Delta r = 0.4$ mm wide between the annular line of coalescence of the drop residue with the target fluid and the growing wall of the cavity and crown is the forming bottom of the cavity with a diameter of $d_c = 5$ mm. The diameter of the crown edge reaches $d_{cr} = 7$ mm at the same time.

The number of spikes decreases with time. Some of them are completely separated from the veil edge. A trough forms in such places (Figure 10c). The heterogeneity of the shape of teeth and spikes indicates the complex three-dimensional nature of fluid flows in the veil. One part of the spikes serves as the continuation of the brightly colored traces of ligaments. The other part is formed as a result of the coalescence of a pair of jets, reaching the thickness of $\delta_j = 0.2$ mm.

Protrusions up to $l_f \sim 1$ mm long are observed at the border of a densely colored domain in the center of the flows. Fast trickles flow out from their tops (Figure 10d). Gradually, as the intermediate layer spreads under the cavity bottom (Figure 10e,f), the background color of the cavity walls becomes more uniform. Triangular elements appear in the flow structure. The heights of the tiers is $l_s = 0.88, 1.24, 1.67$ mm with the thickness of the forming fibers from $0.2 < \delta_s < 0.4$ mm. They are most clearly expressed in the upper part of the cavity in the vicinity of the transition domain where the cavity transfers into the crown, rising above the level of the undisturbed surface (Figure 10e).

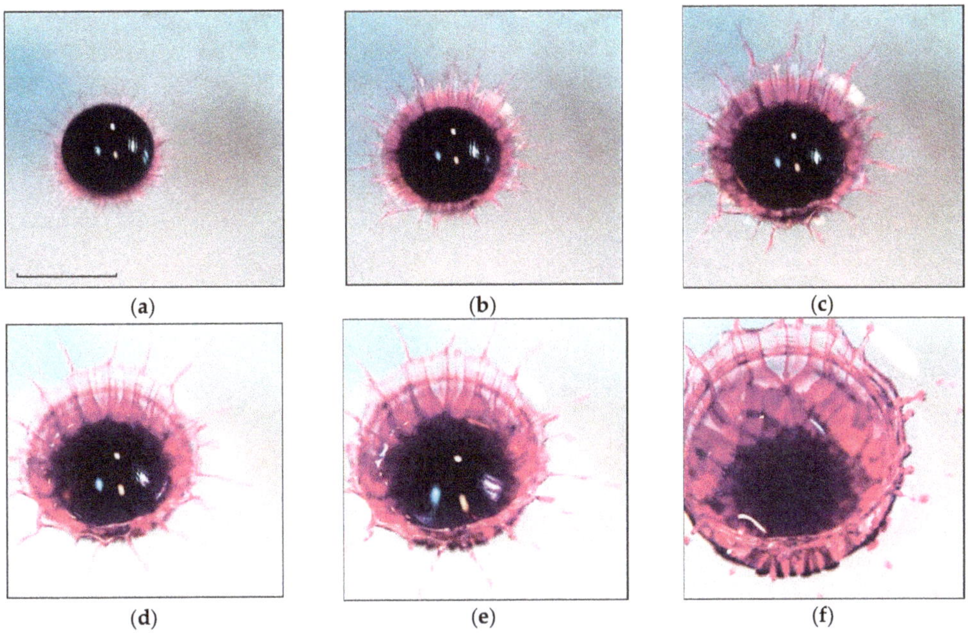

Figure 10. Evolution of the flow pattern as the drop of an aqueous solution of potassium permanganate diluted in a ratio of 1:10 merges with water (D = 4.3 mm, U = 3.1 m/s, E_σ = 4.2 μJ, E_k =200 μJ, Re = 13,300, Fr = 230, We = 570, Bo = 2.5, Oh = 0.0018, $R_{En} = En_k/En_\sigma = 48$, $R_W = 1.66 \cdot 10^{-3}$, the tag length in Figure (**a**) is 5 mm, all frames are presented in the same scale). (**a**) t = 0.25 ms, (**b**) t = 0.5 ms, (**c**) t = 0.75 ms, (**d**) t = 1 ms, (**e**) t = 1.25 ms, (**f**) t = 2.5 ms.

As the cavity and crown grow, the number of triangular grid cells on the bottom and walls of the cavern increases, and boundaries appear in their structure, separating ring sets of cells (Figure 10f). The outer border of the cavity loses its smoothness at the bottom of the image. Here, colored sections of fibers begin to be squeezed into the transparent target liquid, the further evolution of which is traced in [88]. When new tiers of the structure are formed, the size of the central spot (a diffusely colored central domain) decreases. Further, as the crown grows, the central spot grows linearly over time until the cavity reaches its maximum depth. When the cavity collapses, the central spot contracts to the center of the cavern remnant and to the top of the ascending reverse jet.

4.3. Examples of Photographic Registration of the Flow Pattern at the Initial Coalescence Stage of a Free-Falling Drop with the Target Fluid

The indisputable advantage of video recording is the ability to trace the structural connection and the temporal sequence of individual elements of the flow pattern appearance. The benefit of photo registration within the available technical resources is a higher spatial and temporal resolution. We compared the technical characteristics of two cameras. For the Optronis CR 300 × 2 video camera with a Nikon Nikkor 24–85 mm lens, the distance from the front lens of the device to the center of the shooting area is from 10 to 25 cm, and the exposition is from 1/5000 s. For the Canon EOS 350D camera with a Canon EFS 18–55 mm lens, the distance from the front lens of the device to the center of the shooting area is 5–20 cm, and the exposition 1/4000 s. The spatial and temporal resolution in the video frame (at a frame rate of 4000 fps) and the photo is, respectively, from 30 μm/pix and from 10 μm/pix. Under experimental conditions, photo registration allows analyzing the finer elements of the pattern of a rapidly evolving flow on a moving contact surface that continuously changes its location in space and its shape.

In the experiments carried out, the drops of a solution of potassium permanganate, copper sulfate, iron sulfate, or tap water fall into a cuvette $16 \times 16 \times 7$ cm^3 at room temperature. Some physical parameters of solutions are given in Table 1.

Table 1. Physical parameters of saturated solutions.

Solutions	ρ, g/cm^3	σ, g/s^2	$\gamma = \sigma/\rho$, cm^3/s^2	μ, g/s·cm	$\kappa_S \times 10^5$ cm^2/s
copper sulfate	1.206	75	62.2	0.02	0.53
iron sulfate	1.18	75	63.56	0.02	0.38
potassium permanganate	1.04	74	71.15	0.01	4.0

The registration of the flow pattern is carried out with a Canon EOS350D camera with a 12 mm macro ring, ISO 100, D 5.6, with a minimum exposition 1/4000 s. The distance to the object is 15 cm and the spatial resolution is 13 μm/pix. The inclination of the sight line is $\vartheta = 75°$ to the horizon.

4.3.1. Drop Spreading of Saturated Potassium Permanganate Solution in Water

The main components in the finely structured flow pattern formed during the initial contact of the droplet with the target liquid are a veil or spikes and sequences of small droplets (spray) that fly out both from the top of the spikes and directly from the contact line. The presence of "colored tongues" is explained by the difference in the drop shape from the spherical one at the time of the initial contact. The droplet surface is distorted by Rayleigh oscillations and traveling capillary waves [85], and it is also deformed by a radially spreading air jet [48].

The flow photos at the beginning of the coalescence of a drop of potassium permanganate solution with water are shown in Figure 11a (the inclination of the sight line is $\vartheta = 75°$ to the horizon). In the flow pattern, the "tongues" of the veil are expressed in the range of "10–11" and "13–14 o'clock" with a pronounced outer edge and an inner striped structure. The estimation of the velocity of the longest strokes (traces of droplets flying out from the spikes on the tops of the teeth) is measured by the ratio of their length $l_s \sim 5$ mm to the exposure time $\Delta t = 0.25$ ms is $u_s = l_s/\Delta t \sim 22$ m/s (six times the drop velocity $U = 3.8$ m/s). Careful consideration allows distinguishing thin strokes that have a continuation to the tops of the teeth as well as those in contact with troughs at the boundary of the visible contact domain in the upper part of the drop.

In turn, the upper part of the boundary of the colored fluid is not a smooth line, protrusions and troughs with $\Delta\delta_\varphi \sim 0.2$ mm increments are distinguished in it. The transverse dimensions of the fibers on the veil teeth above the drop are $\delta_f \sim 0.05$ mm or less. The question remains open about the coloring nature of the teeth, which may indicate the formation of a continuous inhomogeneously colored surface of the veil with thin trickles lying on its surface or with separate isolated jets located in the space above and below the veil.

The outer wall of the developed spherical cavity, represented in the lower part of Figure 11b, is covered with protruding fibers $\delta_f \sim 0.5$ mm thick, which are at a distance of $0.2 < \Delta\delta_f < 1.2$ mm. Vortex heads of protruding jets appear on some fibers under the pigmented mesh nodes, which transform into loops as the flow evolves [88]. The original linear structure of the distribution of the drop matter is preserved on the wall of the falling crown remnant, which is covered with short capillary waves and lies in the upper part of the cavity.

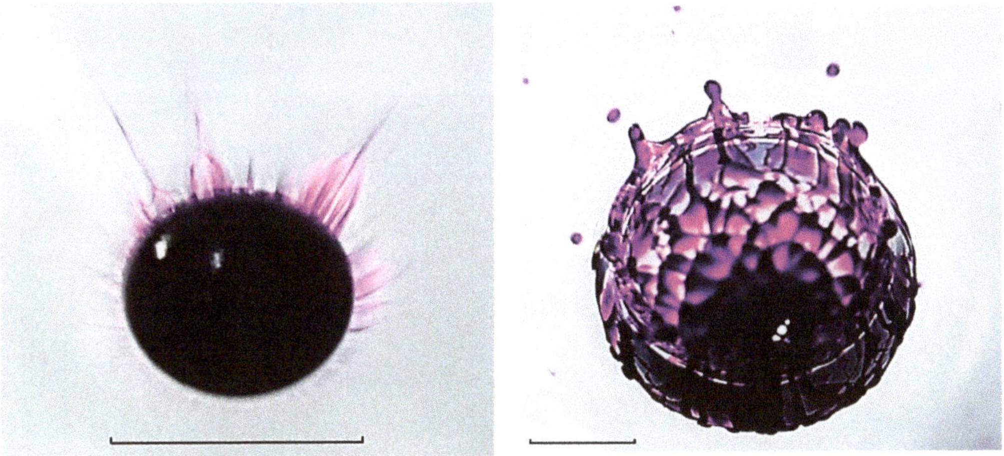

(**a**) (**b**)

Figure 11. A drop of saturated aqueous solution of potassium permanganate falls into the water ($D = 4.3$ mm, $U = 3.8$ m/s, $\rho_d = 1.04$ g/cm^3, $\sigma_d^a = 74$ g/s^2, $\mu = 0.01$ g/cm · s, $En_\sigma = 4.3$ μJ, $En_k = 313$ μJ, $En_k/En_\sigma = 73$, $W_k/W_\sigma = 25 \cdot 10^{-4}$, Re = 17,000, Fr = 350, Bo = 2.55, Oh = 0.0017, We = 880, the tag length is 5 mm). Forms of the flow geometry: (**a**) image of initial contact with annihilation of contacting surfaces, (**b**) developed reticular formation and linear structures on the cavity surface.

At the cavity bottom, fibers with a thickness from $0.1 < \delta_f < 0.4$ mm form a three-tiered mesh with triangular and quadrangular cells of the size from $1.2 < \delta_m < 2.2$ mm. Traces of cells are also visible in the central, more densely colored layer, under which the adjacent intermediate layer is located, which is shown in Figure 9. Dark and light diffuse strips surrounding the cavity in Figure 11b are shadow images of short circular capillary waves [42].

Either a single trickle flows to the tops of the teeth on the upper edge of the crown along its center, or a pair of trickles flow along the outer edge of the tooth, forming a spike, from the top of which sprays fly out.

4.3.2. Drop Spreading of Copper Sulphate Solution in Water

A distinctive feature of the flow during the primary contact of a drop of saturated aqueous solution of copper sulfate with water is also the abundance of rapid small sprays, which is presented in Figure 12a (the inclination of the sight line is $\vartheta = 75°$ to the horizon; registration: Canon EOS 350D with a 12 mm macro-ring, ISO 100, D 5.6).

The maximum length of thin strokes (traces of sprays) reaches $l_s = 4$ mm. Taking the stroke width as the diameter of the ejected droplet, and the length as its movement during exposure time, $\Delta t = 0.25$ ms, it is possible to estimate their velocity by the ratio $u_s = l_s/\Delta t \leq 16$ m/s, which significantly exceeds the droplet velocity $U = 3.8$ m/s. The existence of parallel strokes groups is the evidence of the successive ejection of droplets from the top of the same spike on the edge of the cavity veil, the position of which changes as the crown grows. In this experiment, the spray velocity u_s noticeably exceeds the drop velocity U; their ratio is $u_s/U = 4.2$. The fact indicates the importance of the influence of the processes of converting ASPE into other forms on the dynamics and structure of droplet flows [70,86].

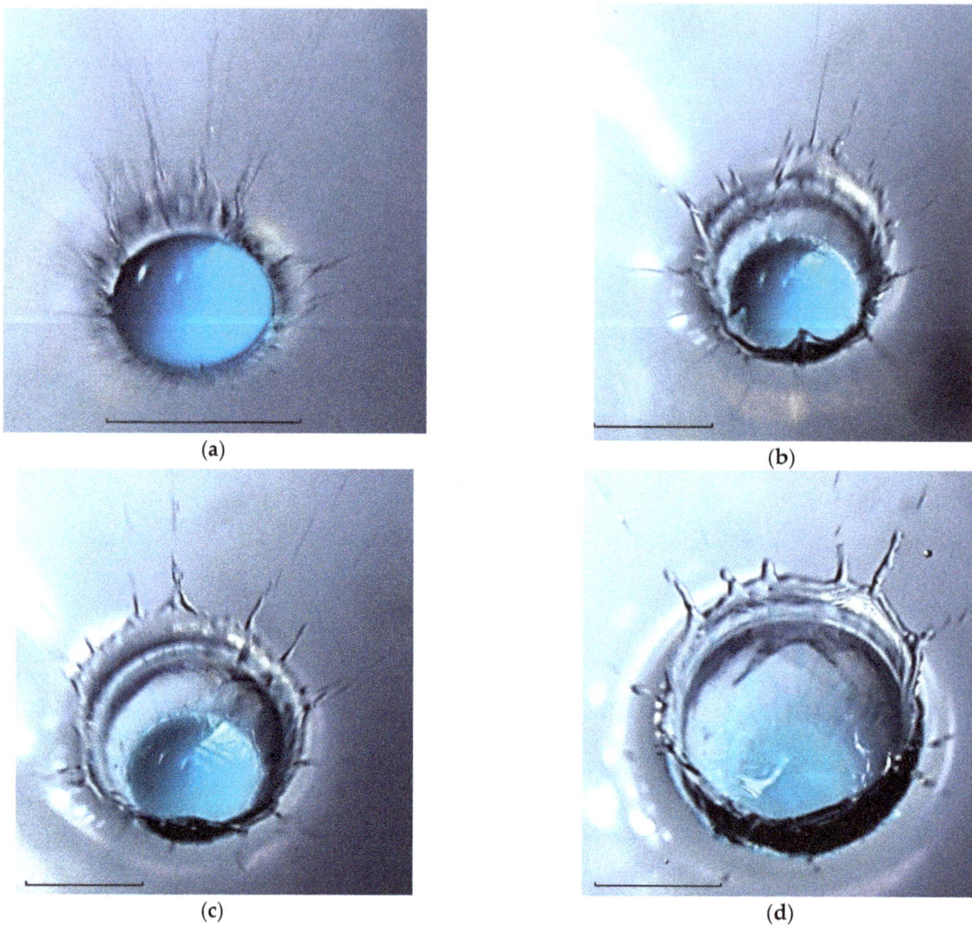

(**a**) (**b**) (**c**) (**d**)

Figure 12. Photos of the copper sulfate solution drop spreading pattern in water ($\rho_d = 1.2$ g/cm^3, $\sigma_d^a = 75$ g/s^2, $\mu = 0.02$ g/cm $\cdot$ s, $D = 4.3$ mm, $U = 3.8$ m/s, $En_\sigma = 4.4$ µJ, $En_k = 363$ µJ, Re $= 9900$, Fr $= 650$, We $= 1000$, Bo $= 2.9$, Oh $= 0.0032$, $R_E = En_k/En_\sigma = 83$, $R_W = 3 \cdot 10^{-3}$, the tag length is 5 mm): Typical flow patterns: (**a**)—the image of the first contact with ejecta formation; (**b**)—growing cavity with drop remnant separated by the complex irregular boundary with circular capillary waves, (**c**)—deepening cavity with forming a growing crown, (**d**)—individual protrusions on the drop remnant boundary and linear coloured fibers on the on the surface of the cavity.

Droplet traces (strokes) are located on a pale background of a thin veil $\Delta\delta_v^r = 1.25$ mm wide, adjacent to a darker annular layer $\Delta\delta_{cr}^r = 0.9$ mm wide, in contact with a light strip $\Delta\delta_{ca}^r = 0.14$ mm (the image of the forming walls and cavity bottom). The density of the strokes indicates the high frequency of the process of the sprays generation on the spike tops at the veil or crown edge.

The drop spreads along the bottom as the cavity grows, and in Figure 12b, it is possible to see the boundary of the contact domain of the merging fluids having a complex irregular shape. Basically, the veil and spikes are inclined outward, but in the lower part, there are two sections of the veil, which are inclined inward.

The droplets from their tops fall on the surface of the merging drop and form short capillary waves (in Figure 12b—in sectors for "6" and "8" o'clock [85]). The complex structure of the fluids' contact domain boundary (colored fibers visualizing thin trickles flowing along the cavity bottom, the crown walls and reaching the spikes on its edge) is preserved throughout the process of droplet coalescence (Figure 12c,d).

Individual protrusions (precursors of the forming vortex loops [88]) are also visible on the outer cavity wall in the lower part of Figure 12b,c.

In the enlarged images of the contact liquids domains shown in Figure 13, fibers adjacent to the contact line and small-scale inhomogeneities of complex shape with a size of $\Delta l_{\varphi} = 0.22$ mm are visible, forming a moving boundary of the coalescence domain. As in other experiments, the fibers containing the droplet matter go along the bottom and the walls of the cavity, reaching the tops of the spikes on the veil edge [86].

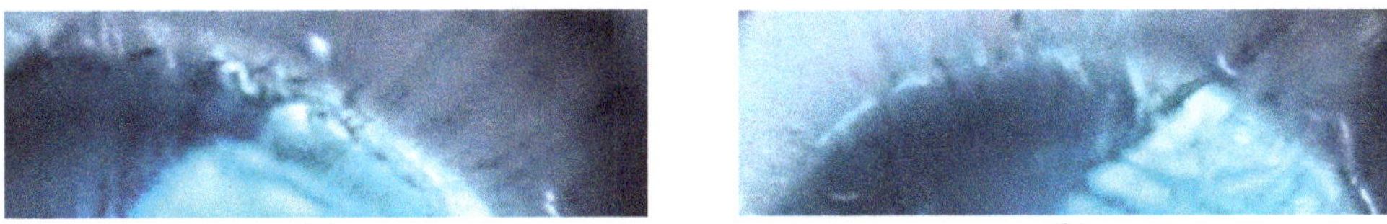

(a) (b)

Figure 13. Enlarged images of the boundary of the droplet coalescence domain; parts of Figure 12b,c. (**a**,**b**) fibers adjacent to the contact line and small-scale inhomogeneities of complex shape.

The colored fibers are traces of fast jets containing droplet matter (ligaments that violate the smoothness of the contact domain boundary of merging liquids). They flow for a long time while maintaining their angular position, leaving connected colored traces on the walls of the cavity and crown.

4.3.3. Spreading of an Iron Sulfate Solution Drop in Water

A large number of fast small droplets (sprays) are also formed when a drop of a saturated solution of iron sulfate $FeSO_4$ merges with water. However, the frequency of their ejection here is less than when a drop of copper sulfate solution merges with water. Splashes fly out from the tops of the spikes, to which thin trickles (ligaments) flow, penetrating the domain boundary where the droplet merges with water. Continuous trickles in Figure 14a can be traced at the cavity bottom, the crown walls, above the crown edge.

However, the fine structure of the flow near the boundary of the fluid coalescence domain is more complex here than in Figure 13. In the flow pattern, several groups of teeth are distinguished, the tops of which are wrapped inside the flow domain—two teeth in the left part, two teeth in the right part, and three teeth at the bottom (the angular position is at "3", "6 and 9 o'clock"). Circular disturbances on the surface of the droplet residue in the lower part of the cavity bottom are systems of short capillary waves formed by the coalescence of sprays flying from the spikes on teeth tops [87].

The coalescence boundary (the enlarged image of which is shown in Figure 15) is uneven.

Fine fibers come out of periodic protrusions and troughs between them. The dark circular line on the cavity wall is the crest of a capillary wave running from the coalescence domain boundary.

Fibers (jets, ligaments, trickles) running from the uneven fluids' contact domain boundary were observed in all experiments conducted with colored droplets in the impact coalescence mode. The high (in comparison with the falling drop velocity) speed of the liquid flow and the density of the additional KEn of the flow in the ligaments were provided by the conversion processes of ASPE, which was released during the fluids coalescence and elimination of free surfaces. The transformed energy was stored in the thin volume in which it was concentrated near the free surface.

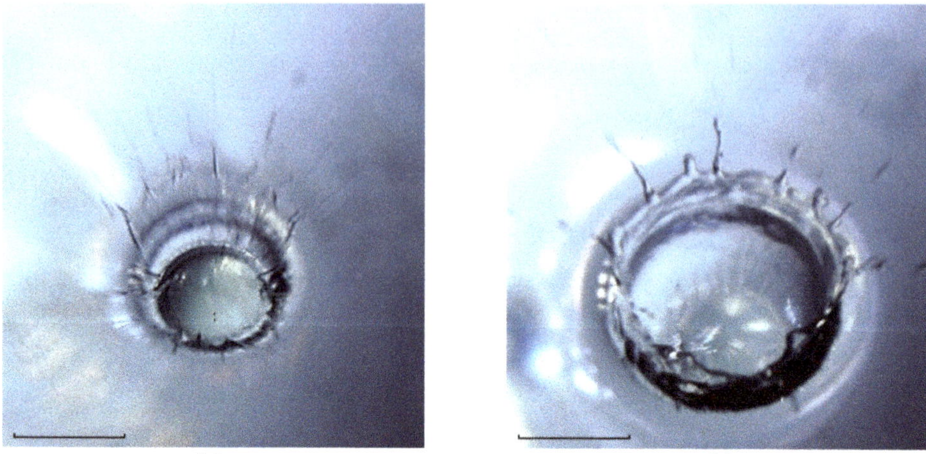

(**a**) (**b**)

Figure 14. A drop of an aqueous solution of iron sulfate falls into the water ($\rho_d = 1.18$ g/cm^3, $\sigma_d^a = 75$ g/s^2, $\mu = 0.02$ g/cm · s, $D = 4.3$ mm, $U = 3.8$ m/s, $En_\sigma = 4.4$ μJ, $En_k = 355$ μJ, $R_E = En_k/En_\sigma = 81$, $W_k/W_\sigma = 2.7 \cdot 10^{-3}$, Re = 9650, Fr = 350, Bo = 2.85, Oh = 0.0032, We = 980, the tag length is 5 mm): (**a**) the complex form of the boundary of the coalescing fluids; (**b**) fine jets of coloured fluid on the cavity surface.

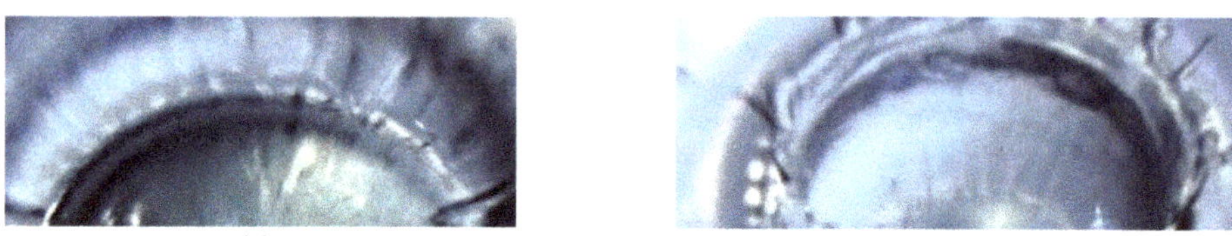

(**a**) (**b**)

Figure 15. Enlarged images of the contact domain boundary of a merging saturated iron sulfate solution drop with water (parts of Figure 13). (**a**) fast jets forming liear sructure on the cavity surface; (**b**) traces of jets—ligaments on the bottom of the cavity.

4.3.4. Natural Shadowgraph Patterns of Water Droplet Spreading in Water

Changing the angles of inclination and the position of the water surface causes a redistribution of light flows and leads to the formation of the flow components' distinctive images. Shadowgraph methods, which have been widely used in the study of fluid flows for almost two centuries [90], can be also employed in the study of droplet flows. The resulting large gradients of the refractive index allows tracing the evolution of a flow fine structure under the cavity water droplet in water by the direct shadowgraphy method [84]. The photos presented in Figure 15 show the possibility of visualizing fibers (traces of trickles) on the surface of a complex-shaped liquid not only by tint but also by observations of the redistribution of image illumination.

In the natural shadowgraph flow image shown in Figure 16a, the merging drop residue in the center of the cavity is outlined with a dark line (the coalescence domain boundary). A light strip, which is wall and the growing cavity image, is penetrated by separate jets: ligaments. Thin jets that are generated in the domain where the droplet merges with the target fluid liquid deform the bottom. One group of jets penetrates through the cavity walls, while another part flows along a cavity bottom and is continued on the crown surface, then in the veil, and finally forms dark spikes protruding from the veil edge. They are especially distinguished in the upper right quarter of the flow. Sprays fly out from the spike tops and are registered as thin strokes, the length of which reaches $l_s = 4.25$ mm. The maximum ratio of splash and drop velocities is $u_s/U = 4.5$. A large number of strokes above the crown and the veil is evidence of a multi-tiered level of sprays distribution flying out from the spikes on the edge of the veil or the crown [86].

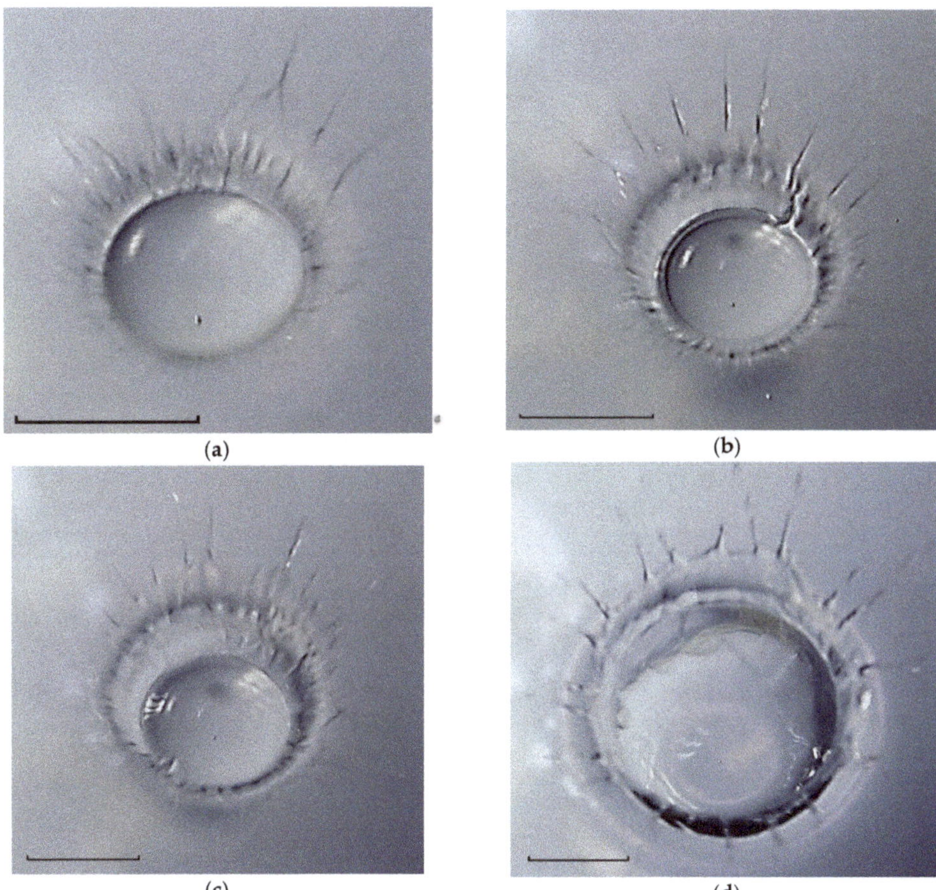

Figure 16. Photos of the drop water coalescence with water in the impact mode ($\rho_d = 1$ g/cm^3, $\sigma_d^a = 73$ g/c^2, $\nu_d = 0.01$ cm^2/s, $D = 4.3$ mm, $U = 3.8$ m/s, $En_\sigma = 4.2$ μJ, $En_k = 240$ μJ, Re = 14,600, Fr = 270, We = 690, Bo = 2.5, Oh = 0.0018, $R_E = En_k/En_\sigma = 57$, $R_W = 2 \cdot 10^{-3}$, the tag length is 5 mm): (**a**) fine sprays, spikes, veil, and jets on the contact line of coalescing fluids; (**b**) spikes, veil, set of jets and sharp contact line; (**c**) growing cavity, crown, veil, and spikes on its outer edge; (**d**) deep cavity with drop remnant, crown, and rare spikes.

The contrast of the flow components' images is determined by the inclination of the contacting surfaces, the location of the light sources and a photocamera [86]. An uneven section of the boundary in the range of angles "10–12 o'clock" allows us to consider the visible boundary as the actual edge of the spreading drop. The boundary is radially stretched by thin-layer flows localized in the contact domain of the cavity shell and the near-surface layer of the target fluid [62].

The resulting flows accelerate at the coalescence domain boundary, where the free surfaces are eliminated and the releasing ASPE is transformed into other forms [70]. In this case, together with the matter of the contacting media, a part of the extensive KEn of the merging drop is captured by sprays. It is the rapid release of the ASPE that is the driver of the flow. At the same time, a noticeable contribution to the fast flow energy is introduced by the energy of the drop mechanical movement.

The spikes on the edge of the cavity can be a continuation of a single trickle (the upper left part of the flow); they can form during the coalescence of two edges of the veil (three spray lines in the upper right part of the figure). The question as to whether the observed structures are continuous jets or a droplets collection consisting of a sequence of small components is still open.

As the drop merges, the number of spikes on the veil border decreases, their thickness increases, and the distribution of illumination in the upper part of the image takes on a diffuse character due to the reconfiguration of a large number of jet elements (Figure 16b). Single jets can be seen more clearly in the lower part of the image. In the upper right corner of the angular position at "2 o'clock", an inverted jet is visible, the top of which falls on the surface of the merging drop. Two clear contours appear in the flow pattern, one of which is the transition from the bottom to the cavity wall, and the second one is the boundary of the spreading drop at the cavity bottom. The images of the fibers in the lower part of the figure emphasize their origin at the contact domain boundary of the droplet residue with the target liquid.

Gradually, the spikes on the upper edge of the veil thicken, and the "strokes" (traces of more and more slowly flying droplets) shorten (Figure 16c). At the same time, the inhomogeneous contrast of dark fibers on the walls of the crown and the cavity is preserved, which is determined by the shape of the surface and lighting conditions. The circular domain boundary where the droplet merges with the deformed liquid surface is visualized as well. In the lower part, the contact of the fibers with the contact line of the merging liquids at the cavity bottom is traced. The sequences of circular lines on the surface of the droplet residue, most pronounced in the sectors at "10" and "14 o'clock", are images of a group of circular capillary waves running to the top of the flow from the expanding boundary of the contact merging fluids domain. A group of capillary waves in an angular position at "7 o'clock" on the surface of the drop residue is a trace of contact of the ejected spikes facing the flow center with the surface of the drop residue [87]. As the flow evolves, the inhomogeneities of the veil structure are smoothed out. The components of the structure on the crown edge and the cavity walls are enlarged.

The formation of additional structural components is accompanied by the emergence of new fine elements in the flow pattern, which, in particular, are expressed as light lines at the cavity bottom in the circular line vicinity of the drop residue coalescence in the "2–5 o'clock" sector. Their geometry, which correlates with the shape of the intermediate layer edge in Figure 4 at $t = 2.75$ ms, confirms the formation of an intermediate layer under the cavity bottom in the impact mode of merging a water drop with water. The layer is created due to the mixing penetrating substance of the drop with the target fluid.

At first glance, the identical physical properties of the contacting liquids in this experiment (tap water) exclude the formation of contrasting images of flow boundaries in the fluid thickness. However, the fact of their reproducible registration in independent experiments [84] indicates the necessity for a more thorough study of the physical nature of the internal boundaries' contrast formation. Perhaps the boundaries reflect large gradients in temperature distributions. Evaporative cooling lowers the surface temperature of a free-falling drop and creates a liquid layer with a changed refractive index, which persists during the flow evolution and is diluted by molecular diffusion processes, as well as the concentration gradient layer in Figures 5 and 8.

A group of light ring strips in the sector of "7–9 o'clock" visualizes capillary waves running along the cavity wall. Radial strokes on the surface of the drop residue are traces of falling sprays. Vertical inhomogeneities of the structure of the cavity side wall in the "10–13 o'clock" sector are the traces of ligaments (thin jets flowing from the fluids coalescence domain boundary to the spikes on the edge of the crown or merged veil, the bases of which thicken over time, and the lengths decrease). Accordingly, the droplet sizes are growing, more and more slowly flying out from their tops.

5. Discussion

From the first experiments up to the present, most of the attention has been paid to the study of either the geometry of the free surface [3,91] or the motion of the forming ring vortices and vortex systems [1,2,22,23] when studying the flows generated by a drop falling into a liquid. To visualize the flow, the methods of "bright dot" or "backlight" illumination and "black and white flow pattern registration" are mainly employed [86,89]. During video and photo registration of the black-and-white flow pattern, clear images of the cavity contour were obtained [14], but the distribution features of the droplet matter in the target fluid were not indicated. The results of single experiments on the color registration of the fibrous pattern of the matter distribution of the fallen drop in shallow [63] and in deep fluids [64] have escaped the attention of theorists. Calculations of the matter transfer are traditionally carried out under the assumption of flow axial symmetry, that is, the axial uniformity of the drop matter distribution on the cavity walls [92].

In modern algorithms for calculating droplet flows, the system of fundamental equations with kinematic and dynamic boundary conditions on a free surface [19–21] is replaced by a model system of constitutive equations [46,93]. They accepted assumptions [94] that allow additional expressions simulating the effect of surface tension to be introduced into the equations. Many references (the number to [94], which is the methodological basis for the constitutive description of the liquid surface geometry in droplet flows [46,93], exceeds 2000) illustrate the popularity of the approach. However, the reduction in the system of fundamental equations, in which only the momentum transfer equation is preserved (and at the same time modified), impoverishes the completeness of the flow description, since it does not allow calculations of the energy and matter transfer.

Gibbs' idea of the existence of a diffuse interface layer of contacting miscible liquids of finite thickness is actively used in numerical calculations of droplet spreading over the surface of a solid body. The dependence of internal energy on the "phase measure" and its gradient was introduced in the first mathematical model of a diffuse layer proposed by Van der Waals at the end of the XIX century, [95]. Later, the representation of internal energy was repeatedly used in the development of drop impact flow calculation programs. The technique [95], modified in [96,97], was later actively employed in the formulation of experiments [98], numerical calculations [99,100], as well as in a comparison of the effectiveness of hydrodynamic and kinetic approaches to describing the structure of a diffuse contact layer when spreading droplets on a solid surface [101] and calculating the cavity shape in a liquid [3].

However, a significant difference in experimental conditions does not allow us to transfer the results [95,101] to the analysis of the coalescence drop pattern with the target fluid at rest. When spreading on a solid impermeable surface, the "no-slip conditions", slippage or non-flow conditions that limit the spatial fluid droplet distribution are met. The solid surface remains stationary and does not lose matter. When the droplet spreads, the contact surface of the target fluid actively moves. The ASPE released at the coalescence transfers into other forms including energy of thin flows and outgoing sprays, which contain both contacting fluids.

In all the experiments carried out, the registration of the flow pattern and the shape of the contact domain boundary of merging miscible liquids is implemented in the impact coalescence mode at $En_k > En_\sigma$. Here, in contrast to the intrusive mode [62], the cavity starts to form from the moment of initial contact. At the same time, the droplet disintegrates into separate thin jets. At the initial stage, some of the jets fly out into the air and form sequences of small droplets, the inclination trajectory angle of which gradually increases to the horizon. The velocity modulus of fine splashes is greater than the drop velocity at the time of primary contact.

Some generated jets cross the boundary of the fluid coalescence domain and flow along the deformable surface of the target fluid. The jets reach the edge of the crown or veil and form the teeth on it. In the center of the teeth there are spikes, from the tops of which new portions of sprays fly out. The size of the drops flying out from the tops of the spikes (the continuations of the jets) grow over time, and the speed decreases rapidly. Some groups of drops flying out from the teeth tilted inward fall on the bottom part of the spreading drop. Since the jet formation sites at the fluids' coalescence domain boundary at the bottom of the growing cavity retain their angular position for a long time, the jets maintain their shape. The remaining colored traces of the ligaments form a linear pattern on the walls of the cavity and crown. At the final stage of the drop spreading, a reticular pattern is formed at the cavity bottom, the size and structure of which depend on the composition of the drop and its shape at the time of initial contact.

Other jets penetrate the bottom of the growing cavity and form a structurally isolated area under its bottom, in which thin fibers containing liquid droplets are separated by layers of the target liquid. Molecular processes of mutual diffusion on the developed contact media surface provide a rapid equalization of the density difference of interpenetrating liquids and form a layer of intermediate density liquid adjacent to the cavity bottom. As the cavity grows, the intermediate layer homogenizes and thins. The rate of density equalization depends on the properties of the medium; for example, the heterogeneity of the distribution of potassium permanganate in it is lost faster than that of alizarin ink. The existence of fine jets at the cavity bottom and the intermediate layer of the cavity is visualized even when a water drop spreads in the water.

The fine structure of the coalescence domain boundary depends on the properties of the fluids and the conditions of the experiment. When saturated solutions of some metal salts (copper and iron sulfate, potassium permanganate) merge with water, the boundary is uneven: there are individual protrusions and troughs, and thin fibers are distinguished in it. The boundary of the ink solution drop coalescence domain is smoother. The complexity of the shape of the boundary is explained by the possibility of forming a large number of multidirectional thin jets, whose flow accelerates the processes of ASPE conversion occurring in a wide range of spatial scales—from the sizes of molecular clusters ($\delta_c \sim 10^{-6}$ cm) to the scales of non-stationary ligaments $\delta_U^\gamma \sim 10^{-2}$ cm and more.

6. Conclusions

A series of experiments was carried out to visualize the fine structure of the flow in the vicinity of the fluid contact domain boundary when a free falling drop merges with stationary water in a cuvette. The beginning of spreading of a free falling drop of water, dilute ink, solutions of potassium permanganate, copper sulfate, and iron sulfate in deep water was visualized.

In the impact flow mode, when the KEn exceeds the ASPE of the droplet, the cavity begins to form from the moment of initial contact of liquids. In all experiments, the drop loses its continuity and breaks up into thin jets that fly out into the air, flow along the fluid surface or penetrate the cavity bottom.

Small fast droplets fly into the air from the tops of thin jets formed directly on the fluid coalescence domain boundary, or they are ejected from the edges of the spikes, which are located on the tops of the teeth of the veil or crown. The droplet sizes grow over time, and the velocity decreases.

Thin, fast jets flowing along the cavity bottom retain the radial direction for a long time. Their traces form linear and reticular structures at the cavity bottom.

The jets penetrating the cavity bottom form an intermediate layer with their own physical properties. The size and shape of the intermediate layer as well as the degree of uniformity of the matter distribution are changing rapidly.

The jet formation causing the loss of continuity of the cavity bottom and the contact line of merging liquids is associated with the rapid transformation of the available potential surface energy and other components of internal energy into other forms—variations in pressure, temperature, redistribution of matter and acceleration of fluid flows acting together with available KEn.

Further improvement of the experimental methodology in terms of expanding the range of lengths of illuminating electromagnetic waves (toward ultraviolet and X-rays), as well as increasing the spatial and temporal resolution of the instruments, will clarify the mechanisms of transmission of energy, momentum and matter when droplets merge with a moving liquid and liquid at rest, which play an important role in modern industrial technologies and environmental processes.

Author Contributions: Conceptualization, Y.D.C.; methodology, Y.D.C. and A.Y.I.; experimental investigation, A.Y.I.; data curation, Y.D.C.; writing—original draft preparation, Y.D.C. and A.Y.I.; writing—review and editing, Y.D.C.; supervision, Y.D.C.; project administration, Y.D.C.; funding acquisition, Y.D.C. All authors have read and agreed to the published version of the manuscript.

Funding: The work was supported by the Russian Science Foundation (project 19-19-00598-P, "Hydrodynamics and energetics of drops and droplet jets: formation, motion, break-up, interaction with the contact surface", https://rscf.ru/en/project/19-19-00598-P/, accessed on 3 August 2023). The experiments were performed at the stands of the Unique Research Facility, Hydrophysical Complex, Ishlinsky Institute for Problems in Mechanics, Russian Academy of Sciences.

Institutional Review Board Statement: Not applicable.

Informed Consent Statement: Not applicable.

Data Availability Statement: Not applicable.

Conflicts of Interest: The authors declare no conflict of interest.

References

1. Rogers, W.B. On the formation of rotating rings by air and liquids under certain conditions of discharge. *Am. J. Sci. Second Ser.* **1858**, *26*, 246–258.
2. Thomson, J.J.; Newall, H.F. On the formation of vortex rings by drops falling into liquids, and some allied phenomena. *Proc. R. Soc. Lond.* **1885**, *29*, 417–436. [CrossRef]
3. Worthington, A.M.; Cole, R.S. Impact with a liquid surface, studied by the aid of instantaneous photography. *Phil. Trans. R. Soc. Lond. A* **1897**, *189*, 137–148. [CrossRef]
4. Thompson, D.W. *On Growth and Form*, 2nd ed.; Dover: Mineola, NY, USA, 1992; 1116p.
5. Alford, R.S.; Emling, J.W. Underwater ambient noise. *J. Mar. Res.* **1948**, *7*, 410–429. Available online: https://elischolar.library.yale.edu/journal_of_marine_research/679 (accessed on 26 September 2023).
6. Franz, G.J. Splashes as sources of sound in liquids. *J. Acoust. Soc. Am.* **1959**, *31*, 1080–1096. [CrossRef]
7. Edgerton, H.E.; Killian, J.R., Jr. *Flash*; Hale, Cushman and Flint: Boston, MA, USA, 1939; 203p.
8. Zhang, L.V.; Toole, J.; Fezzaa, K.; Deegan, R.D. Splashing from drop impact into a deep pool: Multiplicity of jets and the failure of conventional scaling. *J. Fluid Mech.* **2012**, *703*, 402–413. [CrossRef]
9. Lee, J.S.; Weon, B.M.; Park, S.J.; Kim, J.T.; Pyo, J.; Fezzaa, K.; Je, J.H. Air evolution during drop impact on liquid pool. *Sci. Rep.* **2020**, *10*, 5790. [CrossRef]
10. Thoroddsen, S.T.; Etoh, T.G.; Takehara, K. High-Speed Imaging of Drops and Bubbles. *Ann. Rev. Fluid Mech.* **2008**, *40*, 257–285. [CrossRef]
11. Li, E.Q.; Thoraval, M.-J.; Marston, J.O.; Thoroddsen, S.T. Early azimuthal instability during drop impact. *J. Fluid Mech.* **2018**, *848*, 821–835. [CrossRef]
12. Blanken, N.; Saleem, M.S.; Thoraval, M.-J.; Antonini, C. Impact of compound drops: A perspective. *Curr. Opin. Colloid Interface Sci.* **2020**, *51*, 101389. [CrossRef]
13. Moghtadernejad, S.; Lee, C.; Jadidi, M. An introduction of droplet impact dynamics to engineering students. *Fluids* **2020**, *5*, 107. [CrossRef]
14. Lherm, V.; Deguen, R. Velocity field and cavity dynamics in drop impact experiments. *J. Fluid Mech.* **2023**, *962*, A21. [CrossRef]
15. Zhang, L.V.; Brunet, P.; Eggers, K.; Deegan, R.D. Wavelength selection in the crown splash. *Phys. Fluids* **2010**, *22*, 122105. [CrossRef]
16. Chashechkin, Y.D.; Prokhorov, V.E. Drop-impact hydrodynamics: Short waves on a surface of the crown. *Dokl. Phys.* **2013**, *58*, 296–300. [CrossRef]
17. Guilizzoni, M.; Santini, M.; Fest-Santini, S. Synchronized Multiple Drop Impacts into a Deep Pool. *Fluids* **2019**, *4*, 141. [CrossRef]
18. Batchelor, G.K. *An Introduction to Fluid Dynamics*; Cambridge University Press: Cambridge, UK, 1967; 615p.
19. Landau, L.D.; Lifshitz, E.M. *Fluid Mechanics. V.6. Course of Theoretical Physics*; Pergamon Press: Oxford, UK, 1987; 560p.
20. Muller, P. *The Equations of Oceanic Motions*; CUP: Cambridge, UK, 2006; 292p.
21. Vallis, G.K. *Atmospheric and Oceanic Fluid Dynamics*; CUP: Cambridge, UK, 2017; 745p.
22. Okabe, J.; Inoue, S. *The Generation of Vortex Rings, II*; Reports of Research Institute for Applied Mechanics; Kyushu University: Fukuoka City, Japan, 1961; Volume 9, pp. 147–161.

23. Peck, B.; Sigurdson, L. The three-dimensional vortex structure of an impacting water drop. *Phys. Fluids* **1994**, *6*, 564–576. [CrossRef]
24. Sigurdson, L. Flow visualization in turbulent large-scale structure research. *Flow Vis. Soc. Japan Atlas Vis.* **1997**, *3*, 99–113.
25. Sigurdson, L. Atom Bomb/Water Drop. In *A Gallery of Fluid Motion*; Samimy, M., Breuer, K.S., Leal, L.G., Steen, P.H., Eds.; Cambridge University Press: Cambridge, UK, 2003; 128p.
26. Chashechkin, Y.D.; Ilinykh, A.Y. Delay of cavity formation in the intrusive coalescence of a free falling drop with target fluid. *Dokl. Phys.* **2021**, *66*, 20–25. [CrossRef]
27. Ching, B.; Golay, M.W.; Johnson, T.J. Droplet impacts upon liquid surfaces. *Science* **1984**, *226*, 535–537. [CrossRef]
28. Engel, O.G. Crater depth in fluid impacts. *J. Appl. Phys.* **1966**, *37*, 1798–1808. [CrossRef]
29. Engel, O.G. Initial pressure, initial flow velocity, and the time dependence of crater depth in fluid impacts. *J. Appl. Phys.* **1967**, *38*, 3935–3940. [CrossRef]
30. Cai, Y.K. Phenomena of a liquid drop falling to a liquid surface. *Exp. Fluids* **1989**, *37*, 388–394. [CrossRef]
31. Brutin, D. Drop impingement on a deep liquid surface: Study of a crater's sinking dynamics. *Comptes Rendus Mec.* **2003**, *331*, 61–66. [CrossRef]
32. Bisighini, A.; Cossali, G.E.; Tropea, C.; Roisman, I.V. Crater evolution after the impact of a drop onto a semi-infinite liquid target. *Phys. Rev. E* **2010**, *82*, 036319. [CrossRef]
33. Oguz, H.; Prosperetti, A. Surface-tension effects in the contact of liquid surfaces. *J. Fluid Mech.* **1989**, *203*, 149–171. [CrossRef]
34. Oguz, H.; Prosperetti, A. Bubble entrainment by the impact of drops on liquid surfaces. *J. Fluid Mec.* **1990**, *219*, 143–179. [CrossRef]
35. Ray, B.; Biswas, G.; Sharma, A. Bubble pinch-of and scaling during liquid drop impact on liquid pool. *Phys. Fluids* **2012**, *24*, 082108. [CrossRef]
36. Prokhorov, V.E. Underwater gas bubbles produced by droplet impact: Mechanism to trigger volumetric oscillations. *Phys. Fluids* **2023**, *35*, 033314. [CrossRef]
37. Prokhorov, V.E. Acoustics of oscillating bubbles when a drop hits the water surface. *Phys. Fluids* **2021**, *33*, 083314. [CrossRef]
38. Beacham, S.T.; Tilger, C.F.; Oehlschlaeger, M.A. Sound generation by water drop impact on surfaces. *Exp. Therm. Fluid Sci.* **2020**, *117*, 110138. [CrossRef]
39. Prosperetti, A.; Oguz, H.N. The impact of drops on liquid surfaces and the underwater noise of rain. *Ann. Rev. Fluid Mech.* **1993**, *25*, 577–602. [CrossRef]
40. Krechetnikov, R.; Homsy, G.M. Crown-forming instability phenomena in the drop splash problem. *J. Colloid Int. Sci.* **2009**, *331*, 555–559. [CrossRef]
41. Agbaglah, G.; Thoraval, M.-J.; Thoroddsen, S.T.; Zhang, L.V.; Fezzaa, K.; Deegan, R.D. Drop impact into a deep pool: Vortex shedding and jet formation. *J. Fluid Mech.* **2015**, *764*, R1. [CrossRef]
42. Zhu, G.-Z.; Li, Z.-H.; Fu, D.-Y. Experiments on ring wave packet generated by water drop. *Chin. Sci. Bull.* **2008**, *53*, 1634–1638. [CrossRef]
43. Rein, M. The transitional regime between coalescing and splashing drops. *J. Fluid Mech.* **1996**, *306*, 145–165. [CrossRef]
44. Leng, L. Splash formation by spherical drops. *J. Fluid Mech.* **2001**, *427*, 73–105. [CrossRef]
45. Ray, B.; Biswas, G.; Sharma, A. Regimes during liquid drop impact on a liquid pool. *J. Fluid Mech.* **2015**, *768*, 492–523. [CrossRef]
46. Deka, H.; Ray, B.; Biswas, G.; Dalal, A.; Tsai, P.-H.; Wang, A.-B. The regime of large bubble entrapment during a single drop impact on a liquid pool. *Phys. Fluids* **2017**, *29*, 092101. [CrossRef]
47. Thoroddsen, S.T.; Etoh, T.G.; Takehara, K. Air entrapment under an impacting drop. *J. Fluid Mech.* **2003**, *478*, 125–134. [CrossRef]
48. Tran, T.; De Maleprade, H.; Sun, C.; Lohse, D. Air entrainment during impact of droplets on liquid surfaces. *J. Fluid Mech.* **2013**, *726*, R3. [CrossRef]
49. Fezzaa, K.; Wang, Y. Ultrafast X-ray phase-contrast imaging of the initial coalescence phase of two water droplets. *Phys. Rev. Lett.* **2008**, *100*, 104501. [CrossRef]
50. Fink, J.; Gault, D.; Greeley, R. The Effect of Viscosity on Impact Cratering and Possible Application to the Icy Satellites of Saturn and Jupiter. *J. Geoph. Res. Solid Earth* **1984**, *89*, 417–423. [CrossRef]
51. Feifei, J.; Kai, S.; Peng, Z.; Cuicui, Y.; Tianyou, W. Marangoni effect on the impact of droplets onto a liquid-gas interface. *Phys. Rev. Fluids* **2020**, *5*, 073605. [CrossRef]
52. Boruah, M.P.; Randive, P.R.; Pati, S.; Chakraborty, S. Morpho-dynamic evolution due to inertia-mediated impact of a compound drop on a deep liquid pool. *Phys. Fluids* **2022**, *34*, 032106. [CrossRef]
53. Breitenbach, J.; Roisman, I.V.; Tropea, C. From drop impact physics to spray cooling models: A critical review. *Exp. Fluids* **2018**, *59*, 55. [CrossRef]
54. Zhang, Y.J.; Liu, P.Q.; Qu, Q.L.; Hu, T.X. Energy conversion during the crown evolution of the drop impact upon films. *Int. J. Multiph. Flow* **2019**, *115*, 40–61. [CrossRef]
55. Ma, H.; Liu, C.; Li, X.; Huang, H.; Dong, J. Deformation characteristics and energy conversion during droplet impact on a water surface. *Phys. Fluids* **2019**, *31*, 062108. [CrossRef]
56. Gielen, M.V.; Sleutel, P.; Benschop, J.; Riepen, M.; Voronina, V.; Visser, C.W.; Lohse, D.; Snoeijer, J.H.; Versluis, M.; Gelderblom, H. Oblique drop impact onto a deep liquid pool. *Phys. Rev. Fluids* **2017**, *2*, 083602. [CrossRef]
57. Liu, X. Experimental study of drop impact on deep-water surface in the presence of wind. *J. Phys. Oceanogr.* **2018**, *48*, 329–341. [CrossRef]

58. van Hinsberg, N.P.; Charbonneau-Grandmaison, M. Single-drop impingement onto a wavy liquid film and description of the asymmetrical cavity dynamics. *Phys. Rev. E* **2015**, *92*, 013004. [CrossRef]
59. Singh, A.; Kumar, P. Droplet impact dynamics onto a deep liquid pool of wavy free surface. *Phys. Fluids* **2022**, *34*, 022107. [CrossRef]
60. Burzynski, D.A.; Bansmer, S.E. Droplet splashing on thin moving films at high Weber numbers. *Int. J. Multiph. Flow* **2018**, *101*, 202–211. [CrossRef]
61. Burzynski, D.A.; Bansmer, S.E. Role of surrounding gas in the outcome of droplet splashing. *Phys. Rev. Fluids* **2019**, *4*, 073601. [CrossRef]
62. Chashechkin, Y.D.; Ilinykh, A.Y. Intrusive and impact modes of a falling drop coalescence with a target fluid at rest. *Axioms* **2023**, *12*, 374. [CrossRef]
63. Ersoy, N.E.; Eslamian, M. Capillary surface wave formation and mixing of miscible liquids during droplet impact onto a liquid film. *Phys. Fluids.* **2019**, *31*, 012107. [CrossRef]
64. Chashechkin, Y.D.; Ilinykh, A.Y. Banded Structures in the Distribution Pattern of a Drop over the Surface of the Host Liquid. *Dokl. Phys.* **2018**, *63*, 282–287. [CrossRef]
65. Gibbs, J.W. *Elementary Principles in Statistical Mechanics*; Scribner's and Sons: New York, NY, USA, 1902; 207p.
66. Feistel, R. Thermodynamic properties of seawater, ice and humid air: TEOS-10, before and beyond. *Ocean Sci.* **2018**, *14*, 471–502. [CrossRef]
67. Harvey, A.H.; Hrubý, J.; Meier, K. Improved and Always Improving: Reference Formulations for Thermophysical Properties of Water. *J. Phys. Chem. Ref. Data* **2023**, *52*, 011501. [CrossRef]
68. Popov, N.I.; Fedorov, K.N.; Orlov, V.M. *Sea Water*; Nauka: Moscow, Russia, 1979; 330p. (In Russian)
69. Nayfeh, A.H. *Introduction to Perturbation Techniques*; John Wiley & Sons: New York, NY, USA, 2011; 533p.
70. Chashechkin, Y.D. Foundations of engineering mathematics applied for fluid flows. *Axioms* **2021**, *10*, 286. [CrossRef]
71. Chashechkin, Y.D.; Ochirov, A.A. Periodic waves and ligaments on the surface of a viscous exponentially stratified fluid in a uniform gravity field. *Axioms* **2022**, *11*, 402. [CrossRef]
72. Chashechkin, Y.D.; Kistovich, A.V. Classification of three-dimensional periodic fluid flows. *Dokl. Phys.* **2004**, *49*, 183–186. [CrossRef]
73. Eisenberg, D.; Kauzmann, W. *The Structure and Properties of Water (Oxford Classic Texts in the Physical Sciences)*; Oxford University Press: Oxford, UK, 2005; 308p.
74. Malenkov, G.G. Structure and dynamics of surfaces of thin films and water microdroplets. *Colloid J.* **2010**, *72*, 649–659. [CrossRef]
75. Teschke, O.; de Souza, E.F. Water molecule clusters measured at water/air interfaces using atomic force microscopy. *Phys. Chem. Chem. Phys.* **2005**, *7*, 3856–3865. [CrossRef] [PubMed]
76. Bunkin, N.F.; Suyazov, N.V.; Shkirin, A.V.; Ignat'ev, P.S.; Indukaev, K.V. Study of Nanostructure of highly purified water by measuring scattering matrix elements of laser radiation. *Phys. Wave Phenom.* **2008**, *16*, 243–260. [CrossRef]
77. Chashechkin, Y.D.; Ilinykh, A.Y. Drop decay into individual fibers at the boundary of the contact area with a target fluid. *Dokl. Phys.* **2021**, *66*, 101–105. [CrossRef]
78. Chashechkin, Y.D.; Ilinykh, A.Y. Transfer of Drop Material during the Formation of a Primary Cavern. *Dokl. Phys.* **2023**, *68*, 1–10.
79. Chashechkin, Y.D. Conventional partial and new complete solutions of the fundamental equations of fluid mechanics in the problem of periodic internal waves with accompanying ligaments generation. *Mathematics* **2021**, *9*, 586. [CrossRef]
80. Hydrophysical Complex for Modeling Hydrodynamic Processes in the Environment and Their Impact on Underwater Technical Objects, as Well as the Distribution of Impurities in the Ocean and Atmosphere, Institute for Problems in Mechanics RAS. Available online: http://ipmnet.ru/uniqequip/gfk (accessed on 2 August 2023).
81. Bisighin, A.; Cossali, G.E. High-speed visualization of interface phenomena: Single and double drop impacts onto a deep liquid layer. *J. Vis.* **2011**, *14*, 103–110. [CrossRef]
82. Gillot, G.; Derec, C.; Génevaux, J.-M.; Simon, L.; Benyahia, L. A new insight on a mechanism of airborne and underwater sound of a drop impacting a liquid surface. *Phys. Fluids* **2020**, *32*, 062004. [CrossRef]
83. Visser, C.W.; Frommhold, P.E.; Wildeman, S.; Mettin, R.; Lohse, D.; Sun, C. Dynamics of high-speed micro-drop impact: Numerical simulations and experiments at frame to-frame times below 100 ns. *Soft Matter.* **2015**, *11*, 1708. [CrossRef]
84. Castrejón-Pita, A.A.; Castrejón-Pita, J.R.; Hutchings, I.M. Experimental observation of von Kármán vortices during drop impact. *Phys. Rev. E* **2012**, *86*, 045301(R). [CrossRef] [PubMed]
85. Chashechkin, Y.D. Oscillations and Short Waves on a Free Falling Drop Surface (Experiment and Theory). In Proceedings of the Topical Problems of Fluid Mech, Prague, Czech Republic, 20–22 February 2019; pp. 45–52.
86. Chashechkin, Y.D.; Ilinykh, A.Y. Multiple emissions of splashes upon drop impact. *Dokl. Phys.* **2020**, *65*, 384–388. [CrossRef]
87. Chashechkin, Y.D.; Ilinykh, A.Y. Capillary waves on the surface of a droplet falling into a liquid. *Dokl. Phys.* **2015**, *60*, 548–554. [CrossRef]
88. Chashechkin, Y.D.; Ilinykh, A.Y. Formation of a system of inclined loops in the flows of a drop impact. *Dokl. Phys.* **2021**, *66*, 234–242. [CrossRef]
89. Kuhlman, J.M.; Hillen, N.L. Droplet impact cavity film thickness measurements versus time after drop impact and cavity radius for thin static residual liquid layer thicknesses. *Exp. Therm. Fluid Sci.* **2016**, *77*, 246–256. [CrossRef]

90. Settles, G.S. *Schlieren and Shadowgraph Techniques: Visualizing Phenomena in Transparent Media*; Springer: Berlin, Germany, 2001; 376p. [CrossRef]
91. Xu, Z.; Wang, T.; Che, Z. Cavity deformation and bubble entrapment during the impact of droplets on a liquid pool. *Phys. Rev. E* **2022**, *106*, 055108. [CrossRef]
92. Berberovic, E.; Van Hinsberg, N.P.; Jakirlic, S.; Roisman, I.V.; Tropea, C. Drop impact onto a liquid layer of finite thickness: Dynamics of the cavity evolution. *Phys. Rev. E* **2009**, *79*, 036306. [CrossRef]
93. Guilizzoni, M.; Frontera, G. Crater Depth after the Impact of Multiple Drops into Deep Pools. *Fluids* **2022**, *7*, 50. [CrossRef]
94. Brackbill, J.U.; Kothe, D.B.; Zemach, C. A new method for modeling surface tension effects on fluid. *J. Comp. Phys.* **1992**, *100*, 335–354. [CrossRef]
95. van der Waals, J.D. The thermodynamic theory of capillarity flow under the hypothesis of a continuous variation of density. (Verhandel. Konink. Akad. Weten. Amsterdam, 1893). English transl. by Rowlinson, J.S. *J. Statist. Phys.* **1979**, *20*, 197–244. [CrossRef]
96. Cahn, J.W. On spinodal decomposition. *Acta Metall.* **1961**, *9*, 795–801. [CrossRef]
97. Cahn, J.W.; Hilliard, J.E. Free Energy of a Non-Uniform System. Part I. *Interfacial Free Energy J. Chem. Phys.* **1958**, *28*, 258–267. [CrossRef]
98. Šikalo, Š.; Wilhelm, H.-D.; Roisman, I.V.; Jakirlić, S.; Tropea, C. Dynamic contact angle of spreading droplets: Experiments and simulations. *Phys. Fluids* **2005**, *17*, 062103. [CrossRef]
99. Jacqmin, D. Contact-line dynamics of a diffuse fluid interface. *J. Fluid Mech.* **2000**, *402*, 57–88. [CrossRef]
100. Khatavkar, V.; Anderson, P.; Duineveld, P.; Meijer, H. Diffuse-interface modelling of droplet impact. *J. Fluid Mech.* **2007**, *581*, 97–127. [CrossRef]
101. Karim, M.A.; Suszynski, W.J. Physics of dynamic contact line: Hydrodynamics theory versus molecular kinetic theory. *Fluids* **2022**, *7*, 318. [CrossRef]

fluids

MDPI

Article

Numerical Simulation of Dropwise Condensation of Steam over Hybrid Surfaces via New Non-Dimensional Heat Transfer Model

Giulio Croce *,† and Nicola Suzzi †

Dipartimento Politecnico di Ingegneria e Architettura (DPIA), Università Degli Studi di Udine, Via delle Scienze 206, 33100 Udine, Italy; nicola.suzzi@uniud.it

* Correspondence: giulio.croce@uniud.it

† These authors contributed equally to this work.

Abstract: Dropwise condensation (DWC) of steam over hybrid hydrophobic–hydrophilic surfaces is numerically investigated via a phenomenological, Lagrangian model. The full non-dimensionalization of the heat transfer model, needed to determine the droplet growth, allows for generalization of computational results. Hybrid surfaces characterized by recursive geometries are implemented via the introduction of proper boundary conditions. The numerical size distribution of both the large and the small droplet populations, crucial for development of simplified, statistically sound models, is compared with empirical and theoretical correlations. Then, the validation with experimental data involving DWC over an hybrid surface is successfully conducted and the heat flux is enhanced under different operating conditions via hybrid geometry optimization.

Keywords: dropwise condensation; hybrid surface; Lagrangian model

1. Introduction

Heterogeneous steam condensation is involved in engineering applications such as heat exchanger design. In particular, dropwise condensation (DWC) over an inclined surface is known to enhance heat transfer performance, in terms of specific heat flux, of about one order of magnitude, compared with the standard filmwise condensation (FWC) mode. In fact, the continuous surface renewal, due to gravity driven motion of the large droplets, reduces the thermal resistance of the liquid droplets population [1] under DWC regime. Such a mechanism is promoted by hydrophobic surfaces, characterized by static contact angle $\theta > 90°$, and superhydrophobic surfaces, $\theta > 150°$, due to small contact angle hysteresis, which gives smaller departure radii, compared with hydrophilic surfaces [2]. Nanostructure superhydrophobic surfaces promote coalescence-induced jumping of droplets, further reducing the departure radius of the droplets and enhancing heat transfer [2,3]. Recently, hybrid hydrophobic–hydrophilic and patterned surfaces are getting more and more attentions [2,4–8]. In particular, hybrid surfaces often appears as a pattern of recursive, hydrophobic–hydrophilic regions of different geometries [2,4,9]. Appropriate surface configurations allow to control the maximum droplet size over the hydrophobic pattern, where DWC is promoted, while the hydrophilic regions enhance the liquid drainage. For example, Peng et al. [2] conducted an experimental campaign on vapor steam condensation over a vertical hybrid surface, composed by alternate hydrophobic–hydrophilic stripes of width L_D and L_F, looking for the best geometrical configuration in terms of heat flux. Alwazzan et al. [4,5] also studied steam condensation over a patterned cylindrical surface with alternate hydrophobic and less hydrophobic vertical stripes. More complex geometries have also been investigated in the experimental literature, such as inverted V-shaped channels design [9], tested in case of moist air condensation and characterized by alternate hydrophobic and hydrophilic inclined stripes and a vertical hydrophilic region, which enhance condensate removal. The optimization of such hybrid geometries (for example, in terms of optimal hydrophobic width L_D and hydrophilic width L_F for the test case investigated in [2]) plays a crucial role and represents a complex issue, as hydrophilic regions, which enhance heat transfer promoting droplet removal from non-wettable surface, are in turn covered by a continuous film and are thus characterized by lower heat transfer performance.

A deep theoretical investigation on the effect of morphology of superhydrophobic nanograsses surfaces inducing droplet jumping was conducted in ref. [10], with the predictions of the proposed model compared with data from the experimental campaign. Various methods can be used to achieve hydrophobic surfaces [4,11–13]. However, the surface characterization is usually conducted only in terms of static contact angle and contact angle hysteresis, while the determination of the actual density of nucleating sites, which strongly influences heat transfer performance as well as the wettability properties, still remains an open problem. The empirical correlation of Rose [14], which gives the nucleation density as a function of the critical nucleating radius and, thus, as a function of the substrate subcooling, is known to overestimate the actual nucleation density, which usually ranges in $10^9 \div 10^{15}\,\mathrm{m}^{-2}$ for pure steam condensation. An updated theoretical model was proposed in refs. [15,16], where the effect of the hydrophobic coating was introduced. In particular, the effective nucleation density, which must be used in Rose correlation, is greater that the critical one and depends on the substrate subcooling, on the surface wettability and on the coating thermal resistance. However, the effect of surface roughness and morphology is not taken into account and, still, the knowledge of the coating characteristics in terms of thickness and thermal conductivity is not easy to determine in practical problems.

The aim of this work is the numerical investigation of steam condensation over a hybrid surface via a Lagrangian approach. Thus, nucleation, growth (due to condensation process and eventual coalescence), and departure of every single droplet is traced during its whole life. An in-house code, previously developed in FORTRAN90 language to study moist air condensation, is adapted to simulate steam condensation over hybrid surfaces. The steam condensation model is formulated in a non-dimensional form in order to generalize the problem. This allows to reduce the number of computations required to characterize and optimize hybrid hydrophobic–hydrophilic surfaces. Furthermore, the procedure can be used as a postprocessing tool to estimate the nucleation density of the hydrophobic surface from experimental data. The numerical model is firstly validated in terms of droplet size distribution with experimental [17] and theoretical [1] correlations. Then, condensation over an hydrophobic–hydrophilic surface as the one in ref. [2] is investigated. An efficient procedure to characterize the hydrophobic surface in terms of nucleation density, based on the literature experimental data of ref. [2], is proposed. Then, the hybrid geometry is optimized, looking for the best compromise in terms of global heat flux, and the optimal geometrical configuration, numerically obtained, is compared with experimental evidence of [2].

2. Mathematical Model

The evolution of a condensing droplet population over an hydrophobic surface is analyzed via a phenomenological, Lagrangian model. Condensation of pure, saturated vapor steam is considered. Every droplet is modeled as a spherical cap with prescribed contact angle. The main physical mechanisms are:

1. Generation of nuclei at random locations over the computational domain (only dry spots are active);
2. Growth of the whole droplet population due to condensation;
3. Coalescence check (via loop on every droplet of the population, looking for overlap condition with other droplets) and implementation;
4. Hybrid surface implementation as a dedicated boundary condition;
5. Next time step.

The nucleating sites are randomly initialized over the $L_x \times L_y$ computational domain, according to the imposed nucleation density ρ_n. Nucleation is allowed on dry spots only. Thus, a new drop with radius r_n is initialized at each time step on the kth site, with $k = 1, \cdots, \lfloor \rho_n L_x L_y \rfloor$, if the site stands on a dry region. The location of the nucleating sites does not change during the computation. The droplet growth is driven by two combined mechanisms: condensation of pure vapor at the droplet free surface; coalescence with other droplets (i.e., droplets at different locations may interact with each other). The condensation model is derived in a non-dimensional form. The coalescence between two droplets, which is assumed as an instantaneous event, takes place if contact between neighbour droplets is detected. In order to simulate recursive geometries, periodic boundary conditions were eventually implemented, letting two droplets standing at the opposite sides of the computational domain coalesce. The hybrid surface implementation replicates the experimental test case of ref. [2], where a pattern of hydrophobic and hydrophilic stripes on a flat plate is used to enhance droplet regeneration over the hydrophobic surface. The stripes are thin enough to prevent droplet growth above the gravity induced motion threshold: thus, droplet motion due to gravity and/or air shear is not considered and the contact angle is assumed constant and equal to the static contact angle during droplet growth.

An in-house code, previously developed in FORTRAN90 language and validated in case of dropwise condensation of moist air [18], was adapted. The non-dimensional condensation model of vapor steam was implemented. Furthermore, different boundary conditions were added, including periodic conditions and hybrid surface implementation. Parallelization was performed via OpenMP library for a shared memory machine, in order to speed up computations.

2.1. Droplet Growth

Assuming saturation temperature at the droplet free surface, the heat transfer can be modeled via a series of two thermal resistances, related to conduction through coating layer and conduction through liquid droplet:

$$q = \frac{T_s - T_w}{\frac{\delta_c}{\lambda_c A_b} + \frac{\delta_{eq}}{\lambda_l A_b}},\ A_b = \pi r^2 \sin^2\theta,\ F_\theta = \frac{A_b}{r\,\delta_{eq}} = \frac{4\pi \sin\theta}{\theta} \tag{1}$$

where A_b is the droplet base radius, δ_c is the coating thickness, $\delta_{eq} = r\,\theta\,\sin\theta/4$ is the equivalent thickness related to conduction through the droplet core, calculated according to refs. [15,16]. For convenience, we refer to non-dimensional parameter F_θ, which only depends on contact angle, instead of δ_{eq}. Further assuming that condensation drives heat transfer process, the heat flux through the droplet free surface can be expressed as:

$$q = \dot{m}_c\, h_{lv} = 3\,\rho_l\, V_\theta\, h_{lv}\, r^2 \frac{\mathrm{d}r}{\mathrm{d}t},\ V_\theta = \frac{\pi}{3}\,(1-\cos\theta)^2\,(2+\cos\theta) \tag{2}$$

The non-dimensional coefficient V_θ, which depends on the contact angle, is the droplet volume (estimated as the volume of a spherical cap with contact angle θ and radius r) divided by the cube of its radius. Combining Equations (1) and (2) gives the droplet growth rate:

$$3\,\rho_l\, V_\theta\, h_{lv} \frac{\mathrm{d}r}{\mathrm{d}t} = \frac{T_s - T_w}{\frac{\delta_c}{\lambda_c\,\pi\,\sin^2\theta} + \frac{r}{\lambda_l F_\theta}} \tag{3}$$

Further integrating Equation (3) in time yields the droplet radius growth,

$$r(t) = -\ell_c + \sqrt{(\ell_c + r_n)^2 + \frac{2}{3}\frac{\lambda_l\, F_\theta\,(T_s - T_w)}{\rho_l\, V_\theta\, h_{lv}}\,t},\ \ell_c = \frac{\lambda_l\, F_\theta\, \delta_c}{\lambda_c\,\pi\,\sin^2\theta} \tag{4}$$

where r_n is the nucleating radius. We define the scaling length r_0 as the critical coalescence radius and the scaling time t_0 as the time elapse required by a newly nucleated droplet to grow from r_n to r_0:

$$r_0 = \frac{1}{\sqrt{4\rho_n}}, \; t_0 = \left[(\ell_c + r_0)^2 - (\ell_c + r_n)^2\right] \frac{3}{2} \frac{\rho_l V_\theta h_{lv}}{\lambda_l F_\theta (T_s - T_w)} \tag{5}$$

Further introducing the following non-dimensional quantities,

$$\mathbf{x}^* = \frac{\mathbf{x}}{r_0}, \; r^* = \frac{r}{r_0}, \; t^* = \frac{t}{t_0}, \; \xi = \frac{\ell_c}{r_0}, \; \rho_n^* = \rho_n {r_0}^2 \tag{6}$$

where ξ is the ratio between the thermal resistance of the coating and the thermal resistance of a droplet of radius r_0, Equation (4), which gives the droplet growth, can be recast in a non-dimensional form:

$$r^* = -\xi + \sqrt{(r_n^* + \xi)^2 + \left[(1+\xi)^2 - (r_n^* + \xi)^2\right] t^*} \tag{7}$$

Note that the non-dimensional droplet growth, which is shown in Figure 1a at different values of the non-dimensional parameter ξ does not depend on neither the fluid properties nor the substrate subcooling. Furthermore, the non-dimensional nucleation density is not a model parameter as it is always equal to:

$$\rho_n^* = \rho_n {r_0}^2 = \frac{1}{4} \tag{8}$$

where r_0 is defined by Equation (5).

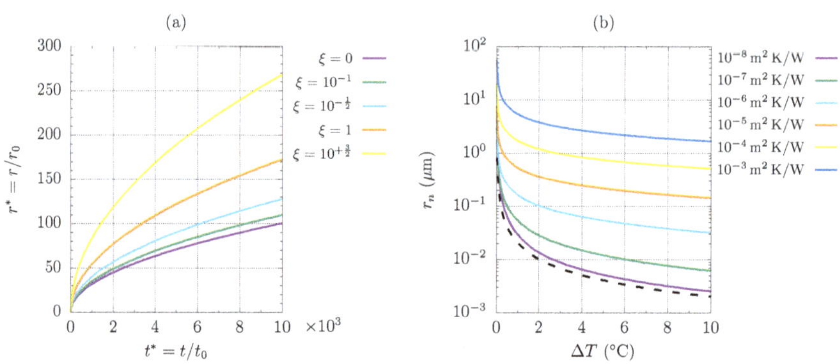

Figure 1. Non-dimensional droplet growth model, Equation (7), at different values of the coating parameter ξ and fixed $r_n^* = 10^{-3}$ (**a**). Nucleating radius as a function of subcooling ΔT and coating thermal resistance $R_{th,c}$; $\theta = 120°$, water properties evaluated at $T_s = 100$ °C (**b**). The dotted line represents the critical nucleating radius, Equation (11).

2.2. Comparison with Literature Growth Model

The condensation model, see Equation (4), was compared with the literature model of ref. [19] in terms of growth of a single, isolated droplet laying on a hydrophobic surface at a given subcooling. If compared with ref. [19], our growth model, Equation (4), neglects the vapor–liquid interface thermal resistance and the interface curvature thermal resistance, defined as [19]:

$$R_{th,i} = \frac{1}{2\,\pi\, r^2\, \alpha_i\, (1-\cos\theta)}, \; \alpha_i = \frac{2\,\sigma_c}{2-\sigma_c}\sqrt{\frac{M}{2\,\pi\, R_g\, T_s}}\,\frac{\rho_v\, {h_{lv}}^2}{T_s} \tag{9}$$

$$R_{th,\sigma} = \frac{2\,\sigma_{lv}\, T_s}{\rho_l\, h_{lv}\, r\, q} \tag{10}$$

where $\sigma_c = 1$ for pure vapor condensation and q is the heat flux through the droplet. It can be verified that $R_{th,i}$ is negligible in case of pure vapor condensation, while $R_{th,\sigma}$ is important only when the droplet radius approaches the critical nucleating radius r_{cr}:

$$r_{cr} = \frac{2\,\sigma_{lv}\, T_s}{\rho_l\, h_{lv}\,(T_s - T_w)} \tag{11}$$

In fact, the model of ref. [19] predicts droplet blow up when $r \le r_{cr}$. Thus, we must impose a realistic value of the effective nucleating radius r_n to get accurate results and perform a proper comparison with our model; see Equation (4). The effective nucleating radius, which is greater than r_{cr} in presence of a hydrophobic coating [15], was derived at different substrate subcooling and coating characteristics via minimization of the availability function $\Psi(r)$, following the procedure proposed by refs. [15,16,19]. The effective nucleating radius of an hydrophobic surface is shown in Figure 1b as a function of the substrate subcooling and of the coating specific thermal resistance, $R''_{th,c} = \delta_c/\lambda_c$, while the dashed line represents the critical nucleating radius. Note that the coating layer plays a crucial role in the determination of the nucleating radius, which, in turn, affects the actual nucleation density [14,16]. Equation (14), presented by ref. [19], which is an ordinary differential equation modeling the droplet growth, was numerically solved by imposing the initial condition $r(t_0) = r_n$ and compared with the proposed analytical model, see Equation (4), in case of water vapor condensation at $T_s = 100$ °C under different substrate subcooling and coating layer properties. The results shows a great agreement between the literature model [19] and the proposed growth model through the whole range of investigated subcooling; see Figure 2.

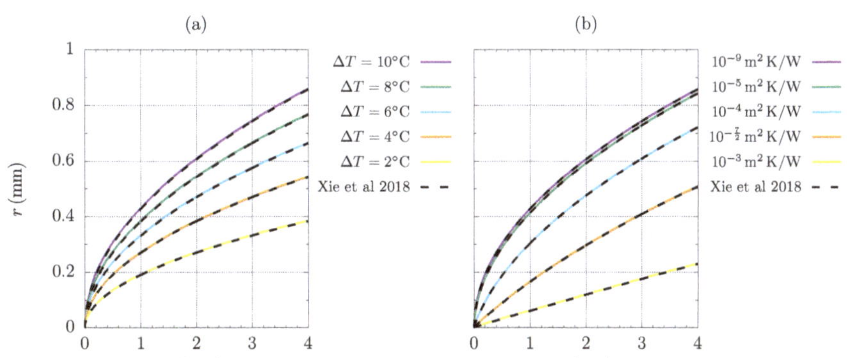

Figure 2. Droplet radius versus time at different substrate subcooling ΔT, fixed $R_{th,c} = 10^{-7}\,\mathrm{m^2\,K/m}$ (**a**), at different coating thermal resistances $R_{th,c}$, fixed $\Delta T = 10$ °C (**b**). $\theta = 120°$, water properties evaluated at $T_s = 100$ °C. The continuous lines represent Equation (4) and the dotted lines are obtained via numerical integration of the droplet growth model by ref. [19].

2.3. Droplet Coalescence

During condensation, droplets also grow due to coalescence. At each time step, a coalescence check is performed for every droplet belonging to the population, looking for an overlap with neighboring droplets. Since it is assumed that the coalescence is an instantaneous process, all the detected coalescences are implemented within the time step. Assuming spherical droplets on a hydrophobic surface, the critical droplet distance between two droplets leading to overlap condition is equal to [18]:

$$\Delta^*_{coa} = \sqrt{(r^*_1 + r^*_2)^2 - (r^*_1 \cos\theta_1 - r^*_2 \cos\theta_2)^2} \tag{12}$$

The coalesced droplet is placed at the center of gravity of the pre-merged droplets and its mass is equal to the sum of the pre-merged droplets mass:

$$\mathbf{x}^*_3 = \frac{m^*_1 \mathbf{x}^*_1 + m^*_2 \mathbf{x}^*_2}{m^*_1 + m^*_2}, \; m^*_3 = m^*_1 + m^*_2, \tag{13}$$

$$m^* = \frac{\pi}{3} r^{*3} (1 - \cos\theta)^2 (2 + \cos\theta) \tag{14}$$

2.4. Hybrid Surface Implementation

The experimental configuration of ref. [2], characterized by vertical hydrophobic–hydrophilic stripes with prescribed width alternating over a vertical plate, was considered. Dropwise condensation takes place on the hydrophobic regions, while the wettable surface portions are characterized by a continuous liquid film. The migration of droplets from hydrophobic regions to hydrophilic regions due to surface tension forces allows for re-generation of the droplets [2,6], while the falling film ensures liquid shedding from the condensing plate through the hydrophilic regions. In order to enhance the heat transfer, the width of the hydrophobic regions must be much lower than the characteristics droplet moving diameter [2].

In presence of an hybrid surface characterized by recursive geometries such as a series of hydrophobic–hydrophilic stripes (as the one of Figure 3), the computational domain is defined by a single frame of hydrophobic region, with proper boundary conditions implemented through the hydrophobic–hydrophilic boundaries. In particular, following experimental evidences [2], when the base surface of a growing droplet reaches the hydrophobic–hydrophilic boundary, the droplet instantaneously migrates to the more wettable region due to surface tension forces and, thus, exits the computational domain. Note that droplet migration is modeled as an instantaneous process as well as coalescence. The effect of droplets eventually moving due to gravity is not considered, as the hydrophobic width is lower than the moving diameter in practical cases.

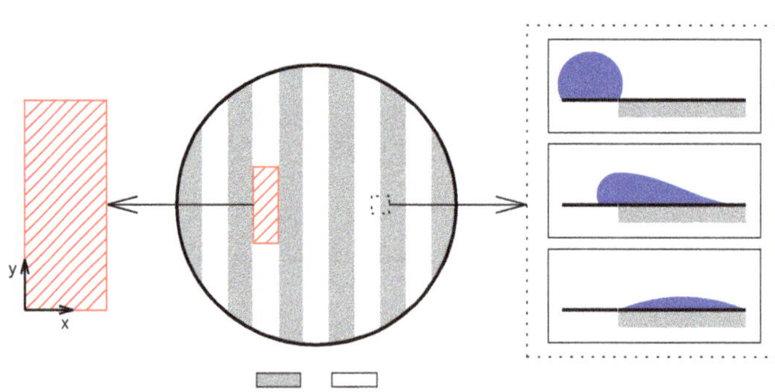

Figure 3. Hybrid surface geometry of ref. [2] and droplet migration (here modeled as an instantaneous process) from a hydrophobic region ($\theta > 90°$) to hydrophilic ($\theta < 90°$) region. The red patterned rectangle denotes the computational domain, where DWC is simulated via Lagrangian model: a periodic condition is applied to both the top and bottom section while a hybrid surface condition is implemented through the lateral boundaries.

3. Result

3.1. Droplet Size Distribution

First, the evolution of a population of droplets condensing on a horizontal plate, of non-dimensional size $L_x^* \times L_y^* = 600 \times 600$, characterized by equilibrium contact angle equal to $\theta = 90°$, was simulated. The eventual presence of a coating layer was neglected, giving $\xi = 0$. A 600×600 computational domain gives 9×10^4 nucleating spots and, thus, ensures that we can get statistically relevant results. After a dependency analysis on the integration time step, it was found that $\Delta t^* \leq 10^{-1}$ does not affect the resulting cumulative size distribution; thus, it was imposed that $\Delta t^* = 5 \times 10^{-2}$. Applying the Rose correlation [14],

$$\rho_n = \frac{0.037}{{r_n}^2} \tag{15}$$

it derives that the non-dimensional nucleating radius is equal to $r_n^* = 2\sqrt{0.037} \simeq 3.84707 \times 10^{-1}$, which was chosen for the current simulation. Periodic conditions were applied through boundaries $x^* = 0, 600$ and $y^* = 0, 600$. The cumulative size distribution of large droplets was computed and validated with the empirical cumulative droplet size distribution, obtained via integration of the well known Le Fevre and Rose empirical correlation [17],

$$N_r''(r) = \int_{r_0}^{r} \frac{r^{-8/3}}{3\,\pi\,{r_{max}}^{1/3}}\,\mathrm{d}r = \frac{{r_0}^{-5/3} - r^{-5/3}}{5\,\pi\,{r_{max}}^{1/3}} \tag{16}$$

where $N_r''(r)$, which has the unit of measure m^{-2}, is the number of droplets per unit surface with a radius ranging in $[r_0, r]$ and r_{max} is the maximum radius of the droplet population. In general, the maximum radius may be the characteristic moving radius [20], in case of gravity and/or shear driven droplet, or it may be derived from geometrical constraints in case of hybrid surfaces [2]. Here, the actual maximum radius of the condensing droplet population was considered. Note that only large droplets, characterized by a radius in the range $r_0 \leq r \leq r_{max}$ were considered to estimate N_r'', because Equation (16) is valid when coalescence drives the droplet growth process rather than condensation. The computed probability of having large droplets with radius in the range $[r_0, r]$, obtained from $N_r''(r)$, was compared with:

$$N_{\%} = \frac{N_r''(r)}{N_r''(r_{max})} = \frac{1 - {r^*}^{-5/3}}{1 - {r_{max}^*}^{-5/3}} \tag{17}$$

Figure 4a shows the droplet population at $t^* = 172$ on the non-dimensional computational domain of size 600×600. The corresponding $N_{\%}$ is compared with Equation (17) in Figure 4b, showing almost a perfect agreement.

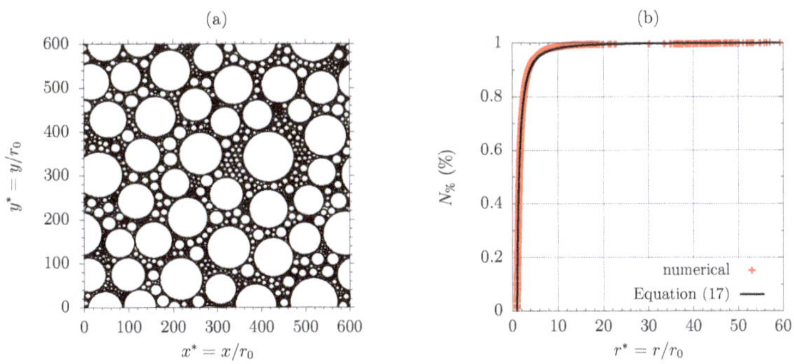

Figure 4. Droplet population at $t^* = 172$ (**a**); comparison between corresponding cumulative droplet size distribution, numerically computed, and empirical correlation, Equation (17) (**b**). $\theta = 90°$, $r_n^* = 2\sqrt{0.037}$, $\xi = 0$.

The small droplet population, characterized by $r_{cr} \le r \le r_0$, in which the growth is driven by a condensation process rather than coalescence, also plays a crucial role in the definition of a simplified, statistically sound model of heat transfer in dropwise condensation [6,20,21]. Due to the tiny size of the small droplets (usually $r < 1\ \mu\text{m}$), poor experimental literature is available, while a correlation, based on the single droplet heat transfer model, was proposed for the small drop size distribution by ref. [1]:

$$n_r'''(r) = N_r'''(r_0)\,\frac{r\,(r_0 - r_{cr})}{r_0\,(r - r_{cr})}\,\frac{A_2\,r + A_3}{A_2\,r_0 + A_3}\,\exp(B_1 + B_2) \tag{18}$$

where $n_r'''(r) = \mathrm{d}n_r''/\mathrm{d}r$ with unit of measure m^{-3}, $A_{2,3}$, $B_{1,2}$, defined according to ref. [1] and r_{cr} is the critical nucleating radius. Equation (18) can be analytically integrated and recast in non-dimensional fashion, yielding the cumulative size distribution of the small droplets $n_r''(r)$, which is the number of droplets per unit surface with a radius lower than r for $r < r_0$:

$$\begin{aligned} n_r''(r)\,{r_0}^2 &= {r_0}^2 \int_{r<r_0}^{r_0} n_r'''(r)\,\mathrm{d}r = {r_0}^3\,N_r'''(r_0)\,\frac{1 - r_{cr}^*}{a_2 + a_3}\left[\exp(B_1 + B_2) - 1\right], \\ B_1 + B_2 &= \frac{a_2}{2}\left(1 - r^{*2}\right) + (a_2\,r_{cr}^* + a_3)\left(1 - r^* + r_{cr}^*\,\log\frac{1 - r_{cr}^*}{r^* - r_{cr}^*}\right) \end{aligned} \tag{19}$$

where the non-dimensional coefficients $a_{2,3}$ are equal to:

$$a_2 = \frac{A_2\,{r_0}^2}{\tau\,A_1},\ a_3 = \frac{A_3\,r_0}{\tau\,A_1} \tag{20}$$

where τ and A_1 are defined in ref. [1]. Further computations were run in order to verify the numerical cumulative size distribution of small droplets with Equation (19). Again, the coating thermal resistance was neglected, leading to $\xi = 0$, while two values of the nucleating radius were imposed: $r_n^* = 2\sqrt{0.037}$, which is consistent with Rose correlation [14]; $r_n^* = 10^{-2}$. A hydrophobic surface with contact angle equal to $\theta = 150°$, of size $L_x^* \times L_y^* = 1000 \times 1000$, was considered. Periodic conditions were applied through the plate boundaries and the integration time step was set to $\Delta t^* = 5 \times 10^{-2}$. The numerical cumulative size distribution (including both small and large drops) was computed from the actual droplet population as,

$$N_r''(r)\,{r_0}^2 = \frac{1}{2\,S^*}\sum_{k=1}^{N_{drop}}\left(1 + \frac{r^* - r_k^*}{|r^* - r_k^*|}\right) \tag{21}$$

where $S^* = 1000 \times 1000$ is the computational domain area and is compared with Equations (16) and (19). Note that, when $R_{th,c} = 0$, the non-dimensional coefficient $a_{2,3}$ in Equation (19) reduces to:

$$a_2 = \frac{11 - 14\,r_{cr}^*}{3},\ a_3 = 0 \tag{22}$$

meaning that the non-dimensional parameter a_2 does not depend on the liquid properties or on the substrate subcooling and contact angle. We further assume that $r_n \sim r_{cr}$ is consistent with the model of ref. [15] in case of negligible coating thermal resistance. As expected, the cumulative size distribution of the large droplet population agrees with the empirical correlation of ref. [17], see Figure 5. The computed, cumulative size distribution of small drops $n_r''(r^*)$ also agrees with the theoretical correlation of ref. [1] in case of $r_n^* = 2\sqrt{0.037}$, while it is lower than the theoretical size distribution for low droplet radius in case of $r_n^* = 10^{-2}$. However, Equation (18) is not reliable in the limit of small r^* as infinite droplet density is predicted when the droplet radius approaches the critical nucleating radius, $\lim_{r \to r_{cr}} n_r'''(r) = +\infty$.

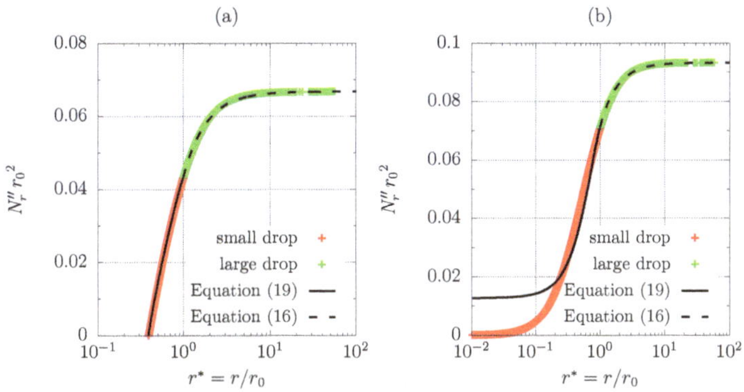

Figure 5. Numerical droplet cumulative size distribution versus empirical correlations for large droplets, Equation (16), and small droplets, Equation (19): $r_n^* = 2\sqrt{0.037}$ (**a**), $r_n^* = 10^{-2}$ (**b**). $t^* = 74$, $\theta = 150°$, $\xi = 0$, 1000×1000 computational domain.

3.2. Hybrid Surface Characterization

The experimental setup of Peng et al. [2] was replicated. Thus, condensation of water vapor at a saturation temperature of $T = 100$ °C (fluid properties listed in Table 1) over a circular, vertical test section of radius $R = 13$ mm, characterized by alternating hydrophilic–hydrophobic vertical stripes of width L_D and L_F was considered, as shown in Figure 3. In particular, condensation over a stripe of hydrophobic surface of non-dimensional width $L_D^* = L_D/r_0$ (white regions in Figure 3) was simulated: periodic conditions were applied through $y^* = 0$, L_y^*, while the presence of hydrophilic regions at $x^* < 0$ and $x^* > L_D^*$ was implemented as described in Section 2.4. It was assumed that the solution of a localized piece of hydrophobic strip is representative of the whole hydrophobic surface, in terms of specific heat flux and migrating flow rate per unit length. The computational domain height L_y^* was set to $3\,L_D^*$ to ensure statistically relevant results. Peng et al. [2] did not provide specific information about the thickness and the thermal conductivity of the self-assembled monolayers (SAM) deposited on the solid substrate to ensure hydrophobicity. We considered $\xi = 0$ in the numerical simulations: such an assumption is consistent with SAMs, characterized by low thermal resistance, of the order of 10^{-7} m^2 K/W [22]. Computations were run at different normalized width L_D^*, while the static contact angle was set to $\theta = 120°$, according to ref. [2]. Once the solution becomes self-similar, the computations were stopped and the non-dimensional heat flux from dropwise condensation (DWC) contribution was estimated from the cumulative condensate volume, which was traced during the whole computation. Furthermore, the volume of liquid migrating through the hydrophobic–hydrophilic boundary due to capillary forces was also traced. As shown in Figure 3, droplets with the base surface touching the hydrophilic stripes instantaneously migrate and merge with the thin film flowing over the hydrophilic side. Snapshots of the droplet population resulting in successive instants from a computation at $L_D^* = 200$ with time integration step of $\Delta t^* = 5 \times 10^{-2}$ are shown in Figure 6a–c. The actual condensing flux $\dot{m}_c$ is shown in Figure 6d, where it can be observed that, after the largest droplet, with a base diameter of about half L_D^*, migrates to the hydrophilic region, $\dot{m}_c$ suddenly increases due to nucleation and growth of several droplets over the dry spot left by the migrating droplet. In fact, droplet growth is faster during the very first stage, as $r^*(t^*) \propto \sqrt{t^*}$, see Equation (7). Thus, a small hydrophobic region ensures an higher sweeping frequency and smaller droplets size, with a benefit in terms of DWC heat flux. On the other hand, reducing the hydrophobic width means that the contribution from FWC, which is about one order of magnitude lower in terms of specific heat flux, becomes more important. The self-similar condition is reached when the sweeping cycles of the largest droplets (characterized by $r^* \sim L_D^*/2$) are observed and, thus, the computed averaged condensing flux stabilizes.

Table 1. Water properties at saturation temperature of $T_s = 100$ °C.

ρ_l (kg/m^3)	μ_l (mPa · s)	λ_l (W/m/K)	ρ_v (kg/m^3)	h_{lv} (kJ/kg)	σ_{lv} (N/m)
958.4	0.2816	0.6772	0.5981	2256	0.5892

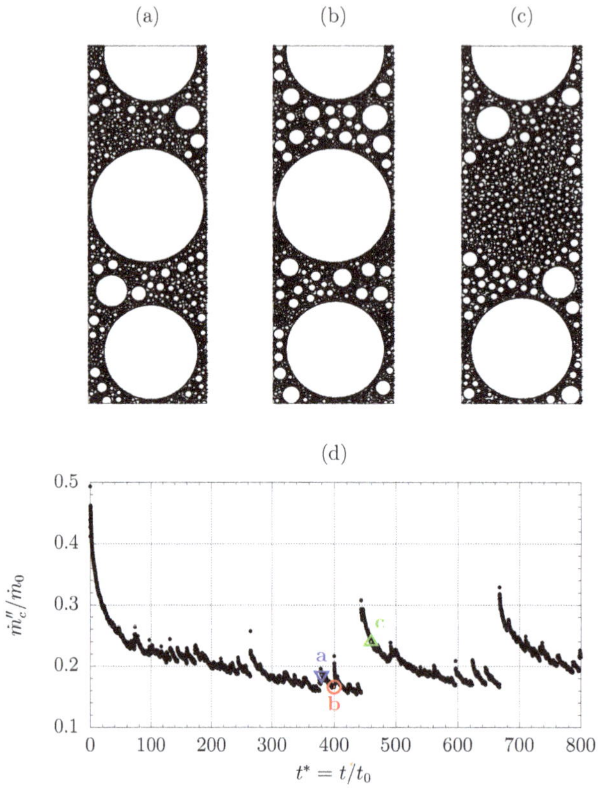

Figure 6. Droplet population at: $t^* = 380$ (**a**); $t^* = 400$ (**b**); $t^* = 460$ (**c**). Non-dimensional condensing flux, with colored markers corresponding to $t^* = 380$, 400, 460 (**d**). $L_D/r_0 = 200$, $\theta = 120°$, $\xi = 0$, $r_n^* = 2\sqrt{0.037}$, $\dot{m}_0 = \rho_l\, r_0/t_0$.

Each simulation represents a family of physical problems, as explained in Section 2.1. Depending on the considered value of ρ_n, the coalescence radius r_0 is univocally defined according to Equation (5). Then, the imposition of the dimensional width L_D of the hydrophobic strip gives the corresponding L_D^*, which is the input parameter of the simulation in the non-dimensional space. Figure 7a explains the procedure to extend a computation at imposed L_D^* to a family of physical test cases, characterized by a different combination of L_D and ρ_n. Further imposing ΔT allows to determine the scaling time t_0, Figure 7b, and, thus, to extend the simulations at any substrate subcooling.

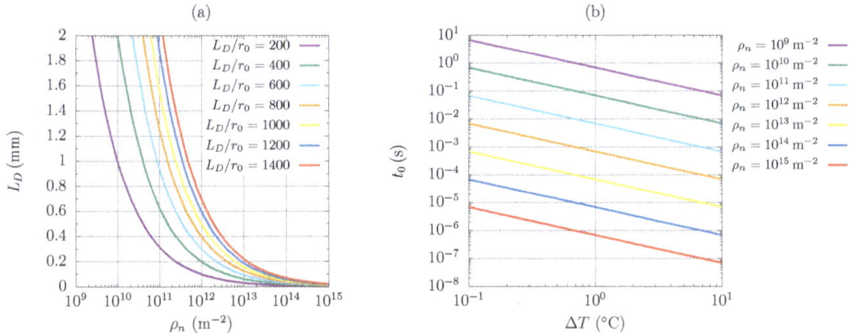

Figure 7. Generalization of results at fixed non-dimensional width L_D^* to different values of nucleation density and hydrophobic width, each line corresponding to a computation in the non-dimensional space (**a**). Scaling time, Equation (5), as a function of subcooling and nucleation density, water properties evaluated at saturation temperature $T_s = 100$ °C (**b**).

The specific heat flux through the hydrophobic region and the migrating flow rate per unit length were estimated as:

$$q''_D = \frac{\rho_l \, h_{lv} \, r_0^3}{L_D \, L_y \, t_0} \, \dot{m}_c^*, \dot{m}'_\sigma = \frac{\rho_l \, r_0^2}{L_y \, t_0} \, \dot{m}_\sigma^* \tag{23}$$

$$\dot{m}_c^* = \frac{V_\theta}{t_n^*} \sum_{i=1}^{n} \left[\sum_{k=1}^{N_{nucl}} r_n^{*3} + \sum_{k=1}^{N_{drop}} \left(r_k^{*3}(t_i^* + \Delta t^*) - r_k^{*3}(t_i^*) \right) \right], \dot{m}_\sigma^* = \frac{V_\theta}{t_n^*} \sum_{i=1}^{n} \sum_{k=1}^{N_{migr}} r_k^{*3} \tag{24}$$

where N_{nucl} is the number of newly nucleated droplets, N_{drop} is the number of droplets, N_{migr} is the number of migrating droplets, and V_θ is defined in Equation (2). Parametric computations were run at different non-dimensional hydrophobic width, with L_D^* ranging $100 \div 800$. Each simulation was run on $L_D^* \times L_y^*$ domain, with the hydrophilic regions placed at $x^* = 0, L_D^*$ and periodic condition imposed at $y^* = 0, L_y^*$, with $L_y^* = 3\,L_D^*$ to ensure statistically relevant information. The time integration step was set in the range $\Delta t = 5 \times 10^{-2} \div 10^{-1}$ and the simulations were stopped after self-similarity was reached.

In order to compute the filmwise condensation (FWC) contribution, the liquid layer solution is required over the hydrophilic regions (grey stripes in Figure 3). It was assumed that a rivulet of width L_F forms over the hydrophilic regions. Applying lubrication theory, a parabolic rivulet profile can be assumed and the flow rate can be estimated by integrating the parabolic velocity profile through the rivulet section:

$$\dot{m} = \frac{16}{105} \, \frac{\rho_l \, (\rho_l - \rho_v) \, g \, L_F}{\mu_l} \, \delta^3 \tag{25}$$

where δ is the maximum rivulet thickness. The liquid flow rate at a distance y from the topmost section of the vertical plate is equal to:

$$\dot{m} = \dot{m}_c + \dot{m}'_\sigma \, y \tag{26}$$

where $\dot{m}_c$ is the condensate flow rate and $\dot{m}'_\sigma$ is the flow rate per unit length, which migrates from the hydrophobic region to the hydrophilic region due to capillary forces. Following Nusselt theory [23], the condensate flow rate is estimated via the energy balance through the liquid rivulet:

$$\frac{\mathrm{d}\dot{m}_c}{\mathrm{d}y} = \frac{\lambda_l \, L_F \, (T_s - T_w)}{\delta \, h_{lv}} \, \frac{\sin\theta_{riv}}{\theta_{riv}}, \tan\theta_{riv} = \frac{4\,\delta}{L_F} \tag{27}$$

Combining Equations (25)–(27) gives an ordinary differential equation,

$$\frac{16}{35}\frac{\rho_l\,(\rho_l-\rho_v)\,g}{\mu_l}\,\delta^2\,\frac{\mathrm{d}\delta}{\mathrm{d}y}=\frac{\lambda_l\,(T_s-T_w)}{\delta\,h_{lv}}\,\frac{\sin\theta_{riv}}{\theta_{riv}}+\frac{\dot{m}'_\sigma}{L_F} \tag{28}$$

which can be analytically solved yielding $\delta(y)$:

$$\frac{u^3}{3}-\frac{3}{2}C\,u^2+3C^2\,u-C^2\log\left(\frac{u}{C}\right)=\frac{11}{6}C^3+\frac{y}{A}, \tag{29}$$

$$u=\delta+C,\;A=\frac{16}{35}\frac{\rho_l\,(\rho_l-\rho_v)\,g\,L_F}{\mu_l\,\dot{m}'_\sigma},\;C=\frac{\lambda_l\,L_F\,(T_s-T_w)}{h_{lv}\,\dot{m}'_\sigma}\,\frac{\sin\theta_{riv}}{\theta_{riv}} \tag{30}$$

Equation (29) has one positive root δ. Following the experimental setup of ref. [2], let us consider water properties at saturation temperature of $T_s=100$ °C, substrate subcooling $\Delta T=10$ °C, and hydrophilic width $L_F=0.45$ mm. The corresponding rivulet height along the vertical plate is plotted in Figure 8a at a different flow rate per unit length $\dot{m}'_\sigma$, migrating from the hydrophobic to the hydrophilic region, and compared with the film thickness from Nusselt theory [23]. The heat flux along the vertical coordinate is also plotted in Figure 8b, showing that the standard Nusselt theory [23] would lead to inaccurate results. The heat flux through a vertical, hydrophilic strip can be simply estimated as $q_F=\dot{m}_c\,h_{lv}$, with $\dot{m}_c$ being the condensate flow rate at the outlet section. As the condensing surface of [2] is a circular plate of radius R, characterized by vertical hydrophilic-hydrophobic stripes, the specific heat flux through the whole hydrophilic surface was evaluated as:

$$q''_F=\frac{L_F}{L_F+L_D}\,\frac{1}{\pi\,R^2}\sum_{k=1}^{n_F}\dot{m}_c(y_k)\,h_{lv} \tag{31}$$

$$n_F=\frac{2\,R}{L_F+L_D},\;y_k=2\,R\sqrt{1-\left[1-(k-1)\,\frac{L_F+L_D}{R}\right]^2} \tag{32}$$

with n_F being the number of hydrophilic stripes and y_k being the kth stripe height.

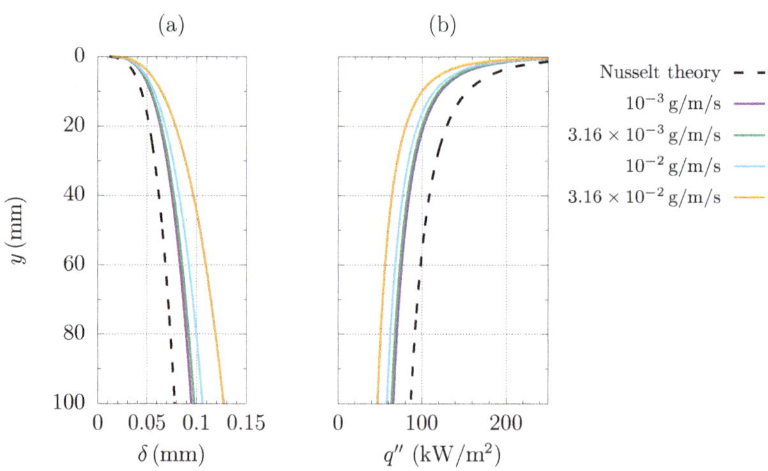

Figure 8. Maximum rivulet thickness δ along the plate at different $\dot{m}'_\sigma$ (**a**); local specific heat flux q'' at different $\dot{m}'_\sigma$ (**b**). Water properties evaluated at $T_s=100$ °C, subcooling $\Delta T=10$ °C, $L_F=0.45$ mm.

Results from DWC computations were combined with the FWC heat flux, which estimation needs $\dot{m}'_\sigma$, derived from postprocessing of DWC data. The global specific heat flux was estimated as:

$$q'' = \frac{q''_F\, L_F + q''_D\, L_D}{L_F + L_D} \tag{33}$$

First of all, we had to characterize the hydrophobic surface. The actual nucleation density is not given in ref. [2]. However, the present numerical model allows us to estimate it for any experimental point. To do so, we considered an experimental data point, which gives the experimental heat flux at a given width of the hydrophobic and of the hydrophilic stripes and at a given substrate subcooling, and we followed the procedure:

- A number of non-dimensional simulations at different L^*_D were carried out, with $L^*_D \in [100, 800]$;
- Each of these non-dimensional computations at different L^*_D may be representative of the experimental hydrophobic width L_D, if ρ_n is determined via Figure 7a;
- The value of ρ_n allows us to deduce the corresponding r_0 and to calculate the DWC heat flux from the non-dimensional condensing flux $\dot{m}^*_c$ via Equations (23) and (24), while the FWC contribution was estimated via Equation (31);
- There will be a single value of ρ_n, which allows the matching between experimental and numerical heat fluxes. This single value is assumed as the actual nucleation density.

The experimental heat fluxes at fixed hybrid surface geometry, given by $L_D = 0.46\,\text{mm}$, $L_F = 0.45\,\text{mm}$, and different subcoolings were used to trace $\rho_n(\Delta T)$. It was found that the nucleation density is nearly independent on the substrate subcooling in the investigated range of ΔT and the value $\rho_n = 4.25 \times 10^{11}\,\text{m}^{-2}$, corresponding to $L^*_D = 600$, allows to accurately reproduce experimental data at $L_D = 0.46\,\text{mm}$, $L_F = 0.45\,\text{mm}$ and $\Delta T = 1 \div 8\,°\text{C}$, as shown in Figure 9a.

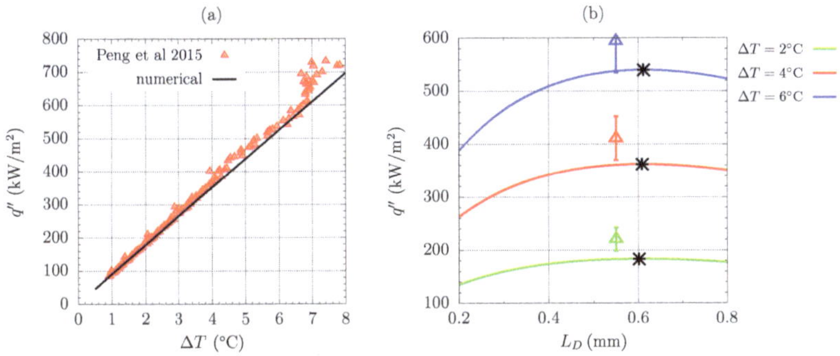

Figure 9. Predicted vs. experimental [2] heat flux through the hybrid surface as a function of substrate subcooling, $L_D = 0.46\,\text{mm}$, $L_F = 0.45\,\text{mm}$ (**a**). Specific heat flux through the hybrid surface as a function of hydrophobic length L_D, at fixed $L_F = 0.45\,\text{mm}$ at $\Delta T = 2, 4, 6\,°\text{C}$ (**b**). Markers with the error bar refer to experimental optimized configuration of [2]. Imposed nucleation density equal to $\rho_n = 4.25 \times 10^{11}\,\text{m}^{-2}$.

The estimated nucleation density was then used in the optimization of the hybrid surface geometry, in order to validate the model with the experimental data of ref. [2] at different hybrid surface geometries, not yet used for characterization of ρ_n and therefore representing different test cases to investigate. In particular, the effect of increasing the hydrophobic width L_D at fixed L_F was investigated. Numerical data at different values of L^*_D were used to trace the DWC heat flux as a function of the hydrophobic width and substrate subcooling. To do so, further computations at $L^*_D > 600$ were run and the available heat flux data were interpolated. The resulting specific heat flux of the hybrid surface is shown as a function of L_D in Figure 9b, at different substrate subcoolings. A maximum of q'' is always observed at $L_D \simeq 0.6$. The experimental optimum observed by

Peng et al. [2] is also identified with markers in Figure 9b, showing a good agreement with numerical data. Furthermore, the predicted heat flux falls inside the uncertainty range of experimental data for $\Delta T = 6\ ^{\circ}\mathrm{C}$.

4. Conclusions

Steam condensation over an hybrid hydrophobic–hydrophilic surface was numerically investigated via a Lagrangian approach, using an in-house code previously developed and adapted to the modelization of hybrid surfaces. The validation of the details of such a numerical model is not trivial, since it is difficult to experimentally evaluate some of its inputs and outputs. Here, the accuracy of the cumulative size distribution of the larger droplets was successfully validated against the well-established empirical correlation of Le Fevre Rose [17], for the case of an unbounded surface. Experimental data on smaller droplet populations is not available, due to difficult access to the microscopic scale of the nucleation phenomenon, but the comparison with cumulative size distribution of small droplets from the theoretical correlation of ref. [1] confirms the robustness of the implemented numerical model.

Furthermore, the simulation of a practical application demonstrates that the code is capable of producing useful information both in terms of surface characterization and as a design tool for optimizing hybrid hydrophilic-hydrophobic surfaces. In particular, by analyzing the experimental data on condensing heat transfer on a hybrid surface, characterized by alternating hydrophobic–hydrophilic vertical stripes, it was possible to estimate the effective nucleation density of the hydrophobic surface, which is not known a priori in most of the practical cases and is quite difficult to determine experimentally. This type of numerical post-processing of experimental data may thus offer an interesting procedure for the practical characterization of the effective ρ_n of real surfaces. Once the surface parameters were known, the numerical simulations were able to define the optimal hybrid surface geometries: the good agreement with experiments of ref. [2] allows for a further validation of the numerical model (note that different data from ref. [2] and different geometries were considered for surface characterization and model validation) and demonstrates that numerical computations are an alternative, robust approach to large experimental campaign for hybrid surface design, while simplified, statistically based DWC models are not able to accurately reproduce hybrid surface behavior.

Computations were run on a 8 core, 64 GB shared memory machine, with the one at the highest L_D^* requiring around 96 h. However, the generalization of the DWC problem via the non-dimensional approach allowed us to investigate different values of nucleation density ρ_n, hydrophobic width L_D, and subcooling ΔT by changing only a non-dimensional model parameter, $L_D^* = L_D/r_0$. Thus, the computational cost of simulations was reduced by about a cubic root compared to a standard modelization of the DWC problem.

A deeper investigation of size distribution of the small droplet population, which is essential to build accurate and simplified, statistical based models [1,16], could further enhance the accuracy of the procedure, as both experimental and theoretical methods showed some limitations when the droplet size approached the nucleating radius. Further developments within the Lagrangian approach could include the modelization of mechanisms such as coalescence-induced jumping, which is promoted by superhydrophobic surfaces, and droplet movement driven by the wettability gradient. Both of them cannot be simply predicted by the definition of a critical departure radius. Thus, the Le Fevre and Rose correlation [17] may not be able to predict the large drop size distribution accurately [10] and computations may help to access useful statistical information.

Author Contributions: Conceptualization, G.C.; Methodology, N.S.; Software, N.S.; Formal analysis, G.C.; Investigation, G.C.; Writing—original draft, N.S.; Writing—review & editing, G.C. All authors have read and agreed to the published version of the manuscript.

Funding: This research received no external funding.

Data Availability Statement: The data presented in this study are available on request from the corresponding author.

Conflicts of Interest: The authors declare no conflict of interest.

References

1. Kim, S.; Kim, K.J. Dropwise Condensation Modeling Suitable for Superhydrophobic Surfaces. *ASME J. Heat Transf.* **2011**, *133*, 081502. [CrossRef]
2. Peng, B.; Ma, X.; Lan, Z.; Xu, W.; Wen, R. Experimental investigation on steam condensation heat transfer enhancement with vertically patterned hydrophobic-hydrophilic hybrid surfaces. *Int. J. Heat Mass Transf.* **2015**, *83*, 27–38. [CrossRef]
3. Miljkovic, N.; Enright, R.; Nam, Y.; Lopez, K.; Dou, N.; Sack, J.; Wang, E.N. Jumping-droplet-enhanced condensation on scalable superhydrophobic nanostructured surfaces. *Nano Lett.* **2012**, *13*, 179–187. [CrossRef] [PubMed]
4. Alwazzan, M.; Egab, K.; Peng, B.; Khan, J.; Li, C. Condensation on hybrid-patterned copper tubes (I): Characterization of condensation heat transfer. *Int. J. Heat Mass Transf.* **2017**, *112*, 991–1004. [CrossRef]
5. Alwazzan, M.; Egab, K.; Peng, B.; Khan, J.; Li, C. Condensation on hybrid-patterned copper tubes (II): Visualization study of droplet dynamics. *Int. J. Heat Mass Transf.* **2017**, *112*, 950–958. [CrossRef]
6. Peng, B.; Ma, X.; Lan, Z.; Xu, W.; Wen, R. Analysis of condensation heat transfer enhancement with dropwise-filmwise hybrid surface: Droplet sizes effect. *Int. J. Heat Mass Transf.* **2014**, *77*, 785–794. [CrossRef]
7. Croce, G.; D'Agaro, P.; Suzzi, N. Optimization of hybrid hydrophilic-hydrophobic surfaces for dropwise condensation enhancement. In Proceedings of the ASME 2019 17th International Conference on Nanochannels, Microchannels, and Minichannels, St. John's, NL, Canada, 23–26 June 2019.
8. Suzzi, N.; Croce, G. Numerical prediction of dropwise condensation performances on hybrid hydrophobic-hydrophilic surfaces. *J. Phys. Conf. Ser.* **2020**, *1599*, 012006.
9. Yang, K.S.; Lin, K.H.; Tu, C.W.; He, Y.Z.; Wang, C.C. Experimental investigation of moist air condensation on hydrophilic, hydrophobic, superhydrophilic, and hybrid hydrophobic-hydrophilic surfaces. *Int. J. Heat Mass Transf.* **2017**, *115*, 1032–1041. [CrossRef]
10. Stevens, K.A.; Crockett, J.; Maynes, D.; Iverson, B.D. Simulation of Drop-Size Distribution During Dropwise and Jumping Drop Condensation on a Vertical Surface: Implications for Heat Transfer Modeling. *Langmuir* **2019**, *35*, 12858–12875. [CrossRef] [PubMed]
11. Zhang, B.J.; Kuok, C.; Kim, K.J.; Hwang, T.; Yoon, H. Dropwise steam condensation on various hydrophobic surfaces: Polyphenylene sulfide (PPS), polytetrafluoroethylene (PTFE), and self-assembled micro/nano silver (SAMS). *Int. J. Heat Mass Transf.* **2015**, *89*, 353–358. [CrossRef]
12. Parin, R.; Martucci, A.; Sturaro, M.; Bortolin, S.; Bersani, M.; Carraro, F.; Del Col, D. Nano-structured aluminum surfaces for dropwise condensation. *Surf. Coat. Technol.* **2018**, *348*, 1–12. [CrossRef]
13. Torresin, D.; Tiwari, M.K.; Del Col, D.; Poulikakos, D. Flow Condensation on Copper-Based Nanotextured Superhydrophobic Surfaces. *Langmuir* **2013**, *29*, 840–848. [CrossRef] [PubMed]
14. Rose, J.W. Further aspects of dropwise condensation theory. *Int. J. Heat Nad Mass Transf.* **1976**, *19*, 1363–1370. [CrossRef]
15. Liu, X.; Cheng, P. Dropwise condensation theory revisited: Part I. Droplet nucleation radius. *Int. J. Heat Mass Transf.* **2015**, *83*, 833–841. [CrossRef]
16. Liu, X.; Cheng, P. Dropwise condensation theory revisited Part II. Droplet nucleation density and condensation heat flux. *Int. J. Heat Mass Transf.* **2015**, *83*, 842–849. [CrossRef]
17. Le Fevre, E.J.; Rose, J.W. A Theory of Heat Transfer by Dropwise Condensation. In Proceedings of the 3rd International Heat Transfer Conference, Chicago, IL, USA, 7–12 August 1966.
18. Suzzi, N.; Croce, G. Numerical simulation of dropwise condensation over hydrophobic surfaces using vapor-diffusion model. *Appl. Therm. Eng.* **2022**, *214*, 118806. [CrossRef]
19. Xie, J.; Xu, J.; Shang, W.; Zhang, K. Dropwise condensation on superhydrophobic nanostructure surface, part II: Mathematical model. *Int. J. Heat Mass Transf.* **2018**, *127*, 1170–1187. [CrossRef]
20. Mirafiori, M.; Abbatecola, A.; Tancon, M.; Bortolin, S.; Del Col, D. A New Model to Study the Life of Millions of Drops During Condensation of Steam. In Proceedings of the 40th UIT International Conference, Assisi, Italy, 26–28 June 2023.
21. Miljkovic, N.; Enright, R.; Wang, E.N. Modeling and Optimization of Superhydrophobic Condensation. *ASME J. Heat Mass Transf.* **2013**, *135*, 111004. [CrossRef]
22. Song, L.; Zhang, Y.; Yang, W.; Tan, J.; Cheng, L. Molecular Structure Effect of a Self-Assembled Monolayer on Thermal Resistance across an Interface. *Polymers* **2021**, *13*, 3732. [CrossRef] [PubMed]
23. Nusselt, W. Die Oberflächenkondensation des Wasserdampfes. *Z. Ver. Dtsch. Ingenieure* **1916**, *60*, 541–546, 569–575.

fluids

MDPI

Article

Influence of Weber Number on Crown Morphology during an Oblique Droplet Impact on a Thin Wall Film

Jonathan Lukas Stober [1,*], Maurizio Santini [2] and Kathrin Schulte [1]

[1] Institute for Aerospace Thermodynamics, University of Stuttgart, 70569 Stuttgart, Germany; kathrin.schulte@itlr.uni-stuttgart.de

[2] Department of Engineering and Applied Sciences, University of Bergamo, 24044 Bergamo, Italy; maurizio.santini@unibg.it

* Correspondence: jonathan.stober@itlr.uni-stuttgart.de

Abstract: Spray impacts can be found in several technical applications and consist of many single droplets, which impact under different trajectories on wetted walls. This study investigates the asymmetric crown morphology resulting from an oblique impact ($\alpha = 60°$) of a single droplet on a horizontal and quiescent wall film of the same liquid. A droplet generator with an accelerated needle releases the droplets ($D = 1.5$ mm) in a controlled trajectory on a thin film ($h_f/D = 0.2$). The impact process is recorded from two perspectives with two synchronized high-speed cameras. Varying the Weber number within the splashing regime reveals distinct crown morphologies, which are described in detail. For $We < 500$, a single central finger develops at the front of the crown, with subsequent detachments of secondary droplets. At higher We (>500), a collision of the crown with the wall film shortly after impact introduces disturbances into the rim, leading to two fingers in the middle of the front crown. A further increase in We (>600) intensifies the crown–film interaction, resulting in an early ejection of tiny droplets and a complete breakup of the front rim. The influence of We on the crown morphology during an oblique impact is also compared to the normal impact (90°). This study paves the way for a classification of impact regimes and a comprehensive picture of the oblique impact process, which deserve more investigation.

Keywords: droplet impact; oblique; thin wall film; horizontal quiescent wall film; Weber number; crown morphology; splashing

1. Introduction

Spray impacts are part of many technical applications and environmental processes, such as the distribution of plant protection agents, coating processes, medical sprays, or soil erosion. A spray consists of many droplets which impact with different trajectories, and thus impact angles, on dry solid walls but also on walls covered with liquid films of different thicknesses. Therefore, a single droplet impact onto a wall film of the same liquid can be regarded as an elementary process in such a spray. While most research focused on the scenario of droplets impacting the wall film at a normal angle (i.e., 90°), the focus of this study is on oblique droplet impact. Its comprehensive physical description and modeling form the basis for optimizing these applications.

During a normal or oblique droplet impact on a wall film, a crown structure is formed, which is characterized by a thicker rim bounding its top (Taylor rim, [1]) due to surface tension forces. As the crown evolves, disturbances within the rim intensify and potentially result in the growth of finger-like structures. Secondary droplets can detach consecutively from the tips of these fingers. The disintegration of the primary droplet with the formation of secondary droplets is called splashing. While a crown splash is described here, it may also occur at other stages of the impact depending on the impact conditions [2–4]. A prompt splash is characterized by the ejection of small droplets directly after impact. A Worthington jet, with the detachment of secondary droplets, results from the break down of a crown leading to a flow back to the center of impact. In case of deposition, on the other hand, the drop merges with the wall film after impact without the ejection of secondary droplets.

Cossali et. al. [2] conducted a large experimental study on normal droplet impacts and derived an empirical correlation that describes the deposition/splashing threshold based on the fluid properties, impact conditions and film thickness. Many other studies continued with the characterization of the morphology after normal impact and the determination of regime thresholds [5–7]. The Weber number, We, which characterizes the ratio of inertial forces to surface tension forces, is one of the significant parameters changing the impact outcome fundamentally and is often used to determine the splashing threshold by $We > We_{crit}$. A higher We generally leads to a larger crown, a thinner crown wall as well as more and smaller secondary droplets [5,8,9]. Further studies focused on detailed investigation of the crown splashing. Roisman et al. [10] performed analytical investigations on the growth of disturbances from bending instabilities, while Agbaglah et al. [11] used numerical simulations to study the transverse instability of the rim. Their findings suggest that the rim breakup is initially driven by Rayleigh–Taylor instabilities, leading to cusps, followed by the Rayleigh–Plateau mechanism, which drives the subsequent breakup of the fingers.

However, an oblique droplet impact onto a wetted wall, which is of great relevance in real applications, changes the impact process considerably compared to a normal impact. Okawa et al. [12] experimentally investigated the ejection of secondary droplets during the impact on thick wall films, $h_f/D > 1.7$. They observed a liquid sheet similar to the shape of a ship's prow, where a liquid column forms and secondary droplets detach. Comparing the splashing process for a normal impact to the oblique impact, they found that the number and size of secondary droplets are highly dependent on the impact angle. Other experimental studies investigated the oblique impact into deep pools [13,14]. The impacts on thin films, $h_f/D < 0.4$, are only investigated in some studies, where the wall film was inclined [15–19]. This either leads to limited impact angles or to a flowing wall film that induces a boundary layer in the films and disturbs the impact. In addition to this experimental work, there are also numerical studies on oblique impacts on thin films [20–23]. However, the investigation of the crown morphology and the disintegration process during an oblique droplet impact onto a quiescent and thin wall film remains to be mostly unexplored. This study starts to close this gap by describing the influence of the Weber number on the splashing process during oblique droplet impact on a thin film.

In order to investigate this, the droplet has to be shot in a trajectory. Previously, different methods were used, for example, a pressure chamber that ejects single droplets [13] or a piezoelectric actuator that produces a droplet chain from which one single droplet is isolated [14]. The latter can reach high velocities, $U < 25$ m/s, but is limited to very small droplets, $D = 0.1$ mm, which makes it impossible to investigate the impact onto a thin wall film, $h_f/D < 0.4$. Okawa et al. [12] generated a flat spray and used an orifice to obtain a single droplet. All the described methods either produce very small droplets, which prevent the investigation of thin wall films, or are limited to low velocities. For this study, a different droplet generator was used, consisting of a needle that is accelerated on a rail to produce millimeter-size droplets by a controlled detachment. The impact velocity and angle can be varied independently from each other while maintaining a constant droplet diameter.

This study is structured the following way. First, the droplet generator itself and the experimental setup with a synchronized two-perspective high-speed camera system is presented. After that, the crown morphology of a normal impact (90°) at different Weber numbers is shown for reference. This is followed by the detailed description of one exemplary oblique droplet impact. Thereafter, the influence of the Weber number on the oblique (60°) impact morphology and splashing dynamics is analyzed. At the end, the crown characteristics and their change with We of the oblique impact and the normal impact are compared.

2. Experimental Method

Figure 1 represents an oblique droplet impact. It is characterized by the droplet diameter D, the total impact velocity U, the impact angle α and the wall film thickness h_f. The fluid of the droplet and wall film is the same, with its properties such as the density ρ, the surface tension σ and the dynamic viscosity μ. To summarize the impact conditions in a dimensionless manner, we consider the Weber number $We = \rho D U^2/\sigma$, the Ohnesorge number $Oh = \mu/\sqrt{\rho D \sigma}$ and the non-dimensional film thickness $\delta = h_f/D$. The dimensionless time after first contact of the droplet and the wall film is defined as $\tau = tU/D$.

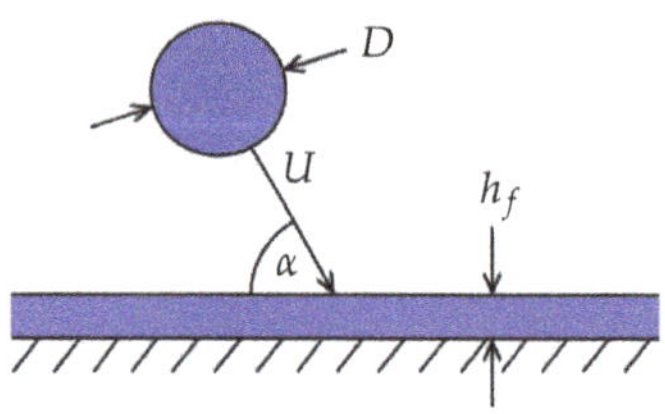

Figure 1. Schematic representation of the impact conditions.

2.1. Oblique Droplet Generator

To achieve an oblique impact, the droplet has to be ejected with a horizontal velocity component. Figure 2A,B show the droplet generator, which was built for that purpose, based on the patent by Santini et al. [24]. A carriage rides along an 80 cm long rail and is powered by a brushless linear motor (H2W Technologies BLDM-B02, Santa Clarita, CA, USA). The magnet assembly for the linear motor is mounted next to the rail and covers the full length, which leads to a usable stroke of 65 cm due to the length of the carriage itself. To monitor the exact position, a linear encoder system (Renishaw RGH41, Warton, UK) is utilized. The motor and the position sensor are connected to a servo drive (Baldor MicroFlex e100, Fort Smith, AK, USA), which is controlled by an in-house Visual Basic program to define the kinematic profile of the carriage.

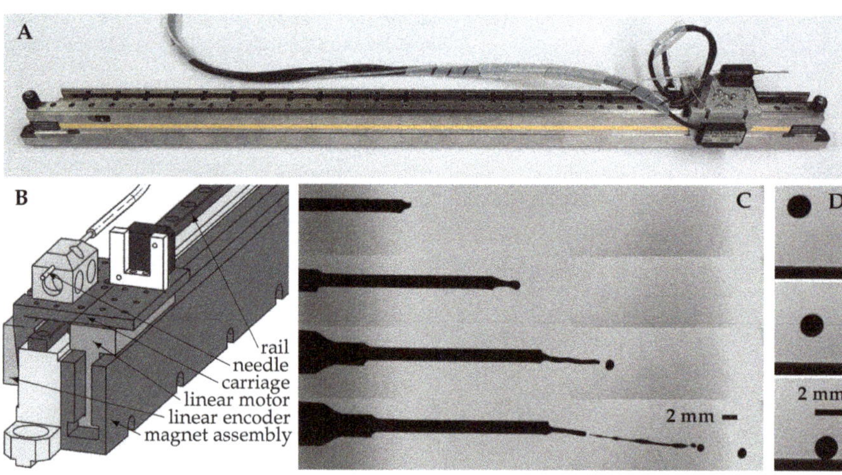

Figure 2. (**A**) Photography of the oblique droplet generator, (**B**) sketch of the droplet generator according to the patent by Santini et al. [24], (**C**) high-speed camera recording of the detachment process, (**D**) high-speed camera recording of droplet prior to impact onto the liquid film.

A blunt syringe needle with an inner diameter of 1.5 mm is fixed on the carriage. Isopropanol is supplied to the needle through a piezoelectric pump (Bartels microtechnik mp6-liq) capable of finely tuning the mass flow rate. To achieve a reproducible and controlled detachment, the needle is driven in a periodic motion. It begins by slowly moving it backwards, followed by a constant acceleration forward to the desired velocity. Then, it decelerates to a stop at the turning point. After that, the process repeats itself. A deceleration of $a_{dec} = 130\ \mathrm{m/s^2}$ was found to be suitable to detach droplets consistently. With needle velocities in the range of $1.5\ \mathrm{m/s} < U_{needle} < 3.3\ \mathrm{m/s}$, the detached droplets show a velocity in the range of $1.3\ \mathrm{m/s} < U_{detach} < 2.9\ \mathrm{m/s}$.

Figure 2C illustrates the detachment process. During the high deceleration phase at the turning point, the fluid within the needle is pushed outwards due to its own inertia and forms a ligament. As more fluid exits the needle, the ligament continues to grow, consecutively detaching droplets from its end. The size and velocity of the droplets decreases as the ligament undergoes the deceleration of the needle. Consequently, only the first droplet, with the highest velocity and diameter, reaches the impact point, while the subsequent droplets follow a shorter trajectory. The diameter of the produced droplets is extremely sensitive to the mass flow rate and requires precise adjustment. Furthermore, the size of the droplets is also dependent on the needle size. In this study, a needle with an inner diameter of 1.5 mm resulted in droplets in the range of $1.2\ \mathrm{mm} < D < 1.7\ \mathrm{mm}$ dependent on the mass flow rate. A thinner needle produces smaller droplets, while a larger one leads to an unstable ligament and irregular droplets. With one fixed set of tuning parameters, the droplet diameter varies in the range of 0.15 mm from one shot to the other. The produced droplets are nearly spherical and do not oscillate as can be seen from Figure 2D, which shows the droplet prior to the impact on the film. Potential oscillations resulting from the forced detachment thus have been damped before impact.

2.2. Test Stand

The experimental setup is illustrated in Figure 3. The imaging system comprises two synchronized high-speed cameras (Photron Fastcam SA-X2, Tokyo, Japan) capturing the process from two perspectives: a side view (blue dash-dotted line) and a front view (red dashed line). For each camera, an LED with a lens acts as a light source for the backlit shadowgraphy. The light for the side view is reflected 90° by a mirror before entering the objective mounted on the camera. Both the side view and front view utilize modular optic systems, consisting of an extension tube, a zoom module and a lower lens (side view: Optem Fusion, front view: Navitar). The cameras record with a frame rate of 12,500 fps and a resolution of 1024 × 1024 pixels. This results in a relative resolution of 24.3 µm/px for the side view and 21.1 µm/px for the front view. For the impact area, a smooth sapphire glass plate with a size of 50 mm × 50 mm and a thickness of 2 mm is used. The liquid film of isopropanol covers the complete glass plate and is held by its surface tension. The size of the glass plate ensures a flat film in the middle of the plate, where the droplet impacts. Its thickness is measured by a confocal chromatic sensor (sensor IFS2405-1, controller IFC2421MP from Micro-Epsilon, Ortenburg, Germany) continuously from below during the experiment to monitor changes due to evaporation. This also ensures that the film is static again after the previous impact and allows us to prove the flatness of the film by moving the sensor horizontally. For the confocal chromatic distance measurement, a white light source is split into different spectra. The light is focused in point but with different focal lengths dependent on the wavelength. A spectrometer measures the spectral intensity of the reflected light from the measurement object. The wavelength of the intensity peaks can be translated back into a distance with the information of the wavelength dependent focal points. For more information, see Ruprecht et al. [25]. The droplet generator is mounted on a separate frame on a different table decoupling it from the test stand to reduce disturbances of the liquid film. Its height, horizontal distance to the impact point and the inclination of the rail can be adjusted, as well as the acceleration profile and hence the detachment velocity. With that, different impact angles and velocities can be realized independently from each other, while the droplets are impacting in the middle of the sapphire plate. With the current setup, impact angles from $\alpha = 30°$ up to $\alpha = 80°$ can be achieved. The maximum impact velocity varies for each angle, mainly due to the limitations of the frame holding the droplet generator. At $\alpha = 80°$, velocities up to $U = 2.6$ m/s can be reached, while for $50° \leq \alpha \leq 70°$, velocities up to $U = 3.4$ m/s can be reached. For lower impact angles, $30° \leq \alpha \leq 40°$, the achievable velocity decreases to $U \approx 2.8$ m/s. Due to small disturbances during the detachment process and in the surrounding air, the trajectories and thus the impact velocity and angle vary in the range of 0.2 m/s and 0.7°, respectively.

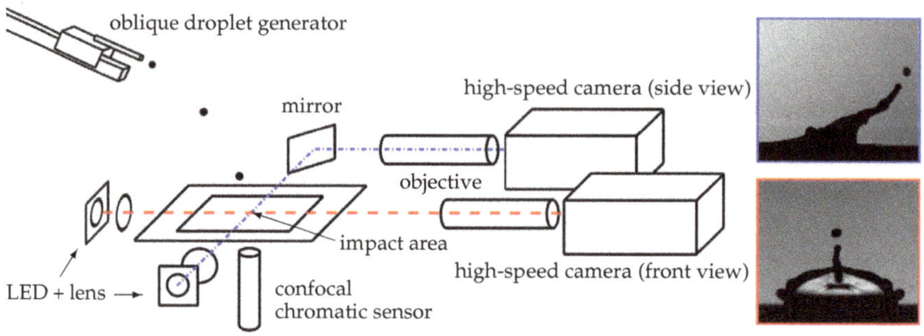

Figure 3. Schematic representation of the experimental setup.

The experimental procedure is the following: First, the oblique droplet generator needs to be tuned in such a way that the desired impact velocity, U; impact angle, α; and droplet diameter, D, are achieved while hitting the middle of the sapphire plate. This is performed iteratively by adjusting the height, the horizontal distance and the inclination of the droplet generator as well as the detachment velocity, which is dependent on the acceleration profile and the mass flow. The impact parameters are determined retrospectively from the camera recording from the side view using an in-house Matlab (R2021a) program. The oblique droplet generator is continuously operated in a periodic motion, releasing droplets approximately every three seconds. This ensures a repeatable detachment and trajectory. At the beginning of a measurement series, the droplets are caught by a trap to apply a thin film of isopropanol on the sapphire plate. Once the film achieves its desired thickness and stability, which is monitored by the sensor, the trap is removed and a single droplet impacts. Both cameras, which are also synchronized to the film thickness measurement, record continuously until the end trigger is pulled manually. With this method, droplets impacting off-center can be sorted out by waiting for the next impact. To ensure a static film after the previous impact, every second droplet is caught with the trap. Due to the impact conditions varying from one shot to another, as described earlier, many experiments have to be conducted.

The droplet diameter is determined utilizing an in-house Matlab program, which analyses the side view of the high-speed camera recordings, assuming a spherical shape. The measurement accuracy of one pixel corresponds to an uncertainty of 1.93%, considering the resolution mentioned earlier. Similarly, the impact velocity and angle are also calculated using the Matlab program evaluating the last ten frames before impact. The velocity is measured with an uncertainty of 1.68%, and the impact angle is measured with an uncertainty of 1.35%. Since the film thickness is measured from below through the sapphire plate, the measured thickness deviates from the actual thickness. To eliminate this error, the thickness was determined with a second sensor from above at the same time to quantify the shift in the measurement. By applying that correction, the film thickness can be determined with an accuracy of 10 µm leading to an uncertainty of 3.3% for the film thickness investigated in this study.

3. Results

This section is structured as follows: First, in Section 3.1, the crown morphology of a normal droplet impact, i.e., $\alpha = 90°$, for different Weber numbers is presented. In the next subsection, the outcome of one oblique impact ($\alpha = 60°$) experiment leading to a single central finger on one side of the crown is shown in detail. After that, the morphology change with different impact Weber numbers is presented and discussed in Section 3.3. At the end, the oblique droplet impact is compared to the normal one in Section 3.4.

In order to investigate the nature of the oblique impact, the impact angle α was not set too close to 90°. However, if the angle is too small ($\leq 45°$), a comparison to normal impacts is not meaningful anymore. Therefore, an impact angle of $\alpha = 60°$ was chosen for this study.

3.1. Crown Morphology for Weber Number Variation at Normal Droplet Impact ($\alpha = 90°$)

Experiments were conducted with $\alpha = 90°$ in a range of $We \approx [300, 700]$, as for the oblique droplet impact. The detailed impact conditions of the experiments shown in this and the following sections are summarized in Table 1. Figure 4 shows the crown development of Experiments 1–4 at the time of maximum crown height (A), during (B) and after the receding phase (C). With an increase of We, the morphology changes the following way. For Experiment 1 ($We = 323$), there is a smooth rim which falls down on the film again without any splashing; see Figure 4(B1). After that, a small Worthington jet [4] forms, C1. With an increase in We (Experiment 2, $We = 481$), the rim remains smooth until it reaches its maximum crown height, and there is still no crown splash. During the receding, waves appear at the rim and an intense Worthington jet can be observed after that, which leads to a detachment of the jet itself and a separation into up to six droplets. Moving on to the third experiment ($We = 618$), an increase in the Weber number results in the emergence of waves in the expansion phase of the crown, A3. During the receding phase, thick fingers start to form and comparably large droplets separate from it, which fall back onto the liquid film, B3. After that, a small Worthington jet is present without further atomization, C3. Finally, as the Weber number is further increased ($We = 708$), fingers appear at the rim before the maximum crown height is reached, and many secondary droplets detach; see Figure 4(A4). The fingers and the secondary droplets at this Weber number are smaller in diameter and fly upwards. Additionally, consecutive detachments of more droplets from these fingers occurs during the receding phase, B4 , which fall down onto the liquid film again, C4.

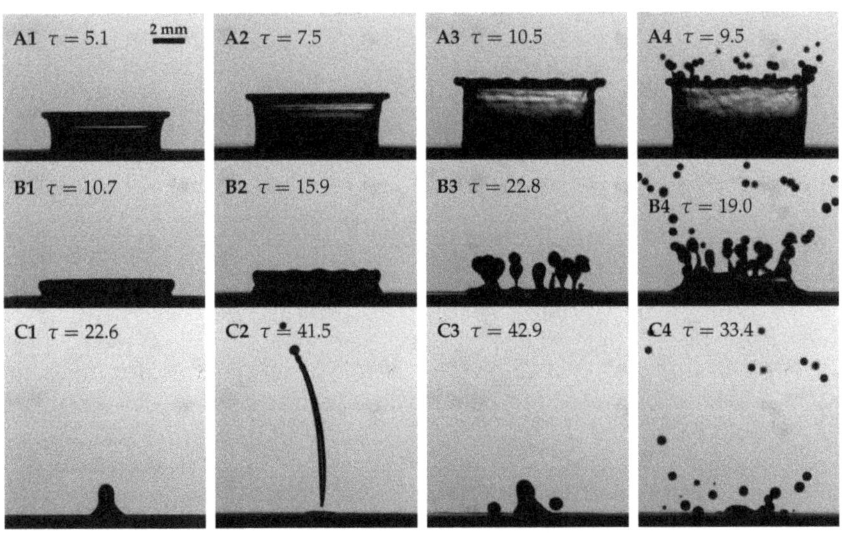

Figure 4. Recording of the crown morphology during a normal droplet impact, i.e., $\alpha = 90°$, for different Weber numbers. (**1**) Exp 1 $We = 323$, (**2**) Exp 2 $We = 481$, (**3**) Exp 3 $We = 618$, (**4**) Exp 4 $We = 708$. Moment of maximum crown height (**A**), receding phase (**B**), after the receding of the crown (**C**).

Table 1. Impact conditions for all experiments shown in this study, as well as fluid properties at 22 °C.

	Exp 1	Exp 2	Exp 3	Exp 4	Exp 5	Exp 6	Exp 7	Exp 8	Exp 9	Exp 10	Exp 11	Uncertainty
droplet diameter D	1.78 mm	1.77 mm	1.74 mm	1.76 mm	1.53 mm	1.47 mm	1.52 mm	1.69 mm	1.45 mm	1.49 mm	1.62 mm	±0.014 mm
impact velocity U	2.23 m/s	2.73 m/s	3.12 m/s	3.32 m/s	2.14 m/s	2.89 m/s	3.05 m/s	3.20 m/s	3.10 m/s	3.35 m/s	3.38 m/s	±0.024 m/s
impact angle α	90°	90°	90°	90°	60.2°	59.7°	60.4°	59.5°	61.7°	59.3°	59.6°	±0.41°
dim.less film thickness δ	0.21	0.22	0.23	0.22	0.21	0.23	0.22	0.17	0.26	0.24	0.22	±0.0038
Weber number We	323	481	618	708	256	448	516	631	508	610	675	*
Ohnesorge number Oh	0.0129	0.0130	0.0131	0.0130	0.0139	0.0142	0.0140	0.0133	0.0143	0.0141	0.0136	±0.0005
density ρ						784.3 kg/m^3						±0.9 kg/m^3
surface tension σ						21.5 mN/m						±0.1 mN/m
dynamic viscosity μ						2.24 mPa/s						±0.07 mPa/s

* uncertainty for comparably low $We_1 = 323 \pm 10$, and high $We_4 = 708 \pm 25$.

These phenomena described here were already observed by others [3,4,7,9]. The threshold derived from the correlation by Cossali et al. [2], at which splashing starts to occur, is $We_{crit} = 485$. They mention a transition region in the vicinity of the threshold, in which both outcomes are possible. At the film thickness and Ohnesorge number we investigated in this study, a splashing from a Worthington jet is observed before the crown splashing. Cossali et al. also investigated this region but did not comment on splashing from a Worthington jet. However, they clearly defined splashing as the formation of secondary droplets and their threshold fits quite well with our observations.

3.2. Single Central Finger Morphology ($\alpha = 60°$)

The Weber number of an oblique impact can be defined either with the total impact velocity, $We = \rho D U^2/\sigma$, or with the velocity component normal to the wall film, $We_n = \rho D U_n^2/\sigma = We\ sin(\alpha)^2$. Using the total Weber number yields the same kinetic energy of the impact as in the normal impact. On the other hand, the normal Weber number scales with the velocity component normal to the wall, which could also be the determining quantity for the splashing characteristic. The normal Weber number is always smaller than the total Weber number; here, for $\alpha \approx 60°$, approximately $We_n \approx We\ 3/4$. In this paper, the Weber number for the oblique impact was calculated with the total impact velocity, meaning that the normal velocity component of the oblique impact is smaller than that of the normal impact at the same We. Nevertheless, as shown later in Sections 3.3 and 3.4, splashing occurs at lower We in the oblique case.

In this section, the outcome of one oblique droplet impact (Experiment 6, see Table 1, $We = 448$, $\alpha = 59.7°$) is shown and explained in detail. Figure 5 presents the development of the impact, captured by the two synchronized high-speed cameras. The left picture shows the morphology from the side view and the right picture from the front view. Both images are scaled to the same size and aligned in height. The difference visible at the film surface is due to a different depths of field and focal points for the two views. The droplet impacts obliquely from the top left; see Figure 5A showing the droplet shortly after first contact with the wall film. In Figure 5B, the early stage after impact reveals that the lamella grows flat and horizontally at the front compared to the higher rising lamella on the sides. Shortly after (see Figure 5C,D), the front lamella turns upwards and reaches higher than the lamella on the side. In picture D, at $\tau = 4.3$, a central finger begins to form on the front side and continues to grow diagonally upwards (Figure 5E–H). A first secondary droplet detaches from the tip of the finger at $\tau \approx 11$, see Figure 5F,G, moving upwards. At this moment, the crown has already reached its maximum height and starts to recede, first at the back and then at the side. Subsequently, the central finger also retracts while a second and a third secondary droplet separate from its tip (Figure 5I,J), which fall back down on the wall film and coalesce with it. At the rim on the side, no fingers form. Instead, one cusp develops on each side of the crown. It starts to form at $\tau = 7.4$, visible in Figure 5E, and becomes evident in both views in Figure 5F,G. On the back side of the crown, the crown is first bent inwards; see Figure 5C. However, a part of the crown, or at least a small finger, flips outwards later, visible in Figure 5E.

Yarin and Weiss [26] provided an explanation for the formation of fingers from the rim bounding the crown during droplet impact. According to their explanation, small disturbances in the Taylor rim [1] are amplified and lead to the formation of cusps. If the kinetic energy is sufficiently high, fingers grow from these cusps and secondary droplets separate from their tips due to the Rayleigh–Plateau instability. This mechanism is also observed in the oblique impact case shown above. However, in this particular case, one single finger grows only at the front side of the crown, which can be attributed to the horizontal velocity component of the impacting droplet. The rim at the side, however, is more stable; see Figure 5D,E. The horizontal velocity component present at the oblique impact probably induces a circumferential flow within the rim. This circumferential flow supplies the central finger and facilitates its growth.

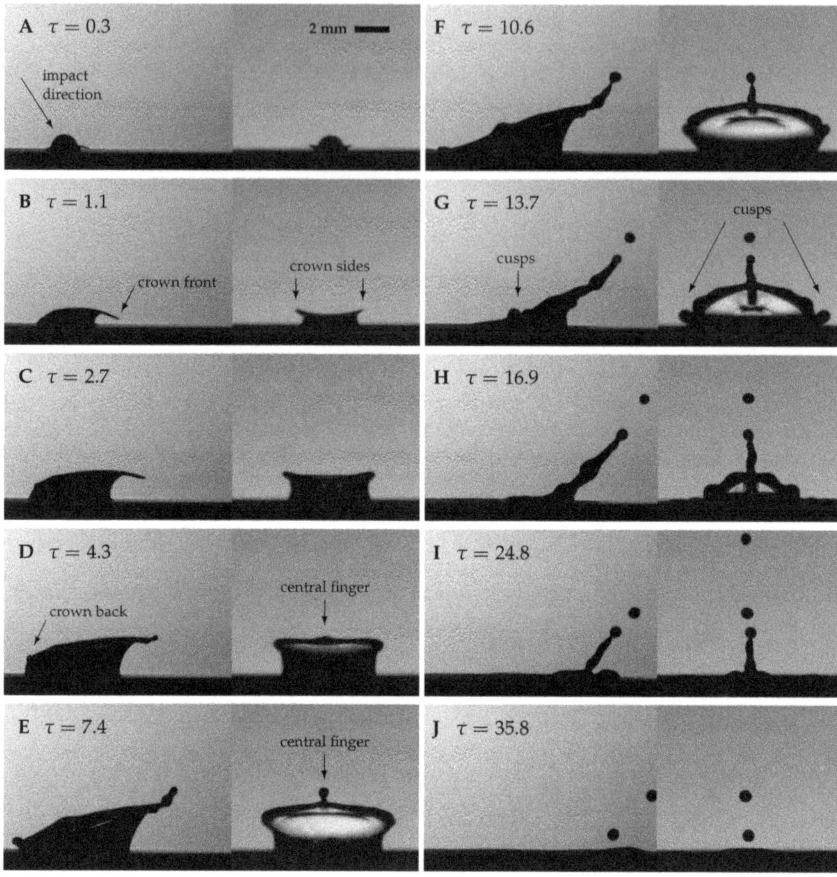

Figure 5. Two-perspective shadowgraphy images of the crown morphology during the oblique droplet impact onto a liquid wall film, Experiment 6 ($We = 448$, $Oh = 0.0142$, $\alpha = 59.7°$); see Table 1. Left picture, side view; right picture, front view. Dimensionless time $\tau = tU/D$. (**A**) droplet shortly after impact; (**B**,**C**) stable crown development; (**D**,**E**) start of formation of central finger; (**F**) first detachment of secondary droplet; (**G**–**I**) crown receding; (**J**) 2nd and 3rd secondary droplet.

To evaluate the repeatability of the experiment, Figure 6 shows the crown morphology from three experiments with the same impact conditions at two moments in time. The first moment is during the early growth phase of the central finger, $\tau_A = 5.0$, shown in the first row. Notably, the height and the shape of the finger are the same across all experiments. Also, the overall shape of the crown is identical. At a later time, $\tau_B = 14.5$, the first secondary droplet has detached from the central finger and the crown has already started descending towards the wall film; see the second row in Figure 6. The shape of the residual crown as well as the position and size of the side cusps are similar in all three experiments. The only observable difference is the position of the secondary droplet, indicating a slight variation in the time of its separation from the finger. A comparison of later times (not shown in this figure) reveals that in all three experiments three secondary droplets detach. Additionally, the crown shape observed in the front view is remarkably symmetric. This demonstrates that the wall film is in fact flat and quiescent prior to droplet impact. Moreover, no noticeable random disturbances, such as oscillations, are introduced by the detachment of the droplet at the droplet generator.

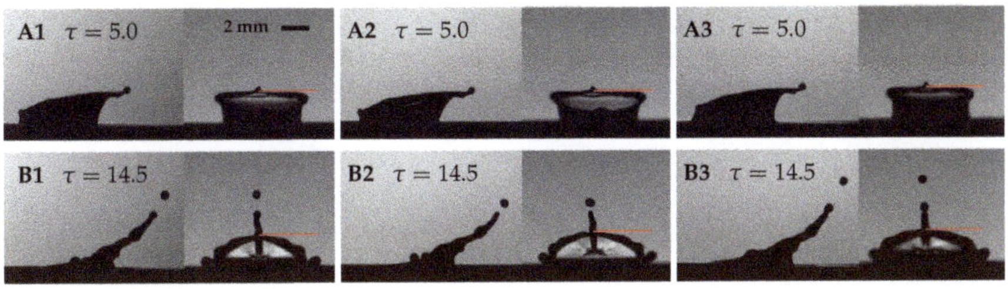

Figure 6. Comparison of the crown morphology of three experiments. (**A**) $\tau = 5.0$, formation of central finger; (**B**) $\tau = 14.5$, crown descending phase; (**1**) Experiment 6 ($We = 448$, $Oh = 0.0142$); (**2**) Experiment 6.1 ($We = 457$, $Oh = 0.0142$); (**3**) Experiment 6.2 ($We = 457$, $Oh = 0.0141$).

Figure 7 presents a comparison of the non-dimensional crown height on the front side h_{cf}/D as function of the non-dimensional time $\tau = tU/D$ for the three experiments under the same conditions (Exp 6, Exp 6.1, Exp 6.2) in order to quantify the repeatability of the experiment. The crown height was measured from the front view, excluding the central finger, as indicated with the red line in Figure 6. The overall development is consistent in all experiments, and they reach their maximum crown height at $\tau = 10$. However, Experiment 6.1 shows a small deviation, with its maximum height being 7% lower compared to the other experiments. It is interesting to note, that the deviation starts at $\tau \approx 5$, which is the time where the central finger starts to grow. The crown height development of Experiment 6 and 6.2 aligns well within the measurement accuracy until the crown starts to recede, $\tau > 14$. During this phase, the rim becomes wavy, as seen in Figure 5H. Small variations in the position of the waves along the receding crown result in changes in the measured height, leading to deviations between the experiments in the receding phase.

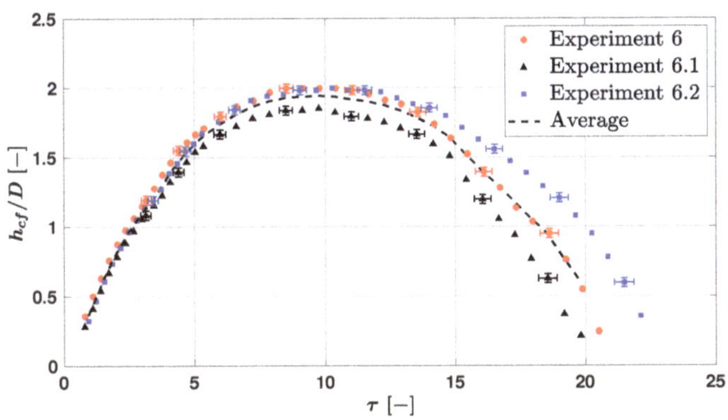

Figure 7. Development of the non-dimensional height of the crown at the front side h_{cf}/D for three experiments with similar impact conditions. Measurement uncertainty marked by error bars for every fourth data point. Average height of three experiments plotted with the dashed line.

3.3. *Variation of the Weber Number ($\alpha = 60°$)*

Figure 8 shows the crown morphology during and oblique droplet impact of four different total Weber numbers, $We = \rho D U^2/\sigma$. It compares Experiment 6 ($We = 448$), which was investigated in the last section, with an experiment with lower Weber number (Experiment 5 $We = 256$) and two experiments with higher We numbers (Experiments 7 + 8, $We = 516$, $We = 631$). The impact angle and the film thickness were kept almost constant; only the droplet diameter was increased for the last experiment to reach higher Weber numbers. The detailed impact conditions are summarized in Table 1. Experiments 6 and 7 lie in the vicinity of the deposition/splashing threshold developed by Cossali et al. [2] for normal droplet impacts on thin wall films, $We_{crit} = 485$. The other experiments represent cases below (Experiment 5) and above it (Experiment 8), which does not mean that they have to lead to deposition and splashing, respectively, because the threshold was not developed for oblique impacts. The left column in Figure 8 shows a moment shortly after the start of finger formation, $\tau_A \approx 6.1$, and the right column a moment where the crown is in the receding phase, $\tau_B \approx 13.9$.

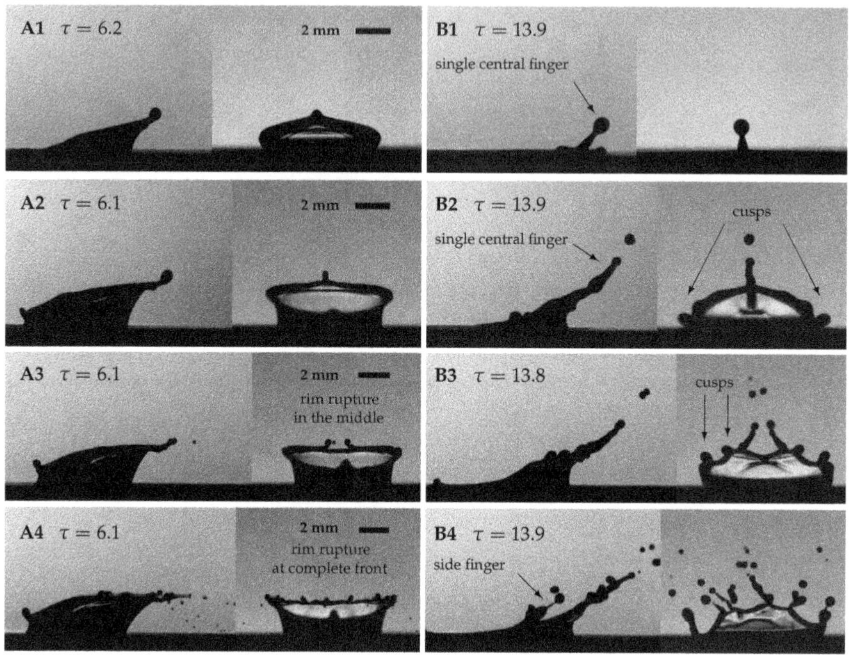

Figure 8. Comparison of the crown morphology for four different impact We numbers. (**A**) $\tau \approx 6.1$, formation of fingers; (**B**) $\tau \approx 13.9$, crown descending phase; (**1**) Exp 5 $We = 256$; (**2**) Exp 6 $We = 448$, (**3**) Exp 7 $We = 516$, (**4**) Exp 8 $We = 631$. For all impact conditions, see Table 1.

As described in the last section, Experiment 6 is characterized by a single central finger forming at the front from which three secondary droplets detach; see Figure 8(A2,B2). At the side of the crown, two cusps can be observed. For Experiment 5, with a lower We number, shown in Figure 8(A1,B1), the crown morphology is similar with a single central finger at the front side of the crown. However, only one secondary droplet detaches in the late receding phase and falls back on the film. Furthermore, the finger is shorter and thicker in this case, and also the detached droplet is larger. If the We number is increased to $We = 516$ (Experiment 7), the crown morphology changes. Here, the rim at the front gets unstable in the middle and two main fingers are visible as well as a tiny secondary droplet which is generated very early; see Figure 8(A3). Secondary droplets detach from the fingers at the front. The fingers are thinner than the single central finger for Experiment 6 and the secondary droplets are smaller. In the receding phase, similar to Experiment 6, cusps can be found at the side of the rim. In this case, however, there are two at each side, which are larger and can be already described as small fingers. In Experiment 8, the impact velocity and the droplet diameter were increased further to reach $We = 631$. Now, the rim is unstable along the complete front, many tiny droplets are present and many fingers are formed already during the initial finger formation; see Figure 8(A4). Droplets separate from those fingers at a later point in time. Furthermore, also at the rim on the side, not only cusps, but fingers are visible and droplets detach from them.

On the one hand, the disturbances in the early rim for Experiments 7 + 8 are explained by the increase in the ratio of kinetic energy to surface tension energy, as it can be also observed for the normal droplet impact. On the other hand, here additionally a collision of the early crown with the wall film is the cause. To better observe the early phase after impact, and to confirm this statement, the experiments were reproduced with a higher magnification for the side view. Figure 9 shows the reproduced experiments which result in the outcome of a single central finger (Exp. 9) as well as an early growth of many fingers (Exp. 10 + 11); see Table 1. In the case of a single central finger, see Figure 9 first row (A1/B1), the crown does not touch the wall film at any time in the early phase after impact. However, with an increase in *We* number, the crown interacts with the wall film in the early time; see Figure 9(A2). Note that, only the middle part of the front rim touches the film as observed from the front view perspective. This seems to introduce disturbances in this part of the rim, which lead to a partial early rupture and four fingers in the middle of the crown's front. The third row in Figure 9(A3/B3) shows a collision of the complete front part of the crown with the wall film. This interaction is more intense and leads to the ejection of tiny droplets. The rim is unstable on the complete front of the crown, which later leads to several fingers and the detachment of many droplets.

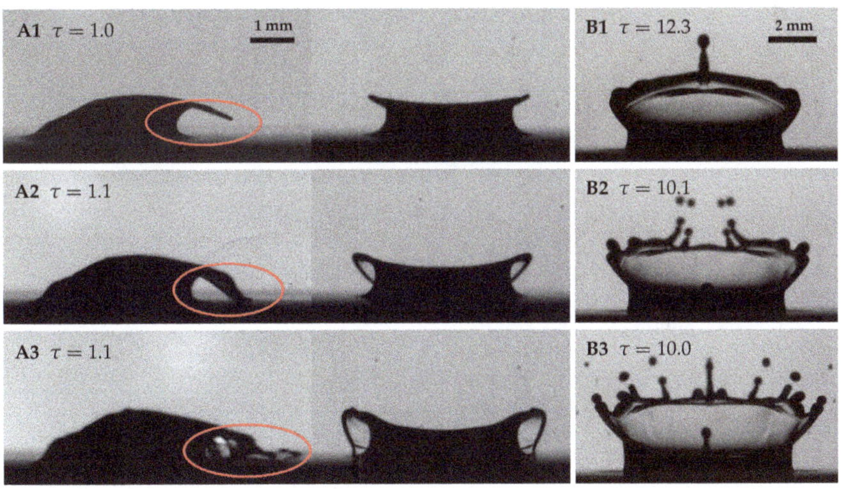

Figure 9. Early crown morphology side and front view (**A**) and late crown morphology front view (**B**). (**1**) Experiment 9 $We = 508$ (first row); (**2**) Experiment 10 $We = 610$ (second row); (**3**) Experiment 11 $We = 675$ (third row). For all impact conditions, see Table 1.

In summary, the Weber number has a big influence on the impact process and changes its outcome fundamentally. Three different morphologies can be distinguished with increasing *We*. The first is a single central finger with the detachment of between one and four secondary droplets. Starting at low *We*, the central finger is comparably thick and only one secondary droplet detaches during the end of the receding phase, which falls down on the film again. The rim at the side of the crown is stable without any cusps or waves on it. With an increase in *We*, the central finger gets longer and thinner. The first secondary droplet detaches earlier, flies upwards and is smaller. Also, a second, a third and a fourth secondary droplet detach from the central finger. With increased *We*, the rim at the side starts to show small waves and cusps. The second morphology is characterized by a partial early rim rupture in the middle with two (up to four) fingers. This is caused by a collision of the middle of the early front crown with the wall film. The fingers are thinner than the single central finger of the first morphology. Also, the trend continues so that the secondary droplets detach earlier and are smaller than at lower *We*. The rim at the side becomes increasingly unstable showing more cusps. A third morphology was observed consisting of a complete early rim rupture with several (>4) fingers. Here, the early rupture is not only limited to the middle of the front crown, and it is caused by an intense collision of the complete front side of the early crown with the wall film.

These different outcomes are summarized in Table 2 with the *We* range in which they were observed. The ranges overlap slightly, which is also reported in other studies that separate regimes of droplet impact, e.g., Cossali et al. [2]. The *We* numbers separating the regime of the single central finger from the partial early rim rupture and the complete early rim rupture result to $We \approx 500$ and $We \approx 600$, respectively. Deposition, e.g., an impact without the separation of any secondary droplets, was not observed in the investigated *We* range. Table 3 shows the splashing characteristics in from the three different morphologies by giving the numbers of fingers and secondary droplets.

Table 2. Summary of the Weber range in which the three different morphologies were observed for $\alpha \approx 60°$, $\delta \approx 0.22$ and $Oh = 0.014$.

Morpholgy Observed	We Range	Exp #
single central finger	[249, 515]	5, 6, 9
partial early rim rupture	[480, 610]	7, 10
complete early rim rupture	[592, 675]	8, 11

Table 3. Detailed description of observed morphologies and their splashing characteristic (number of fingers and secondary droplets).

Morpholgy Observed	We Range	Exp #	Number of Fingers	Number of Sec. Droplets
single central finger	[249, 420]	5	1	1–2
	[390, 500]	6, 9	1	3
	[450, 515]		1	4
partial early rim rupture	[480, 610]	7, 10	2–4	8–15
complete early rim rupture	[592, 675]	8, 11	5–12	10–25

In total, 80 experiments were conducted in a range of $249 \leq We \leq 675$ for the $\alpha \approx 60°$ impact case with $\delta = 0.22$. At the points of a change in morphology ($We = 500 \pm 20$, $We = 600 \pm 20$), more than ten experiments were conducted.

Table 4 gives examples of three phenomena from the literature for oblique droplet impacts, which are similar to the morphologies observed in this study. The impact conditions are listed in the table. Okawa at al. described the formation of a prow-like structure and the detachment of secondary droplets from it. This splashing mechanism is similar to the single central finger observed in this study. The Weber number was the same as in this study and the impact angle was smaller. The most significant difference is the $\delta = 3.8$, which can be already described as a deep pool impact. At larger *We* and larger impact angles, Okawa et al. [12] observed an early detachment of many tiny droplets, called prompt splash. Gielen et al. [14] investigated the impact on deep pools ($\delta = \infty$) and determined the Weber number at which splashing starts to occur at the front side in the 60° to $We = 370$.

Table 4. Comparable phenomena observed in the literature.

Morphology Observed	We	Impact Conditions
splash forms prow-like structure (single central finger) [12]	293	$\alpha = 41°$ $\delta = 3.8$ $Oh = 0.0044$
prompt splash (early detachment of many tiny droplets) [12]	378	$\alpha = 66°$ $\delta = 3.6$ $Oh = 0.0043$
splash on front side of crown [14]	≥370	$\alpha = 60°$ $\delta = \infty$ $Oh \approx 0.01$

3.4. Comparison of the Oblique Impact with the Normal Impact

Comparing the oblique impact with $\alpha = 60°$ to the normal impact, the following differences in morphology become evident. The most obvious difference is that the oblique impact has only one symmetry plane and the crown is asymmetric from the side view perspective, while the normal impact is axisymmetric. Consequently, the occurrence of different phenomena, e.g., a single central finger on the front and cusps at the side, are unique features of the oblique impact. Also, a collision of the early crown with the wall film can only be observed during an oblique impact. Furthermore, the crown recedes first at the back and last at the front, whereas the crown collapses at once in the $\alpha = 90°$ case. This explains why a Worthington jet can only occur in the normal droplet impact. Another difference is that at the oblique impact splashing can be observed already at $We = 249$, while splashing from a Worthington jet and crown splash only happens at $We \approx 400$ and $We \approx 600$, respectively, in the conducted normal impact experiments. In addition to that, the trajectory of the secondary droplets follows all radial directions, while for the oblique impact, the trajectories are only directed in the direction of impact.

In addition to these differences, there are also various observations, which are similar. The progression of phenomena with increasing Weber number is the same. First, only waves and cusps form, and then fingers and the detachment of secondary droplets, which fall down on the film again. This starts to occur in the late receding phase first and shifts to earlier times if We is increases further, which then leads to a ejection of secondary droplets, that fly upwards. The only difference is that this process happens for the whole crown at once for the normal impact, whereas at the oblique impact, it progresses individually for the different parts of the crown, e.g., first at the front and then for the rim at the sides. Another similarity is that the diameter of the fingers and secondary droplet decreases with an increase in We.

The practical implication of the splashing characteristic of the oblique droplet impact is a significant lower splashing threshold. In other words, secondary droplets are ejected during an oblique impact at conditions where no secondary droplets are generated at a normal impact. Furthermore, the secondary droplets do not fly in all radial directions as observed in the normal impact but are ejected only in the tangential direction of the impact. These differences can be crucial for spray impact applications.

4. Conclusions and Outlook

The oblique and normal impact of a single droplet onto a thin horizontal and quiescent wall film was investigated experimentally for different Weber numbers ($We = \rho D U^2/\sigma \approx$ [250; 700]) and the following conclusions can be drawn.

The method and the design of the oblique droplet generator, which was used in this study to shoot the droplets in a trajectory, was explained in detail, including the detachment process of the droplets from the needle tip and its tuning parameters. This setup allows us to vary the impact velocity and the impact angle independently from each other within a wide range. Furthermore, droplet impacts on deep pools as well as thin and thick wall films can be investigated in the future.

The study revealed a significant change in the morphology of the crown with increasing We and thus in the disintegration of the initial droplet and the formation of secondary droplets. Three different morphologies could be identified: The first one shows a single central finger on the front of the crown with one to four secondary droplets detaching from it. The second morphology resulting from an increase in We is characterized by a partial rupture of the early rim at the front, which is caused by a collision with the film. Two to four fingers are formed at the front. If the Weber number further increases, the interaction of the crown with the wall film is intensified. The rim breaks up also at the sides of the crown and several fingers form (third morphology). The critical Weber numbers separating the regime of the single central finger from the partial early rim rupture and the complete early rim rupture are $We \approx 500$ and $We \approx 600$, respectively.

A comparison to normal impacts revealed the important differences between both impact conditions. Obviously, an asymmetric crown which results in the formation of one single central finger at the front can be observed only during an oblique impact. Also, the collision of the crown with the wall film in the early time after impact only occurs during oblique impact. A Worthington jet, on the other hand, can only be found during normal impact. The progression from waves and cusps to fingers and the detachment of secondary droplets with an increase in We is similar for the oblique and normal impact scenario. In addition to the different morphologies, the splashing threshold in the oblique impact case is significantly lower. Hence, secondary droplets are ejected in conditions where no secondary droplets are generated during a normal impact. Furthermore, they fly only in the tangential direction of the impact velocity, not in all radial directions as in the normal droplet impact.

This study lays the foundation for a multidimensional regime map, which is valuable for the understanding of the fundamentals of a spray process and its optimization. Future investigations varying the impact angle and the film height will enhance the described regime separation and will provide a comprehensive picture of the process.

Author Contributions: Conceptualization, all authors; methodology, all authors; validation, J.L.S.; visualization, J.L.S.; investigation, J.L.S.; writing—original draft, J.L.S.; writing—review and editing, all authors.; visualization, J.L.S.; supervision, K.S. and M.S.; project administration, K.S. and M.S.; funding acquisition, K.S. and M.S. All authors have read and agreed to the published version of the manuscript.

Funding: The authors kindly acknowledge the financial support of this work by the Deutsche Forschungsgemeinschaft (DFG) in the frame of the International Research Training Group "Droplet Interaction Technologies" (GRK2160/2: DROPIT).

Data Availability Statement: The data presented in this study are available on request from the corresponding author.

Acknowledgments: Many thanks to my student assistant Simon Weitling, who helped in conducting the experiments.

Conflicts of Interest: The authors declare no conflict of interest. The funders had no role in the design of the study; in the collection, analyses, or interpretation of data; in the writing of the manuscript; or in the decision to publish the results.

References

1. Taylor, G.I. The dynamics of thin sheets of fluid. III. Disintegration of fluid sheets. *Proc. R. Soc. Lond.* **1959**, *253*, 313–321. [CrossRef]
2. Cossali, G.E.; Coghe, A.; Marengo, M. The impact of a single drop on a wetted solid surface. *Exp. Fluids* **1997**, *22*, 463–472. [CrossRef]
3. Deegan, R.D.; Brunet, P.; Eggers, J. Complexities of splashing. *Nonlinearity* **2008**, *21*, C1–C11. [CrossRef]
4. Worthington, A. *A Study of Splashed*; Longmans, Green, and Co, London: London, UK, 1908.
5. Vander Wal, R.L.; Berger, G.M.; Mozes, S.D. Droplets splashing upon films of the same fluid of various depths. *Exp. Fluids* **2005**, *40*, 33–52. [CrossRef]
6. Rioboo, R.; Bauthier, C.; Conti, J.; Voué, M.; De Coninck, J. Experimental investigation of splash and crown formation during single drop impact on wetted surfaces. *Exp. Fluids* **2003**, *35*, 648–652. [CrossRef]
7. Geppert, A.; Chatzianagnostou, D.; Meister, C.; Gomaa, H.; Lamanna, G.; Weigand, B. Classification of Impact Morphology and Splashing/Deposition Limit for n-Hexadecane. *At. Sprays* **2016**, *26*, 983–1007. [CrossRef]
8. Macklin, W.C.; Metaxas, G.J. Splashing of drops on liquid layers. *J. Appl. Phys.* **1976**, *47*, 3963–3970. [CrossRef]
9. Cossali, G.; Marengo, M.; Coghe, A.; Zhdanov, S. The role of time in single drop splash on thin film. *Exp. Fluids* **2004**, *36*, 888–900. [CrossRef]
10. Roisman, I.V.; Horvat, K.; Tropea, C. Spray impact: Rim transverse instability initiating fingering and splash, and description of a secondary spray. *Phys. Fluids* **2006**, *18*, 102104. [CrossRef]
11. Agbaglah, G.; Josserand, C.; Zaleski, S. Longitudinal instability of a liquid rim. *Phys. Fluids* **2013**, *25*, 022103. [CrossRef]
12. Okawa, T.; Shiraishi, T.; Mori, T. Effect of impingement angle on the outcome of single water drop impact onto a plane water surface. *Exp. Fluids* **2008**, *44*, 331–339. [CrossRef]

13. Leneweit, G.; Koehler, R.; Roesner, K.G.; Schäfer, G. Regimes of drop morphology in oblique impact on deep fluids. *J. Fluid Mech.* **2005**, *543*, 303. [CrossRef]
14. Gielen, M.V.; Sleutel, P.; Benschop, J.; Riepen, M.; Voronina, V.; Visser, C.W.; Lohse, D.; Snoeijer, J.H.; Versluis, M.; Gelderblom, H. Oblique drop impact onto a deep liquid pool. *Phys. Rev. Fluids* **2017**, *2*, 83602. [CrossRef]
15. Liang, G.; Guo, Y.; Yang, Y.; Zhen, N.; Shen, S. Spreading and splashing during a single drop impact on an inclined wetted surface. *Acta Mech.* **2013**, *224*, 2993–3004. [CrossRef]
16. Chen, J.J.; Liu, Q.F.; Zhang, F.; Zhao, F.L.; Qu, J.Y. Experimental Investigation on Water Droplet Impacting Liquid Film on Inclined Wetted Surface. *IOP Conf. Ser. Earth Environ. Sci.* **2021**, *701*, 012005. [CrossRef]
17. Che, Z.; Deygas, A.; Matar, O.K. Impact of droplets on inclined flowing liquid films. *Phys. Rev. E* **2015**, *92*, 23032. [CrossRef]
18. Adebayo, I.T.; Matar, O.K. Droplet impact on flowing liquid films with inlet forcing: The splashing regime. *Soft Matter* **2017**, *13*, 7473–7485. [CrossRef]
19. Kittel, H. Drop Impact onto a Wall Wetted by a Thin Film of Another Liquid. Ph.D. Thesis, Technische Universität Darmstadt, Darmstadt, Germany, 2019.
20. Cheng, M.; Lou, J. A numerical study on splash of oblique drop impact on wet walls. *Comput. Fluids* **2015**, *115*, 11–24. [CrossRef]
21. Brambilla, P.; Guardone, A. Assessment of dynamic adaptive grids in Volume-Of-Fluid simulations of oblique drop impacts onto liquid films. *J. Comput. Appl. Math.* **2015**, *281*, 277–283. [CrossRef]
22. Guo, Y.; Lian, Y. High-speed oblique drop impact on thin liquid films. *Phys. Fluids* **2017**, *29*, 082108. [CrossRef]
23. Chen, Z.; Shu, C.; Wang, Y.; Yang, L. Oblique drop impact on thin film: Splashing dynamics at moderate impingement angles. *Phys. Fluids* **2020**, *32*, 033303. [CrossRef]
24. Santini, M.; Cossali, G.E.; Marengo, M. Device and Method for Drops Generation. WO 2010/021004 A1, PCT/IT2008/000554, 22 August 2008.
25. Ruprecht, A.K.; Pruss, C.; Tiziani, H.J.; Osten, W.; Lucke, P.; Last, A.; Mohr, J.; Lehmann, P. Confocal micro-optical distance sensor: Principle and design. In *Proceedings of the Optical Measurement Systems for Industrial Inspection IV*; SPIE: Munich, Germany, 2005; Volume 5856, pp. 128–135. [CrossRef]
26. Yarin, A.L.; Weiss, D.A. Impact of drops on solid surfaces: Self-similar capillary waves, and splashing as a new type of kinematic discontinuity. *J. Fluid Mech.* **1995**, *283*, 141–173. [CrossRef]

fluids

MDPI

Article

Dynamics of a Water Droplet Impacting an Ultrathin Layer of Oil Suspended on a Pool of Water

Amir Dehghanghadikolaei, Bilal Abdul Halim, Ehsan Khoshbakhtnejad and Hossein Sojoudi *

Department of Mechanical, Industrial, and Manufacturing Engineering, The University of Toledo, Toledo, OH 43606, USA
* Correspondence: hossein.sojoudi@utoledo.edu

Abstract: This study investigates water droplets impacting a two-layered pool, consisting of a deep pool of water above which an ultrathin a suspended layer of silicone oil is present. Initially, the difference between the impact dynamics of water droplets on ultrathin and thick layers of oil were studied. It was found that the existence of an ultrathin layer of oil changes the impact characteristics such how aggressively the jet rises, how the dimensions of the impact impression change, and how the jets are broken down on their tops. Then, in a series of experiments on ultrathin layers of oil, the droplet size, the velocity of the droplets upon impact, and the viscosity of the oil layers were changed to observe and measure the characteristic dimensions of the formed craters and the jets. It was observed that when the viscosity of oil layers decreased to a minimum of 1 (cSt), the jet height and crater sizes increased to their maximum value. In addition to the effect of the oil viscosity, it was found that the droplet size and the release heights of the droplets were in the next orders of significance in determining the impact dynamics. The impacts were also characterized qualitatively by specifically looking into the crown and crater formations, pinch-off modes in jets, and number of formed secondary droplets. As well as the quantitative conclusion, it was found that the major affecting parameter in changing each of these qualities was the viscosity of the suspended oil layer.

Keywords: oil-water droplet dynamics; crater formation; jet break up

1. Introduction

Droplet impact on surfaces can be seen in a wide range from daily life to industrial applications such as inkjet printers, pesticide sprayers, irrigation systems, fuel injection in engines, fire extinguishers, metal quenching systems, and plasma spraying, among others [1–5]. The surfaces that experience impacting droplets could be a solid, a pool of liquid, or a thin liquid film laid either on a solid substrate or suspended on a pool of another immiscible liquid [6–8]. Properties of the impacting droplets (e.g., size, velocity, viscosity, surface tension, etc.) and characteristics of the pool/thin film liquids (e.g., thickness, viscosity, surface tension, etc.) and/or solid surfaces (e.g., surface energy, roughness, polarity, wetness) all influence the outcome of the impact [9,10]. For instance, modulating roughness and energy of solid surfaces has resulted in bouncing/repelling of the impacting droplets and development of so-called superhydrophobic surfaces [11–13]. Other potential outcomes of such impacts include homogenous wetting, splashing, and impalement/sticking [14,15]. On the other hand, droplets impacting miscible or immiscible liquids (ultrathin film or pool) might lead to outcomes such as crater and jet formation depends on liquid thickness, temperature, viscosity, surface tension, etc. [16]. Effects of droplet size, the number of falling droplets, the impact velocity, and condition of the impacted pool of liquid (temperature, viscosity, stationary/moving, etc.) have all been investigated [17,18]. These studies have mainly focused on the formation and dimensions of the craters upon impact, the retraction of the craters, and the consequent jet formations [19–21].

Inertia, capillary, viscous, and surface tension forces are all involved in studying droplet dynamics. Competing effects of these forces have been studied through various dimensionless numbers such as Reynolds (*Re*), Weber (*We*), Froude (*Fr*), Ohnesorge (*Oh*), Bond, and capillary [22–24]. Among those, the *We* and the *Re* numbers have been used mostly to explain various phenomena involved in droplet dynamics. For instance, Manzello et al. defined a critical *We* for jet break up that was independent of the pool depth [25]. Castillo-Orozco et al. found that the jet height increases as the *We* number increases and that the chances of the jet break up significantly decreases in higher droplet viscosity [26]. Ding et al. studied dynamics of viscous droplets impacting a less viscous pool of liquid (a mixture of water and glycerol with the oil content changing from 50 to 100 (wt.%), providing various dynamic viscosity values). They observed two successive jet formations after the impact in which the first jet corresponds to climbing behavior of the more viscous droplet and the second one is the so-called Worthington's jet [27]. Che et al. showed that the miscibility of two interacting liquids (i.e., droplet and pool of liquid) play a major role in determining outcome of the impact. They found that when a water droplet impacts an oil layer a compound jet forms while when an oil droplet impacts a water layer, the droplet spreads quickly [28]. Jain et al., investigated the effect of viscous oil droplets impacting on a pool of water, showing a correlation between the droplet viscosity and the crater depth. They showed that as the viscosities of the droplets and the pool liquid becomes closer, shallower craters are formed [29]. One of the few studies on the impact of droplets on multi/double layered pools was conducted by Kim et al., where oil droplets impact a suspending oil layer on a pool of water. With different droplet diameters and suspended layer thicknesses, they found out that with even a very thin layer of oil, the jet height and the crater depth are smaller when compared to the case in which no suspended oil layer is present [30]. In a similar study by Zhang et al., a water droplet impacted a suspended heptane layer on a deep pool of water. The study concluded that with changing the impact velocity (*We* number) and the thickness of the suspended oil layer, four types of impact outcomes observed (single crown-jet, single crown-bubble, double crown-bubble, and double crown-jet). They also showed that the existence of the water-oil interface substantially reduces the crater dimensions [31]. There are other works focusing on the impact of droplets on solid surfaces with engineered properties and thin films of liquid spread on solid substrates. For instance, Raiyan et al. used macrotextures to reduce contact time of viscous liquids impacting superamphiphobic surfaces [32]. On the other hand, Yeganehdoust et al. used a three phase flow model to study air entrapment during droplet impact on a thin film of immiscible liquid [33].

More recently droplet impacts on multilayers of immiscible liquids have been studied. Among various parameters, thickness of each layer have shown major effect on the characteristics of the crater and jet formation [31,34]. When a droplet impacts a multilayer of immiscible liquids, relative thickness of the outmost liquid (suspended layer) to the droplet size plays a major role in characteristics of the impact. The physics becomes more complex when this relative thickness becomes ultrathin [30]. Despite previous reports, there are no major efforts that study dynamics of droplets impacting ultrathin layer of a liquid suspended/floating on a pool of another immiscible liquid.

Here, dynamics of water droplets impacting an ultrathin layer of silicone oil suspended on a deep pool of water is investigated. Effect of oil viscosity and droplet size to oil thickness ratio on impact dynamics are studied. Characteristics of crater, crown, jet, and secondary droplets and their dependence to droplet/bilayer liquid (ultrathin silicone oil and pool of water) properties are studied in detail. It is found that an increase in the droplet size leads to an increase in jet height and crater dimensions (i.e., depth and width). In addition, higher *We* and *Re* numbers lead to formation of tall crowns and jets with top pinch-off modes, respectively. Furthermore, it is found that the oil viscosity plays a major role in defining the dynamics of the impact. Building upon these insights and drawing from the work of Yongsheng Wu et al. [35] we recognize that environmental factors such as rainfall must be considered in the context of oil spills. Their study underscores that rain-induced turbulence significantly affects oil droplet size and dispersion rates, similar to the turbulence caused by wind-induced waves. This phenomenon is critical for understanding the maximum droplet sizes and dispersion fractions in varying conditions, including the use of chemical dispersants. Their findings suggest that heavy rainfall can influence the dispersion of light and heavy oils to varying degrees, which should be considered when developing strategies to mitigate oil spill. Integrating this knowledge, our study will further elucidate the potential spread of oil spills over oceans due to rain, providing insights for better handling of such environmental crisis.

2. Results and Discussions

Droplet dynamics involves competition of inertia, viscous, surface tension, and gravity forces, among others. To understand significance of each force, various dimensionless numbers are employed. Weber (*We*) number represents ratio of inertial force to surface tension force and is defined as $We_e = \frac{\rho_w V^2 d}{\sigma_e}$ (effective impact Weber number) while Reynolds (*Re*) number represents competing effects of the inertial force to viscous resistance and is defined as $Re = \frac{Vd}{\nu_w}$. Here $V\ (\frac{m}{s})$ is the impact velocity, $\rho_w\ (\frac{kg}{m^3})$ is the droplet density, $d\ (m)$ is the droplet diameter, $\nu_w\ (\frac{kg}{m.s})$ is the droplet kinematic viscosity, and $\sigma_e\ (\frac{mN}{m})$ is the effective surface tension of the pool liquid (ultrathin silicone oil suspended on water or just water). As shown in their mathematical forms, these numbers vary by droplet viscosity, droplet size, impact velocity, equivalent surface tension, and the viscosity of the liquids [36]. Existence of an ultrathin layer of oil suspended on a pool of water changes some of the properties mentioned above. For instance, the active surface tension of the impact changes to interfacial tension since at the same time the water droplet is in interaction with air, and an oil layer suspended on a pool of water [30]. Figure 1C shows three different combinations of pool surface in which the oil thickness (*H*) varies from 0 to ultrathin layer ($H \ll d$) to comparable to water droplet diameter ($H \geq d$). In the case that there is no oil layer present ($H = 0$), the interfacial tension would be equal to that of air-water surface tension. When the oil layers are added (i.e., from ultrathin to thicknesses close to droplet diameter), the interfacial tension between the water droplet, air, and oil layer would become equal to a combination of oil-air and oil-water surface tensions. Finally, when the oil layer is thick enough (i.e., $H > d$), the interfacial tension changes to just the surface tension of oil-water without considering the term for air-water or air-oil surface tensions [37,38]. By adjusting the surface tensions (interfacial tensions in the case of ultrathin oil layer), the effective surface tension (σ_e) is calculated based on each experimental combination (equations shown in Figure 1C are adopted with modifications from Kim et al. [30]). Knowing the effective surface tension, the effective We_e number is calculated and used for data analysis. Since the only changing parameter in the ultrathin oil layer cases is the equivalent surface tension, the rest of the dimensionless numbers are kept and used as they were. Figure 1 shows the overall experimental setup and steps and Table 1 summarizes the liquid properties used in this study and the ranges of their values [39].

When a water droplet impacts an ultrathin layer of silicone oil suspended on a pool of water, its kinetic energy leads to deformation in the droplet and the pool liquid. Figure 2 shows different stages of the impact dynamics. Upon impact, the liquid pool surface deforms forming a crater with a depth of *D* and width of *W* significantly larger than the droplet diameter (*d*). This phenomenon is also called the crater evolution (step 2). Next, the crater retracts and pushes a column of liquids (water and silicone oil) upward forming a varying-height jet which is known as jet evolution and build up (step 3). In some cases, as the jet height reaches to its peak (based on initial impact conditions), the jet breaks up and forms secondary/child droplets (step 4). This moment is exactly when the jet peaks in its height and starts to descent. Generally, this region is called pinch-off or jet break-up and it is where the secondary/child droplets form from the descending liquid column (step 5). In some instances, as the jet descends the weight of that liquid column forms another crater upon impact to the pool forming a secondary crater (step 6). This might lead to formation of a secondary jet which is not typically tall and ultimately results in a liquid dimple and a tertiary droplet (step 7). However, from the point that the jet is descending, the process of droplet impact moves toward a stable state. At this stage, the droplet initially gets trapped in the suspended oil layer and eventually it passes the oil layer and meets with the deep pool of water, mixing with it (step 8). Movie S1 illustrates all the introduced stages of the impact. The video corresponds to a droplet impact falling from height of $Z = 750$ (mm) with a diameter of $d \approx 4.1$ (mm) on an ultrathin layer of silicone oil with kinematic viscosity of $\nu = 50$ (cSt) suspended on a deep pool of water.

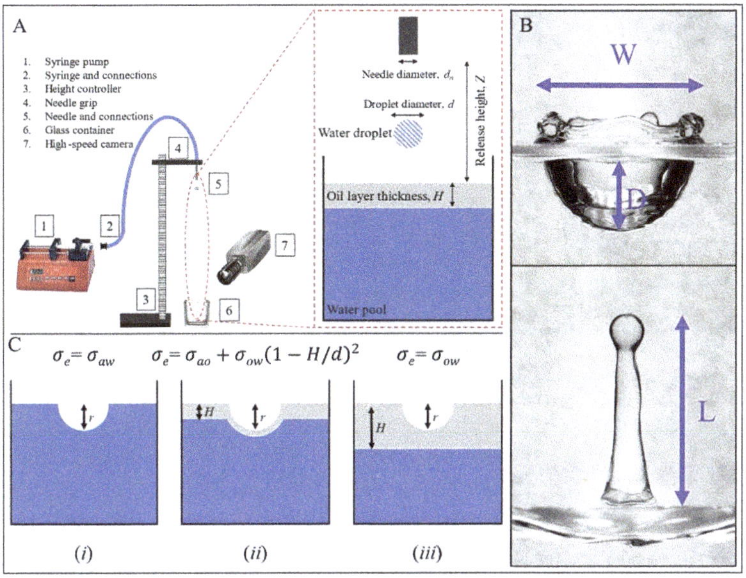

Figure 1. Schematic of the setup, measured features, and different thicknesses of silicone oil layer on pool of water. (**A**) Shows the components of the setup; a syringe pump (1) is used to pump the water with a set of syringes and connections (2) which are fixed on an adjustable height controller (3). The needle is fixed at the grip location (4). After releasing the droplets (5) their impact on the pool (6) are recorded using a high-speed camera (7). (**B**) Shows the crater dimensions (width of W and depth of D) in addition to the jet height (L). (**C**) Shows thicknesses of silicone oil on pool of water where in (*i*) there is no oil layer present ($H = 0$), in (*ii*) an ultrathin oil layer is present ($0 < H \ll d$), and in (*iii*) a thick oil layer is present ($H \geq d$), in addition to the equations used to calculate the effective surface tensions, σ_e. Note $d = 2r$.

Table 1. List of liquid properties, dimensionless numbers, and range of their values.

Variable	Symbol	Range
Water droplet diameter (mm)	d	2.4–4.1
Water droplet release height (mm)	Z	250–750
Water density (kg/m^3)	ρ_w	997
Water kinematic viscosity (cSt)	ν_w	1
Water surface tension (mN/m)	σ_w	72.2
Effective Weber number	We_e	145–704
Reynolds number	Re	5195–15216
Oil density (kg/m^3)	ρ_o	962–964
Oil kinematic viscosity (cSt)	ν_o	1–100
Oil surface tension (mN/m)	σ_o	19.8–21.3

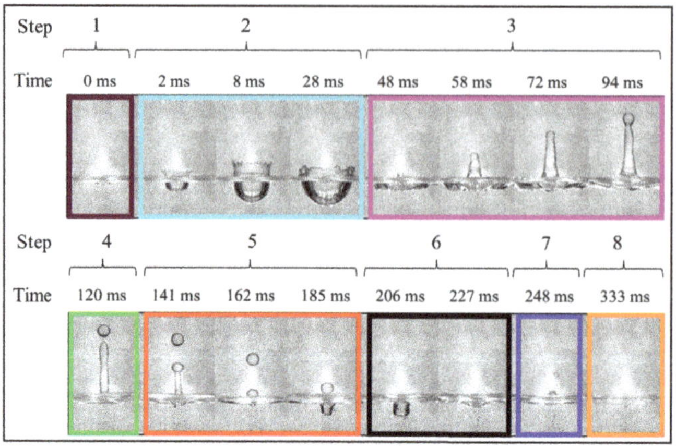

Figure 2. Different steps when a water droplet impacts on an ultrathin oil layer suspended/floating on a deep pool of water and their associated timeline. (**1**) A moment before impact, (**2**) Crater formation and evolution (with or without crown formation), (**3**) Jet evolution and build up (formation), (**4**) Jet break-up/pinch-off and formation of a secondary/child droplet, (**5**) Descending jet and secondary/child droplet, (**6**) Secondary crater formation, (**7**) Dimple and tertiary droplet formation, and (**8**) Steady state pool surface where the droplet finally meets and mixes with the deep pool of water (the sequences correspond to a water droplet with diameter of $d \approx 4.1$ (mm) falling from a height of $Z = 750$ (mm) on an ultrathin oil layer with viscosity of $\nu_o = 1$ (cSt)).

2.1. Effect of Oil Layer Thickness

Figure 3A represents three different silicone oil thicknesses employed in this study. The oil thicknesses of $H = 5$ and 8 (mm) are considered thick layers ($H > d$), the oil thickness of $H = 3$ (mm) is considered a thin layer (nearly equal to the water droplet diameter or $H \approx d$) and finally, the ultrathin oil layers in which $H \ll d$. When the oil is ultrathin ($H \approx 0$) it does not resist the impacting water droplets much. This leads to formation of deep craters regardless of oil viscosity. However, when the suspended oil becomes thicker ($H = 5$ and 8 (mm)) it resists the water droplet upon impact, leading to a decrease in the depth of formed craters. However, the kinetic energy of the impacting droplets leads to creation of wider craters instead. This can be seen when plotting the normalized depth (D') vs. normalized width of craters (W') (Figure 3B). It can be hypothesized that an increase in the normalized jet height (L') means less energy dissipation during the droplet impact. Therefore, it can be concluded that the wider crater formation leads to more energy dissipation than the formation of a deeper one. Difference between water droplets impacting an ultrathin vs. a thick oil layer is evident in characteristics of impact dynamics. With no research on dynamics of a water droplet impacting an ultrathin layer of oil suspended on a pool of water, characteristics of that impact such as formation of crater, crown, jet, and secondary droplets are studied in detail and presented here. Movie S2 compares the characteristics of impacts on ultrathin and thicker layers of suspended oil on a pool of water. The pool surface conditions are with no oil, ultrathin oil layer, and thicker oil layers ($H = 3, 5$, and 8 (mm) and $\nu = 50$ (cSt)). The droplet size is $d \approx 2.9$ (mm) released from $Z = 750$ (mm). A 45° view of the same impact is presented in Movie S3.

Figure 3C shows the difference between ratio of crater dimensions (in terms of $\frac{W}{D}$) for impacts on ultrathin layers oil in comparison to similar impacts on thicker oil layers. It can be seen that for impacts on ultrathin layers of suspended silicone oil, the majority of the impacts form a crater with $\frac{W}{D} < 1$ while for impacts on thicker suspended oil layers the crater dimension ratio changes to $\frac{W}{D} > 1$. Although the impacts on $\nu = 50$ (cSt) and $H = 8$ (mm) falls below the margin of thin and thick oil layers, it can be safely anticipated that the impacts on thick oil layers are different in nature.

Finally, Figure 3D shows that despite dissipation of the kinetic energy upon droplet impact, an increase in inertial forces leads to enhanced retainment of the energy leading to a jet formation with increased normalized jet heights (L' which will be discussed in next sections). However, this trend is less intensified when the water droplets impacted a thicker layer of oil suspended on a pool of water ($H > d$). This can be explained due to surface tension mismatch between the water and the silicone oil and the corresponding damping nature. As expected, when the oil is ultrathin, a linear trend between the impact We number and the normalized jet height (L') is observed. In addition, a decrease in oil viscosity leads to an increase in the normalized jet height, especially where an ultrathin layer of oil is suspended on a deep pool of water. Overall, at a given number a decrease in oil viscosity or its thickness leads to taller jet formation.

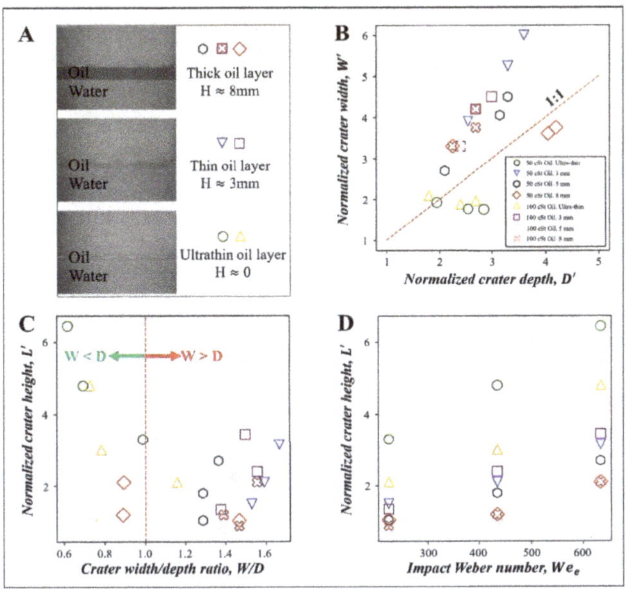

Figure 3. Crater and jet characteristics based on oil thickness. (**A**) Three different thicknesses of suspended silicone oil layer on a pool of water defining thick ($H > d$), thin ($H \approx d$), and ultrathin layers ($0 < H \ll d$) discussed in this work. (**B**) Shows the difference between the normalized crater depth, D', and normalized crater width, W', of impacts between impacts on ultrathin and thicker oil layers. (**C**) Shows the margin of normalized crater $\frac{width}{depth}$ ($\frac{W'}{D'}$). It is evident that the impacts on ultrathin oil layers cause relatively deeper craters than wider when compared to the instances of impact on thicker oil layers. (**D**) Shows the normalized jet height as a function of impact We numbers for different oil thicknesses and kinematic viscosities. In all impacts, water droplets with diameters of $d \approx 3.4$ (mm) were released from height of $Z = 250$ (mm), 500 (mm), and 750 (mm) on oil thicknesses of $H \approx 0$, 3, 5, and 8 (mm). The oil kinematic viscosity was $\nu = 50$ (cSt) or $\nu = 100$ (cSt).

2.2. *Vertical Displacement of Liquid and Formation of Crowns*

As the falling water droplets impact the pool surface (water only or ultrathin oil on water), a deformation occurs in both the falling droplets and the pool surface [40,41]. The impact leads to liquid deformation and consequently to formation of craters. However, the liquid retracts due to surface tension forces resulting in its vertical displacement [42,43]. The extent of vertical displacement in the liquid depends on the impact condition such as We number. In addition, the effect of pool surface viscosity should always be considered as surface/interfacial tension and viscosity forces are in constant interaction during liquid deformation and motion. Four different impact conditions and the extent of vertical pool liquid displacement, up to formation of a full crown, are shown in Figure 4A–D. As the impact conditions change, it can be seen that the viscosity is playing an important role.

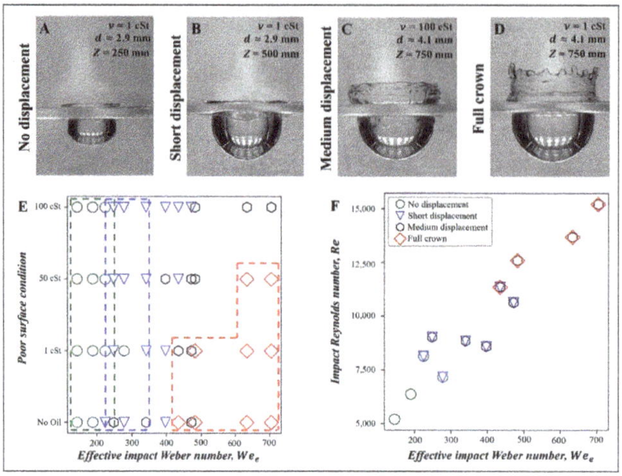

Figure 4. Vertical displacement of the pool liquid (ultrathin layer of oil suspended on water) under various impact conditions. The extent of displacement is based on the impact conditions such as droplet release height, impacting velocity, and droplet size, while oil kinematic viscosity has its own effect, as well. Highest vertical displacements result in formation of a full crown (as shown in (**D**)). The corresponding impact conditions are shown in each section (**A**–**D**). (**E**) Shows the effect of pool condition (with or w/o oil) on the extent of vertical displacement at a given impact We_e. It can be seen that the full crowns are more likely to form at lower oil viscosity. (**F**) Shows the the relation between the impact We_e and Re numbers. As the dimensionless numbers get larger (larger droplet sizes or higher release heights), higher chances of observing tall crowns are expressed.

When the effective Weber number is small ($150 \leq We_e \leq 230$), in most impact instances, no vertical displacement occurs regardless of existence and/or viscosity of the suspended ultrathin oil (see Figure 4E). With an increase in the effective Weber number ($230 \leq We_e \leq 400$), slight/short vertical displacement in the liquid is observed during the crater formation. This is expected given the increase in the kinetic energy of the impacting droplets. Further increase in the Weber number ($470 \leq We_e \leq 710$) leads to medium and large vertical displacements in the liquid. However, at higher Weber numbers adding an ultrathin layer of silicone oil on pool of water leads to an overall decrease in the extent of vertical displacement. This can be due to the surface tension mismatch and immiscibility difference between the silicone oil and the water. In addition, the viscosity of the silicone oil can also play a role. At higher Weber numbers ($470 \leq We_e \leq 710$), an increase in the viscosity of the ultrathin silicone oil from 1 to 100 (cSt) leads to an overall decrease in the extent of vertical displacement in the liquid and consequently crown formation. It is hypothesized that more viscous oil leads to enhanced energy dissipation during the impact and less likelihood of large vertical displacements at high Weber numbers. When the vertical displacement of the liquid is large it leads to so-called crown formation which is visible at the tips of the displaced liquid (Figure 4E). It has been shown that the crown becomes more enhanced when the water droplet impacts the pool of water without any suspended ultrathin layer of oil [44].

Figure 4F shows the behavior of the impact regarding its We_e and Re numbers. As these dimensionless numbers increase, there is a higher chance in observing a full crown formation. Based on the impact Re numbers, three distinct regions of no displacement, short displacements, and full crown formations could be observed over the ranges of $5000 \leq Re \leq 8000$, $8000 \leq Re \leq 11{,}000$, and $11{,}000 \leq Re \leq 15{,}500$, respectively. However, the instances of medium deformations were available over a wide range of impact *Re* numbers ($8000 \leq Re \leq 15{,}500$) which was expected as these deformations were transition modes between a short deformation and a full crown formation. Although it is possible to distribute this category to short displacements and full crowns, they are kept in order to have the highest distinction between all the observed categories. The We_e and *Re* numbers include the terms for impact velocity and the droplet size themselves. Thus, it can be concluded that as the release heights and droplet sizes increase, a linear increasing trend will be observed for chances of having full crown formations.

2.3. Crater Formation and Its Characteristics

Upon droplet impact on a pool of liquid (ultrathin silicone oil suspended on water), deformations occur both in the droplet and the liquid pool leading to formation of craters with various depths and widths 44. As discussed earlier, the craters are wider than deeper ($\frac{W}{D} > 1$) due to more resistance droplets experience in the impact direction (see Figure 3B,C). However, as the We_e number increases, the difference between the width and depth of the craters fades away. This means that at impacts of larger droplets from higher release heights, the craters are formed closer to a hemisphere shape rather than oval-shaped ones which occurs during impact of smaller droplets released from lower heights. It can be also seen that larger craters form when the impacts are more aggressive (increase rate of D is higher than that of W). Thus, such impacts result in increased depth of craters rather than width which enhances the retraction and formation of full crowns. This trend is mostly seen where there are comparable values of W and D ($W \approx D$) (see Figure 5B). Furthermore, an increase in droplet size and its release height collectively leads to an increase in impact We_e number. As expected, the enhancement in the kinetic energy leads to larger crater formation as evident by an increase in the overall dimension ($W + D$) of the formed craters. This consequently leads to enhancement in the vertical displacement of the liquid, resulting in increased occurrence of full crown formation. Figure 5C,D represent the effect of droplet size and release height on enhanced chances of full crown formations. For larger droplets released from a higher height (larger impact We_e), most of the impact instances results in a full crown formation. A real snapshot of the impact instances with droplet sizes 1–4 are shown in Figure 5A where in similar impact conditions, increasing the droplet size causes an increase in the crater dimensions and consequently formation of taller jets which will be discussed in the following sections.

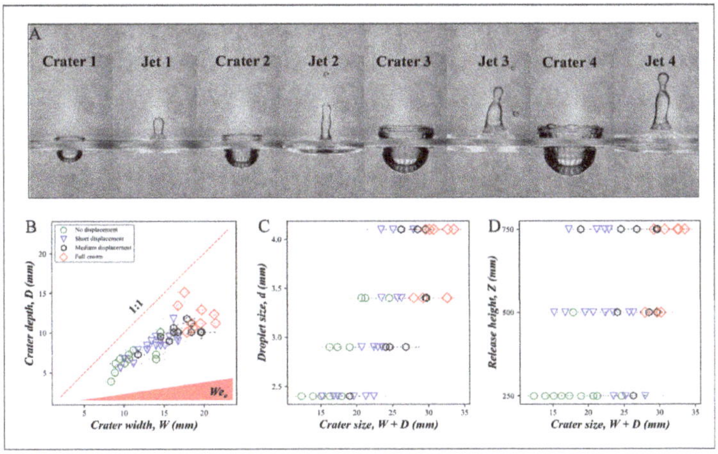

Figure 5. Crater characteristics based on size and velocity of droplets impacting an ultrathin layer of silicone oil suspended on a pool of water. (**A**) shows the sequences of crater formation and jet formation

for impacts of droplets in different diameters in the same conditions resulting in different impact characteristics. (**B**) shows the variations of crater depth (D) versus crater width (W) from smaller to larger impact We_e numbers. (**C**) shows the specific effect of droplet size d) on the crater dimensions ($W + D$). And (**D**) shows the effect of release height (Z) on overall crater dimensions ($W + D$). For a more appealing look, the droplet sizes were coded as 1, 2, 3, and 4 corresponding to the actual droplet sizes of ≈2.4, 2.9, 3.4, and 4.1 (mm), respectively. All the snapshots are for droplets size 1–4 released from height of $Z = 500$ (mm) on suspended ultrathin oil layer of $\nu = 50$ (cSt) oil (crater 1/jet 1–crater 4/jet 4).

2.4. Jet Pinch-Off Modes

Consequent to crater formation and its retraction, the excessive energy of the impacting droplets leads to vertical displacement of the liquid forming a jet [45]. Depending on the extent of energy dissipation during the impact, liquid jets with different heights form some of which experience a disintegration and a pinch-off on top of the liquid pillar. This occurs due to different velocities in a pillar of the liquid that moves upward [46,47]. The process starts with elongation of a column of liquid followed by necking and partial separation, as shown in Figure 6A. If observed pinch-off modes are categorized as low, medium, and top depending on their height as shown in Figure 6B. Figure 6C shows that with an increase in the impact We_e number an increase in the normalized jet height (L') is observed. As shown in Figure 5D, with increased release heights, the crater size increases which translates to more severe crater retraction since more liquid is displaced. As the retraction became more aggressive, a higher normalized jet height was occurred, consequently leading to enhancement in top pinch-off modes. However, at any given release height, there were several instances of top pinch-off which means the release height alone is not a well-defined indicative of the pinch-off modes.

On the other hand, the size of the droplets is considered as the other impacting parameter in changing the impact We_e number. As shown in Figure 6D, with increased droplet sizes, a higher number of top pinch-off instances were observed. Similar to the droplets released from a higher height, when the larger droplets impact the pool surface, higher kinetic energy is converted and larger craters form. Depending on the properties of the droplet and pool liquid properties (such as viscosity), they will deform to different extents. For instance, if the pool surface has a higher equivalent viscosity, the impacting droplets deform more easily. In general, the impacts made wider craters rather than deeper ones which can be obviously seen in Figure 5C,D but with an increased droplet size and release height, this difference became smaller (higher impact We_e numbers in Figure 5B). Overall, increase in effective We number leads to formation of craters with larger depth and width. This leads to enhanced vertical displacement of the liquid, ultimately forming crown. Upon crater retraction the height of the formed jets depends on the size of craters and extent of vertical displacement in the liquid. Observed increase in the depth and width of the formed craters ultimately lead to taller jet formations that experience top pin-off modes.

Figure 6E shows that as the viscosity of the ultrathin oil layer increases the dissipation of kinetic energy increases during the impact. This leads to formation of smaller craters and consequently shorter jets that result in low or no pinch-offs. For instance, an impact on ν = 100 (cSt) silicone oil is less severe compared to the same impact on a lower viscosity (e.g., ν = 1 (cSt)) as in the former, the resisting viscous forces of the pool surface are larger. On the other hand, surface tension plays a key role when comparing the impacts on silicone oil of ν = 1 (cSt) and pool of water with the same viscosity. Since the oil layers are ultrathin ($H/d \ll 1$), the effective surface tension ($\sigma_e \approx 82\ (\frac{mN}{m})$) is roughly equal to the sum of oil-air ($\sigma_{oa} \approx 27.2\ (\frac{mN}{m})$) and oil-water ($\sigma_{ow} \approx 55.2\ (\frac{mN}{m})$) surface tensions and greater than the water-air surface tension ($\sigma_{wa} \approx 72.8\ (\frac{mN}{m})$) [30]. Upon impact, both water surface and oil layer deform. However, on the oil layer, the crater faces a resistance caused by higher surface tension and retracts more aggressively. As a result, the formed jets are taller for the instances of impacts on $\nu = 1$ (cSt) oil layer and the chances of observing top pinch-offs increase [29].

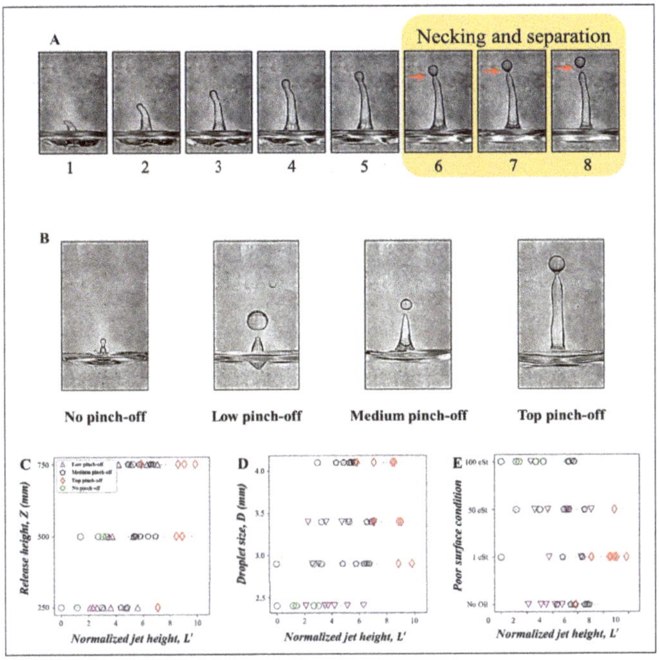

Figure 6. Sequences of pinch-off process, different types of pinch-off modes, effect of processing parameters on pinch-off modes. **(A)** sequence of jet evolution by an impact corresponding to $d \approx 2.9$ (mm), $Z = 750$ (mm), and $\nu = 1$ (cSt) kinematic viscosity. The white arrows in 6, 7, and 8 show necking, partial separation, and pinch-off, respectively. **(B)** shows the different types of pinch-off modes from no pinch-off to top pinch-off. **(C–E)** are the corresponding plots to the specific effects of release height, droplet size, and pool surface conditions on pinch-off modes.

Based on the presented data, it can be concluded that for larger droplets and higher release heights impacting less viscous ultrathin oil layers, it is highly expected to observe deeper craters, taller jets, and higher chances of top pinch-off (see discussion on Figure 6E). Figure 7A shows that the minimum $\frac{W}{D}$ ratio occurs at impacts on $\nu = 1$ (cSt) oil layer and the highest number of instances with top pinch-off occurs at this condition. As a result, it can be concluded that with a deeper (or less wide) crater, the chances of experiencing a top pinch-off increases. Since the instances of top pinch-off occur nearly in any given impact conditions (a combination of release height and droplet size in Figure 7B), it is apparent that the major affecting parameter in defining the pinch-off modes is the viscosity of the target surface followed by its surface/interfacial tension. Nearly all the impacts on the low viscosity oil layer resulted in top pinch-off modes, justifying this finding. Although the droplet size and the release height are the other important contributing factors to categorize the pinch-off modes (Figure 6C,D), the major effect is reflected by the pool surface conditions. Figure 7C represents impacts on different pool surfaces from different heights, corresponding to the presented discussion. When water droplets impact a pool of water only instead of oil, the formed craters disappear faster due to lower viscosity of water when compared to oil. This results in less energy dissipation during the impact and formation of taller jets with potential occurrence of a pinch-off. Adding an oil with a high viscosity to the pool of water results in formation of craters with increased width to depth ration ($\frac{W}{D}$) due to high shear forces leading to shorter jet formation that do not experience top pinch-offs. However, for the impacts on a pool of water and $\nu = 1$ (cSt), the ruling factor seems to be interfacial tension of oil-water-air. It is hypothesized that the suspended oil layer acts as a cushion but due to its negligible thickness it does not provide any significant resistance to deformation and further facilitates the impingement of the falling droplet. Although a deeper crater formation is expected for impacts on a pool of water, an opposite trend was observed. This is a distinct observation when compared to the studies on conventionally thicker oil layers [29,48].

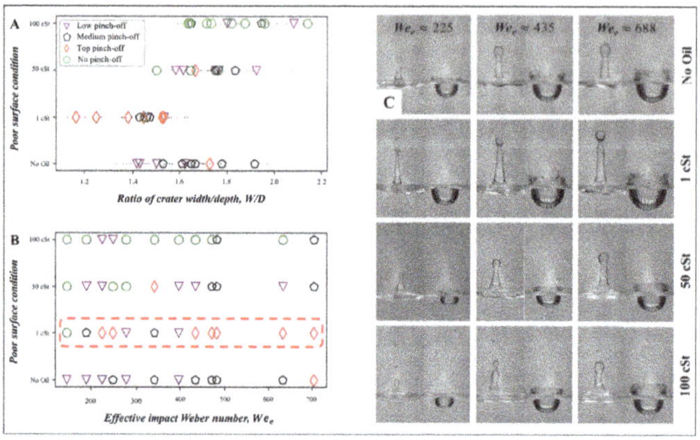

Figure 7. Various pinch off modes upon impact on different pool surfaces. (**A**) shows the effect of pool surface conditions on ratio of crater dimensions ($\frac{W}{D}$). (**B**) shows a wide range of impact We_e numbers on experiencing top pinch-off instances which mostly occurred in presence of $\nu = 1$ (cSt) silicone oil on the pool surface. (**C**) represents the different snapshots of craters and jets formed for impacts of different release heights and pool surfaces. The sequences are obtained from a droplet size of $d \approx 3.4$ (mm) impacting different oil layers ($\nu = 1$, 50, and 100 (cSt) and pool of water), released from Z = 250, 500, and 750 (mm), respectively.

2.5. Secondary Droplet Formation

When jet pinch-off occurs, the liquid body (jet) divides into a main descending column and a secondary droplet. Different pinch-off modes (low, medium, and top) result in different types of secondary droplet formations [49]. A single droplet or a series of droplets (two or more) might form during jet pin-offs depending on the impact condition [50–52]. Here, the results are categorized into three distinct groups of: no secondary droplets, single secondary droplet, and double/multiple secondary droplets as shown in Figure 8A.

When a water droplet with a threshold effective Weber number ($We_e \geq 190$) impacts a pool of water, the resulting jet always experiences a pinch-off with one or more secondary droplets formation. This is valid even after adding an ultrathin layer of silicone oil with kinematic viscosities of $\nu = 1$ and 50 (cSt) to the pool of water. However, when the silicone oil viscosity increases further to $\nu = 100$ (cSt) the formed jet does not necessarily experience a pinch-off, thus no formation of secondary droplets. With a $\nu = 100$ (cSt) silicone oil suspended on pool of water, at higher We_e numbers a single secondary droplet form in rare occasions of impacts. However, adding a less viscous ($\nu = 1$ or 50 (cSt)) ultrathin layer of oil to the pool of water leads to enhanced chances of jet pinch-offs, forming single or multiple secondary droplets (see Figure 8B). Comparing impacts on a deep pool of water and a suspended oil layer of $\nu = 1$ (cSt), at higher impact We_e numbers ($We_e \geq 430$), the impacts on pool of water mostly resulted in single secondary droplet formation while for lower We_e numbers a greater number of secondary droplets were formed. Since the overall air-water surface tension is less than the sum of interfacial tensions between oil and air and oil and water, the retraction of the crater formed upon impact on the pool of water was less abrupt compared to that of 1 (cSt) oil layer (see discussion on Figure 6E). In addition, higher impact We_e numbers result in a thick compound jet where a higher volume of liquid is displaced compared to singular jets (lower We_e numbers). As a result, the necking is less likely to happen for cases of compound jets and lower chances of pinch-off and formation of multiple secondary droplets will exist (see Movie S4).

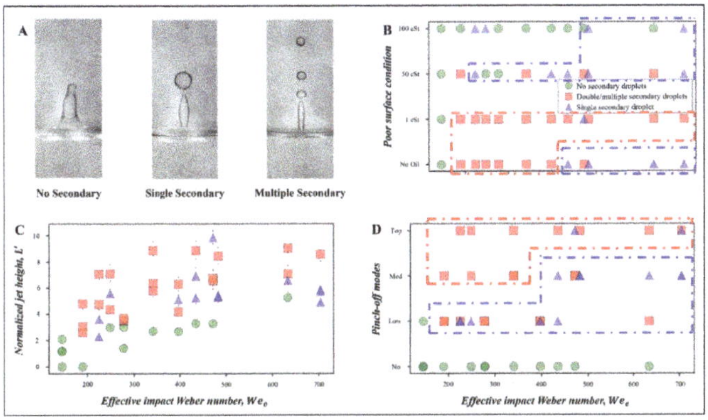

Figure 8. Relation of different processing parameters on the characteristics of impact and secondary droplet formation. (**A**) shows three different impacts with different numbers of secondary droplets. (**B**) shows the relation between the different pool surfaces and the combined effect of impact velocity and droplet size, We_e, with identified categories. (**C**) shows the relation of the normalized jet height, L', with We_e. (**D**) categorizes the different number of secondary droplets based on pinch-off modes (Top = top pinch-off, Med = medium pinch-off, Low = low pinch-off, and No = no pinch-off) and the impact We_e numbers. The sequences shown in (**A**) (from left to right) are results of impacts as follows: ν = 100 (cSt) oil layer, $d \approx 3.7$ (mm), Z = 750 (mm) − ν = 100 (cSt) oil layer, $d \approx 4.1$ (mm), Z = 750 (mm), and ν = 1 (cSt) oil layer, $d \approx 4.1$ (mm), Z = 250 (mm), respectively.

Figure 8C shows that over a wide range of We_e numbers and normalized jet heights different numbers of secondary droplets form. Once a threshold $We_e \approx 190$ number is reached, double/multiple secondary droplets are observed during pinch-off of even short jets. This shows that the secondary droplet formation is not exactly following the increasing trend observed for the crown formation as a function of the impact We_e numbers. However, there is a threshold for the normalized jet height ($L' \approx 3$) that is needed to lead to formation of single/double/multiple secondary droplets. Overall, a combined condition of increase in effective Weber number and normalized jet height lead to formation of single/double/multiple secondary droplets during jet pin-offs.

Figure 8D presents the data for secondary droplet formations based on pinch-off modes. At different pinch-off modes (Low, Med, and Top), the instances of multiple secondary droplets increase over any given impact We_e number. Most likely for any instance of top pinch-off and regardless of the impact We_e numbers, there would be multiple secondary droplets formed. For medium (Med) pinch-offs there is a higher chance of multiple secondary droplet formation by critical $We_e \leq 350$ but for higher We_e values mostly single secondary droplets form. Lower We_e numbers reflect smaller impacting droplets released from a lower height which are more likely to form thin singular jets. This leads to a more feasible necking and pinch-off leading to formation of multiple secondary droplets, compared to thicker compound jets formed at higher impact We_e numbers. It can be concluded that viscosity of the suspended oil layer is the major determining factor in qualitative analysis of secondary droplet formation. Table 2 summarizes how variations in Weber number and oil layer thickness influence the observed jet pinch-off modes and various liquid displacement during crater formation.

While this study has provided valuable insights into droplet dynamics on ultrathin oil layers, we acknowledge certain limitations. Our experiments were conducted under controlled laboratory conditions, which may not fully capture the complexities of natural environments where factors such as wind, temperature fluctuations, and surface heterogeneity can play a role. Furthermore, we focused on a specific range of droplet sizes and velocities limiting the range of studied We number which can be broadened in future studies.. In addition, oil with varying surface tension and thickness can be explored further to better mimic crude oil that spills over oceans/seas.

Table 2. Impact Dynamics by *We* Number, Oil Thickness, and Jet Pinch-Off Modes

We Number Range	Oil Layer Thickness	Jet Pinch-Off	Crater Formation
Low (150–230)	Ultrathin	No/low pinch-off	No displacement
	Thick	Slight pinch-off	No displacement
Medium (230–400)	Ultrathin	Low to medium pinch-off	Short displacement
	Thick	Medium pinch-off	Short to medium displacement
High (470–710)	Ultrathin	Medium to high pinch-off	Medium to full crown
	Thick	Less frequent high pinch-off	Medium, full crown less likely

3. Conclusions

In the current study, it was found that with different impact conditions, different impact dynamics are observed. Initially, it was observed that droplet impacts on ultrathin and thicker suspended silicone oil layers express different impact characteristics; impacts on ultrathin oil layers relatively result in deeper crater formation followed by taller jets. After observing the significant difference in dynamics of impacts on ultrathin oil layers, the experiments were focused on investigating the different characteristics of such impacts. In that regard, it was found that as the impact We_e increases, the chances of formation of full crowns increase as well. Most of the instances of full crown formations occurred in impacts on a pool of water or $\nu = 1$ (cSt) silicone oil. Addition of more viscous oil layers reduced the chances of full crown formation. Furthermore, the size of the formed craters enhanced with increase in droplet size and release height (higher impact We_e numbers). In addition, at higher impact We_e numbers, the crater dimension ratio ($\frac{W}{D}$) decreased, shifting the morphology of the craters to near-sphere shapes. Jet pinch-off modes were observed to be directly tied to droplet size and release height while the major effect was reflected by the pool surface conditions.

Going forward, it was observed that as the oil viscosity decreases, the chances of observing top pinch-offs increased. Comparing the case of a similar viscosity ($\nu = 1$ (cSt) silicone oil and a pool of water), the effective surface tension plays the major role by increasing the depth of the craters and resulting in taller jets and more instances of top pinch-offs for impacts on $\nu = 1$ (cSt) ultrathin oil layer. Other than that, the impacts on this viscosity oil resulted in formation of deeper craters compared to the similar impacts on a pool of water. This is a key characteristic difference compared to the results of previous studies that focused on thicker oil layers. By different behaviors in pinch-off modes, it was expected to have a notably different number of secondary droplet formations. It was observed that secondary droplets can form over a wide range of impact We_e numbers. However, the highest chance of observing multiple secondary droplets is at lowest oil viscosity of $\nu = 1$ (cSt), followed by impacts on a pool of water. The number of secondary droplets is directly tied to the dynamics of jet formation and pinch-off modes.

It was observed that for impacts on a pool of water, as the impact We_e numbers increase, the compound jets form (relatively thicker compared to singular jets) which decreases the chances of necking and pinch-off. As a result, lower number of secondary droplets were formed at this specific impact condition. For a similar reason, the impacts resulted in medium pinch-off form multiple secondary droplets at lower impact We_e numbers while most top and low pinch-off instances resulted in multiple and single secondary droplet formation, respectively.

Future studies can consider impact of water salinity to better mimic ocean water. There is also a great need for studying the impacts of different types of crude oils to better understand impact of rain droplets on crude oil spill over oceans. It can also help in drawing generalized inferences for other industrial applications and environmental conditions by extending the range of the *We* number using wider range of droplets sizes and impacting velocities/release heights.

4. Materials and Methods

4.1. Experimental Setup

Figure 1A shows a schematic of the setup used in this study and the corresponding parameters. Experiments are conducted in a controlled environment (i.e., temperature of 21–23 °C, and relative humidity of 75–80%). A large transparent container (100 mm × 100 mm × 110 mm) is made of polycarbonate and filled with deionized water (Millipore Sigma, Burlington, MA, USA). The container is made large enough to minimize any wall effects [26]. Silicone oils with kinematic viscosities of $\nu_o = 1, 50$, and 100 (cSt) (μ MicroLubrol, Clifton, NJ, USA) and thicknesses of $H = 3, 5$, and 8 (mm) are poured on the deep pool of water. Another set of experiments are conducted in which droplets of the silicone oil (with various viscosities as listed above) are placed on the deep pool of water, forming a suspended ultrathin layer of oil ($H \ll d$). Water droplets are generated with a syringe pump (NE-300 Just Infusion, Braintree, MA, USA) at a flowrate of $Q = 50\frac{ml}{hr}$ and using four different needle diameters, ($d_n \approx 0.6, 0.9, 1.6$, and 2.1 (mm)) to obtain droplets with different diameters (d). Based on the needle sizes, the droplet diameters (d) are mathematically calculated to be around $2.4, 2.9, 3.4$ and 4.1 (mm), respectively. These numbers are also verified through acquired images. The water droplets are released from different heights, (Z = 250, 500, and 750 (mm)) to obtain different impact velocities upon reaching the pool liquid. A high-speed camera (Olympus i-speed, Digital Solutions Americas, Bethlehem, PA, USA) is used to image the impact dynamics at 1500 fps. The silicon oils are dyed with Sudan III (Sigma-Aaldrich, Burlington, MA, USA) to improve contrast and to obtain better quality videos/images.

To accurately replicate real-world conditions in our experiment, we used droplets with varying sizes and release heights. This partially enables findings of this study to be applicable for real worlds scenarios such rain droplets and industrial spraying. Such a varying range for the *We* number allows us to explore droplet dynamics comprehensively, identifying regime transitions and clarifying the underlying mechanisms. Our focus on specific droplet parameters guarantees that our findings are attributable to the interplay of inertial and surface tension forces, offering critical insights into droplet impact physics relevant for environmental and industrial advancements.

4.2. Video Processing

The acquired videos are processed to obtain characteristics of various features (i.e., formation of craters, crowns, and jet and their breakup) observed during the impact dynamics. Dimensions of various features are obtained based on the ratio of the image size to the actual size of the droplets using pixel counting. The reported values are based on averages of three experiments/measurements. Figure 1B shows representative images of the crater and jet formation during the impact and their characteristic dimensions (W and D are the width and the depth of craters, respectively and L is the jet height). Their corresponding nondimensional values are obtained by dividing them over the size of the impacting water droplets ($W' = \frac{W}{d}$, $D' = \frac{D}{d}$, and $L' = \frac{L}{d}$). In addition, characteristics of other features observed during the impact dynamics such as crown height, jet pinch-off modes, and number of secondary formed droplets are obtained from the videos, qualitatively.

Supplementary Materials: The following supporting information can be downloaded at: https://www.mdpi.com/article/10.3390/fluids9040082/s1, The supporting information files of this work include discussion about the different stages of an impact of a falling droplet to a target pool in addition to representing the impact characteristics of similar impacts on a pool of water, on an ultrathin layer, a thin layer, and two thicker layers of suspended silicone oil on a deep pool of water from different angles of view (front view and 45° top view). Furthermore, a comparison of two different jet types is presented where their different characteristics result in different impact dynamics. Movie S1:

Represents different impact stages of a falling water droplet on an ultrathin layer of silicone oil suspended on a pool of water. The pool of water is large and deep enough to consider it semi-infinite pool. A droplet in size of d $\approx$ 4.1 (mm) was released from Z = 750 (mm) on an ultrathin layer of silicone oil (ν = 1 (cSt)) suspended on a deep pool of water; Movie S2: Represents different impact characteristics for different experimental conditions. (A) impact occurs on a deep pool of water without any suspended oil layer (H = 0). (B) impact occurs on an ultrathin layer of silicone oil (ν = 50 (cSt), H $\approx$ 0) suspended on a deep pool of water. (C) impact occurs on a thin layer of silicone oil (ν = 50 (cSt), H = 3 (mm)) suspended on a deep pool of water. (D) impact occurs on a thick layer of silicone oil (ν = 50 (cSt), H = 5 (mm)) suspended on a deep pool of water. (E) impact occurs on a thicker layer of silicone oil (ν = 50 (cSt), H = 8 (mm)) suspended on a deep pool of water. The difference between the impact characteristics is evident as the thickness of the oil layer decreases and reaches to an ultrathin layer; Movie S3: Represents a similar impact from a 45° view angle. The impact conditions are identical as well; Movie S4: Represents the two different types of jet formations, singular jets, and compound jets and how the jet type affects the pinch-off modes and consequently the number of formed secondary droplets. The singular jet is resulted from an impact with relatively lower impact Wee numbers, but it has a higher number of secondary droplets formed. Evidently, the thicker jet needs more upward velocity/energy to onset necking and pinch-off as it contains a larger volume of liquid, compared to that of the singular jet.

Author Contributions: Conceptualization, A.D. and B.A.H.; methodology, A.D.; data curation, A.D. and E.K.; writing—original draft preparation, A.D.; writing—review and editing, E.K., B.A.H. and H.S.; supervision, H.S. All authors have read and agreed to the published version of the manuscript.

Funding: This research received no external funding.

Data Availability Statement: The data presented in this study are available on request from the corresponding author due to (specify the reason for the restriction).

Acknowledgments: The authors cordially thank the University of Toledo machine shop and John Jaegly for his help in preparing the setup for the experiments.

Conflicts of Interest: The authors declare no conflict of interest.

References

1. Pan, Z.; Weibel, J.A.; Garimella, S.V. Influence of Surface Wettability on Transport Mechanisms Governing Water Droplet Evaporation. *Langmuir* **2014**, *30*, 9726–9730. [CrossRef] [PubMed]
2. Liang, G.; Mudawar, I. Review of Pool Boiling Enhancement by Surface Modification. *Int. J. Heat Mass Transfer* **2019**, *128*, 892–933. [CrossRef]
3. Charalampous, G.; Hardalupas, Y. Collisions of Droplets on Spherical Particles. *Phys. Fluids* **2017**, *29*, 103305. [CrossRef]
4. Luo, K.; Shao, C.; Chai, M.; Fan, J. Level Set Method for Atomization and Evaporation Simulations. *Prog. Energy Combust. Sci.* **2019**, *73*, 65–94. [CrossRef]
5. Pasandideh-Fard, M.; Aziz, S.D.; Chandra, S.; Mostaghimi, J. Cooling Effectiveness of a Water Drop Impinging on a Hot Surface. *Int. J. Heat Fluid Flow* **2001**, *22*, 201–210. [CrossRef]
6. Josserand, C.; Thoroddsen, S.T. Drop Impact on a Solid Surface. *Annu. Rev. Fluid Mech.* **2016**, *48*, 365–391. [CrossRef]
7. Lohse, D. Fundamental Fluid Dynamics Challenges in Inkjet Printing. *Annu. Rev. Fluid Mech.* **2022**, *54*, 349–382. [CrossRef]
8. Yu, X.; Zhang, Y.; Hu, R.; Luo, X. Water Droplet Bouncing Dynamics. *Nano Energy* **2021**, *81*, 105647. [CrossRef]
9. Biroun, M.H.; Rahmati, M.; Tao, R.; Torun, H.; Jangi, M.; Fu, Y. Dynamic Behavior of Droplet Impact on Inclined Surfaces with Acoustic Waves. *Langmuir* **2020**, *36*, 10175–10186. [CrossRef]
10. Hu, Z.; Wu, X.; Chu, F.; Zhang, X.; Yuan, Z. Off-centered Droplet Impact on Single-ridge Superhydrophobic Surfaces. *Exp. Therm. Fluid Sci.* **2021**, *120*, 110245. [CrossRef]
11. Breitenbach, J.; Roisman, I.V.; Tropea, C. From Drop Impact Physics to Spray Cooling Models: A Critical Review. *Exp. Fluids* **2018**, *59*, 1–21. [CrossRef]
12. Weisensee, P.B.; Tian, J.; Miljkovic, N.; King, W.P. Water Droplet Impact on Elastic Superhydrophobic Surfaces. *Sci. Rep.* **2016**, *6*, 30328. [CrossRef] [PubMed]
13. Schutzius, T.M.; Jung, S.; Maitra, T.; Graeber, G.; Köhme, M.; Poulikakos, D. Spontaneous Droplet Trampolining on Rigid Superhydrophobic Surfaces. *Nature* **2015**, *527*, 82–85. [CrossRef] [PubMed]
14. Xu, M.; Wang, C.; Lu, S. Water Droplet Impacting on Burning or Unburned Liquid Pool. *Exp. Therm. Fluid Sci.* **2017**, *85*, 313–321. [CrossRef]
15. Raiyan, A.; Mohammadian, B.; Sojoudi, H. Droplet Dynamics and Freezing Delay on Nanoporous Microstructured Surfaces at Condensing Environment. *Coatings* **2021**, *11*, 617. [CrossRef]

16. Hasegawa, K.; Nara, T. Energy Conservation during Single Droplet Impact on Deep Liquid Pool and Jet Formation. *AIP Adv.* **2019**, *9*, 085218. [CrossRef]
17. Blanken, N.; Saleem, M.S.; Thoraval, M.-J.; Antonini, C. Impact of Compound Drops: A Perspective. *Curr. Opin. Colloid Interface Sci.* **2021**, *51*, 101389. [CrossRef]
18. Cheng, X.; Sun, T.-P.; Gordillo, L. Drop Impact Dynamics: Impact Force and Stress Distributions. *Annu. Rev. Fluid Mech.* **2022**, *54*, 57–81. [CrossRef]
19. Pasandideh-Fard, M.; Qiao, Y.M.; Chandra, S.; Mostaghimi, J. Capillary Effects during Droplet Impact on a Solid Surface. *Phys. Fluids* **1996**, *8*, 650–659. [CrossRef]
20. Zhang, G.; Quetzeri-Santiago, M.A.; Stone, C.A.; Botto, L.; Castrejón-Pita, J.R. Droplet Impact Dynamics on Textiles. *Soft Matter* **2018**, *14*, 8182–8190. [CrossRef]
21. Liu, Q.; Lo, J.H.Y.; Li, Y.; Liu, Y.; Zhao, J.; Xu, L. The Role of Drop Shape in Impact and Splash. *Nat. Commun.* **2021**, *12*, 3068. [CrossRef]
22. Moghtadernejad, S.; Lee, C.; Jadidi, M. An Introduction of Droplet Impact Dynamics to Engineering Students. *Fluids* **2020**, *5*, 107. [CrossRef]
23. Attinger, D.; Moore, C.; Donaldson, A.; Jafari, A.; Stone, H.A. Fluid Dynamics Topics in Bloodstain Pattern Analysis: Comparative Review and Research Opportunities. *Forensic Sci. Int.* **2013**, *231*, 375–396. [CrossRef]
24. Li, X.J.; Zhou, Y. *Microfluidic Devices for Biomedical Applications*; Woodhead Publishing: Sawston, UK, 2021.
25. Manzello, S.L.; Yang, J.C. An Experimental Study of a Water Droplet Impinging on a Liquid Surface. *Exp. Fluids* **2002**, *32*, 580–589. [CrossRef]
26. Castillo-Orozco, E.; Davanlou, A.; Choudhury, P.K.; Kumar, R. Droplet Impact on Deep Liquid Pools: Rayleigh Jet to Formation of Secondary Droplets. *Phys. Rev. E* **2015**, *92*, 053022. [CrossRef]
27. Ding, Q.; Wang, T.; Che, Z. Two Jets during the Impact of Viscous Droplets onto a Less-viscous Liquid Pool. *Phys. Rev. E* **2019**, *100*, 053108. [CrossRef]
28. Che, Z.; Matar, O.K. Impact of Droplets on Immiscible Liquid Films. *Soft Matter* **2018**, *14*, 1540–1551. [CrossRef]
29. Jain, U.; Jalaal, M.; Lohse, D.; van der Meer, D. Deep Pool Water-impacts of Viscous Oil Droplets. *Soft Matter* **2019**, *15*, 4629–4638. [CrossRef] [PubMed]
30. Kim, D.; Lee, J.; Bose, A.; Kim, I.; Lee, J. The Impact of an Oil Droplet on an Oil Layer on Water. *J. Fluid Mech.* **2021**, *906*, A5. [CrossRef]
31. Zhang, X.; Wang, C.; Rui, S. Phenomena of Single Water Droplet Impacting a Heptane Layer on Water Pool. *Eur. J. Mech. B/Fluids* **2022**, *92*, 65–74. [CrossRef]
32. Raiyan, A.; McLaughlin, T.S.; Annavarapu, R.K.; Sojoudi, H. Effect of Superamphiphobic Macrotextures on Dynamics of Viscous Liquid Droplets. *Sci. Rep.* **2018**, *8*, 15344. [CrossRef] [PubMed]
33. Yeganehdoust, F.; Attarzadeh, R.; Karimfazli, I.; Dolatabadi, A. A Numerical Analysis of Air Entrapment During Droplet Impact on an Immiscible Liquid Film. *Int. J. Multiph. Flow* **2020**, *124*, 103175. [CrossRef]
34. Wang, W.; Ji, C.; Lin, F.; Wei, X.; Zou, J. Formation of Water in Oil in Water Particles by Drop Impact on an Oil Layer. *Phys. Fluids* **2019**, *31*, 037107. [CrossRef]
35. Wu, Y.; Hannah, C.G.; Thupaki, P.; Mo, R.; Law, B. Effects of Rainfall on Oil Droplet Size and the Dispersion of Spilled Oil with Application to Douglas Channel, British Columbia, Canada. *Mar. Pollut. Bull.* **2017**, *114*, 176–182. [CrossRef] [PubMed]
36. Ribeiro, D.F.S.; Silva, A.R.R.; Panão, M.R.O. Insights into Single Droplet Impact Models Upon Liquid Films Using Alternative Fuels for Aero-Engines. *Appl. Sci.* **2020**, *10*, 6698. [CrossRef]
37. Camp, D.W.; Berg, J.C. The Spreading of Oil on Water in the Surface-Tension Regime. *J. Fluid Mech.* **1987**, *184*, 445–462. [CrossRef]
38. Razavi, S.; Hernandez, L.M.; Read, A.; Vargas, W.L.; Kretzschmar, I. Surface Tension Anomaly Observed for Chemically-Modified Janus Particles at the Air/Water Interface. *J. Colloid Interface Sci.* **2020**, *558*, 95–99. [CrossRef] [PubMed]
39. Mohamed, M.H.; Shahbaznezhad, M.; Dehghanghadikolaei, A.; Haque, M.A.; Sojoudi, H. Deformation of Bulk Dielectric Fluids Under Corona-Initiated Charge Injection. *Exp. Fluids* **2020**, *61*, 1–12. [CrossRef]
40. Guo, Y.; Wei, L.; Liang, G.; Shen, S. Simulation of Droplet Impact on Liquid Film with CLSVOF. *Int. Commun. Heat Mass Transfer* **2014**, *53*, 26–33. [CrossRef]
41. Trujillo, M.F.; Lee, C.F. Modeling Crown Formation Due to the Splashing of a Droplet. *Phys. Fluids* **2001**, *13*, 2503–2516. [CrossRef]
42. Bernard, R.; Baumgartner, D.; Brenn, G.; Planchette, C.; Weigand, B.; Lamanna, G. Miscibility and Wettability: How Interfacial Tension Influences Droplet Impact onto Thin Wall Films. *J. Fluid Mech.* **2021**, *908*, A36. [CrossRef]
43. Bernard, R.; Vaikuntanathan, V.; Weigand, B.; Lamanna, G. On the Crown Rim Expansion Kinematics during Droplet Impact on Wall-Films. *Exp. Therm. Fluid Sci.* **2020**, *118*, 110168. [CrossRef]
44. Ersoy, N.E.; Eslamian, M. Phenomenological Study and Comparison of Droplet Impact Dynamics on a Dry Surface, Thin Liquid Film, Liquid Film and Shallow Pool. *Exp. Therm. Fluid Sci.* **2020**, *112*, 109977. [CrossRef]
45. Santra, S.; Das, S.; Chakraborty, S. Electric Field-Induced Pinch-Off of a Compound Droplet in Poiseuille Flow. *Phys. Fluids* **2019**, *31*, 062004. [CrossRef]
46. Almohammadi, H.; Amirfazli, A. Droplet Impact: Viscosity and Wettability Effects on Splashing. *J. Colloid Interface Sci.* **2019**, *553*, 22–30. [CrossRef]

47. Sayyari, M.J.; Naghedifar, S.A.; Esfahani, J.A. Pinch-Off Location and Time during 2D Droplet Impact onto a Wetted Stationary Cylinder Using the Lattice Boltzmann Method. *J. Braz. Soc. Mech. Sci. Eng.* **2020**, *42*, 1–13. [CrossRef]
48. Blanco–Rodríguez, F.J.; Gordillo, J.M. On the Jets Produced by Drops Impacting a Deep Liquid Pool and by Bursting Bubbles. *J. Fluid Mech.* **2021**, *916*, A37. [CrossRef]
49. Pan, K.-L.; Hung, C.-Y. Droplet Impact upon a Wet Surface with Varied Fluid and Surface Properties. *J. Colloid Interface Sci.* **2010**, *352*, 186–193. [CrossRef]
50. Huang, K.-L.; Pan, K.-L.; Josserand, C. Pinching Dynamics and Satellite Droplet Formation in Symmetrical Droplet Collisions. *Phys. Rev. Lett.* **2019**, *123*, 234502. [CrossRef]
51. Samenfink, W.; Elsäßer, A.; Dullenkopf, K.; Wittig, S. Droplet Interaction with Shear-Driven Liquid Films: Analysis of Deposition and Secondary Droplet Characteristics. *Int. J. Heat Fluid Flow* **1999**, *20*, 462–469. [CrossRef]
52. Liu, D.; Tan, H.-W.; Tran, T. Droplet Impact on Heated Powder Bed. *Soft Matter* **2018**, *14*, 9967–9972. [CrossRef] [PubMed]

fluids

MDPI

Article

Understanding the Application of Emulsion Systems for Bacterial Encapsulation and Temperature-Modulated Release

Nur Suaidah Mohd Isa [1,2,*], Hani El Kadri [1], Daniele Vigolo [1,3,4], Nur Farra Adlina Mohamed Zakhari [2] and Konstantinos Gkatzionis [1,5,*]

1 School of Chemical Engineering, University of Birmingham, Birmingham B15 2TT, UK; dabigribosome@gmail.com (H.E.K.); daniele.vigolo@sydney.edu.au (D.V.)
2 Faculty of Fisheries and Food Science, Universiti Malaysia Terengganu, Kuala Terengganu 21030, Malaysia; farra.adliina@gmail.com
3 School of Biomedical Engineering, The University of Sydney, Sydney, NSW 2006, Australia
4 The University of Sydney Nano Institute, The University of Sydney, Sydney, NSW 2006, Australia
5 Department of Food Science and Nutrition, University of the Aegean, Metropolite Ioakeim 2, 81400 Myrina, Lemnos, Greece
* Correspondence: n.suaidah@umt.edu.my (N.S.M.I.); kgkatzionis@aegean.gr (K.G.)

Abstract: The encapsulation of bacteria in emulsion droplets offers various advantages over other conventional methods of encapsulation, such as improvements in bacterial viability, and may serve as microenvironments for bacterial growth. Nevertheless, changes in temperature may affect bacterial viability and droplet stability. In this study, the encapsulation of bacteria in single water-in-oil (W/O) and double water-in-oil-in-water ($W_1/O/W_2$) emulsions under cold storage and temperature-modulated release were investigated. The microencapsulation of bacteria in emulsion droplets was achieved by using a flow-focusing microfluidic device. Droplet stability was determined by measuring changes in droplet size and creaming behaviour at different temperatures. The thermal properties of the samples were determined by using differential scanning calorimetry, while the release of bacteria with changes in temperature was determined by measuring the colony form unit (CFU) of the released bacteria and conducting fluorescence microscopy. Higher bacterial viability was observed for encapsulated samples compared to free cells, indicating the ability of the emulsion system to improve bacterial viability during cold-temperature storage. The crystallisation temperature was lowered in the presence of bacteria, but the melting temperature was similar with or without bacteria. Storage in freezing temperatures of −20 °C and −80 °C led to extensive droplet destabilisation, with the immediate release of encapsulated bacteria upon thawing, where the temperature-modulated release of encapsulated bacteria was achieved. This study provides an overview of the potential application of emulsion droplets for bacterial encapsulation under cold-temperature storage and the controlled release of encapsulated bacteria mediated by changes in temperature, which is beneficial for various applications in industries such as food and pharmaceuticals.

Keywords: microfluidics; bacterial release; W/O/W emulsion; stability; crystallisation

1. Introduction

Emulsion droplets have been extensively studied to support the encapsulation of materials such as bacteria, bioactive compounds, and drugs [1–5]. These materials usually require storage at low temperatures to maintain their stability, especially for long-term storage. The storage of bacteria at low temperatures is a common practice for bacterial preservation to maintain their viability and is used for, for example, lactic acid bacteria in the dairy industry and in biotechnology applications. The freezing of cells to temperatures such as −20 °C and −80 °C helps to maintain not only their viability but also their functionality over a long storage period [6,7]. The encapsulation of bacteria in emulsions, especially food-based emulsions such as milk and ice cream, has been reported to improve bacterial viability during freezing, depending on the nature of the emulsion [8–11]. Nevertheless, further studies are still required not only to assess the viability of bacteria encapsulated in emulsion droplets but also to determine the effect of bacteria on emulsion stability during storage at cold temperatures.

The successful application of emulsion droplets for bacterial encapsulation depends on their stability. The stability of emulsions during cold-temperature storage is highly affected by changes in the phase behaviour of the oil and aqueous phases, as reported in many related studies [12–15]. The crystallisation of the oil phase followed by the aqueous phase in oil-in-water (O/W) emulsions may cause a partial coalescence of the oil droplets whereby, in loosely packed emulsions, the collision between the still-liquid oil droplets and crystallised oil droplets with protruding crystals may cause membrane rupture, resulting in the droplets with still-liquid oil to flow out, forming a linkage between the two droplets as they freeze. This mechanism causes droplet flocculation, which eventually leads to complete coalescence and oiling-off upon thawing [12,16]. A different destabilisation mechanism for multiple emulsions, namely, water-in-oil-in-water ($W_1/O/W_2$) emulsions, has been proposed, which involves the external coalescence of the inner W_1 phase with the outer W_2 phase, which occurs during the thawing process, while the W_1 inner phase remains intact during the freezing process [17,18].

Although emulsion destabilisation is unfavourable for food emulsions as it may affect product quality, it has the potential to be applied for the controlled and immediate release of materials, particularly from double-emulsion droplets. The targeted release of microbials from double emulsions was discussed recently by Zhang et al. [19], who showed the promising application of this system whereby release can be triggered by various factors, which may help to enable applications in various food and drug environments. Previous studies also show an interesting application of emulsions for the controlled release of encapsulated materials with changes in temperature, which has good potential to be applied in industries such as food and pharmaceuticals [17,18,20,21]. However, it is crucial to understand the mechanism behind the release to widen the applications of this system for controlled release. This study focused on the application of emulsion droplets for bacterial encapsulation under cold-temperature storage. It aimed to investigate the encapsulation of bacteria within single W/O and double $W_1/O/W_2$ emulsion droplets, focusing on their stability and viability during cold storage. A flow-focusing microfluidic device was utilised for droplet formation, where the effects of temperature variations on droplet stability, bacterial viability, and the controlled release of bacteria were explored. The detrimental effects of cold temperature on bacterial viability and droplet destabilisation during storage remain a challenge. Therefore, this study is crucial and may provide valuable information on the suitability of emulsion droplets for bacterial encapsulation under cold-temperature storage. In addition, temperature-induced bacterial release from double water-in-oil-in-water emulsions was explored to further determine its potential in the controlled release of encapsulated materials.

2. Materials and Methods

2.1. Materials and Bacterial Cultures

Microfluidic device fabrication was performed using a Polydimethylsiloxane (PDMS) preparation set (Sylgard 184, Dow-corning, Midland, MI, USA), which included the curing agent and pre-polymer. The oil-soluble surfactant, polyglycerol polyricinoleate (PGPR), was obtained from Danisco (Copenhagen, Denmark). Soybean oil (Alfa Aesar, Lancashire, UK) was used as the oil phase for both single water-in-oil (W/O) and double water-in-oil-in-water ($W_1/O/W_2$) emulsion droplets. For bacterial culture preparation, the materials used were nutrient agar, Luria Bertani broth (LB broth), and phosphate-buffered saline (PBS), all from Oxoid Ltd. (Hampshire, UK). *Escherichia coli* strain SCC1 (MG1655-GFP mutation) expressing green fluorescent protein (*E. coli*-GFP) stock culture was obtained from the Biochemical Engineering Laboratory, University of Birmingham, United Kingdom.

2.2. Bacterial Cell Preparation

Bacterial cultures used for encapsulation in emulsion droplets were prepared by culturing *E. coli*-GFP on nutrient agar at 37 °C for 24 h. The cultured bacteria were kept at 4 °C prior to the experiment. The bacterial cells were then inoculated into 50 mL of Luria Bertani broth (LB Broth) in a shaking incubator at 37 °C and 150 rpm for 24 h, and then sub-cultured in fresh LB broth (1:50), followed by incubation for another 2 h. The bacterial culture was then centrifuged (10,000× *g*, 10 min) and washed two times with 50 mL of PBS. After centrifugation, the supernatant was discarded and replaced with 50 mL of fresh LB broth or DIW to re-suspend the bacterial cells for encapsulation. The bacterial cell concentration was prepared to 10^8 CFU/mL.

2.3. Microfluidic Device Fabrications

The microfluidic devices used for droplet generation were prepared using the standard soft lithography technique [22]. Device designs were based on specifications provided by Bauer et al. [23]. Patterned moulds were produced by printing the design onto high-resolution photomasks (Micro Lithography Services Limited, Essex, UK). Silicon wafers (Si-Mat, Kaufering, Germany) were spin-coated with SU-8 photoresist (SU-8, Microchem) and then exposed to UV light through the photomasks with a mask aligner (Canon PLA-501FA mask aligner, Canon, Tokyo, Japan). The PDMS mixture, consisting of PDMS and curing agent prepared at a recommended ratio of 1:10, was then poured onto the prepared mould, degassed, and cured in an oven at 70 °C for 1 h. The device was then cut out of the mould, and the inlet and outlet holes were punched, followed by corona discharge treatment (Relyon, PZ2) for approximately 30 s, which bonds the device onto a glass slide and seals the channels. The prepared device was then left on a hot plate for approximately 15 min at 100 °C. A new device was prepared for every experiment to minimise contamination.

As part of microfluidic device preparation, a partial hydrophilic surface treatment was performed on double-emulsion droplet devices to ease droplet formation [23]. Polyelectrolyte multilayers (PEM) consisting of alternating sequence of poly(allylamine hydrochloride) (PAH) and poly(sodium 4-styrenesulfonate) (PSS) solutions in 0.5 M of aqueous sodium chloride solution (0.1% *w*/*v*) was loaded into a Tygon tube (Cole-Parmer instrument Co. Ltd., Cambridge, UK) with 0.1 M of sodium chloride (NaCl) in deionised water as the washing solution. The PEM was then flushed through the lower part of the microfluidic device while the upper part was blocked with a stream of deionised water. At the end of the process, the lower part of the microfluidic channel was coated with PEM, rendering the channel walls hydrophilic while the upper part remained hydrophobic.

2.4. Microfluidic Encapsulation of E. coli-GFP in W/O and W_1/O/W_2 Droplets

The prepared culture of *E. coli*-GFP in deionised water (DIW) was then used as the aqueous phase of W/O and the inner aqueous phase of W_1/O/W_2 droplet. The oil phase for both types of emulsion consisted of soybean oil with 1.5% *w*/*v* PGPR surfactant, while for W_1/O/W_2, the outer aqueous phase consisted of DIW with 1% *w*/*v* Tween 80 surfactant. The bacterial encapsulation was conducted using a flow-focusing microfluidic device, whereby a single junction device was used for generating W/O droplets, while the encapsulation into W_1/O/W_2 was performed using a double junction device. The devices were connected to a pressure controller (OB1 MK3, Elveflow, Paris, France) to produce monodispersed droplets of approximately 50 µm in diameter for W/O. To generate a double emulsion, the W/O droplets were further emulsified with an outer aqueous phase, producing dispersed oil globules of approximately 100 µm in diameter. The produced droplets were collected in Eppendorf tubes for further analysis.

2.5. Storage of Samples at Different Temperatures

The prepared samples of emulsion droplets with or without *E. coli*-GFP were stored at different temperatures, 25 °C (control), 5 °C (refrigeration temperature), −20 °C (freezing temperature), and −80 °C (freezing temperature), in laboratory-based refrigerators and freezers, with the temperature monitored throughout the storage period. Samples were stored for 24 h and thawed at 25 °C for 1 hour prior to further tests. Samples of *E. coli*-GFP suspended in sterilised deionised water (DIW) were also prepared and kept at the designated temperature as non-encapsulated controls.

2.6. Bacterial Viability Determination

The viability of *E. coli*-GFP was determined before and after the thawing process of samples stored in different temperatures. Encapsulated samples were centrifuged at 15,800× *g* for 1 min to break the emulsion and release the entrapped cells. Serial dilutions of the samples were performed using phosphate-buffered saline (PBS), and viable cells were counted immediately after sample preparation and after storage using the Miles and Misra method [24].

2.7. Microscopic Observation of Droplet Destabilisation

Droplet destabilisation was observed in relation to changes in temperature. The samples were prepared for microscopy by placing approximately one drop of sample onto the glass slide. The prepared glass slide was then placed on a Peltier temperature-controlled stage (Model PE-120, LINKAM scientific instruments, Surrey, UK), where the temperature of the stage was controlled using Link software (Linksys32, version 2.4.3). An Eheim circulation pump (ECP) was used for water circulation throughout the stage to maintain the desired temperature. The temperature-controlled stage was fitted on a Nikon Eclipse Ti microscope equipped with a Pco. Edge 5.5 sCMOS camera, where the samples were cooled and thawed on the microscope stage while photomicrographs of the samples were acquired. The sample was prepared at an initial temperature of 25 °C and was cooled down to −25 °C. The sample was then thawed to the initial temperature of 25 °C. For fluorescence microscopy, the encapsulated *E. coli*-GFP was observed at 509 nm emission.

2.8. Determination of Droplet Size and Phase Separation

The effect of different storage temperatures on droplet size was determined by measuring the diameter of the droplet immediately after sample preparation and after the thawing process at 25 °C. Single and double-emulsion samples without *E. coli*-GFP were prepared as controls. Photomicrographs of the droplet were taken before and after the storage period, and changes in the size of the dispersed aqueous phase for single-emulsion droplets and oil globule size for double-emulsion droplets were measured using MATLAB software (version 9.6, R2019a) and the circular Hough transform [25].

Droplet stability was also determined by measuring the amount of free water (for W/O emulsion) and free oil (for $W_1/O/W_2$ emulsion) with or without the presence of *E. coli*-GFP. Phase separation measurement was performed by collecting 1 mL of sample into graduated syringes immediately after droplet preparation. The samples were then stored at different temperatures in an upright position for 24 h. The volume of the free water and oil were measured from the graduated syringe immediately after the thawing process, and the percentage of free water and oil were determined as follows:

$$Percentage\ of\ separated\ water/oil\ (\%) = \frac{V_{water/oil}}{V_{emulsion}} \times 100$$

where $V_{water/oil}$ is the volume of separated water or oil measured from the graduated syringe, while $V_{emulsion}$ is the total volume of the emulsion sample.

2.9. Measuring the Release of Bacteria from $W_1/O/W_2$ Droplet

The release of bacteria was determined by measuring the number of cells in the outer aqueous phase immediately after sample preparation and after the thawing process of the double-emulsion samples ($W_1/O/W_2$). Samples stored in Eppendorf tubes were kept upright to allow the separation of the samples into a cream layer (oil globules) and serum phase (W_2) due to the difference in density between the phases. The serum was then carefully withdrawn from the sample using a pipette and serially diluted in PBS. Plate counts were then performed according to the Miles and Misra [24] method, where the encapsulation efficiency was calculated following El Kadri et al. [26], and the percentage of released cells was quantified relative to the overall viable cells in the samples after storage as follows:

$$Encapsulation\ efficiency\ (\%) = \left(\frac{N_0 - N}{N_0}\right) \times 100$$

$$Bacterial\ release\ (\%) = \frac{Log_{10}N_1 - Log_{10}N_t}{Log_{10}N_1} \times 100$$

where N is the number of viable cells counted in the serum phase immediately after emulsion preparation and N_0 is the overall viable cell count prior to encapsulation in $W_1/O/W_2$ droplets. N_1 is the overall viable cells after storage, and N_t is the viable cell count in the serum phase after storage at temperature t.

2.10. *Differential Scanning Calorimetry*

Differential scanning calorimetry (DSC) was performed to determine the thermal properties of emulsions samples and bulk aqueous solutions. The crystallisation and melting behaviours were characterised using a differential scanning calorimeter (Mettler Toledo, model 822e, Mettler Scientific Instruments, Giessen, Germany) with liquid nitrogen as the cooling substance. Approximately 5 mg of sample was weighed and hermetically sealed in aluminum pans together with an empty pan as reference. Samples were then cooled from 25 °C to −70 °C and held at −70 °C for 5 min before being heated back to 25 °C at a rate of 1 °C/min. The cooling and heating curves were obtained using STARe software version 9.0 (Mettler Toledo, Mettler Scientific Instruments, Germany), whereby positive curves resulted from an exothermic reaction, while an endothermic reaction created negative curves. The crystallisation and melting points of the tested samples were determined from the peak onset temperatures of the curves using the same software.

2.11. *Statistical Analysis*

All experiments were conducted with three replicates ($n = 3$). The data obtained were analysed with Excel (Microsoft corp., Washington, DC, USA) to determine the mean, standard deviation (SD), and standard error of means (SEM). A one-way ANOVA with Tukey HSD was conducted to compare several means, while a Student's t-test was conducted to compare two means using IBM SPSS Statistical software, version 23. Differences between means were considered significant at $p < 0.05$.

3. Results and Discussion

3.1. *The Effect of Storage on Droplet Size Change*

Monodisperse droplets for both single- and double-emulsion droplets were observed for all samples, with or without *E. coli*-GFP, with a CV value of approximately 9% for single-emulsion droplets and 7% for double-emulsion droplets prior to the storage test, as observed in Figure 1A,C,E,G and Table 1. The results indicate the ability of the flow-focusing microfluidic devices to produce highly monodisperse droplets, and the presence of *E. coli*-GFP did not affect droplet formation. For samples of single-emulsion droplets, no significant change ($p > 0.05$) in droplet size was observed after storage at 25 °C and 5 °C (Table 1) for both samples with and without bacteria. The droplets remained monodisperse, with only a 0.1–0.2% change in the value of the coefficient of variation (CV). However, a significant change ($p < 0.05$) in droplet size and distribution was observed after 24 h of storage for samples stored at freezing temperatures of −20 °C and −80 °C for both samples with and without bacteria. The greatest change was observed for samples stored at −80 °C, where the CV value increased from 9.6% to 28.3% for samples without *E. coli*-GFP, indicating a polydisperse distribution of droplets (CV values greater than 25% are regarded as polydisperse). Comparing samples with or without *E. coli*-GFP, the change in droplet stability was minimised in the presence of *E. coli*-GFP, whereby a smaller change in average droplet size and CV was observed with samples containing *E. coli*-GFP compared to samples without *E. coli*-GFP.

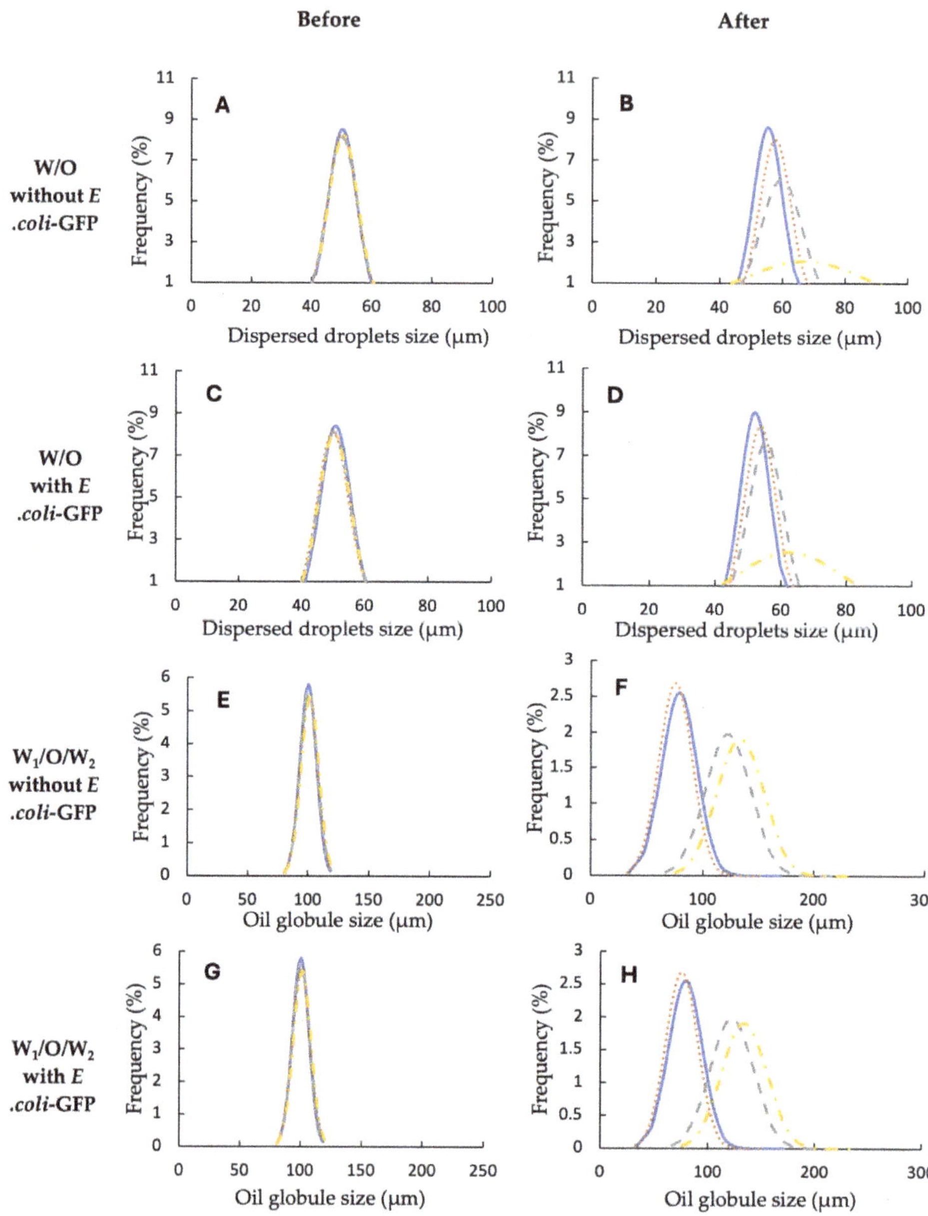

Figure 1. Droplet size distribution for single W/O (**A–D**) and double $W_1/O/W_2$ (**E–H**) emulsions at 25°C (—), 5°C (····), −20°C (----), and −80°C (- · -) before and after 24 h of storage. Droplet size measurements were taken after the samples were allowed to reach 25 °C following storage at cold temperatures.

Table 1. Summary of droplet size and distribution measured before and after 24 h of storage. Samples were prepared with or without *E.coli*-GFP for W/O and W1/O/W2 droplets. Data represent the mean ± standard deviation taken from three independent experiments with N = 900 droplets. The coefficient of variation (%) was measured by dividing the standard deviation by the mean droplet diameter. Mean comparisons indicated by small letters and capital letters represent significant differences ($p < 0.05$) within each sample across temperatures and before and after storage at each temperature, respectively. Data were analysed using one-way ANOVA and Student's *t*-test.

Samples	Temperature (°C)	Average Droplet Diameter (μm)		Coefficient of Variation (%)	
		Before	After	Before	After
W/O without *E. coli*-GFP	25	50.4 ^{aA} ± 4.7	55.3 ^{aA} ± 5.2	9.3	9.4
	5	50.4 ^{aA} + ± 4.8	57.7 ^{aA} ± 5.6	9.5	9.7
	−20	50.3 ^{aA} ± 4.9	59.6 ^{cB} ± 6.6	9.7	11.1
	−80	50.7 ^{aA} ± 4.8	67.2 ^{dB} ± 19.0	9.6	28.3
W/O with *E. coli*-GFP	25	50.8 ^{aA} ± 4.7	52.1 ^{aA} ± 4.9	9.3	9.4
	5	50.0 ^{aA} ± 4.9	53.7 ^{aA} ± 5.3	9.8	9.9
	−20	50.4 ^{aA} ± 4.9	55.3 ^{bB} ± 5.4	9.7	9.8
	−80	50.5 ^{aA} ± 5.0	62.8 ^{cB} ± 15.6	9.9	24.8
$W_1/O/W_2$ without *E. coli*-GFP	25	100.3 ^{aA} ± 6.9	79.0 ^{aB} ± 15.6	6.9	19.7
	5	100.7 ^{aA} ± 7.3	75.4 ^{bB} ± 14.9	7.2	19.8
	−20	100.6 ^{aA} ± 7.1	122.3 ^{cB} ± 20.2	7.1	16.5
	−80	101.8 ^{aA} ± 7.4	133.8 ^{dB} ± 20.9	7.3	15.6
$W_1/O/W_2$ with *E. coli*-GFP	25	100.9 ^{aA} ± 7.1	79.4 ^{aB} ± 15.6	7.0	19.6
	5	100.3 ^{aA} ± 7.7	76.3 ^{bB} ± 14.7	7.7	19.3
	−20	100.4 ^{aA} ± 7.1	116.7 ^{cB} ± 18.5	7.1	15.9
	−80	100.4 ^{aA} ± 7.4	126.3 ^{dB} ± 18.9	7.4	14.9

A significant change ($p < 0.05$) in the size of oil globules after 24 h of storage was observed for all double $W_1/O/W_2$ emulsion samples, with a decrease in average oil globule size observed for samples stored at 25 °C and 5 °C. However, the presence of large droplets up to 170 μm in size was also observed (Figure 1F,H). For samples stored at freezing temperatures of −20 °C and −80 °C, an increase ($p < 0.05$) in droplet size after storage was noted (Table 1). Samples containing *E. coli*-GFP demonstrated better droplet stability compared to those without, indicating the ability of bacterial cells to maintain droplet stability during cold-temperature storage. Complete external coalescence occurred between the inner W_1 droplet and the outer W_2 phase after 24 h of storage, forming a single oil-in-water (O/W) emulsion.

The storage of single W/O samples at −20 °C resulted in the solidification of the aqueous phase and partial crystallisation of the oil phase, whereas storage at −80 °C completely solidified the emulsions. With a broad crystallisation temperature, as opposed to pure water, which exhibits a sharp crystallisation curve, it is quite difficult to determine the exact crystallisation temperature of soybean oil due to the presence of lipids and a mixture of molecules. It has been reported previously that soybean oil is partially crystallised at a temperature between −10 °C and approximately −20 °C, whereby below this temperature, soybean oil is most likely to become solid [27–30].

Several destabilisation mechanisms of single W/O emulsions during the freeze–thaw process have been proposed, depending on droplet arrangements in the emulsion [16,31–34]. The destabilisation of loosely packed W/O emulsion droplets is due to collision-mediated coalescence [16]. This process is triggered by the uneven crystallisation of polydisperse water droplets in the emulsion, whereby the collision between the smaller, still-liquid water droplet and protruding ice crystals of the larger frozen droplet punctures the membrane surrounding the smaller, still-liquid water droplet. This leads to heterogeneous nucleation, with smaller droplets coalescing to form larger droplets upon thawing [16,35]. However, this may not be the case for a single W/O droplet generated using the microfluidic device, as it is highly monodispersed and therefore minimises the heterogeneous crystallisation of the water droplet and the effect of collision-mediated coalescence. Although the effect of collision-mediated coalescence was minimised, the destabilisation of W/O droplets was still observed, which is mainly attributed to the static and upright storage of the samples during freezing that caused the inevitable sedimentation of the water droplets, making them closely packed. For densely packed emulsions, the destabilisation mechanism is induced by the direct breakage of interfacial films and emulsion inversion [31,32]. The direct breakage of the droplet interfacial film occurred as neighbouring droplets crystallised and expanded, causing destabilisation. Droplet expansion pressed the droplets closely together, while crystallised droplets punctured the interfacial film of neighbouring still-liquid droplets, leading to droplet rupture [14]. This caused the content of the ruptured droplet to flow out and form a link between the two frozen droplets, resulting in flocculation, as observed in this study in Figure 2b [36]. During the thawing process, the crystal network collapsed, and the two partially coalesced droplets merged together, forming a larger droplet (complete coalescence), as observed in Figure 2d [12]. Droplet crystallisation and expansion also resulted in the thinning of the oil and surfactant layer around the aqueous phase, which accelerates droplet coalescence during the melting process [35]. Emulsion inversion, as reported by Aronson and Petko [31], involves the rearrangement of ice crystals and the still-liquid oil, which leads to the presence of distinct oil droplets within the ice structure. However, this may not be the case for samples tested in this study, as it occurs mainly in densely packed emulsions containing distinctively high-volume fractions of the dispersed phase [12].

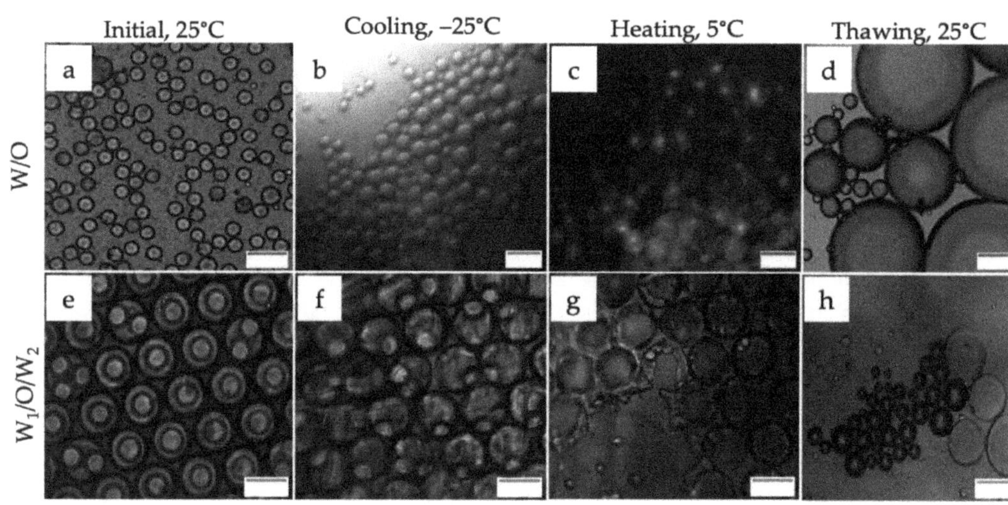

Figure 2. Photomicrographs of W/O (**a–d**) and $W_1/O/W_2$ (**e–h**) emulsion droplets with change in temperature. The cooling and thawing processes were conducted on the temperature-controlled microscope stage. Scale bar represents 100 μm.

It has been reported previously that the stability of water-in-oil droplets during the freeze–thaw process depends on the freezing sequence of the oil and aqueous phases and also the type of surfactants used [35]. The freezing sequence between the oil and aqueous phase may depend on the type of oil used, as some oils may have a lower or higher freezing point than the aqueous phase. Ghosh and Coupland [14] reported that for single water-in-oil emulsions, instability is mostly evident in samples where the oil phase crystallises first prior to the aqueous phase and when the emulsifier is in a liquid state during the freezing process. For emulsions containing soybean oil, the oil phase crystallises at a much lower temperature than the aqueous phase, minimising droplet destabilisation. However, significant changes ($p < 0.05$) in droplet size distribution were observed for samples stored at −80 °C. At −80 °C, the gradual crystallisation of the oil phase occurs due to differences in the melting fraction of fats in soybean oil, as the high-melting fraction crystallises first, followed by the low-melting fraction [30], resulting in ice crystals being forced into the still-liquid oil phase forming a region with highly concentrated ice crystals [14,37]. During thawing, as the oil melts, the ice crystals coalesce immediately, as a further increase in temperature causes the aqueous droplets to melt and fuse together [38]. A previous study reported the separation of the W_1 phase from double-emulsion droplets, where the separated W_1 phase was retained in the W_2 phase supported by a very thin and complex layer of oil and surfactants [39]. This also caused the reduction in oil globule size as the W_1 phase escaped into the W_2 phase. Similar behaviour was also seen for samples with double emulsions kept at a lower temperature of 5 °C. However, for double-emulsion samples kept at a much lower temperature of −20 °C and −80 °C, a significant ($p < 0.05$) increase in the average oil globule size and change in the droplet size distribution was observed. Interestingly, the inner W_1 phase remained intact in the $W_1/O/W_2$ during the freezing process, as shown in Figure 2f, and the droplet separation process was not observed.

The destabilisation mechanism of the double-emulsion droplets was primarily attributed to external coalescence between the W_1 and W_2 phases, a process distinct from the freeze–thaw destabilisation of single emulsions [17,40]. In contrast to single-emulsion droplets, whereby the droplet destabilisation process occurs throughout the freeze and thawing processes, the external coalescence of double emulsions occurred during the thawing process of the emulsions, while the inner W_1 phase remained intact during the freezing process [17]. The thawing process caused deformation of the inner W_1 phase followed by the complete burst of the droplet due to the external coalescence, and this process is mainly affected by the size of the inner W_1 phase and surfactant concentration. According to Rojas and Papadopoulos [17], the determined threshold of the W_1 inner aqueous phase size to the oil globule size ratio is 0.3, above which coalescence occurs upon thawing. In addition, an emulsion droplet with a thick surfactant layer creates a repulsive force between the W_1 inner phase and the O/W_2 interface that improves droplet stability against external coalescence [41]. A similar mechanism was observed in this study, whereby the inner W_1 droplet remained intact during the freezing process and rapid freezing prevented the droplet from splitting (Figure 2e–f). Emulsion destabilisation only occurred during the thawing process for both samples kept at −20 °C and −80 °C. As the frozen oil melts at a lower temperature during the thawing process of samples kept at freezing temperatures, the liquid oil flows through the cracks and gaps created by the gradual thawing of the aqueous phase (Figure 2g), leading to the absence of the oil layer that separates the inner W_1 phase from the outer W_2 phase. This accelerates the external coalescence of the aqueous phase, as it melts completely with an increase in temperature, forming a single oil-in-water (O/W) emulsion. Moreover, the droplet distribution graph that tails towards the larger droplet size with sharp distribution in the smaller droplet size region shows evidence of oil globule coalescence after the thawing process, which contrasts with droplet instability that was induced by Ostwald ripening, whereby it tends to form a sharp droplet distribution in the larger region with tailing in the smaller droplet size region [31].

The presence of bacteria also affected droplet stability, especially for single-emulsion droplets. As the PGPR surfactant remained liquid during the freezing process [35], it is most likely retained in the oil phase and withdrawn from the ice crystals as the aqueous phase freezes. This accelerates the coalescence process of the aqueous phase as the emulsions warm up. However, the presence of *E. coli*-GFP on the interface minimises droplet coalescence during the thawing process as it is most likely crystallised on the interface. *E.*

coli-GFP was observed to have a high affinity towards the soybean oil, as determined by the BATH assay (Figure S2). Changes in bacterial characteristics due to storage in freezing temperatures ease the attachment of the bacteria onto the interface, making the droplet less susceptible to rupture. The stabilisation effect of bacteria may be similar to emulsion stabilisation due to surfactant crystallisation and the presence of particles on the droplet interface forming Pickering emulsions, as reported previously [34,41]. It has been reported previously that the crystallisation of glycerol monostearate (GMS) at 25 °C creates a crystalline shell around the water droplet that helps prevent crystallisation damage between the adjacent water droplets and reduces coalescence during the thawing process, which is in contrast to emulsions containing molten PGPR [35]. According to Zhu et al. [42], the addition of heated soy and whey protein in emulsions improves droplet stability against freeze/thaw cycling due to Pickering stearic stabilisation. However, further studies are still required to confirm the effect of bacteria on droplet stability in freezing conditions.

Recent studies applied numerical and computational models to understand the formation and stability of emulsion droplets in micro-fluidic channels without bacteria [43,44]. Furthermore, since bacteria have an impact on emulsion formation and stability, in future, numerical and computational studies can simulate this complex interaction to optimise the formation of emulsions encapsulating bacteria by predicting conditions that lead to the desired size and stability and to develop an efficient and scalable system for various applications.

3.2. The Effect of Storage on Phase Separation of Emulsions

Droplet destabilisation of W/O droplets resulted in the presence of a free water layer due to droplet coalescence, while the destabilisation of $W_1/O/W_2$ droplets led to the presence of a free oil layer due to the external coalescence between the inner and outer aqueous phases (forming an O/W emulsion) and coalescence between large oil droplets that resulted in the presence of an oil layer on top of the sample. Figure 3 shows the percentage ratio of free water and oil measured after 24 h of storage with respect to the total emulsion. Given the results obtained, it was determined that phase separation occurred for samples of both single and double emulsions kept at −20 °C and −80 °C, with the highest amount of free water and free oil observed for samples kept at −80 °C. Emulsion stability for samples stored at 25 °C was maintained during the storage period, as there was no observable free water or oil. However, a small percentage of free water and oil (approximately 2%) was observed for samples without *E. coli*-GFP at 5 °C, and this is mainly attributed to the closely packed water droplets during storage and the absence of *E. coli*-GFP that helped in improving droplet stability. An overall comparison between samples with or without the presence of *E. coli*-GFP shows that phase separation was minimised with the presence of *E. coli*-GFP.

Droplet destabilisation occurred for W/O emulsions kept at −20 °C and −80 °C due to partial coalescence of the droplets, as the crystallisation of the dispersed phase resulted in film rupture that connected neighbouring droplets, leading to complete coalescence upon thawing [12,45]. This eventually led to bulk water separation that was minimised with the presence of *E. coli*-GFP in the inner W_1 phase. During the consecutive crystallisation of the oil phase, the dispersed ice crystals were forced into the still-liquid region of the oil phase, and a further reduction in temperature led to the withdrawal of crystallised oil from this region. A combination of these processes led to complete droplet coalescence and extensive aqueous phase separation from the emulsion during the thawing process [37,38]. The partially crystallised oil phase of samples stored at −20 °C prevented the complete withdrawal of the oil phase, as some regions within the emulsion contained liquid oil, thus minimising extensive droplet coalescence.

The direct destabilisation process of the double-emulsion droplet that occurs immediately after the thawing process leads to oil phase separation and the external coalescence of W_1 and W_2 phases. As shown in Figure 2g, the movement of liquid oil through the cracks of the outer aqueous phase during the thawing process immediately separates the oil and aqueous phases, forming a bulk oil layer due to the coalescence of bigger oil droplets, while some oil droplets remain in a layer of O/W emulsion. The unfrozen W_1 phase will remain in the W_2 phase and eventually fuses with the W_2 phase as it melts (Figure 2h). This process leads to the complete release of the W_1 phase into the W_2 phase and extensive separation of the oil phase from the emulsion. Complete $W_1/O/W_2$ emulsion destabilisation due to external coalescence during the thawing phase of the oil layer was also reported previously by Rojas and Papadopoulos [17], whereby it led to the complete release of the W_1 phase into the W_2 phase, forming O/W emulsions.

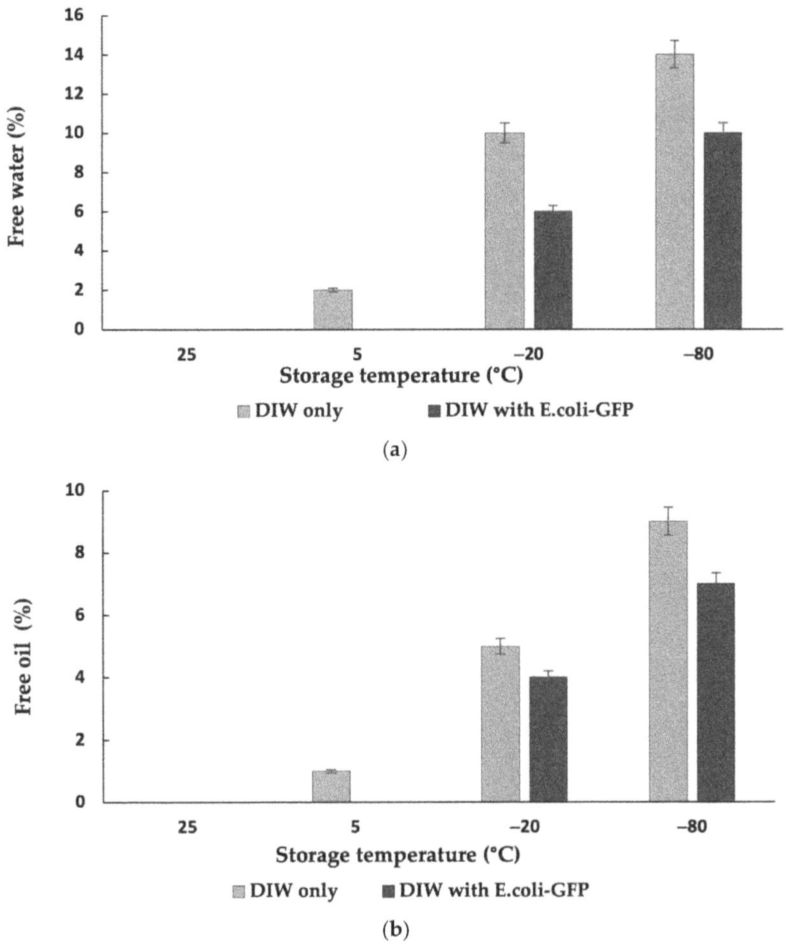

Figure 3. The percentage of (**a**) free water measured for single-emulsion samples and (**b**) free oil for double-emulsion samples. The volume of free water and free oil was measured with respect to the total volume of emulsions after 24 h of storage at different temperatures. Bars represent the mean ± SEM obtained from 3 independent experiments.

The presence of crystallised surfactants and Pickering particles has been reported to improve droplet stability, as it minimises droplet coalescence during the thawing process by creating a protective layer that prevents droplet fusion [14,42,46]. A similar stabilisation effect may occur with the presence of *E. coli*-GFP on the interfacial layer. Under freezing conditions, *E. coli*-GFP undergoes changes in membrane conformation and cell shrinkage [47], making it easily embedded in the interfacial layer. A previous study reported that the presence of *E. coli*-GFP in single W/O emulsion droplets helps in maintaining the stability of the droplet during five days of storage at 25 °C, whereby the presence of *E. coli*-GFP, particularly dead cells, helps in reducing interfacial tension due to its attachment on the interfacial layer [48]. The bacterial affinity for the interface is due to its hydrophobic nature, as the dead *E. coli*-GFP cells exhibit a higher affinity for the oil phase compared to live cells. The BATH assay of *E. coli*-GFP with soybean oil reveals its affinity for the oil phase (Figure S2); thus, this may help to reduce the interfacial tension and create a protective barrier that prevents extensive droplet coalescence during the thawing process. Moreover, the presence of bacteria in the W_2 phase due to its release as the $W_1/O/W_2$ emulsion destabilises may help prevent coalescence between the oil droplets, as reported previously by Firoozmand and Rousseau [49], whereby the presence of bacterial cells in the aqueous phase of the O/W emulsions improves droplet stability due to the attachment of bacterial cells onto the oil droplets, creating a boundary layer that protects the droplet against coalescence and phase separation.

3.3. *Thermal Properties of Emulsions Determined Using Differential Scanning Calorimetry (DSC)*

The thermal properties of the bulk aqueous solution (Figure 4) were characterised as opposed to emulsified samples of single and double emulsions in soybean oil (Figure 5). The crystallisation temperature of the samples was determined from the onset temperatures of the crystallisation peaks in which the temperature was recorded at the beginning of the crystallite formation. The onset temperature was determined by using Star e software version 9.0 (Mettler Toledo, Mettler Scientific Instruments, Giessen, Germany). Referring to the crystallisation peaks in Figure 4a, it was determined that the crystallisation of *E. coli*-GFP suspension in DIW occurred at a lower temperature compared to pure DIW. Pure DIW crystallises at −12 °C and melts at 0 °C, which is around the same temperatures as reported previously in several studies [13,35,50]. For the *E. coli*-GFP suspension, the crystallisation temperature was recorded at −17 °C, while the melting of the *E. coli*-GFP suspension occurred around the same temperature as the pure DIW. The reduction in the crystallisation temperature with the presence of *E. coli*-GFP is due to the freezing point depression, where the presence of *E. coli*-GFP as a solute tends to lower the freezing point.

Figure 5 shows that the emulsified samples freeze at a much lower temperature compared to bulk solutions. A broader crystallisation curve was observed for samples of W/O emulsions as compared to $W_1/O/W_2$ emulsions. This is mostly attributed to the presence of a large percentage of oil in the continuous phase as compared to the aqueous phase for W/O emulsions. The presence of long chains of lipids together with a mixture of other molecules in soybean oil results in a broader crystallisation temperature as opposed to pure water, making it difficult to determine the exact freezing point of the soybean oil. It has been reported that pure soybean oil crystallises at approximately −10 °C to −20 °C, where it is most likely to solidify below these temperatures, while soybean oil emulsions solidify at a much lower temperature than pure oil [28,30]. In this study, the single W/O emulsion crystallised at −20 °C for samples without *E. coli*-GFP, while samples with *E. coli*-GFP crystallised at −28 °C. The melting point of both emulsions was around 0 °C. The difference in crystallisation temperature is mainly attributed to the destabilisation of emulsions without *E. coli*-GFP, where the presence of free water due to phase separation increases the crystallisation temperature of the samples. In addition, the similar melting point of the emulsions as that of bulk solutions also indicates the loss of emulsion stability due to an increase in the percentage of free water during the thawing process [34,50]. It has been reported previously that the W/O emulsions solidify at a lower temperature than that of bulk water as they overcome the free energy barrier due to homogeneous nucleation [12]. Water droplets in emulsions solidify at a lower temperature than the temperature required to solidify bulk water (−20 °C) and are much lower than the equilibrium melting point of water of 0 °C [16].

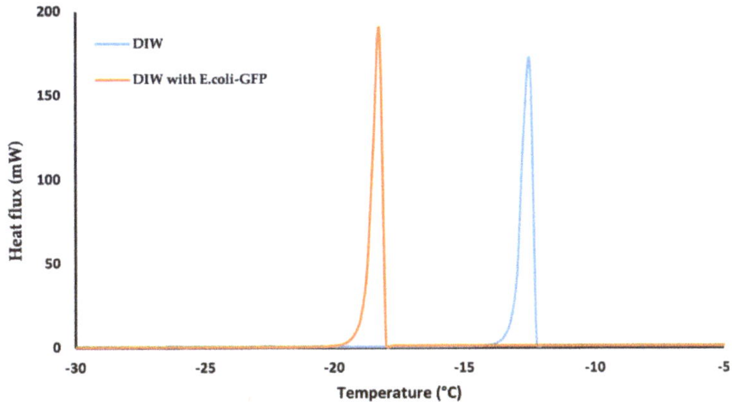

(**a**)

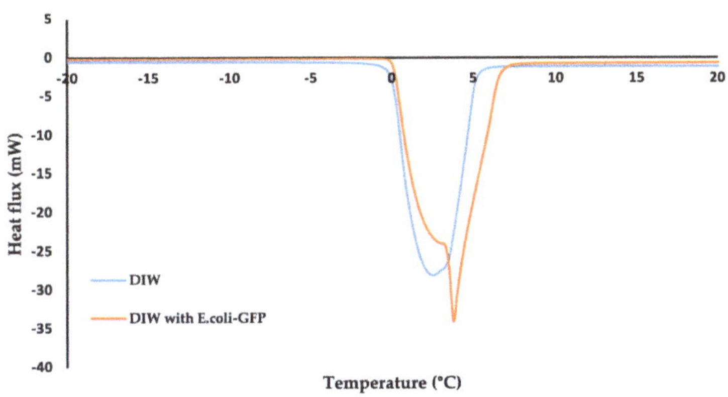

(**b**)

Figure 4. DSC thermograms showing the cooling curves (**a**) and heating curves (**b**) of bulk aqueous samples with or without *E. coli*-GFP.

For double $W_1/O/W_2$ emulsions, a small difference in crystallisation temperature was observed between samples with or without *E. coli*-GFP, as samples containing *E. coli*-GFP crystallised at −21 °C while samples without *E. coli*-GFP crystallised at −23 °C. Only one peak was observed, which represents the crystallisation peak of the W_2 phase as opposed to the two peaks reported by Kovács et al. [51]. The absence of the second peak indicates the loss of the inner W_1 phase due to emulsion breakdown [51,52]. However, this may not be the case in this study, as the inner W_1 droplets were intact during the freezing process, as observed in Figure 2f, and the loss of the inner W_1 phase occurred during the thawing/melting process, which explains the presence of only one peak for the melting curve with a melting temperature of approximately 0 °C for all samples. Schuch, Köhler and Schuchmann [52] also described that it is possible to distinguish between the inner W_1 phase and the outer W_2 phase from the cooling curves due to the presence of two peaks that result from the difference in crystallisation temperatures, but it is impossible to distinguish between these two phases from the melting curves of the $W_1/O/W_2$. Another possible explanation for the presence of only one cooling curve is due to the highly monodispersed distribution of the droplets that were prepared by using the microfluidic method with a single and large size W_1 core and a thin oil layer separating the W_1 phase from the W_2 phase. A combination of these factors may have caused the outer W_2 phase and the inner W_1 phase to crystallise simultaneously within the same temperature range. It has been reported previously that the crystallisation of emulsion droplets may depend on the size of the droplet [52]. However, further studies are still required to understand the effect of the droplet size distribution on the crystallisation temperature of multiple emulsions.

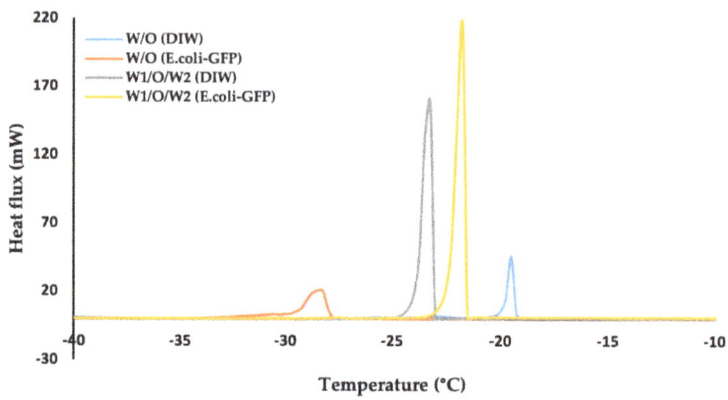

(**a**)

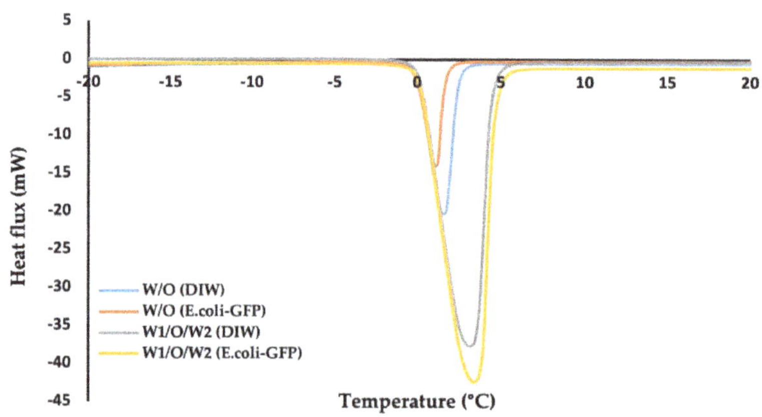

(**b**)

Figure 5. DSC thermograms of (**a**) cooling curves and (**b**) heating curves of emulsified samples with or without *E. coli*-GFP.

3.4. Bacterial Viability During Storage

Referring to Figure 6, a large reduction in the *E. coli*-GFP cell count was observed for samples kept at −20 °C and −80 °C compared to samples kept at 25 °C and 5 °C, whereby, comparing the non-encapsulated samples, a 1.2 log CFU/mL and 1.3 log CFU/mL reduction in viable cell count was observed for samples kept at 25 °C and 5 °C, respectively, while reductions of 2.3 log CFU/mL and 2.7 log CFU/mL were observed for samples kept at −20 °C and −80 °C, respectively. *E. coli*-GFP encapsulated in W/O droplets showed a smaller percentage of bacterial reduction compared to samples encapsulated in $W_1/O/W_2$ droplets and control samples of free *E. coli*-GFP cells in sterilised DIW.

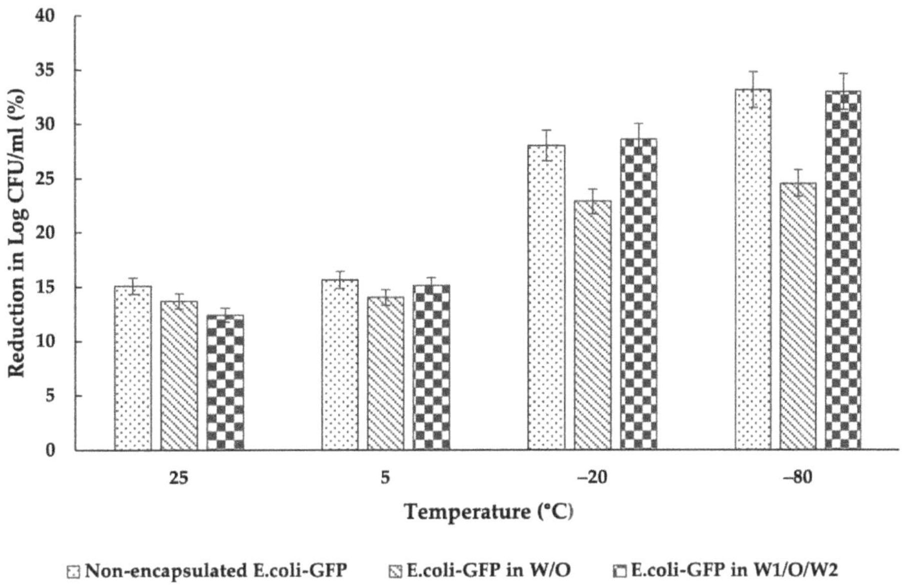

Figure 6. Reduction in Log CFU/mL of free *E. coli*-GFP in sterilised DIW and encapsulated in W/O or $W_1/O/W_2$ emulsion droplets. Bars represent mean SEM taken from 3 independent experiments.

It has been reported previously that freeze–thawing may cause detrimental effects on bacterial cells that lead to a reduction in bacterial viability [47,53–55]. Cell damage during the freeze–thaw process occurs due to various physico-chemical reactions [7,56]. One of the factors responsible for cell damage during storage at low temperatures is the nature of the cells [7,57], whereby different bacterial species exhibit different resistance to cell damage depending on the cell structure. Furthermore, the use of cryoprotectants with various formulations [7] and the freeze–thawing conditions [6,58] of the samples may also affect the degree of cell damage during storage.

The effect of water crystallisation is one of the most common factors responsible for cell damage [47,58,59]. The detrimental effect of water crystallisation on bacterial viability depends on the freezing rate, whereby the decrease in bacterial viability is minimised at a low freezing rate compared to a high freezing rate. This is due to the cryoconcentration effect, as the crystallisation of the external medium resulted in cell dehydration, thus preventing lethal intracellular crystallisation. Crystallisation of the extracellular medium forces the cells to be concentrated in the unfrozen region of the medium, whereby a continuous decrease in temperature causes the region to be increasingly concentrated. This leads to cell dehydration as the cells are exposed to a highly concentrated solution during the freezing process. Meanwhile, the rapid freezing of bacterial suspension resulted in extensive supercooling that crystallised intracellular water, as it was not able to flow out of the cells fast enough, causing extensive cell damage [56]. These findings indicate that the degree of cell damage during the freezing process depends on the availability of intra- and extracellular water. Referring to the DSC thermograms (Figure 5), the broader crystallisation temperature of the emulsion compared to the bulk bacterial solution (Figure 4) indicates that the encapsulation of bacteria in emulsion droplets promotes slow freezing and thus minimises cell damage during the freezing process. Encapsulation in W/O emulsion droplets shows a smaller reduction in cell count compared to samples encapsulated in $W_1/O/W_2$ emulsion droplets and free cells in DIW due to the thick oil layer and the low water ratio that minimises cell damage due to water crystallisation.

The effect of encapsulation in improving bacterial viability during the freeze–thaw process has also been reported previously [8,9,60]. The microencapsulation of *Lactobacillus acidophilus* in the self-assembled polyelectrolyte layers of chitosan and carboxymethyl cellulose has been reported to protect bacteria not only from the adverse effects of the simulated GI tract but also during the freezing and freeze-drying processes [60]. Furthermore, the microencapsulation of *Bifidobacterium longum* in milk proteins and sugar alcohols aids in improving the protection effect of cryoprotectants such as glycerol, resulting in better bacterial viability and function after freezing and freeze-drying processes [9].

Moreover, the freezing of emulsion-based products such as milk has also been reported to cause small changes in the viability of bacteria, indicating the ability of the emulsion structure to protect bacteria against extensive cell damage. A study conducted by Sánchez et al. [61] shows that the freezing of goat milk at −20 °C or even after extended storage at −80 °C does not significantly affect the viability of *E. coli* as opposed to cow milk due to differences in milk composition. A similar result was also reported by Nurliyani, Suranindyah, and Pretiwi [10], whereby the frozen storage of Ettawah Cross-bred goat milk samples for 60 days did not cause significant changes in the total bacteria, while changes in emulsion stability were observed after 30 days of storage.

Therefore, the encapsulation of bacteria in emulsion droplets, especially W/O, may help in improving bacterial viability during the freezing process due to the protective effect of the emulsion structure, as it reduces the available water ratio and prevents the lethal crystallisation effect of the water phase.

3.5. The Release of Bacteria from Double Emulsions

Although the storage of samples at freezing temperatures leads to unfavourable phase separation, the immediate destabilisation of the $W_1/O/W_2$ emulsion triggered by the thawing process may be beneficial for the controlled release of bacteria for various applications. A high encapsulation efficiency of approximately 99% was achieved by using microfluidics for bacteria encapsulation in $W_1/O/W_2$, indicating the successful encapsulation of bacteria prior to cold storage. The increase in cell counts in the W_2 phase, therefore, resulted from the release of bacteria from the W_1 phase due to the emulsion destabilisation process during storage, as presented in Table 2.

Table 2. The effect of storage temperature on the release of bacteria from $W_1/O/W_2$ droplets. Samples were prepared with 99.9% encapsulation efficiency, and the release of bacteria into the W_2 phase was determined after 24 h of storage. Data represent mean ± standard deviation taken from 3 independent experiments.

Temperature (°C)	Overall Viability (log CFU/mL)	Released (log CFU/mL)	Released (%)
25	6.42 ± 0.03	2.75 ± 0.02	42.80 ± 0.51
5	6.18 ± 0.01	2.77 ± 0.01	44.89 ± 0.23
−20	5.19 ± 0.02	5.10 ± 0.01	98.36 ± 0.18
−80	4.93 ± 0.03	4.90 ± 0.01	99.35 ± 0.40

Complete bacterial release into the W_2 phase was observed for samples kept in freezing temperatures of −20 °C and −80 °C, with 5.10 log CFU/mL of viable bacteria observed in the W_2 phase out of the overall viable cell count of 5.19 log CFU/mL (for samples kept at −20 °C) and 4.90 log CFU/mL of viable bacteria observed in the W_2 phase out of the overall viable cell count of 4.93 log CFU/mL (for samples kept at −80 °C). Only 2.75 log CFU/mL of cells were released into the W_2 phase out of the overall viable cell count of 6.42 log CFU/mL for samples kept at 25 °C, and 2.77 log CFU/mL of cells were released out of the overall viable cell count of 6.18 log for samples kept at CFU/mL at 5 °C.

The release of bacteria from samples kept at 25 °C and 5 °C was mainly attributed to the rupture of the thin film that separates the W_1 phase from the W_2 phase, induced by the change in the osmotic balance of the droplet due to the production of bacterial by-products during the storage period. Conversely, the complete destabilisation of the

$W_1/O/W_2$ emulsion droplet stored at frozen temperatures is due to the freeze–thawing process that leads to phase separation. This process resulted in the complete release of the encapsulated bacteria as the emulsion destabilises into a layer of the bulk oil phase (top), O/W emulsion (middle), and the aqueous phase (bottom) containing the released bacteria.

As the inner W_1 phase remains intact during the freezing process (Figure 2f), the frozen $W_1/O/W_2$ emulsion may serve as an alternative for bacterial microencapsulation, whereby the controlled release of the encapsulated bacteria at the desired time can be achieved by simply thawing the frozen emulsion samples, causing immediate and complete release of the bacteria into the W_2 phase. Emulsion destabilisation induced by the freeze–thawing process has been used extensively in various applications, such as in emulsion liquid membrane (ELM) processing [62,63] and waste emulsion treatment, such as in oil sludge demulsification [33,34].

Its application in pharmaceuticals was also discussed previously by Bernal-Chávez et al. [21]. Moreover, a study by Rojas and Papadopoulos [17] also revealed the potential application of $W_1/O/W_2$ emulsion droplets for the controlled release of encapsulated materials triggered by the thawing of the oil phase. Similar to the results obtained in this study, the internal coalescence of the inner W_1 phase and the external coalescence of the outer W_2 phase were not observed during the freezing stage but occurred extensively during the thawing of the oil phase, which led to the complete release of the encapsulated material (Figure 7).

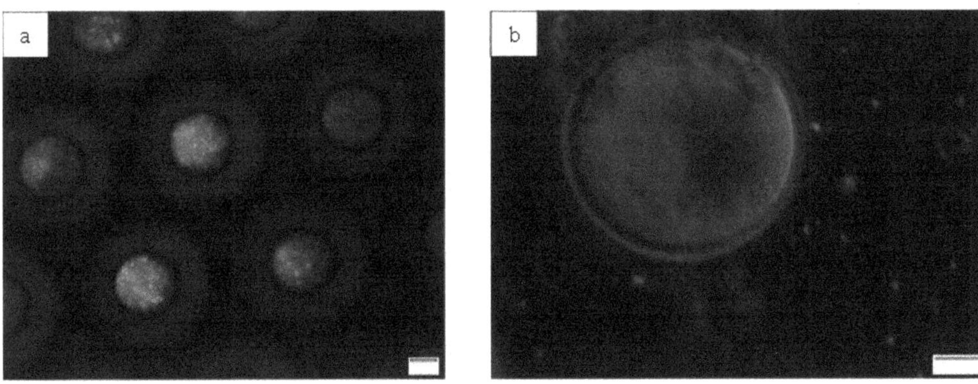

Figure 7. Photomicrographs of double-emulsion droplets before (**a**) and after (**b**) storage at −80 °C for 24 h. Scale bar represents 20 µm.

Further studies were also conducted on its potential application to the controlled release of protein from bulk emulsion systems [18]. As expected, similar results were also obtained in the bulk emulsion systems, whereby the stability of the droplet was maintained during the freezing process, while the thawing of the oil phase led to an instant release of the encapsulated protein.

4. Conclusions

This study revealed the feasibility of using emulsion droplets for bacterial encapsulation and storage not only at ambient but also at cold temperatures. The findings demonstrate that emulsion-based encapsulation enhances bacterial viability at low temperatures and that temperature changes can modulate the release of bacteria. This study provides a better understanding of the stability of emulsions in the presence of bacteria in cold temperatures and demonstrates the benefits of encapsulation in maintaining the viability of the bacteria by providing protection against the lethal effect of the freezing process.

These characteristics are beneficial for the production of emulsions containing bacteria for laboratory-based and industrial applications. In addition, it also demonstrates the potential use of the freeze–thaw technique for the controlled release of not only bacteria but also for other applications using temperature-sensitive compounds, such as in drug delivery.

Nevertheless, further studies are still required in order to compare the freeze–thaw mechanism of highly monodispersed $W_1/O/W_2$ droplets containing a single W_1 inner phase with those containing multiple W_1 inner phases, as it may exhibit a different mechanism, as observed in the absence of a second peak in the cooling curves of the double-emulsion droplet. In addition, further studies are also required to investigate the process underlying the ability of bacterial cells to maintain droplet stability during cold-temperature storage and to improve bacterial stability during the freezing process, especially for samples encapsulated in double $W_1/O/W_2$ droplets.

Supplementary Materials: The following supporting information can be downloaded at: https://www.mdpi.com/article/10.3390/fluids9120274/s1, Figure S1: Microscopic observation of emulsion destabilisation with the change in temperature; Figure S2: The bacterial adherence to soybean oil assay for live and dead *E. coli*-GFP at different soybean oil volumes.

Author Contributions: Conceptualisation, N.S.M.I.; Formal analysis, N.S.M.I.; Investigation, N.S.M.I.; Methodology, N.S.M.I., H.E.K., D.V. and K.G.; Project administration, D.V. and K.G.; Resources, D.V. and K.G.; Supervision, D.V. and K.G.; Validation, H.E.K. and D.V.; Visualisation, H.E.K. and N.F.A.M.Z.; Writing—original draft, N.S.M.I. and H.E.K.; Writing—review and editing, D.V., N.F.A.M.Z. and K.G. All authors have read and agreed to the published version of the manuscript.

Funding: This research was funded by the Ministry of Higher Education Malaysia (MOHE) as part of a doctorate programme.

Data Availability Statement: The original contributions presented in the study are included in the article and Supplementary Materials; further inquiries can be directed to the corresponding author.

Acknowledgments: The authors acknowledge the Ministry of Higher Education (MOHE), Malaysia, for funding the research as part of a doctorate programme. The authors would like to thank Iqram Fauzi, Putu Virgina Partha Devanthi, and Elaine Mitchell for their technical support. The PGPR was kindly provided by Danisco (Denmark).

Conflicts of Interest: The authors declare no conflicts of interest.

References

1. Chavarri, M.; Maranon, I.; Carmen, M. Encapsulation Technology to Protect Probiotic Bacteria. In *Probiotics*; Rigobelo, E.C., Ed.; IntechOpen: London, UK, 2012; pp. 502–529.
2. Devanthi, P.V.P.; Linforth, R.; El Kadri, H.; Gkatzionis, K. Water-in-oil-in-water double emulsion for the delivery of starter cultures in reduced-salt moromi fermentation of soy sauce. *Food Chem.* **2018**, *257*, 243–251. [CrossRef] [PubMed]
3. Peng, Q.; Meng, Z.; Luo, Z.; Duan, H.; Ramaswamy, H.S.; Wang, C. Effect of Emulsion Particle Size on the Encapsulation Behavior and Oxidative Stability of Spray Microencapsulated Sweet Orange Oil (*Citrus aurantium* var. *dulcis*). *Foods* **2023**, *12*, 116. [CrossRef] [PubMed]
4. Sobti, B.; Kamal-Eldin, A.; Rasul, S.; Alnuaimi, M.S.K.; Alnuaimi, K.J.J.; Alhassani, A.A.K.; Almheiri, M.M.A.; Nazir, A. Encapsulation Properties of *Mentha piperita* Leaf Extracts Prepared Using an Ultrasound-Assisted Double Emulsion Method. *Foods* **2023**, *12*, 1838. [CrossRef] [PubMed]
5. Dluska, E.; Markowska-Radomska, A.; Metera, A.; Tomaszewski, W. Drug-Core Double Emulsions for Co-Release of Active Ingredients. *Int. J. Chem. Eng. Appl.* **2017**, *7*, 428–432. [CrossRef]
6. Fonseca, F.; Béal, C.; Corrieu, G. Operating conditions that affect the resistance of lactic acid bacteria to freezing and frozen storage. *Cryobiology* **2002**, *43*, 189–198. [CrossRef]
7. Wang, G.; Yu, X.; Lu, Z.; Yang, Y.; Xia, Y.; Lai, P.F.-H.; Ai, L. Optimal combination of multiple cryoprotectants and freezing-thawing conditions for high lactobacilli survival rate during freezing and frozen storage. *LWT* **2019**, *99*, 217–223. [CrossRef]
8. Goderska, K.; Czarnecki, Z. Influence of microencapsulation and spray drying on the viability of Lactobacillus and Bifidobacterium strains. *Pol. J. Microbiol.* **2008**, *57*, 135–140.
9. Dianawati, D.; Mishra, V.; Shah, N.P. Survival of *Bifidobacterium longum* 1941 microencapsulated with proteins and sugars after freezing and freeze-drying. *Food Res. Int.* **2013**, *51*, 503–509. [CrossRef]
10. Nurliyani Suranindyah, Y.; Pretiwi, P. Quality and Emulsion Stability of Milk from Ettawah Crossed Bred Goat During Frozen Storage. *Procedia Food Sci.* **2015**, *3*, 142–149. [CrossRef]

11. de Farias, T.G.S.; Ladislau, H.F.L.; Stamford, T.C.M.; Medeiros, J.A.C.; Soares, B.L.M.; Arnaud, T.M.S.; Stamford, T.L.M. Viabilities of *Lactobacillus rhamnosus* ASCC 290 and *Lactobacillus casei* ATCC 334 (in free form or encapsulated with calcium alginate-chitosan) in yellow mombin ice cream. *LWT* **2019**, *100*, 391–396. [CrossRef]
12. Vanapalli, S.A.; Palanuwech, J.; Coupland, J.N. Stability of emulsions to dispersed phase crystallization: Effect of oil type, dispersed phase volume fraction, and cooling rate. *Colloids Surf. A Physicochem. Eng. Asp.* **2002**, *204*, 227–237. [CrossRef]
13. Cramp, G.L.; Docking, A.M.; Ghosh, S.; Coupland, J.N. On the stability of oil-in-water emulsions to freezing. *Food Hydrocoll.* **2004**, *18*, 899–905. [CrossRef]
14. Ghosh, S.; Coupland, J.N. Factors affecting the freeze-thaw stability of emulsions. *Food Hydrocoll.* **2008**, *22*, 105–111. [CrossRef]
15. Tippetts, M.; Martini, S. Effect of cooling rate on lipid crystallization in oil-in-water emulsions. *Food Res. Int.* **2009**, *42*, 847–855. [CrossRef]
16. Lin, C.; He, G.; Li, X.; Peng, L.; Dong, C.; Gu, S.; Xiao, G. Freeze/thaw induced demulsification of water-in-oil emulsions with loosely packed droplet. *Sep. Purif. Technol.* **2007**, *56*, 175–183. [CrossRef]
17. Rojas, E.C.; Papadopoulos, K.D. Induction of instability in water-in-oil-in-water double emulsions by freeze-thaw cycling. *Langmuir* **2007**, *23*, 6911–6917. [CrossRef]
18. Rojas, E.C.; Staton, J.A.; John, V.T.; Papadopoulos, K.D. Temperature-induced protein release from water-in-oil-in-water double emulsions. *Langmuir* **2008**, *24*, 7154–7160. [CrossRef]
19. Zhang, L.; Hu, Y.; Jiang, L.; Huang, H. Microbial Encapsulation and Targeted Delivery Mechanisms of Double Emulsion Loaded with Probiotics—A State-of-Art Review. *Food Rev. Int.* **2023**, *40*, 1731–1755. [CrossRef]
20. Guan, X.; Li, Y.; Jiang, H.; Tse, Y.L.S.; Ngai, T. Temperature-responsive Pickering double emulsions stabilized by binary microgels. *Chem.—Asian J.* **2023**, *18*, e202300587. [CrossRef]
21. Bernal-Chávez, S.A.; Romero-Montero, A.; Hernández-Parra, H.; Peña-Corona, S.I.; Del Prado-Audelo, M.L.; Alcalá-Alcalá, S.; Cortés, H.; Kiyekbayeva, L.; Sharifi-Rad, J.; Leyva-Gómez, G. Enhancing chemical and physical stability of pharmaceuticals using freeze-thaw method: Challenges and opportunities for process optimization through quality by design approach. *J. Biol. Eng.* **2023**, *17*, 35. [CrossRef]
22. Kim, P.; Kwon, K.W.; Park, M.C.; Lee, S.H.; Kim, S.M.; Suh, K.Y. Soft lithography for microfluidics: A review. *Biochip. J.* **2008**, *2*, 1–11.
23. Bauer, W.A.C.; Fischlechner, M.; Abell, C.; Huck, W.T.S. Hydrophilic PDMS Microchannels for High-Throughput Formation of Oil-in-Water Microdroplets and Water-in-Oil-in-Water Double Emulsions. *Lab Chip* **2010**, *10*, 1814–1819. [CrossRef] [PubMed]
24. Miles, B.Y.A.A.; Misra, S.S.; Irwin, J.O. The Estimation of The Bactericidal Power of The Blood. *Epidemiol. Infect.* **1938**, *38*, 732–749. [CrossRef] [PubMed]
25. Find Circles Using Circular Hough Transform—MATLAB Imfindcircles. Available online: https://www.mathworks.com/help/images/ref/imfindcircles.html (accessed on 8 October 2020).
26. El Kadri, H.; Overton, T.; Bakalis, S.; Gkatzionis, K. Understanding and Controlling the Release Mechanism of Escherichia Coli in Double W 1 /O/W 2 Emulsion Globules in the Presence of NaCl in the W 2 Phase. *RSC Adv.* **2015**, *5*, 105098–105110. [CrossRef]
27. Tieko Nassu, R.; Gonçalves, L.A.G. Determination of Melting Point of Vegetable Oils and Fats by Differential Scanning Calorimetry (DSC) Technique. *Grasas y Aceites* **1999**, *50*, 16–21. [CrossRef]
28. Harada, T.; Yokomizo, K. Demulsification of Oil-in-Water Emulsion under Freezing Conditions: Effect of Crystal Structure Modifier. *J. Am. Oil Chem. Soc.* **2000**, *77*, 859–864. [CrossRef]
29. Mezzenga, R.; Folmer, B.M.; Hughes, E. Design of Double Emulsions by Osmotic Pressure Tailoring. *Langmuir* **2004**, *20*, 3574–3582. [CrossRef]
30. Ishibashi, C.; Hondoh, H.; Ueno, S. Influence of Morphology and Polymorphic Transformation of Fat Crystals on the Freeze-Thaw Stability of Mayonnaise-Type Oil-in-Water Emulsions. *Food Res. Int.* **2016**, *89*, 604–613. [CrossRef]
31. Aronson, M.P.; Petko, M.F. Highly Concentrated Water-in-Oil Emulsions: Influence of Electrolyte on Their Properties and Stability. *J. Colloid Interface Sci.* **1993**, *159*, 134–149. [CrossRef]
32. Aronson, M.P.; Ananthapadmanabhan, K.; Petko, M.F.; Palatini, D.J. Origins of Freeze-Thaw Instability in Concentrated Water-in-Oil Emulsions. *Colloids Surf. A Physicochem. Eng. Asp.* **1994**, *85*, 199–210. [CrossRef]
33. He, G.; Chen, G. Lubricating Oil Sludge and Its Demulsification. *Dry. Technol.* **2002**, *20*, 1009–1018. [CrossRef]
34. Chen, G.; He, G. Separation of Water and Oil from Water-in-Oil Emulsion by Freeze/Thaw Method. *Sep. Purif. Technol.* **2003**, *31*, 83–89. [CrossRef]
35. Ghosh, S.; Rousseau, D. Freeze-Thaw Stability of Water-in-Oil Emulsions. *J. Colloid Interface Sci.* **2009**, *339*, 91–102. [CrossRef]
36. van Boekel, M.A.J.S.; Walstra, P. Stability of Oil-in-Water Emulsions with Crystals in the Disperse Phase. *Colloids Surf.* **1981**, *3*, 109–118. [CrossRef]
37. Degner, B.M.; Olson, K.M.; Rose, D.; Schlegel, V.; Hutkins, R.; McClements, D.J. Influence of Freezing Rate Variation on the Microstructure and Physicochemical Properties of Food Emulsions. *J. Food Eng.* **2013**, *119*, 244–253. [CrossRef]
38. Degner, B.M.; Chung, C.; Schlegel, V.; Hutkins, R.; Mcclements, D.J. Factors Influencing the Freeze-Thaw Stability of Emulsion-Based Foods. *Compr. Rev. Food Sci. Food Saf.* **2014**, *13*, 98–113. [CrossRef]
39. Mohd Isa, N.S.; El Kadri, H.; Vigolo, D.; Gkatzionis, K. Optimisation of Bacterial Release from a Stable Microfluidic-Generated Water-in-Oil-in-Water Emulsion. *RSC Adv.* **2021**, *11*, 7738–7749. [CrossRef]

40. Magdassi, S.; Garti, N. Formation of Water/Oil/Water Multiple Emulsions with Solid Oil Phase. *J. Colloid Interface Sci.* **1987**, *120*, 537–539. [CrossRef]
41. Hou, W.; Papadopoulos, K.D. Stability of Water-in-Oil-in-Water Type Globules. *Chem. Eng. Sci.* **1996**, *51*, 5043–5051. [CrossRef]
42. Zhu, X.F.; Zhang, N.; Lin, W.F.; Tang, C.H. Freeze-Thaw Stability of Pickering Emulsions Stabilized by Soy and Whey Protein Particles. *Food Hydrocoll.* **2017**, *69*, 173–184. [CrossRef]
43. Chen, Z.A.; Huang, F.; Tsai, P.A.; Komrakova, A. Numerical Study of Microfluidic Emulsion Dynamics under the Influence of Heterogeneous Surface Wettability. *Int. J. Multiph. Flow* **2022**, *147*. [CrossRef]
44. Malakshah, V.M.; Darabi, M.; Sattari, A.; Hanafizadeh, P. Numerical Investigation of Double Emulsion Formation in Non-Newtonian Fluids Using Double Co-Flow Geometry. *Chem. Eng. Res. Des.* **2024**, *203*, 165–177. [CrossRef]
45. Boode, K.; Walstra, P.; de Groot-Mostert, A.E.A. Partial Coalescence in Oil-in-Water Emulsions 2. Influence of the Properties of the Fat. *Colloids Surf. A Physicochem. Eng. Asp.* **1993**, *81*, 139–151. [CrossRef]
46. Marefati, A.; Rayner, M.; Timgren, A.; Dejmek, P.; Sjöö, M. Freezing and Freeze-Drying of Pickering Emulsions Stabilized by Starch Granules. *Colloids Surf. A Physicochem. Eng. Asp.* **2013**, *436*, 512–520. [CrossRef]
47. Souzu, H. Studies on the Damage to Escherichia Coli Cell Membrane Caused by Different Rates of Freeze-Thawing. *BBA Biomembr.* **1980**, *603*, 13–26. [CrossRef]
48. Mohd Isa, N.S.; El Kadri, H.; Vigolo, D.; Gkatzionis, K. The Effect of Bacteria on the Stability of Microfluidic-Generated Water-in-Oil Droplet. *Micromachines* **2022**, *13*, 2067. [CrossRef]
49. Firoozmand, H.; Rousseau, D. Microbial Cells as Colloidal Particles: Pickering Oil-in-Water Emulsions Stabilized by Bacteria and Yeast. *FRIN* **2016**, *81*, 66–73. [CrossRef]
50. Clausse, D.; Gomez, F.; Pezron, I.; Komunjer, L.; Dalmazzone, C. Morphology Characterization of Emulsions by Differential Scanning Calorimetry. *Adv. Colloid. Interface Sci.* **2005**, *117*, 59–74. [CrossRef]
51. Kovács, A.; Csóka, I.; Kónya, M.; Csányi, E.; Fehér, A.; Er\Hos, I. Structural Analysis of w/o/w Multiple Emulsions by Means of DSC. *J. Therm. Anal. Calorim.* **2005**, *82*, 491–497. [CrossRef]
52. Schuch, A.; Köhler, K.; Schuchmann, H.P. Differential Scanning Calorimetry (DSC) in Multiple W/O/W Emulsions: A Method to Characterize the Stability of Inner Droplets. *J. Therm. Anal. Calorim.* **2013**, *111*, 1881–1890. [CrossRef]
53. Souzu, H. Changes in Chemical Structure and Function in Escherichia Coli Cell Membranes Caused by Freeze-Thawing. I. Change of Lipid State in Bilayer Vesicles and in the Original Membrane Fragments Depending on Rate of Freezing. *BBA Biomembr.* **1989**, *978*, 105–111. [CrossRef] [PubMed]
54. Souzu, H.; Sato, M.; Kojima, T. Changes in Chemical Structure and Function in Escherichia Coli Cell Membranes Caused by Freeze-Thawing. II. Membrane Lipid State and Response of Cells to Dehydration. *BBA Biomembr.* **1989**, *978*, 112–118. [CrossRef] [PubMed]
55. Strocchi, M.; Ferrer, M.; Timmis, K.N.; Golyshin, P.N. Low Temperature-Induced Systems Failure in Escherichia Coli: Insights from Rescue by Cold-Adapted Chaperones. *Proteomics* **2006**, *6*, 193–206. [CrossRef] [PubMed]
56. Simonin, H.; Bergaoui, I.M.; Perrier-Cornet, J.M.; Gervais, P. Cryopreservation of Escherichia Coli K12TG1: Protection from the Damaging Effects of Supercooling by Freezing. *Cryobiology* **2015**, *70*, 115–121. [CrossRef] [PubMed]
57. O'Brien, K.V.; Aryana, K.J.; Prinyawiwatkul, W.; Ordonez, K.M.C.; Boeneke, C.A. Short Communication: The Effects of Frozen Storage on the Survival of Probiotic Microorganisms Found in Traditionally and Commercially Manufactured Kefir. *J. Dairy Sci.* **2016**, *99*, 7043–7048. [CrossRef]
58. Powell-Palm, M.J.; Preciado, J.; Lyu, C.; Rubinsky, B. Escherichia Coli Viability in an Isochoric System at Subfreezing Temperatures. *Cryobiology* **2018**, *85*, 17–24. [CrossRef]
59. Mazur, P. Freezing of Living Cells: Mechanisms and Implications. *Am. J. Physiol.-Cell Physiol.* **2017**, *247*, C125–C142. [CrossRef]
60. Priya, A.J.; Vijayalakshmi, S.P.; Raichur, A.M. Enhanced Survival of Probiotic Lactobacillus Acidophilus by Encapsulation with Nanostructured Polyelectrolyte Layers through Layer-by-Layer Approach. *J. Agric. Food Chem.* **2011**, *59*, 11838–11845. [CrossRef]
61. Sánchez, A.; Contreras, A.; Jiménez, J.; Luengo, C.; Corrales, J.C.; Fernández, C. Effect of Freezing Goat Milk Samples on Recovery of Intramammary Bacterial Pathogens. *Vet. Microbiol.* **2003**, *94*, 71–77. [CrossRef]
62. Hirai, T.; Hodono, M.; Komasawa, I. Preparation of Spherical Calcium Phosphate Fine Particles Using an Emulsion Liquid Membrane System. *Langmuir* **2000**, *16*, 955–960. [CrossRef]
63. Hirai, T.; Hariguchi, S.; Komasawa, I.; Davey, R.J. Biomimetic Synthesis of Calcium Carbonate Particles in a Pseudovesicular Double Emulsion. *Langmuir* **2002**, *13*, 6650–6653. [CrossRef]

fluids

MDPI

Article

Instability of a Film Falling Down a Bounded Plate and Its Application to Structured Packing

Giulio Croce *,† and Nicola Suzzi †

Dipartimento Politecnico di Ingegneria e Architettura, University of Udine, Via delle Scienze 206, 33100 Udine, Italy; nicola.suzzi@uniud.it
* Correspondence: giulio.croce@uniud.it
† These authors contributed equally to this work.

Abstract: The instability of a film falling down a vertical plate with lateral walls, which is the base configuration describing the structured packing geometry, is numerically investigated via the lubrication theory. The solid substrate wettability is imposed through the disjoining pressure, while the assumption of a tiny, precursor film thickness allows for modelling a moving contact line. Contact angles up to 60°, which falls in the range of structured packing applications, are investigated, thanks to the full implementation of the capillary pressure instead of the small slope approximation. Parametric computations are run for a film falling down a vertical plate bounded by lateral walls, changing the plate width and the flow characteristics. An in-house, finite volume method (FVM) code, previously developed in FORTRAN language and validated in the case of film instability and rivulet flow, is used. The number of observed rivulets, triggered by the instability induced by the lateral walls, is traced for each computation. The numerical results suggest that rivulets with a given wavelength, equal to the one provided by the linear stability analysis, are generated, but only those characterized by a wavelength greater than a minimum threshold, which depends on the substrate wettability, induce partial dewetting of the domain. This allowed for the development of a simplified, statistically based model to predict the effective interface area and the rivulet holdup (required to estimate the mass transfer rate in absorption/distillation applications). Compared to the literature models of the structured packing hydrodynamics, which usually assume a continuous wetting layer, the influence of the flow pattern (continuous film or ensemble of rivulets) on the liquid holdup and on the interfacial area is introduced. The predicted flow regime is successfully verified with evidence from the literature, involving a flow down a corrugated sheet.

Keywords: film instability; lubrication; hydrophilic surface; structured packing

1. Introduction

Liquid film hydrodynamics presents complex issues, with a number of physical discontinuities caused by instability phenomena and moving contact lines. Indeed, different liquid flow patterns may be observed: still/moving droplets, ensemble of rivulets, and continuous liquid layer (or a combination of them). Furthermore, the transition from a given flow pattern to another may also occur: a one-dimensional advancing film front transforms into an ensemble of rivulets due to finger instability, eventually inducing a partial dewetting of the domain if a receding contact line is observed through the valley between neighbor rivulets [1]; if one of these rivulets flowing over a non-wettable surface is narrow enough, it may break into moving droplets [2]; small perturbations applied to an initially uniform wetting layer may induce the formation of large dry patches [3,4].

 https://doi.org/10.3390/fluids10020030

Several numerical and analytical studies investigate the evolution of a liquid film assuming the lubrication approximation [5–13], which allows us to reduce the 3D free boundary physical problem to a 2D mathematical problem. Oron et al. (1997) [14] presented a unified mathematical model of the evolution of thin liquid films over a solid substrate based on the long-wave theory. They performed a linear stability analysis of a continuous liquid film for several cases (including films driven by constant surface tension and gravity, films subjected to Van der Waals forces and constant surface tension, thermocapillary driven films) and investigated the eventual rupture of the wetting layer with the generation of dry holes. An extensive review of various film instability phenomena is also carried out in [15], including the complex patterns observed during surfacant-induced spreading. The finger instability of a film front falling down a vertical surface, perturbed by spanwise disturbances, was numerically investigated by Zhao and Marshall (2006) [16]: they modeled the effect of the surface heterogeneity via a non-uniform surface wettability. The depinning instability of a liquid ridge on a heterogeneous surface was investigated in [9]. An extensive review of the microscopic theories beyond wetting phenomena is carried out in [17], including the disjoining pressure model and the dynamics driving the formation of a precursor film thickness. Schwartz and Eley (1998) [18] numerically investigated the contact line spreading and dewetting over a solid substrate via lubrication theory introducing the disjoining pressure concept, which allows for implicitly implementing the surface wettability and obviates the need for explicitly tracking the moving contact line.

The wealth of literature references on the evolution of a thin liquid layer reflects its great interest from an engineering point of view: as an example, in the absorption and distillation processes inside packing columns, a liquid flows down a collection of corrugated sheets, while a countercurrent gas flow goes up through the interspace formed between neighbor sheets. Knowing the liquid flow pattern is crucial here since it allows for predicting both the liquid holdup and the effective interfacial area, the former essential to estimating the flooding threshold (i.e., liquid blockage and blow up of the gas pressure drop) and the latter correlated to the actual mass transfer performances. A number of empirical correlations have been proposed to predict the interfacial area based on the liquid flow characteristics and on the structured packing geometry [19–25]. For example, Onda et al. (1968) [19] developed an empirical correlation to estimate the interfacial area for different types of random packing and used it to predict the mass transfer coefficient in the case of gas absorption in packed columns. However, the range of validity is often limited to the particular test case the correlation was derived from. Indeed, it was shown in [26] that two different correlations may give completely different values of the interfacial area (one coherent with a continuous film, and the other one coherent with a rivulet pattern) under the same operative condition. Furthermore, in the full modelization of the mass transfer process (including the calculation of the liquid holdup, gas pressure drop, flooding threshold and effective interfacial area), a continuous wetting layer is typically assumed on the corrugated sheets. For example, the semi-empirical model proposed by Rocha et al. (1993) [23,24] describes structured packing as a flow channel whose side length and inclination depend on the packing geometry, while the liquid holdup inside structured packing and the liquid–gas interfacial area are estimated assuming that a continuous film flows through the walls of the flow channel. On the other hand, various numerical studies [26–29] show a much more complex behavior, with the eventual formation of a rivulet flow pattern. In [26], a fully 3D CFD approach, with the liquid–gas interface solved via the volume of fluid (VOF) method, was used to investigate the hydrodynamics of liquid water flowing down two adjacent sheets of structured packing. The numerical results show the presence of both dry patches and wet regions, with an effective area fraction, defined as the ratio between the liquid–gas interfacial area and structured packing area, lower than one. It was also proved in [26] that empirical literature correlations are not able to accurately predict the effective interface area in the range of the investigated water/air superficial velocities. A similar study was conducted in [28], where the flow down a drilled corrugated sheet was investigated via a 3D CFD analysis, which includes using the VOF method to trace the liquid–gas interface. In particular, the performances of drilled or micro-textured surfaces and the effect of the cross junctions between adjacent sheets were analyzed in terms of the liquid–gas interfacial area, and it was proved that the efficiency of the existing packing can be further increased. However, only a few test cases were studied in [28] due to the inherently high computational cost related to 3D simulations, with some of the simulations performed in large parallel computer cluster nodes. A simplified configuration, characterized by a liquid flow down an inclined plate bounded by lateral walls, which is representative of the hydrodynamics inside structured packing, was numerically investigated in [27,29], again adopting a fully 3D approach: the CFD analysis was validated with experimental evidence and the complex case of a three-phase distillation process was also studied in [27], while Singh et al. (2016) [29] proved that a rivulet pattern can be observed in practical applications involving a absorption/distillation process through structured packing, where a highly viscous solvent promotes the formation of a rivulet pattern. Indeed, different fluids, used as liquid solvents for carbon dioxide absorption, were numerically investigated in [29] and a scaling theory based on the liquid Kapitza number was carried out to estimate the effective interfacial area, correlated to the mass transfer performances. A numerical approach based on the lubrication theory, which allows for the reduction of the 3D physical problem to a 2D mathematical formulation with no sensible loss of information regarding the liquid flow, was proposed in [30–32], with the main goal of strongly reducing the computational cost compared to the standard 3D CFD approach. In particular, preliminary studies showed the capability of the 2D approach to accurately predict (with the same precision as the fully 3D CFD modelization) the wetting layer behavior in the case of rivulet generation down an inclined plate.

Here, a discontinuous film flowing down a vertical plate bounded by lateral walls is numerically investigated through the lubrication theory, aiming at mapping the possible flow configurations as a function of plate geometry and injected liquid flow rate. Such numerical mapping can, thus, be used to build a statistically based modelization, able to predict the liquid pattern inside structured packing and to estimate relevant mean quantities such as the liquid holdup and the effective interfacial area, which directly affects the mass transfer performances in absorption and distillation processes. Following [11,16,18], the assumption of a thin precursor film and the implementation of the disjoining pressure allow for modelling the substrate wettability. The exact formulation of the film curvature is implemented, as done in [31,33]: this allows for widening the range of possible applications, since previous literature studies adopt the standard small free surface slope approximation, which is accurate only for low surface wettabilities, $\theta_s < 30^\circ$ [34]. An in-house finite volume method code, previously developed in FORTRAN language, successfully validated with experimental evidence and verified with analytical and numerical data for static contact angles up to $\theta_s = 60^\circ$ [31], predicts the instability of a thin film flowing down a vertical plate bounded by lateral walls, which is the simplest configuration describing structured packing [29]. Compared to [33], where the instability of an advancing film on an indefinitely large domain was triggered by a non-uniform surface wettability, a bounded film on a homogeneous surface is investigated here since the instability caused by the lateral walls is correlated to the one that eventually induces the dewetting of a corrugated sheet. This requires a dedicated boundary condition to simulate a slip wall (while periodic conditions were used in [33] to mimic an indefinitely large domain). First of all, the dynamic contact angle at the apparent contact line of a 1D falling film front is computed as a function of the undisturbed flow characteristics, defined by the non-dimensional film capillary number, for different surface wettabilities and compared with the well-known Hoffman–Voinov–Tanner law. Then, parametric computations are run in the case of a film falling down a plate with lateral walls, characterized by a static contact angle equal to $\theta_s = 60^\circ$, which is consistent with structured packing applications. The liquid pattern (continuous film, ensemble of rivulets), which is correlated to the effective interfacial area, is traced as a function of the injected flow characteristics and of the plate geometry. Such information is used to develop a new, statistically based modelization of the hydrodynamics in structured packing, which is able to estimate the liquid holdup and the interfacial area as a function of the effective flow pattern.

2. Mathematical Model

Let us consider a thin film of thickness h flowing down a vertical plate, driven by gravity. Applying the lubrication theory and assuming a Newtonian fluid, the Navier–Stokes equations can be integrated across the film thickness, reducing the 3D physical problem to a 2D mathematical problem. Further assuming a low film Reynolds number, a parabolic velocity profile can be assumed. Integrating the velocity profile across the film thickness and substituting to the continuity equation yields the governing lubrication equation,

$$\frac{\partial h}{\partial t} + \nabla \cdot \left(\frac{\rho_l\, g\, h^3}{3\,\mu_l}\,\hat{\mathbf{j}} - \frac{\nabla p}{3\,\mu_l}\, h^3 \right) = 0, \; p = -2\,\sigma_{lg}\,\kappa - D \tag{1}$$

with $\hat{\mathbf{j}}$ being the versor identifying the plate's downhill direction, with p being the sum of the capillary pressure and disjoining pressure. The full curvature implementation,

$$2\,\kappa = \frac{\frac{\partial^2 h}{\partial x^2}\left[1 + \left(\frac{\partial h}{\partial y}\right)^2\right] + \frac{\partial^2 h}{\partial y^2}\left[1 + \left(\frac{\partial h}{\partial x}\right)^2\right] - 2\,\frac{\partial h}{\partial x}\frac{\partial h}{\partial y}\frac{\partial^2 h}{\partial x \partial y}}{\left[1 + \left(\frac{\partial h}{\partial x}\right)^2 + \left(\frac{\partial h}{\partial y}\right)^2\right]^{3/2}} \tag{2}$$

allows for accurately investigating contact angles up to $\theta_s = 60^\circ$ [31], while the standard small slope approximation, $2\,\kappa \simeq \nabla^2 h$, usually assumed within lubrication theory, is not accurate for $\theta_s \geq 30^\circ$ [31,34]. The disjoining pressure allows for modelling the solid substrate wettability [16,18],

$$D = B\left[\left(\frac{\delta}{h}\right)^n - \left(\frac{\delta}{h}\right)^m\right], \; B = f\,\frac{(n-1)(m-1)}{n-m}\,\frac{\sigma_{lg}}{\delta}\,(1 - \cos\theta_s) \tag{3}$$

with θ_s being the static contact angle, $\delta \ll h_0$ being the precursor film thickness and n, m being positive constants with $n > m > 1$. The coefficient f, introduced by [16] as a function of δ/h_0, takes into account the imposed precursor film thickness, usually higher than its physical value ($\sim 10 \div 100\,\mathrm{nm}$) to reduce the computational cost.

To generalize the problem, let us introduce the following non-dimensional quantities,

$$\mathbf{X} = \frac{\mathbf{x}}{L_0}, \; H = \frac{h}{h_0}, \; T = \frac{t}{(L_0/u_0)}, \; K = \frac{\kappa}{\left(h_0/{L_0}^2\right)}, \; \Pi = \frac{D}{\left(\sigma_{lg}/h_0\right)} \tag{4}$$

with h_0 being the undisturbed film thickness, u_0 the undisturbed film velocity and L_0 the characteristic length scale:

$$u_0 = \frac{\rho_l\, g\, {h_0}^2}{3\,\mu_l}, \; L_0 = \left(\frac{\sigma_{lg}\, h_0}{\rho_l\, g}\right)^{1/3} \tag{5}$$

Further defining the film capillary number as $\mathrm{Ca} = \mu_l\, u_0/\sigma_{lg}$, the governing lubrication equation, Equation (1), can be recast in non-dimensional form:

$$\frac{\partial H}{\partial T} + \nabla \cdot \left[(\hat{\jmath} - \nabla P)\, H^3 \right] = 0,\ P = -2\,K - (3\,\mathrm{Ca})^{-2/3}\,\Pi \tag{6}$$

Equation (6) is numerically solved on a structured grid via a finite volume method in-house code, previously developed in FORTRAN language. The in-house code was validated in [31] for contact angles up to $\theta_s = 60°$ for the dry patch generation on a continuous wetting layer and for the generation of an isolated rivulet via the injection of liquid through a confined inlet section. The second-order centered scheme of [35] was implemented for the spatial discretization, while the alternating-direction implicit (ADI) approximate factorization of [36] was used for marching in time. To limit the error of the numerical solution, an adaptive time stepping was implemented limiting the allowed increment in the film thickness H at each time step,

$$\Delta T^n : \max\left(\frac{\Delta T^n}{H_k^n} \left| \frac{\mathrm{d}H_k^n}{\mathrm{d}T} \right| \right) \leq 10^{-2} \tag{7}$$

with n being the time step and k being the mesh element index. This allowed for reaching a time step up to 10^8, the Newmann limit (i.e., the maximum step for stability of the forward Euler scheme, given by $\Delta T \sim \Delta X^4$) during slow shape evolution, when a spatial discretization step of $\Delta X \leq \delta/h_0$ is used. In order to speed up the computations, the open source programming interface OpenMP is used to parallelize the source code for shared memory machines. A detailed explanation of the implemented numerical method is provided by [31,33,35,36].

3. Problem Statement

Let us consider a gravity-driven film flowing down a vertical plate of dimension $L_X \times L_Y$ bounded by two lateral walls, such as the one shown in Figure 1. According to [27,29], this is the simplest configuration describing the hydrodynamics through structured packing since it allows for replicating the same instability phenomena that eventually induce partial dewetting of the corrugated sheets inside structured packing.

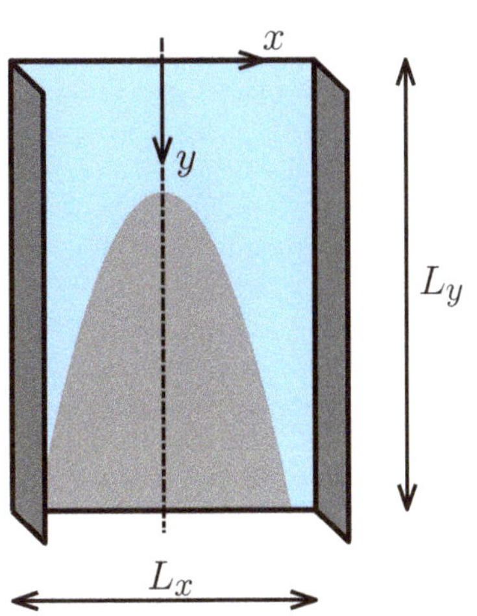

Figure 1. Investigated setup: liquid film down a vertical plate bounded by lateral walls.

The undisturbed film thickness and the undisturbed film flux are imposed through the inlet section, located at the topmost section of the plate, i.e., $Y = 0$,

$$H|_{Y=0} = 1, \ \mathbf{Q} \cdot \hat{\mathbf{n}} = 1 \tag{8}$$

with $\mathbf{Q} = (\hat{\mathbf{j}} - \nabla P)\, H^3$ being the non-dimensional film flux and $\hat{\mathbf{n}}$ being the normal inward direction to the boundary. The fully developed flow condition is imposed through the outlet section, located at the bottommost section of the plate, i.e., $Y = L_Y$,

$$\nabla H \cdot \hat{\mathbf{n}} = 0, \ \nabla P \cdot \hat{\mathbf{n}} = 0 \tag{9}$$

Following [31], a slip wall with a prescribed contact angle is implemented through the lateral boundary $X = L_X/2$,

$$\nabla H \cdot \hat{\mathbf{n}} = -\frac{\sqrt{1 + (3\,\mathrm{Ca})^{2/3} (\nabla H \cdot \hat{\mathbf{t}})^2}}{(3\,\mathrm{Ca})^{1/3} \tan\theta_s}, \ \mathbf{Q} \cdot \hat{\mathbf{n}} = 0 \tag{10}$$

with $\hat{\mathbf{t}}$ being the boundary tangential direction. Taking advantage of the problem symmetry, only half of the domain is simulated, and the symmetry condition is imposed through the section $X = 0$,

$$\nabla H \cdot \hat{\mathbf{n}} = 0, \ \mathbf{Q} \cdot \hat{\mathbf{n}} = 0 \tag{11}$$

A dry domain is always initialized at the first stage of the computation. Thus, the solid surface is initially wetted by the tiny precursor film of thickness δ at $T = 0$:

$$H(X,Y)|_{T=0} = \frac{\delta}{h_0} \tag{12}$$

4. Results

4.1. Dynamic Contact Angle

First of all, preliminary computations were run to investigate a 1D falling film: the dynamic contact angle of the advancing liquid front was traced as a function of the imposed film capillary number at different substrate wettabilities and compared with the Tanner–Voinov–Hoffman law [37]. The reduction to a 1D physical problem was achieved considering a computational domain of width $L_X = 1$ and imposing symmetry conditions through both $X = 0$ and $X = L_X$. As described in Section 3, the liquid was injected to an initially dry domain through the topmost section $Y = 0$, while the fully developed flow condition was imposed at $Y = L_Y$. Since imposing non-physical initial conditions may affect the prediction of film instability at low capillary numbers, the computational domain is initialized with an advancing film front (with the apparent contact line located at $Y = 10$) using the computed solution at a slightly higher capillary number and assuming a dry surface downstream of it, i.e., $H|_{Y>10} = \delta/h_0$. Static contact angles ranging from 30° to 60° were investigated. The exponents in Equation (3) were set to $m = 2$, $n = 3$ according to [11,16,18], while the precursor film thickness was set to $\delta/h_0 = 5 \times 10^{-2}$ [16,18]. A grid dependency analysis was carried out for a 1D falling film in the case of $\theta_s = 60°$ and $\mathrm{Ca} = 2 \times 10^{-2}$. Progressively increasing the grid accuracy, it was found that $\Delta X = 5 \times 10^{-2}$ ensures grid independency since further decreasing ΔX to 2.5×10^{-2} gives less than 1% deviation on the computed dynamic contact angle. Thus, the spatial discretization step was set to $\Delta X = 2 \times 10^{-2}$, while Equation (7) was used to dynamically adjust the time step.

Computations were run for film capillary numbers ranging in $\mathrm{Ca} \in [10^{-3}, 1]$ and three values of the static contact angle, $\theta_s = 30°, 45°, 60°$: both the full curvature formulation, Equation (2), and the small slope approximation were considered, in order to show that the full curvature implementation is required for an accurate prediction of the free surface profile. The dynamic contact angle was estimated as the highest value of the free surface slope close to the apparent contact line of the self-similar advancing film front. For each of the investigated surface wettabilities, the capillary number, related to the imposed liquid flow rate, was lowered until the film rupture was observed, i.e., the film breaks inducing the formation of self-similar droplets falling down the vertical plate, such as the one shown in Figure 2. As mentioned before, imposing consistent initial conditions is crucial to properly predict the instability threshold. Thus, the film profile derived from the solution at a slightly higher capillary number is initialized at $T = 0$, with the apparent contact line located at a distance $Y = 10$ from the inlet section. For each computation, the dynamic contact angle θ_d is estimated as the maximum value of the detected film slope:

$$\tan\theta_d = (3\,\mathrm{Ca})^{1/3} \max\Big(|\nabla H|\Big), \; |\nabla H| = \left|\frac{\partial H}{\partial Y}\right| \tag{13}$$

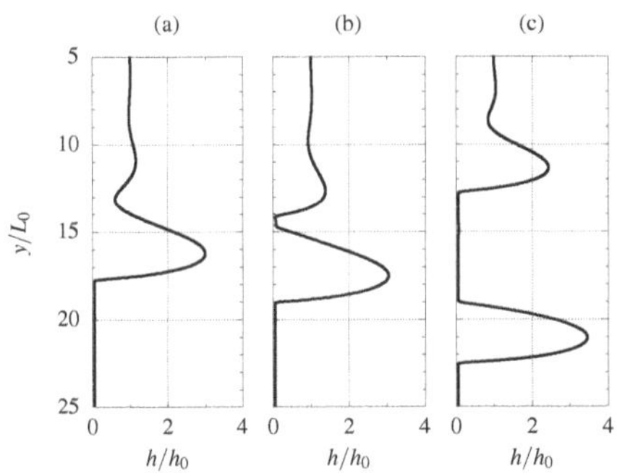

Figure 2. Free surface evolution at successive instances (time elapse $\Delta T = 2$) in the case of film breaking into self-similar falling droplets: $T = 8$ (**a**); $T = 9$ (**b**); $T = 10$ (**c**). A 1D film falling down a vertical plate, $\theta_s = 60°$, $\mathrm{Ca} = 1.33 \times 10^{-2}$.

In Figure 3a, the computed dynamic contact angle of the moving contact line is compared with the one predicted by the well-known Hoffman–Voinov–Tanner law [17,37],

$$\theta_d{}^3 = \theta_s{}^3 + 9\,\mathrm{Ca}\,\log\left(\frac{c\,h_0}{\delta}\right) \tag{14}$$

with δ being the microscopic length scale (i.e., the precursor film thickness), h_0 being the macroscopic one (i.e., the undisturbed film thickness) and c being an unknown constant, which has to be estimated by fitting the numerical data. The numerical points obtained via the full curvature implementation, colored markers in Figure 3a, are in good agreement with Equation (14) for all the investigated static contact angles, while Figure 3b proves that the small slope approximation does not give a consistent estimate of the dynamic contact angle for $\theta_s = 45°, 60°$, since θ_d does not even approach the static contact angle at a low contact line velocity (i.e., at a low capillary number). For each θ_s, the lowest capillary number (left-most markers in Figure 3) corresponds to the instability threshold since further decreasing Ca would give film instability (i.e., film rupture with formation of self-similar droplets): increasing the static contact angle (i.e., lowering the surface wettability) promotes the instability occurrence, with the breaking of the 1D film into falling droplets observed at lower capillary numbers ($\mathrm{Ca} = 1.671 \times 10^{-2}$ for $\theta_s = 60°$, $\mathrm{Ca} = 2.103 \times 10^{-3}$ for $\theta = 30°$).

Note that the small slope approximation, which does not provide an accurate estimate of the free surface profile (as proved by the computed dynamic contact angle), is still able to predict the onset of film instability at the same Ca.

The effect of changing the surface wettability on the computed profile of the advancing film front is shown in Figure 4a (with a magnification of the free surface profile shown close to the apparent contact line), in the case of full curvature implementation: as expected, increasing the imposed static contact angle at fixed capillary number $\mathrm{Ca} = 3 \times 10^{-2}$ yields a higher ridge forming close to the apparent contact line. On the other hand, Figure 4b presents the influence of the capillary number: the height of the liquid ridge surrounding the contact line increases, lowering the capillary number at fixed $\theta = 60^\circ$.

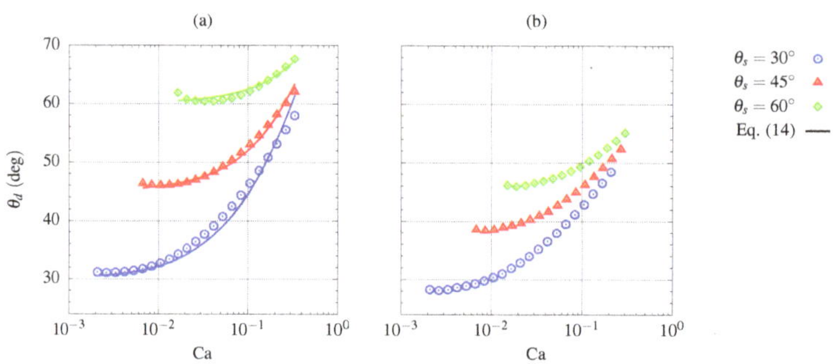

Figure 3. Dynamic contact angle as a function of the film capillary number at different substrate wettabilities: computed (markers) vs. Equation (14) (continuous lines). Full curvature implementation (**a**); small slope approximation (**b**).

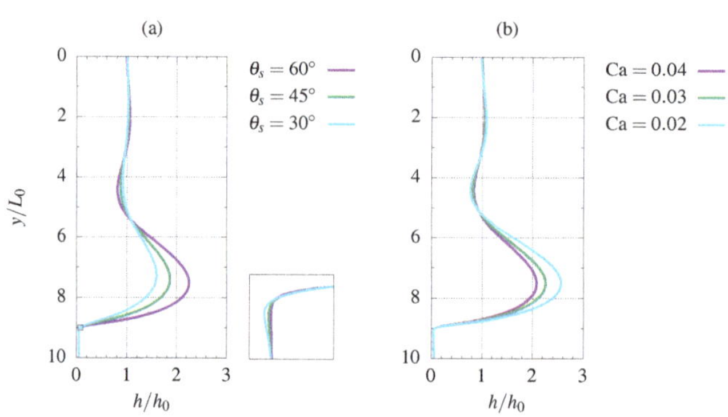

Figure 4. Self-similar free surface profile: influence of the surface wettability, $\mathrm{Ca} = 3 \times 10^{-2}$ (**a**); influence of the film capillary number, $\theta = 60^\circ$ (**b**).

4.2. Finger Instability Validation

Apart from the validation in terms of the dynamic contact angle of an advancing film, the in-house code had been already validated with the experimental evidence of [3,4] in the case of dry patch generation, induced via a small perturbation applied on a uniform wetting layer [31]. Furthermore, isolated rivulets flowing down an inclined plate were generated in [31] injecting the liquid through a confined inlet section at the top of the plate: the computed rivulet interfacial area was successfully compared with numerical data of [29], obtained via a fully 3D modelization, for contact angles up to 60°; the shape of the fully developed rivulet was successfully verified with the analytical solution.

To extend the code validation to problems involving the transition from a continuous film regime to a rivulet pattern, we investigated the finger instability of an advancing film front down a vertical plate. Thus, the liquid is injected through the topmost section of the plate, located at $Y = 0$, while the fully developed flow condition, Equation (9), is applied through the bottom-most section, at $Y = L_Y$. The symmetry condition, Equation (11), is applied though the lateral boundaries, located at $X = 0$, L_X. The liquid layer is initialized at $T = 0$ with the precursor film thickness (i.e., an initially dry domain is considered) according to Equation (12). To perturb the advancing film front, a horizontal stripe of width L_0, characterized by a non-uniform contact angle, was implemented at $Y = 10$, while a uniform substrate wettability was applied to the rest of the computational domain. Computations were run for different values of the capillary number, with Ca ranging between 10^{-3} and 1. Since the boundary effects may affect the number of rivulets, the plate width $L_X = L/L_0$ must be large enough to ensure a sufficient number of rivulets to form. Two domains of different sizes were investigated: $L_X \times L_Y = 100 \times 150$ and $L_X \times L_Y = 90 \times 150$. The precursor film thickness was set to $\delta/h_0 = 5 \times 10^{-2}$, and the exponents of Equation (3) were set to $m = 2$, $n = 3$. According to the grid dependency analysis, a spatial discretization step of ΔX, $\Delta Y = 5 \times 10^{-2}$ was imposed. An adaptive time step size was used applying Equation (7). The static contact angle was set to $\theta_s = 10°$ over the homogeneous region of the computational domain, while a random distribution with $\theta_s \in [0°, 20°]$ was implemented along the heterogeneous stripe of width L_0 and located at $Y = 10$. The investigated θ_s is consistent with the value reported in [1], in which the authors experimentally investigated the finger instability of different liquids flowing down an inclined plate. After the advancing liquid front meets the heterogeneous stripe, the induced spanwise perturbances propagate and finger instability is observed: the number of rivulets N and, thus, the average rivulet spacing (i.e., the wavelength λ) were traced. The computations were stopped when the first advancing rivulet front reaches the outlet section. Figure 5a shows the contact line evolution for Ca $= 10^{-2}$, $L_X \times L_Y = 100 \times 150$: eight rivulets are observed, corresponding to an average wavelength of $\lambda = L/N = 12.5\,L_0$. Figure 5c compares the computed wavelength with the most unstable wavelength $\lambda_{max} = 13.09\,L_0$, provided by the linear stability analysis of [16]: the imposed capillary number does not affect the number of detected rivulets; the numerical wavelength is close to the expected value λ_{max} ($\lambda/L_0 = 12.87$ for the $L_X \times L_Y = 90 \times 150$ domain). Small discrepancies between the numerical data and stability analysis are coherent with the boundary effects, determined by a computational domain of finite width L_X.

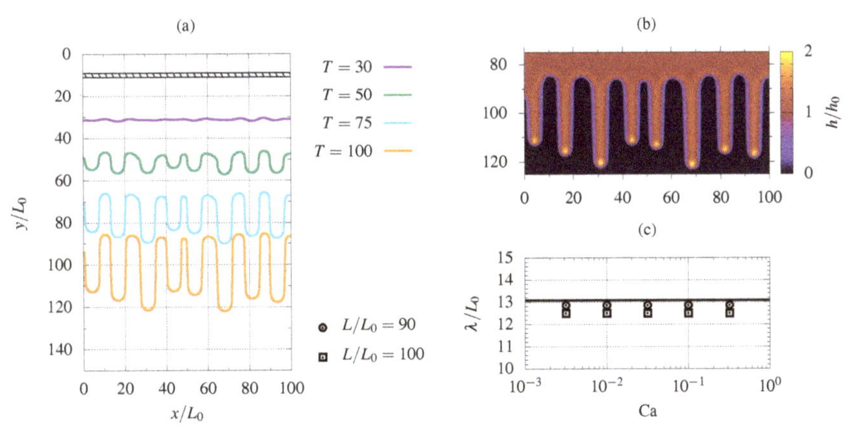

Figure 5. Contact line evolution at successive instants for Ca $= 10^{-2}$, $L_X \times L_Y = 100 \times 150$ (the patterned stripe is characterized by a random contact angle distribution) (**a**). Detail of the liquid layer solution at $T = 100$ (**b**). Computed rivulet wavelength (black markers) versus linear stability analysis (continuous line) of [16] (**c**).

Figure 6 compares the numerical points with the experimental results of [1], in which the authors investigated the finger instability of four different fluids (each one characterized by a particular glycerin-water mixture) flowing down an inclined plate with a slope ranging between 7.2° and 90°. The computed wavelengths λ/L_0 are made dimensional for the fluid A: $\rho_l = 1075\,\mathrm{kg/m^3}$, $\sigma_{lg} = 6.9 \times 10^{-2}\,\mathrm{N/m}$. To do so, the film thickness h_0 is deduced from the capillary number, while the scaling length L_0 is calculated via Equation (5). It is worth pointing out that reducing the numerical data to the other fluids would result in points laying on the same line. The experimental wavelengths of [1] are well predicted by the numerical data, which fall inside the errorbar of most of the experimental points. Furthermore, Figure 6 proves that the numerical results can be extended to any Newtonian fluid, provided that the effect of liquid inertia is negligible (i.e., low film Reynolds number).

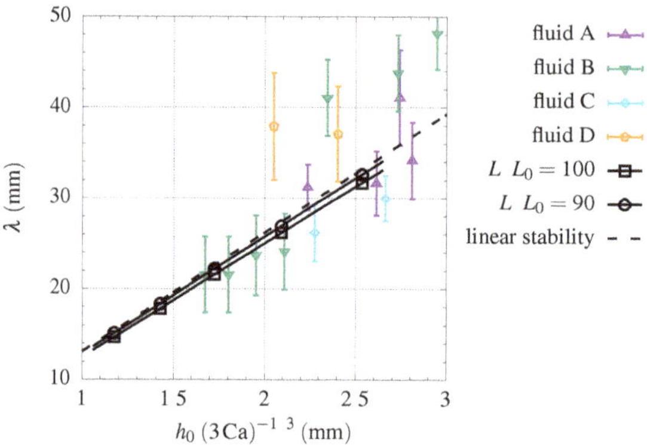

Figure 6. Experimental wavelengths of [1] for different fluids flowing down an inclined plate (colored markers with errorbar) versus numerical data (black markers) and linear stability analysis (dashed line).

4.3. Rivulet Onset

A parametric analysis was conducted for a film flowing down a vertical plate bounded by lateral walls as in Figure 1. Such geometry is chosen as the simplest configuration describing the hydrodynamics inside structured packing [27,29]. The flow pattern was traced as a function of the film capillary number, Ca, related to the liquid flow rate; the non-dimensional wall distance, L/L_0, which defines the problem geometry. The boundary conditions and the initial condition are described in Section 3: the liquid is injected through the inlet section, located at the top of the vertical plate; Equation (10) allows the implementation of the lateral wall at $X = L_X/2$, with $L_X = L/L_0$ being the non-dimensional wall distance, while symmetry conditions are applied through $X = 0$; the solid surface is initially dry, meaning that the precursor film thickness is initialized over the computational domain according to Equation (12). Following reference [16,18,31], the precursor film thickness was set to $\delta = 5 \times 10^{-2}\,h_0$ and the exponents in Equation (3) were set to $m = 2$, $n = 3$. The spatial discretization step was set to ΔX, $\Delta Y = 2 \times 10^{-2}$ for all the computations, according to the grid dependency analysis described in Section 4.1, while the time step was dynamically adjusted using Equation (7). The computations were stopped after the steady state solution was reached. A static contact angle of $\theta_s = 60°$, which is consistent with structured packing applications, was chosen. For each computation, the number of rivulets, eventually induced by the disturbance due to the presence of the solid walls, was traced until the steady state solution was reached.

The maximum number of rivulets at the onset of the instability (let us call it N_{max}), detected during the first stage of the simulations, is shown in Figure 7a. It is interesting highlighting that a change in the number of rivulets always occurs at fixed values of the non-dimensional wall distance L/L_0 (i.e., N_{max} depends on the parameter L/L_0 rather than Ca). This is consistent with the linear stability analysis of [16], where the instability of an advancing 1D film front to spanwise perturbances was analyzed assuming lubrication approximation. It was found in [16] that, for a fully wettable substrate (i.e., $\theta_s = 0°$), the most unstable wavelength is $\lambda_{max}/L_0 = 13.09$, with a weak influence of the static contact angle. This means that, when finger instability is observed, the spanwise distance between neighbor rivulets is expected to be equal to λ_{max}, as experimentally observed by [1]. Figure 7a shows that rivulets at a distance smaller than λ_{max} are induced by the presence of the lateral walls. It is important to mention that rivulet spacing lower than λ_{max} were numerically observed also by [16] in the case of finger instability of a film down a heterogeneous substrate, characterized by a non-uniform surface wettability. However, increasing the wall distance (i.e., decreasing the influence of boundary effects), the number of rivulets detected during the first stage of the computations becomes proportional to the most unstable wavelength:

$$\lim_{L \to \infty} N_{max} = \frac{L}{\lambda_{max}} = 7.64 \times 10^{-2} \frac{L}{L_0} \tag{15}$$

On the other hand, the number of stationary rivulets N (corresponding to the steady state solution), see Figure 7b, looks much more difficult to predict since some of the dry patches forming between neighbor rivulets just drain away, while some other dry patches eventually move upwards inducing partial dewetting of the computational domain. However, it can be clearly observed that it is always verified that $N \leq N_{max}$.

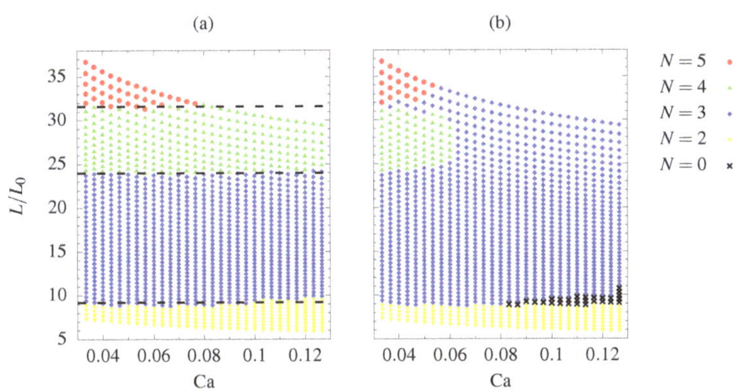

Figure 7. Number of rivulets as a function of the capillary number and non-dimensional wall distance: maximum number of rivulets observed during transient (**a**); steady state solution (**b**). $\theta_s = 60°$.

As an example, Figure 8a shows the computed liquid layer thickness after $T = 12$ for $\mathrm{Ca} = 3.67 \times 10^{-2}$ and $L/L_0 = 32.6$, while the corresponding steady state solution is shown in Figure 8b. Note that all the five rivulets formed during the first stage of the computation (three rivulets inside the computational domain and two half rivulets along the boundary walls) survive in the steady state solution, $N = N_{max} = 5$.

Decreasing the non-dimensional wall distance yields a reduction in the number of rivulets, as proved by Figures 9a and 10a, referring to $L/L_0 \simeq 30$, with four rivulets detected at the onset of the instability. Again we observe that $N = N_{max}$ in Figure 9b, while only three stationary rivulets, i.e., $N < N_{max}$, are observed for a higher film capillary number in Figure 10b, meaning that the dry patch apex enclosed between the central rivulets moves downward so that the two central rivulets coalesce into a single one. As pointed out, N_{max} depends only on the geometrical parameter L/L_0 according to the linear stability analysis, while the number of stationary rivulets N changes with both L/L_0 and Ca. To better understand such a complex behavior, the width of the steady rivulets was computed, looking for a possible threshold based on the minimum dimension that a rivulet can assume.

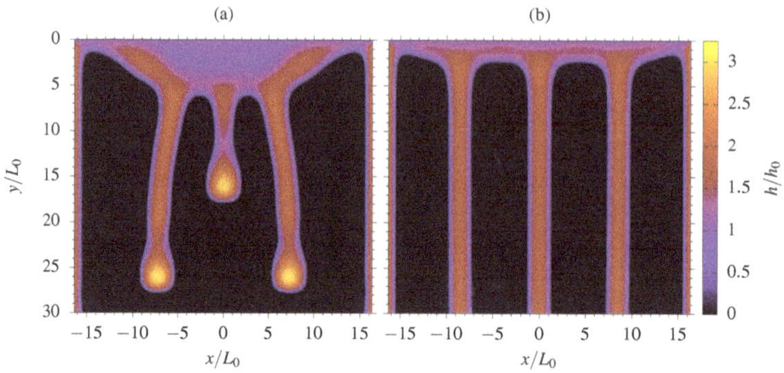

Figure 8. Liquid layer solution at $T = 12$ (**a**); steady state solution (**b**). $\theta_s = 60°$, Ca $= 3.67 \times 10^{-2}$, $L/L_0 = 32.6$.

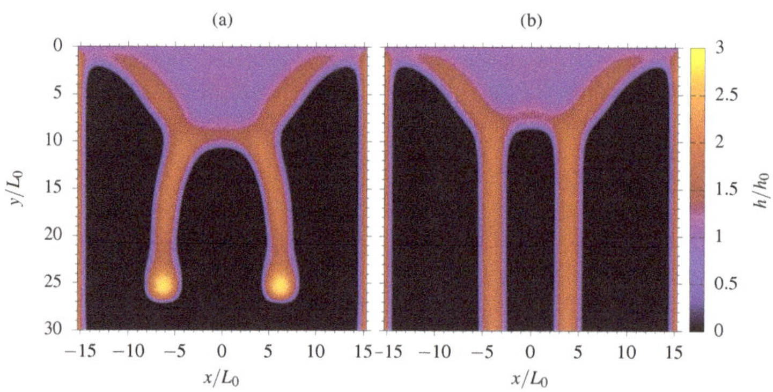

Figure 9. Liquid layer solution at $T = 12$ (**a**); steady state solution (**b**). $\theta_s = 60°$, Ca $= 5.33 \times 10^{-2}$, $L/L_0 = 30.6$.

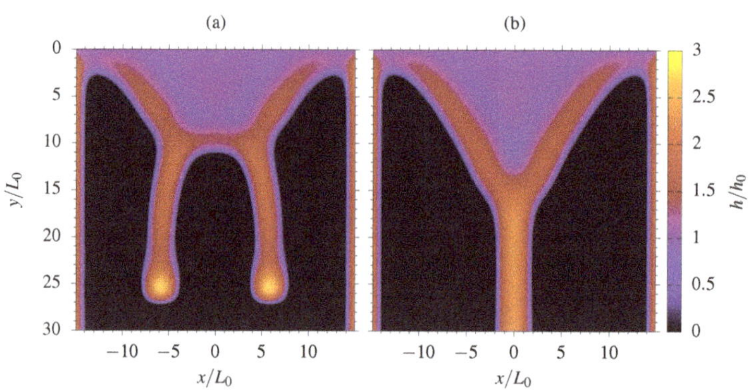

Figure 10. Liquid layer solution at $T = 12$ (**a**); steady state solution (**b**). $\theta_s = 60°$, Ca $= 6.33 \times 10^{-2}$, $L/L_0 = 29.9$.

Figure 11 shows, for each simulation, the computed width of the narrowest stationary rivulet in the core domain (i.e., the eventual rivulets adjacent to the lateral wall are not considered). Results for $L/L_0 > 24$ are shown: we have $N = N_{max}$ if four or five stationary rivulets are observed, while a reduction in the number of rivulets with $N < N_{max}$ is observed when the number of stationary rivulets is $N = 3$. It can be pointed out that the rivulet width is always higher than a critical value, which depends on the capillary number. This is consistent with the linear stability analysis of a continuous film initially wetting the whole domain, subjected to capillary pressure and surface tension, perturbed with a periodic disturbance. Following [14], a uniform film of thickness h_0 flowing down a vertical plate is perturbed with a small disturbance $h = h_0 + h^*$ periodic in x, which is substituted in the governing lubrication equation. Seeking a seperate solution $h^* = h_0 \exp(i\,k\,x + s\,t)$ and linearizing the lubrication equation around $h = h_0$ yields the characteristic equation for s. Imposing zero growth rate for h^*, i.e., $\mathcal{R}(s) = 0$, gives the critical wavelength:

$$\lambda_{cri} = \frac{2\,\pi}{k_{cr}} = h_0 \sqrt{f\,(1 - \cos\theta_s)\,\frac{(m-1)(n-1)}{n-m}\left[m\left(\frac{\delta}{h_0}\right)^{m-1} - n\left(\frac{\delta}{h_0}\right)^{n-1}\right]} \tag{16}$$

If the disturbance wavelength is higher than λ_{cri}, the instability induces the film break into rivulets of spacing equal to the wavelength. Thus, we assume that the minimum width L_{cri} that a rivulet can assume is proportional to λ_{cri},

$$\frac{L_{cri}}{L_0} = \frac{c\,h_0}{L_0} = c\,(3\ \mathrm{Ca})^{1/3} \tag{17}$$

with c being a constant, which depends on the substrate wettability (i.e., on the static contact angle θ_s). Comparing Equation (17) with the numerical data (see Figure 11) shows a good agreement since the lowest rivulet width derived via the postprocess of the computed liquid distribution is well predicted for each of the investigated capillary numbers.

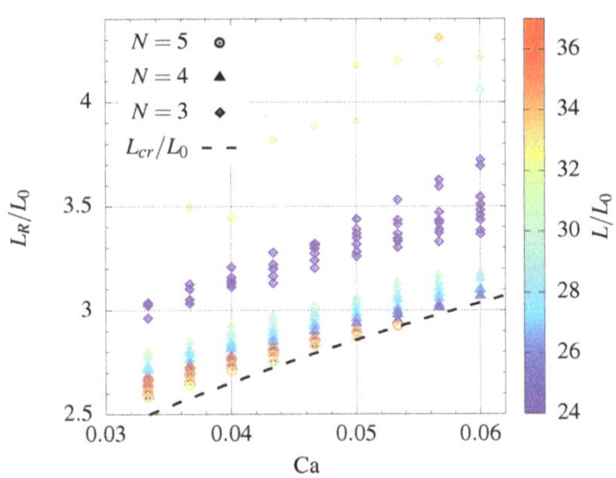

Figure 11. Computed width of the narrowest stationary rivulet in the core domain as a function of the film capillary number (markers) and minimum rivulet width predicted via Equation (17) imposing $c = 5.37$ (dashed lines). Distance between lateral walls is within the range $L/L_0 \in [24, 38]$, $\theta_s = 60°$.

4.4. *Statistically Based Modelization of the Hydrodynamics in Structured Packing*

Since the investigated setup is the simplest configuration describing the hydrodynamics inside structured packing, we can use the numerical results to predict the flow pattern down the corrugated sheets and to estimate the effective liquid–gas interfacial area.

Modeling the structured packing geometry shown in Figure 12 as an inclined flow channel having a squared section with the side length equal to the corrugation length l_c [23], we can correlate l_c with the lateral walls distance (which is the geometrical parameter defining the investigated numerical setup). Knowing the liquid superficial velocity u_{ls} inside the structured packing, which is the liquid velocity calculated as though the whole structured packing was filled with the liquid phase, the theoretical thickness of the continuous film flowing down the corrugated sheets of the structured packing can be derived via the Nusselt theory, which yields [23],

$$h_0 = \left(\frac{3\,\mu_l\,u_{ls}\,l_c}{4\,\rho_l\,g\,\sin\alpha\,\epsilon} \right)^{1/3} \tag{18}$$

with ϵ being the void fraction inside the structured packing and α being the corrugation angle. The film thickness given by Equation (18) was used in [23,24] to estimate the liquid holdup (i.e., the liquid volume fraction inside the structured packing, a crucial parameter for the prediction of the flooding occurrence) as $h_t = 4\,F_t\,h_0/l_c$, with F_t being an empirical correction factor. However, a continuous film may not be representative of the effective liquid pattern inside structured packing, as proved by [26,29]. If we assume a rivulet pattern inside structured packing, we can correlate the liquid superficial velocity u_{ls} to the averaged rivulet characteristics (i.e., spacing, height and width). This allows the estimate of both the average liquid thickness (which gives the liquid holdup) and the effective interfacial area (required to estimate the mass transfer rate). Assuming that each rivulet, which collects the liquid from a portion λ_R of undisturbed film, has a circular section with prescribed contact angle θ_s, the liquid flow rate through a single rivulet can be expressed as,

$$\dot{m}_R = \rho_l\,u_0\,h_0\,\lambda_R = \frac{{\rho_l}^2\,g\,\sin\alpha\,{L_R}^4}{24\,\mu_l\,\sin^4\theta_s} \int_0^{\theta_s} (\cos\phi - \cos\theta_s)^3 \cos\phi\,\mathrm{d}\phi \tag{19}$$

with u_0 being the reference film velocity. Equation (19) allows for correlating the rivulet wavelength to the average rivulet width, both of which are required to estimate the interfacial area. The rivulet wavelength λ_R is estimated as the maximum between the one coming from the linear stability analysis of [16] (i.e., $\lambda_{max} = 13.09\,L_0$, with L_0 being the scaling length) and the one corresponding to a rivulet width L_R equal to the threshold value L_{cri}, defined by Equation (17). The definition of the undisturbed film velocity and of the scaling length, introduced in Equation (5) and strictly valid for a film on a vertical plate, must be extended to a film flowing down an inclined plate with the slope equal to the corrugation angle α:

$$u_0 = \frac{\rho_l\,g\,{h_0}^2\,\sin\alpha}{3\,\mu_l},\quad L_0 = \left(\frac{\sigma_{lg}\,h_0}{\rho_l\,g\,\sin\alpha} \right)^{1/3} \tag{20}$$

Dividing the corrugation length l_c by the calculated rivulet spacing λ_R gives a sharp criterion to determine the liquid pattern: a continuous film is expected if $l_c/\lambda_R < 1$; an ensemble of rivulets is expected if $l_c/\lambda_R \geq 1$. Such a sharp transition between a continuous film and an ensemble of rivulets would give a discontinuity in the predicted liquid holdup, which is what was observed on a small portion of structured packing, as the one numerically investigated. However, in a large domain including several corrugated sheets, small fluctuations of the injected liquid flow rate induce wet regions locally covered by a continuous film and partially dry regions filled with rivulets, as proved by the numerical evidence of [26,28]. To account for this effect, we define $u_{ls}^* = u_{ls}\,\cos\alpha$, which is correlated to the maximum flow rate that can cross the corrugations transversely. Then, we compute the corresponding rivulet spacing $\lambda_R^{\pm}$ for $u_{ls} \pm u_{ls}^*$. If the condition $\lambda_R^- \leq l_c \leq \lambda_R^+$

is verified, the liquid holdup and the effective interfacial area are estimated as a weighted arithmetic mean between the ones derived for a film and the ones derived for a rivulet,

$$h_t = \eta_R \frac{4\bar{h}_R}{l_c}\frac{L_R}{\lambda_R} + \eta_F \frac{4h_0}{l_c}, \quad \frac{a_e}{a_p} = \eta_R \frac{L_R}{\lambda_R}\frac{\theta_s}{\sin\theta_s} + \eta_F \tag{21}$$

$$\begin{cases} \lambda_R^+ < l_c : \ \eta_R = 1, \ \eta_F = 0 \\ \lambda_R^- \le l_c \le \lambda_R^+ : \ \eta_R = \dfrac{l_c - \lambda_R^-}{\lambda_R^+ - \lambda_R^-}, \ \eta_F = 1 - \eta_R \\ \lambda_R^- > l_c : \ \eta_R = 0, \ \eta_F = 1 \end{cases} \tag{22}$$

with a_p being the total packing surface area per unit volume and $\bar{h}_R$ being the averaged rivulet height:

$$\bar{h}_R = L_R \frac{\theta_s - \sin\theta_s \cos\theta_s}{4\sin^2\theta_s} \tag{23}$$

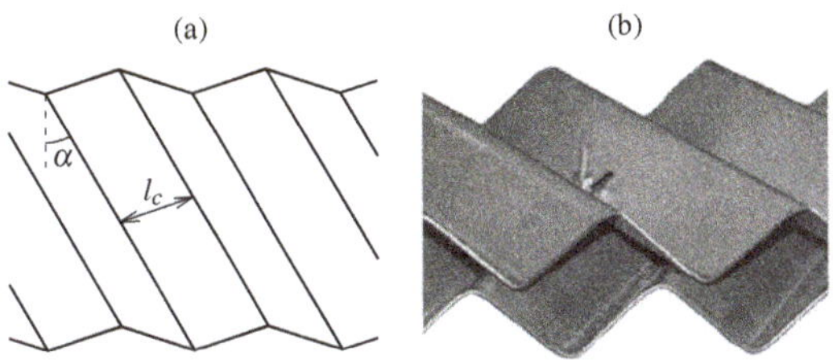

Figure 12. Structured packing geometry: corrugation length l_c and corrugation angle α of the corrugated sheet (**a**); packing element (**b**).

To verify the consistency of the proposed modelization, the effective interfacial area, calculated via Equation (21), is compared in Figure 13 with the empirical correlations [22–25] listed in Table 1 and with the numerical data of [26]. Liquid water, $\rho_l = 997\,\mathrm{kg\,m^{-3}}$, $\mu_l = 8.89 \times 10^{-4}\,\mathrm{kg\,m^{-1}s^{-1}}$ and $\sigma_{lg} = 0.0728\,\mathrm{N\,m^{-1}}$, flowing down the Gempak-2A structured packing, $l_c = 18\,\mathrm{mm}$, $a_p = 223\,\mathrm{m^{-1}}$ and $\alpha = 45°$, was considered according to [26]. The static contact angle was set to $\theta_s = 60°$, which is close to the value reported in [26] and is consistent with water on metal surfaces.

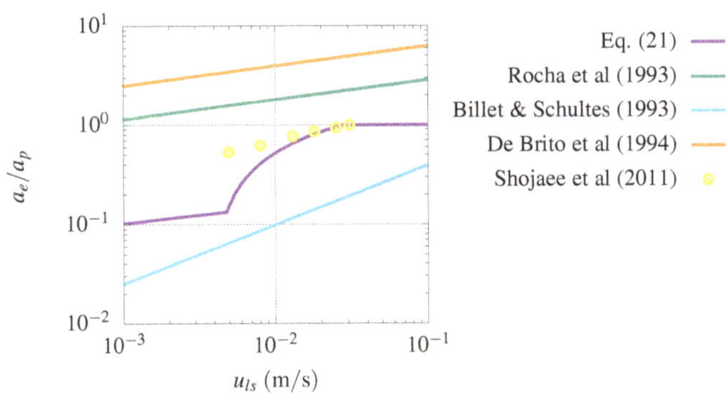

Figure 13. Predicted interfacial area as a function of the liquid superficial velocity: Equation (21) versus empirical models of [22,23,25] and numerical data (markers) of [26]. Liquid water in Gempak-2A structured packing.

Table 1. Literature models for the prediction of the effective interfacial area inside structured packing.

Reference	Correlation
Rocha et al. (1996) [23,24]	$\frac{a_e}{a_p} = F_{SE}\,\frac{29.12\,(\mathrm{We}_l\,\mathrm{Fr}_l)^{0.15}\,l_c^{\,0.359}}{\mathrm{Re}_l^{\,0.2}\,\epsilon^{0.6}\,(1-0.93\,\cos\theta_s)\,(\sin\alpha)^{0.3}}$
Billet and Schultes (1993) [22]	$\frac{a_e}{a_p} = 1.5\,(a_p\,l_c)^{-0.5}\,\mathrm{Re}_l^{\,-0.2}\,\mathrm{We}_l^{\,0.75}\,\mathrm{Fr}_l^{\,-0.45}$
De Brito et al. (1994) [25]	$\frac{a_e}{a_p} = 0.465\left(\frac{\rho_l\,u_l}{\mu_l\,a_p}\right)^{0.3}$

Figure 13, which plots the effective interfacial area as a function of the liquid superficial velocity, shows a good agreement of the proposed statistically based model with the numerical data of [26]. Indeed, the transition from a rivulet to a continuous film is successfully predicted. However, the current modelization seems to underestimate the effective interfacial area predicted by [26] under the rivulet regime: this is probably due to the assumption of a circular rivulet cross section, which is strictly valid only for a liquid layer flowing down a vertical plate (i.e., the effect of the hydrostatic pressure is neglected when $\alpha < 90°$). On the other hand, the available literature models give different trends: the empirical correlation of [25] predicts an interfacial area much greater that one, which is not consistent with small waves and ripples forming on the free surface of a continuous film, while the theoretical correlation of [22], valid for both structured and random packing, gives a regime of partial wetting, consistent with a rivulet pattern, over the whole range of the investigated u_{ls}. The empirical correlation of [24], derived from the analysis of [20] and used in [23,24] to estimate the mass transfer performances of different structured packing geometries including Gempak-2A for water at $u_{ls} \geq 7\,\mathrm{mm\,s^{-1}}$, predicts a continuous film pattern, but it does not deal with the eventual maldistribution of liquid in the packed bed [24]. For a full validation of the statistically based modelization, the liquid layer model should be coupled with the core gas flow solution (or, at least, with a proper channel flow model): this would give an estimate of the gas pressure drop, allowing for a direct comparison with experimental data of [23,24]. However, the modelization of the gas core flow is beyond the scope of the present study, where the consistency of the liquid hydrodynamics modelization has been verified with the data of [26].

To show the practicality of the proposed modelization, Equation (21) is used to calculate the liquid holdup as a function of the liquid superficial velocity and of the structured packing geometry: the liquid holdup is a crucial operative parameter since it provides a direct prediction of the flooding occurrence (observed when $h_t \geq 1$) and allows for estimating the gas pressure drop; the liquid superficial velocity is one of the input parameters for the utilization of a packed column. The 1-methylepiperazione, which is used as a liquid solvent for CO_2 absorption in structured packing [29], is chosen: $\rho_l = 981.31\,\mathrm{kg\,m^{-3}}$, $\mu_l = 3.642 \times 10^{-2}\,\mathrm{kg\,m^{-1}s^{-1}}$, $\sigma_{lg} = 3.84 \times 10^{-2}\,\mathrm{N\,m^{-1}}$. The static contact angle is set to $\theta_s = 60°$. Figure 14a provides a chart to determine h_t from u_{ls} for different corrugation geometries. The colored markers identify the transition between the rivulet pattern to a continuous film and, thus, correspond to the minimum liquid superficial velocity, which gives the maximum interfacial area, i.e., $a_e/a_p = 1$. The definition of such a threshold for u_{ls} is crucial for proper structured packing design since it allows for minimizing the liquid holdup (i.e., reduce the risk of flooding) and enhance the interfacial area (i.e., augment the mass transfer rate). The minimum superficial velocity giving $a_e/a_p = 1$ is shown in Figure 14b as a function of the corrugation length for two values of the corrugation angle.

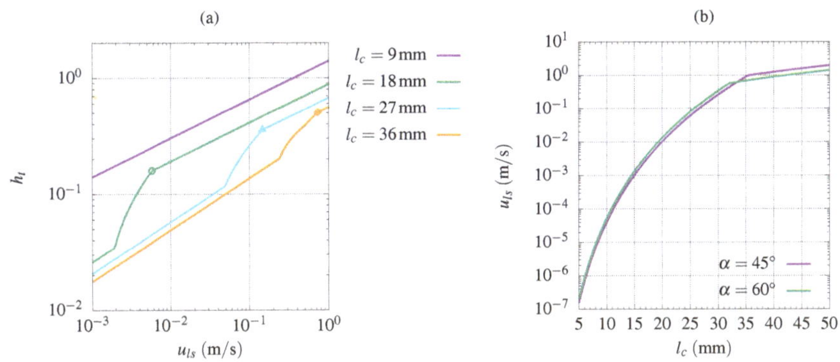

Figure 14. Liquid holdup as a function of the liquid superficial velocity, $\alpha = 60°$: markers identify the transition from rivulet pattern to a continuous film (**a**). Minimum liquid superficial velocity, which gives a continuous film, as a function of the corrugation length (**b**). $\epsilon = 0.95$, $\theta_s = 60°$, 1-methylepiperazine.

5. Conclusions

An in-house code was used to study the hydrodynamics down a portion of a corrugated sheet of structured packing, modeled as an inclined plate with lateral walls. The model was successfully validated by comparing the dynamic contact angle of a 1D falling film front with the Hoffman–Voinov–Tanner law for a static contact angle up to $\theta_s = 60°$. Such comparison demonstrated that the present full implementation of the film curvature accurately predicts the self-similar profile of the advancing film close to the apparent contact line for $\theta_s > 30°$, well beyond the limits of standard small slope approximation, which shows an error on the computed dynamic contact angle up to 25% for $\theta_s = 60°$. This confirms that the details of the free surface evolution cannot be described in the range of structured packing applications using the small slope approximation. The numerical model was further validated with experimental data giving the rivulet wavelength at different flow conditions in the case of finger instability of an advancing film down a vertical plate.

The parametric computations of a discontinuous film down a bounded plate at different corrugation geometries and flow characteristics (the latter being defined by the film capillary number) showed a complex behavior, with different flow patterns observed for slightly different setups. It was found that the linear stability analysis is still able to predict the number of rivulets at the onset of the instability, even if the lateral walls promote the finger instability at wavelengths lower than the one provided by the linear stability. On the other hand, the linear stability analysis does not give information regarding the steady state solution and, thus, the wetting regime (continuous film or ensemble of rivulets, number of rivulets). Indeed, postprocessing numerical results shows that the number of steady rivulets may be lower than the one detected at the instability onset, with rivulets thinner than a threshold value just draining down the vertical plate without inducing partial dewetting. The information provided by computations were used to implement a new statistically based model, able to give a simplified (but still consistent) description of the effective flow pattern inside the structured packing, with the aim of predicting relevant mean parameters, such as the liquid holdup and the effective interfacial area, that allow a direct estimate of the mass transfer performances under given operative conditions. Comparing the statistically based model with the numerical data of [26] shows a good agreement, while the available literature correlations, often at odds with each other, are not able to give an accurate estimate of the interfacial area. This proves that a consistent and accurate prediction of the mass transfer in structured packing, which is crucial for a proper design and utilization, requires the implementation of the effective liquid pattern features (i.e., continuous film or ensemble of rivulets, average rivulet spacing and size). In perspective, more complex geometries should be numerically investigated, such as a periodic curved surface. This would provide a better description of the wetting behavior of a liquid flowing on a corrugated sheet, allowing to eventually include the effect of other geometrical parameters, such as the corrugation height, in the simplified modelization.

Author Contributions: Conceptualization, G.C.; methodology, G.C. and N.S.; software, N.S.; investigation, N.S.; writing—original draft, G.C. and N.S.; writing—review and editing, G.C.; supervision, G.C. All authors have read and agreed to the published version of the manuscript.

Funding: This research received no external funding.

Data Availability Statement: The raw data supporting the conclusions of this article will be made available by the authors upon request.

Acknowledgments: We wish to acknowledge NextGenerationEU for supporting the present work, performed within the project Interconnected Nord-Est Innovation Ecosystem (iNEST), placed within the context of the National Recovery and Resilience Plan (PNRR).

Conflicts of Interest: The authors declare no conflicts of interest.

Glossary

The following abbreviations are used in this manuscript:

Abbreviations

ADI	alternating-direction implicit
CFD	computational fluid dynamics
FVM	finite volume method
VOF	volume of fluid

Nomenclature

Greek symbols

α	corrugation angle
δ	precursor film thickness
ϵ	void fraction
θ	contact angle
κ	free surface curvature
λ	wavelength
μ	dynamic viscosity
ρ	density
σ	surface tension

Other symbols

a_p	packing surface area per unit volume
a_e	effective interfacial area per unit volume
Ca	capillary number, $\mu_l\, u_0/\sigma_{lg}$
D	disjoining pressure
g	gravity acceleration
h	film thickness
h_t	liquid holdup
L	distance between lateral walls
L_0	scaling length
l_c	corrugation length

L_R	rivulet width
$\dot{m}_R$	mass flow rate through a rivulet
N	number of rivulets
p	pressure
t	time
u	film velocity
u_{ls}	liquid interfacial velocity
$\mathbf{x}$	Cartesian coordinates
Subscripts	
0	undisturbed film
g	gas
l	liquid

References

1. Johnson, M.F.G.; Schluter, R.A.; Miksis, M.J.; Bankoff, S.G. Experimental study of rivulet formation on an inclined plate by fluorescent imaging. *J. Fluid Mech.* **1999**, *394*, 339–354. [CrossRef]
2. Suzzi, N.; Croce, G. Numerical modelization of contact angle hysteresis of falling droplet under enhanced lubrication approximation. *J. Phys. Conf. Ser.* **2022**, *2177*, 012043. [CrossRef]
3. Podgorski, T.; Flesselles, J.M.; Limat, L. Dry arches within flowing films. *Phys. Fluids* **1999**, *11*, 845–852. [CrossRef]
4. Rio, E.; Limat, L. Wetting hysteresis of a dry patch left inside a flowing film. *Phys. Fluids* **2006**, *18*, 032102. [CrossRef]
5. Weidner, D.E.; Schwartz, L.W.; Eres, M.H. Simulation of coating layer evolution and drop formation on horizontal cylinders. *J. Colloid Interface Sci.* **1997**, *187*, 243–258. [CrossRef]
6. Duffy, B.R.; Moffatt, H.K. A similarity solution for viscous source flow on a vertical plane. *Eur. J. Appl. Math.* **1997**, *8*, 37–47. [CrossRef]
7. Wilson, S.K.; Duffy, B.R.; Davis, S.H. On a slender dry patch in a liquid film draining under gravity down an inclined plane. *Eur. J. Appl. Math.* **2001**, *12*, 233–252. [CrossRef]
8. El-Genk, M.S.; Saber, H.H. Minimum thickness of a flowing down liquid film on a vertical surface. *Int. J. Heat Mass Transf.* **2001**, *44*, 2809–2825. [CrossRef]
9. Beltrame, P.; Knobloch, E.; Hanggi, P.; Thiele, U. Rayleigh and depinning instabilities of forced liquid ridges on heterogeneous substrates. *Phys. Rev. E* **2011**, *83*, 016305. [CrossRef]
10. Kay, E.D.; Hibberd, S.; Power, H. Inertial effects at moderate Reynolds number in thin-film rimming flows driven by surface shear. *Phys. Fluids* **2013**, *25*, 102108. [CrossRef]
11. Sellier, M. Modelling the wetting of a solid occlusion by a liquid film. *Int. J. Multiph. Flow* **2015**, *71*, 66–73. [CrossRef]
12. Suzzi, N.; Croce, G. Numerical Simulation of Thin Film Breakup on Nonwettable Surfaces. *J. Phys. Conf. Ser.* **2017**, *796*, 012038. [CrossRef]
13. Magnini, M.; Khodaparast, S.; Matar, O.K.; Stone, H.A.; Thome, J.R. Dynamics of long gas bubbles rising in a vertical tube in a cocurrent liquid flow. *Phys. Rev. Fluids* **2019**, *4*, 023601. [CrossRef]
14. Oron, A.; Davis, S.H.; Bankoff, S.G. Long-scale evolution of thin liquid films. *Rev. Mod. Phys.* **1997**, *69*, 931–980. [CrossRef]
15. Craster, R.V.; Matar, O.K. Dynamics and stability of thin liquid films. *Rev. Mod. Phys.* **2009**, *81*, 1131–1198. [CrossRef]
16. Zhao, Y.; Marshall, J.S. Dynamics of driven liquid films on heterogeneous surfaces. *J. Fluid Mech.* **2006**, *559*, 355–378. [CrossRef]
17. Bonn, D.; Eggers, J.; Indekeu, J.; Meunier, J.; Rolley, E. Wetting and spreading. *Rev. Mod. Phys.* **2009**, *81*, 739–805. [CrossRef]
18. Schwartz, L.W.; Eley, R.R. Simulation of droplet motion on low-energy and heterogeneous surfaces. *J. Colloid Interface Sci.* **1998**, *202*, 173–188. [CrossRef]
19. Onda, K.; Takeuchi, H.; Okumoto, Y. Mass transfer coefficients between gas and liquid phases in packed columns. *J. Chem. Eng. Jpn.* **1968**, *1*, 56–62. [CrossRef]
20. Shi, M.G.; Mersmann, A. Effective interfacial area in packed columns. *Ger. Chem. Eng.* **1985**, *8*, 87.
21. Spiegel, L.; Meier, W. Correlations of the performance characteristics of the various Mellapak types (capacity, pressure drop and efficiency). *Inst. Chem. Eng. Symp. Ser.* **1987**, *62*, A203–A215.
22. Billet, R.; Schultes, M. Predicting mass transfer in packed columns. *Chem. Eng. Technol.* **1993**, *16*, 1–9. [CrossRef]
23. Rocha, J.A.; Bravo, J.L.; Fair, J.R. Distillation columns containing structured packings: A comprehensive model for their performance. 1. Hydraulic models. *Ind. Eng. Chem. Res.* **1993**, *32*, 641–651. [CrossRef]
24. Rocha, J.A.; Bravo, J.L.; Fair, J.R. Distillation columns containing structured packings: A comprehensive model for their performance. 2. Mass-transfer model. *Ind. Eng. Chem. Res.* **1996**, *35*, 1660–1667. [CrossRef]

25. De Brito, H.M.; Von Stockar, U.; Bangerter, A.M.; Bomio, P.; Laso, M. Effective mass-transfer area in a pilot plant column equipped with structured packings and with ceramic rings. *Ind. Eng. Chem. Res.* **1994**, *33*, 647–656. [CrossRef]
26. Shojaee, S.; Hosseini, S.H.; Rafati, A.; Ahmadi, G. Prediction of the effective area in structured packings by computational fluid dynamics. *Ind. Eng. Chem. Res.* **2011**, *50*, 10833–10842. [CrossRef]
27. Hoffmann, A.; Ausner, I.; Repke, J.U.; Wozny, G. Fluid dynamics in multiphase distillation processes in packed towers. *Comput. Chem. Eng.* **2005**, *29*, 1433–1437. [CrossRef]
28. Subramanian, K.; Wozny, G. Analysis of hydrodynamics of fluid flow on corrugated sheets of packing. *Int. J. Chem. Eng.* **2012**, *2012*, 838965. [CrossRef]
29. Singh, R.K.; Galvin, J.E.; Sun, X. Three-dimensional simulation of rivulet and film flows over an inclined plate: Effects of solvent properties and contact angle. *Chem. Eng. Sci.* **2016**, *142*, 244–257. [CrossRef]
30. Suzzi, N.; Croce, G. Numerical simulation of rivulet build up via lubrication equations. *J. Phys. Conf. Ser.* **2017**, *923*, 012020. [CrossRef]
31. Suzzi, N.; Croce, G. Numerical simulation of film instability over low wettability surfaces through lubrication theory. *Phys. Fluids* **2019**, *31*, 122106. [CrossRef]
32. Suzzi, N.; Croce, G. Numerical simulation of film instability over a corrugated sheet. *J. Phys. Conf. Ser.* 2024, *in press*.
33. Suzzi, N.; Croce, G. Numerical bifurcation analysis of a film flowing over a patterned surface through enhanced lubrication theory. *Fluids* **2021**, *6*, 405. [CrossRef]
34. Perazzo, C.A.; Gratton, J. Navier-Stokes solutions for parallel flow in rivulets on an inclined plane. *J. Fluid Mech.* **2004**, *507*, 367–379. [CrossRef]
35. Diez, J.A.; Kondic, L. Computing three-dimensional thin film flows including contact lines. *J. Comput. Phys.* **2002**, *183*, 274–306. [CrossRef]
36. Witelski, T.P.; Bowen, M. ADI schemes for higher-order nonlinear diffusion equations. *Appl. Numer. Math.* **2003**, *45*, 331–351. [CrossRef]
37. Blake, T.D. The physics of moving wetting lines. *J. Colloid Interface Sci.* **2006**, *299*, 1–13. [CrossRef]

fluids

MDPI

Article

Computational Modeling of Droplet-Based Printing Using Multiphase Volume of Fluid (VOF) Method: Prediction of Flow, Spread Behavior, and Printability

Rauf Shah [1] and Ram V. Mohan [2,*]

[1] Department of Nanoengineering, North Carolina A&T State University, Greensboro, NC 27401, USA; rgshah@aggies.ncat.edu
[2] Department of Mechanical Engineering, North Carolina A&T State University, Greensboro, NC 27401, USA
* Correspondence: rvmohan@ncat.edu; Tel.: +1-(336)285-2867

Abstract: The evolution of droplets during the printing process is modeled using the volume of fluid (VOF) method, which involves solving the Navier–Stokes and continuity equations for incompressible flow with multiple immiscible phases on a finite volume grid. An indicator function tracks the interfaces and calculates surface tension forces. A grid independence study confirmed the convergence and efficacy of the solutions. The computational model agreed well with experimental data, accurately capturing the impact, spreading, and recoiling of droplets on a solid surface. Additionally, the model validated the interaction of droplets with hydrophilic and hydrophobic surfaces for both constant and dynamic contact angles. Key non-dimensional numbers (Re, We, Oh) were considered to study the interplay of forces during droplet impact on a solid surface. The final print quality is influenced by droplet dynamics, governed by body forces (surface tension, gravity), contact angle, dissipative forces due to motion, and material properties. Computational studies provide insights into the overall process performance and final print quality under various process conditions and material properties.

Keywords: droplet dynamics; volume of fluid (VOF) method; spreading factor; droplet-based printing; print quality

1. Introduction

The dynamics of droplet behavior and its spreading mechanism have been extensively explored through experimental and computational methods. The droplets interact with the printing substrate in various ways, including spreading, rebounding, recoiling, and splashing, as reviewed by Yarin [1]. These interactions vary depending on the process conditions and material properties. Despite significant research on droplet spreading simulations, there has been less emphasis on studying the recoiling behavior. This study aims to fill that gap by providing a comprehensive three-dimensional (3D) analysis of droplet impact, spreading, recoiling, and interaction with different surfaces.

The deposition process is characterized by the 'Spreading Factor', which is defined as the ratio of the maximum spread diameter to the initial drop diameter. This factor directly reflects the printability and quality of droplet-based printing. Experimenting with droplet-based printing can be challenging, costly, and cumbersome. However, computational fluid dynamics (CFD) analyses offer a practical and efficient alternative for exploring the potential applications of droplet-based printing in additive manufacturing. Although numerical methods for accurately representing the interfaces between different phases

https://doi.org/10.3390/fluids10050123

are still evolving, they hold promise for investigating droplet behavior and predicting the outcomes of droplet-based printing.

Limited research has been conducted on the oscillatory behavior of droplets after colliding with a solid surface. Elliot and Ford [2] studied how droplets spread and recoil upon collision, observing changes in contact angles until equilibrium is reached. Similarly, Fukai et al. [3] used both experimental and theoretical models to analyze the fluid mechanics of droplet impact and spreading on a solid surface. While some authors have reported empirical and semi-empirical relations for the spreading diameter based on fluid properties and process conditions, these studies often exclude surface characteristics and dynamic contact angles [4–8]. Pasandideh-Fard et al. [9] and Feng [10] presented improved empirical relations incorporating advancing contact angles and energy conservation factors for predicting the final spreading diameter.

Various modeling techniques are available for numerical investigations of free surface flow with deformations, including explicit interface tracking and implicit Eulerian interface capturing methods. The implicit Eulerian methods, particularly the volume of fluid (VOF) model, are highly effective in capturing flows with deformations, including interface breakups. Most numerical studies employ the VOF model alongside the solution of Navier–Stokes equations and a dynamic contact angle model to accurately depict droplet impact and spreading. For instance, Adeniyi et al. [11] used a coupled Euler–Lagrange approach to model droplet-to-film interactions in aeroengine bearing chambers, highlighting the challenges of modeling droplet impact phenomena in finite-volume CFD techniques.

Pasandideh-Fard et al. [9] conducted simulations using VOF and finite difference Navier–Stokes solvers, concluding that dynamic contact angles yield more accurate results than constant or equilibrium contact angles. Yokoi et al. [12] used the finite volume method and VOF model, incorporating a switching function for dynamic contact angles based on the velocity of the interface normal near the wall. Gujjula [13] employed similar methods with different dynamic contact angle models, comparing their results to those of Pasandideh-Fard et al. [9] and finding that the spread diameter was underestimated. Wang et al. [14] introduced a morphology-adaptive multifield two-fluid model to simulate annular flow, proposing a new droplet entrainment model based on the shear-off entrainment mechanism at the gas–liquid interface.

The present study also examines the influence of various forces (surface tension, gravitational forces, contact angle forces, and other dissipative forces), material properties (density, viscosity, and surface tension), and processing conditions (droplet size, impact velocity, and nozzle-substrate distance) on the spreading and recoiling behavior of droplets, which dictate the final printing resolution. Building on previous work by Shah and Mohan [15], the present work extends the computational domain to three dimensions. It investigates the effects of relevant non-dimensional numbers (Reynolds, Weber, and Ohnesorge numbers) on droplet dynamics to analyze the interplay of different forces in droplet-based printing. In related work, Vontas [16] explored the practical applications of droplet impact in fuel injection systems, using enhanced VOF-based CFD simulations to model the interaction of fuel spray droplets with solid surfaces in combustion chambers.

During continuous printing, the print nozzle moves in a traverse direction depending on the print speed and deposits the subsequent droplets. The print setup and traverse speed decide the position of consequent droplets deposited on the substrate. It is essential to study the dynamics of droplets being printed, with the droplet interaction with the substrate and other droplets. The droplet–droplet interaction and the dynamics of coalescence determine the final print resolution for continuous droplet-based printing. The coalescence mechanism involves rupture, deformation, and interface merging of the droplets and needs further investigation, focusing on droplet-based printing.

Applications:

Droplet-based printing has various applications across various fields due to its precision and versatility. In biomedical engineering, it is used for bioprinting to fabricate complex 3D structures with living cells for tissue engineering and regenerative medicine, and to create drug delivery systems with controlled release properties [17,18]. In microfluidics, droplet-based printing is employed to fabricate lab-on-a-chip devices for biochemical analyses and to create monodisperse emulsions for chemical reactions and material synthesis [19,20]. In flexible electronics, it produces printed sensors for wearable devices and prints conductive inks for flexible circuits and displays [21,22]. In the aerospace and automotive industries, droplet-based printing is utilized to manufacture lightweight components, improve thermal management systems, and optimize fuel injection processes [23–25]. These applications highlight technology's potential to enhance manufacturing processes, improve product performance, and enable innovative solutions across various sectors.

This paper is structured to provide a detailed exploration of droplet dynamics in the context of droplet-based printing. The main content is organized as follows: Section 2 outlines the numerical methods employed, including the governing equations and computational framework. Section 3 presents the results of the computational model, validated against experimental studies, and discusses the effects of key non-dimensional numbers (Re, We, Oh) on droplet dynamics. Section 4 delves into the fluid–fluid interaction and final print resolution prediction, examining droplets' coalescence behavior during continuous printing. Finally, Section 5 offers a comprehensive discussion of the findings, highlighting the implications for additive manufacturing and potential areas for future research. The conclusions summarize the key insights gained from the study and propose directions for further investigation.

2. Numerical Method

2.1. Governing Equations

The Navier–Stokes equations describe the behavior of an incompressible fluid and elucidate the dynamics of a droplet impinging upon a solid surface. This is complemented by the volume fraction function, denoted as the indicator function α. The pertinent equations governing this isothermal modeling framework are delineated as follows:

Continuity equation:

$$\nabla \cdot U = 0 \tag{1}$$

Momentum equation:

$$\rho \frac{DU}{Dt} = -\nabla p + \rho g + \nabla \cdot \mu (\nabla U + \nabla U^T) + F_\sigma \tag{2}$$

Conservation of phase fraction:

$$\frac{D\alpha}{Dt} = 0 \tag{3}$$

where U is the velocity vector, t is time, p is the lumped pressure, g is the acceleration due to gravity, and F_σ are the body (surface tension and gravity) forces

In this computational framework, the gas–liquid interface is conceptualized as a dispersed layer with a thickness commensurate with a finite volume cell. This interface is characterized by the indicator function, wherein the gradient of the volume fraction α delineates the phase discontinuity at the interface. The volume fraction, continuously distributed within the domain ($0 \le \alpha \le 1$) except at the interface, is instrumental in determining phase properties such as distributed density and viscosity. In fluid modeling, the distributed density is typically calculated using the volume fraction of the phases involved. For a two-phase system, the density and viscosity at any point in the domain are a weighted average of the densities and viscosities of the individual phases, respectively, based on their corresponding volume fractions. Mathematically, this can be expressed as follows:

$$\rho = \rho_d \alpha + \rho_g (1 - \alpha) \tag{4}$$

$$\mu = \mu_d \alpha + \mu_g (1 - \alpha) \tag{5}$$

where α is '1' for the liquid phase, '0' for the gas/air phase, and takes an intermediate value at the interface.

As the resolution of the interface is equal to the size of the finite volume cell, it needs to be sharpened in the momentum equation. The '*interfoam*' solver (OpenFOAM® v2021) improves the resolution of the interface using an additional compression velocity term in the conservation equation of the phase fraction (Equation (3)) [26]. The surface tension is considered as body force in the momentum equation and modeled as follows:

$$F_\sigma = \sigma k n = \sigma (\nabla n) \cdot n = \sigma \nabla \left(\frac{\nabla \alpha}{|\nabla \alpha|} \right) \tag{6}$$

where σ is the surface tension coefficient, k is the interfacial curvature, and 'n' is the interfacial normal.

The transient terms in the Navier–Stokes equations are discretized using the Crank–Nicolson scheme, a second-order implicit method known for its stability and accuracy in time integration. The time step is determined based on the specified Courant–Friedrichs–Lewy (CFL) number, ensuring numerical stability by maintaining the CFL condition [16]. Meanwhile, the spatial terms are discretized using a second-order central differencing scheme, which provides a balance between accuracy and computational efficiency by approximating derivatives at the cell centers. This method is particularly effective in minimizing numerical diffusion and maintaining sharp gradients.

The pressure gradient is solved using the conjugate gradient method, an efficient iterative solver for large, sparse linear systems typically encountered in computational fluid dynamics (CFDs). This method leverages the symmetry and positive definiteness of the pressure matrix to converge rapidly. Pressure–velocity coupling is achieved through the pressure implicit with splitting of operators (PISO) method. The PISO method is an extension of the semi-implicit method for pressure-linked equations (SIMPLE), designed to enhance the convergence rate and accuracy of the solution. In the PISO algorithm, the corrected velocities are used to update the pressure field iteratively. This iterative process ensures that the pressure and velocity fields satisfy the momentum and continuity equations, achieving a consistent and accurate solution. The PISO method addresses the challenge of not having a separate equation for the pressure term in the Navier–Stokes equations. Instead, it updates the pressure field after correcting the velocities to satisfy the continuity equation, followed by further corrections using the momentum equation. This iterative correction process ensures the pressure–velocity coupling is accurately resolved.

Furthermore, the transport and volume of fluid (VOF) equations are solved using the Gauss–Seidel algorithm, an iterative method that solves linear systems by successively updating the solution vector. This algorithm is particularly effective for solving the discretized transport equations in CFD.

The multidimensional universal limiter for explicit solution (MULES) technique is employed to achieve interface compression in the VOF method [26]. MULES is designed to maintain a sharp interface between immiscible fluids by limiting numerical diffusion and ensuring that the volume fraction remains between 0 and 1. This technique is crucial for accurately capturing the interface dynamics in multiphase flow simulations.

By incorporating these advanced numerical methods and algorithms, the solver ensures robust and accurate solutions to the Navier–Stokes equations, enabling detailed and reliable fluid dynamics simulations in various engineering applications. Several researchers have utilized the '*interFoam*' VOF solver provided by the open-source CFD software OpenFOAM® in their studies [26]. This solver offers numerous benefits, including enhanced representation of the interface, an efficient advection scheme, and the ability to handle large density ratios. These features make '*interFoam*' particularly suitable for simulating multiphase flows with complex interface dynamics.

Fujimoto et al. [27] achieved good agreement between their numerical simulations and experimental data by utilizing a three-dimensional finite volume solver with the VOF method. Their study incorporated a switching dynamic contact angle approach, similar to that used by Yokoi et al. [12]. This approach allows for more accurate modeling of contact angle hysteresis, which is crucial for simulating droplet behavior on solid surfaces. The dynamic contact angle model incorporates the contact line velocity as a critical parameter, thereby accounting for the influence of flow velocity and accurately capturing the behavior of droplets on both hydrophilic and hydrophobic surfaces. Feng [10] also employed the finite volume method with the VOF model using the '*interFoam*' solver. Their results were comparable to empirical correlations, demonstrating the solver's capability to accurately predict fluid dynamics in multiphase systems. The study highlighted the solver's robustness in handling complex flow phenomena, such as droplet impact and spreading.

The present work focuses on analyzing the behavior of droplets upon impact and spreading using the '*interFoam*' solver provided by OpenFOAM®. The main objective is to investigate droplets' spreading and recoiling behavior, particularly in the context of additive manufacturing, and to predict the final spread diameter. Such analysis involves detailed numerical simulations to capture the intricate dynamics of droplet deformation, spreading, and recoiling, which are critical for optimizing processes in additive manufacturing.

By leveraging the advanced capabilities of the '*interFoam*' solver, this study aims to provide deeper insights into the fluid mechanics of droplet interactions with surfaces. The findings are expected to contribute to developing more efficient and precise additive manufacturing techniques, where control over droplet behavior is essential for achieving high-quality prints.

2.2. Computational Domain

The computational domain and associated boundary conditions are depicted in Figure 1. The domain consists of a rectangular block representing the air phase, while the dark blue region indicates the water droplet. The bottom wall of the domain represents a solid substrate with 'no slip' boundary conditions, ensuring that the fluid velocity at the wall is zero.

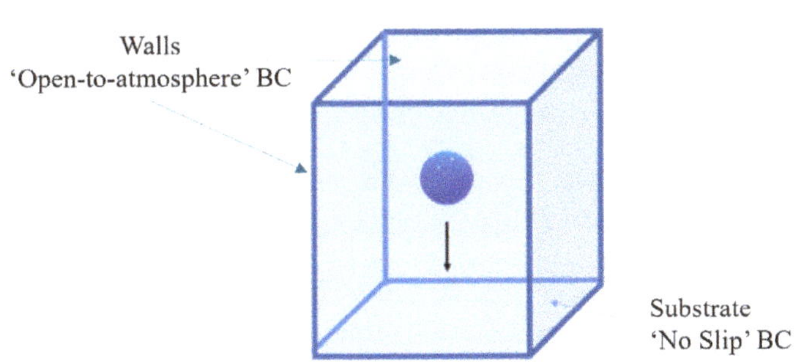

Figure 1. Schematic showing the computational domain and associated boundary conditions for a droplet impinging on a solid surface.

Constant and dynamic contact angle boundary conditions are applied at the bottom wall to accurately capture the fluid–solid interaction. The dynamic contact angle model from the OpenFOAM® library is utilized, which is described by the following equation:

$$\theta = \theta_e + (\theta_a - \theta_r)\tanh\left(\frac{U_w}{U_\theta}\right) \tag{7}$$

where θ_e, θ_a, and θ_r represent the equilibrium, advancing, and receding contact angles, respectively, U_w denotes the normal velocity of the interface near the wall. and U_θ is the characteristic velocity scale. Feng [10] validated with the experimental data that the model successfully predicts the contact angle evolution on various surfaces, including oscillations and other dynamic behaviors. The current study follows the empirical relation reported by Feng [10] with a simplified approach by specifying a static contact angle (θ_0) at the contact line with $\theta_a = \theta_0 + 5^\circ$, $\theta_r = \theta_0 - 5^\circ$, and $U_w = 1\ m/s$.

The top wall and four vertical walls are treated as open to the atmosphere faces with the '*PressureInletOutletVelocity*' boundary conditions for velocities and pressure, as specified by the OpenFOAM CFD solver [26]. The pressure settings are maintained at constant atmospheric pressure, ensuring that the pressure at the inlets/outlets remains stable and does not fluctuate. The velocity settings are zero uniform velocities, meaning there is no net flow through these boundaries. This setup allows for handling the open-to-atmosphere boundary conditions, allowing flow to freely enter and exit the domain.

The rectangular domain is of size 10 × 10 × 10 mm, which is divided into 125 × 125 × 75 finite volume cells after a convergence study is conducted for the present numerical scheme. A domain independence study is conducted to ensure that the computational results are unaffected by the dimensions of the computational domain. This study evaluated multiple domain sizes and systematically compared outcomes to verify that the selected dimensions yield accurate and stable solutions. Given that the initial droplet diameter ranged from 2.5 to 3.5 mm, we assessed various domain sizes to capture critical aspects of droplet dynamics, including maximum diameter spread and droplet velocity. The grid independence study confirmed that a domain size of 10 × 10 × 10 mm is adequate to encapsulate the fundamental dynamics of droplet behavior, ensuring reliable and precise computational results. A comprehensive grid independence study was conducted to ensure the accuracy and reliability of the simulation results, involving the evaluation of computational performance and solution convergence across multiple grid resolutions. A detailed summary is provided in Appendix A. The governing equations and required numerical methods are well described in Section 2.1.

3. Results

The computational method is initially validated by comparing the results with previously reported experimental studies by Kim and Chun [28], which investigated the interaction of water droplets with polycarbonate surfaces. Their study provided a comprehensive analysis of droplet dynamics, including spreading and recoiling behaviors, using high-speed video systems to capture the intricate details of the interactions.

We have conducted a validation study and performed a parametric analysis for a more comprehensive understanding of droplet dynamics by incorporating additional relevant non-dimensional numbers, such as the Reynolds (Re), Weber (We), and Ohnesorge (Oh) numbers, and their effects on droplet dynamics. These dimensionless parameters are crucial for characterizing the fluid dynamics of droplet impact and spreading, providing valuable insights into the underlying mechanism. By incorporating these non-dimensional numbers, the study aims to provide a more detailed and accurate prediction of droplet dynamics, particularly in additive manufacturing. The analysis focuses on droplets' spreading and recoiling behavior, aiming to predict the final spread diameter and understand the influence of various fluid properties and impact conditions.

In Section 3.1, we present the validation studies for the droplet impact on a hydrophilic and hydrophobic surface, implicating the efficacy of the model to capture the complex physics of the droplet impact and spreading.

3.1. Flow and Spreading Behavior Method Validation: Impact of a Water Droplet on a Hydrophilic Surface

3.1.1. Method Validation: Impact of a Water Droplet on a Hydrophilic Surface

The numerical results were validated by comparing them with the experimental study reported by Kim and Chun [28]. In their study, a water droplet with a diameter of 3.6 mm was made to impact a polycarbonate surface at an initial velocity of 0.77 m/s. The experimental setup consisted of a solid surface onto which liquid droplets were projected using an inkjet printer at varying velocities. The recoiling behavior of the droplets upon impact was recorded using high-speed cameras and subsequently analyzed using advanced image processing techniques. Detailed descriptions of the experimental procedures can be found in the work published by Kim and Chun [28].

The equilibrium contact angle between the water and polycarbonate surface was measured to be $\theta_0 = 87.4^\circ$. The dynamic contact angles, including the advancing and receding angles, were specified as $\theta_0 + 5^\circ$ and $\theta_0 - 5^\circ$. These contact angles are critical parameters for accurately modeling the fluid–solid interaction and predicting the droplet dynamics.

The results of the droplet shape evolution over time were compared with the experimental observations, as illustrated in Figure 2. This comparison demonstrated good agreement between the numerical simulations and the experimental data, validating the accuracy of the computational model. The results highlight the importance of accurately capturing the contact angle dynamics to predict the spreading and recoiling behavior of droplets upon impact.

To further validate the computational results, Figure 3 presents the superimposed images of the simulated droplet shape extracted at the same time intervals as the experimental images. This overlay allows for a direct visual comparison, highlighting discrepancies or agreements between the simulation and experimental results. The validation study demonstrated a strong agreement between the present numerical simulations and the experimental data reported by Kim and Chun [28]. The superimposed droplet shapes from the simulations closely matched the experimental images at various time steps, indicating the accuracy and fidelity of the computational model. These findings validate the effectiveness of the '*interFoam*' solver in capturing the complex dynamics of droplet impact and spreading, providing a robust foundation for further investigations in this domain.

As illustrated in Figures 2 and 3, the simulation accurately captures both the spreading and recoiling phases during the shape evolution of a water droplet impinging on a polycarbonate surface. The spreading phase is characterized by a rapid expansion, reaching the maximum spread diameter at 11.6 ms, and is marked by the formation of a rim at the furthest point from the droplet's center. Conversely, the recoiling phase is slower and exhibits an oscillatory behavior in the spread diameter before stabilizing at equilibrium at 39.8 ms.

This study effectively captures the droplet's detailed shape evolution and dynamic behavior throughout the impact process. The qualitative analysis of the shape evolution is complemented by a quantitative analysis of the droplet's spread diameter over time. For the quantitative analysis, we use normalized base and normalized time as defined below:

Normalized base: $D_b^* = \frac{D_b}{D_0}$

Normalized time: $t^* = \frac{t\sqrt{\sigma}}{\sqrt{\rho D_0^3}}$

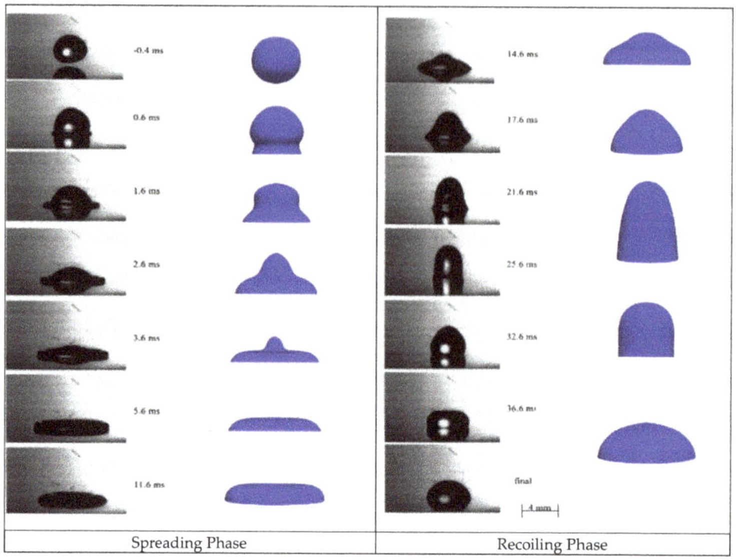

Figure 2. Comparison of experimental and numerical results for a water droplet impinging on a polycarbonate surface. Experimental results are shown in black-and-white images [28], while the numerical results are blue, representing the evolution of the water droplet shape. Experimental results are reproduced with permission from [28]. Copyright 2001 AIP Publishing LLC.

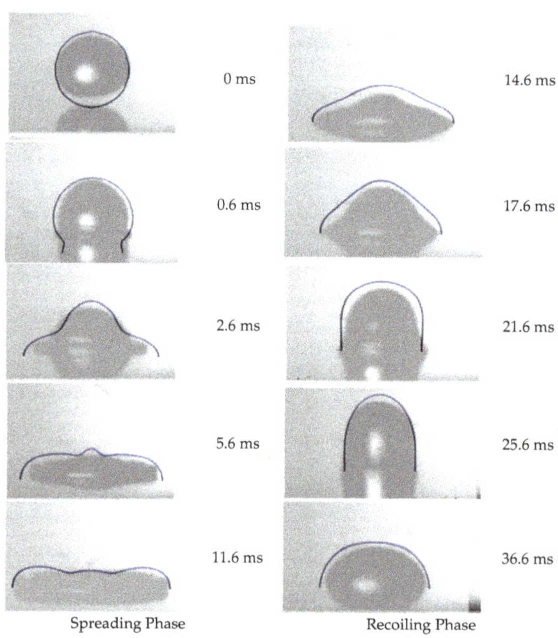

Figure 3. Comparison of shape evolution of the droplet: The surface of the droplet captured in the numerical study is compared with experimental results. Experimental results are shown in gray area images [28], while the numerical results are blue contours representing the water droplet shape evolution. Experimental results are reproduced with permission from [28]. Copyright 2001 AIP Publishing LLC.

The normalized base is plotted against the normalized time to further elucidate the dynamics. This plot provides insights into the orders of magnitude changes in the spreading diameter over time, highlighting the intricate interplay between inertial, viscous, and surface tension forces during the droplet's impact and subsequent evolution. Figure 4 below shows the variation in the normalized base with normalized time for constant and dynamic contact angle models for comparison with the experimental results reported by Kim and Chun [28].

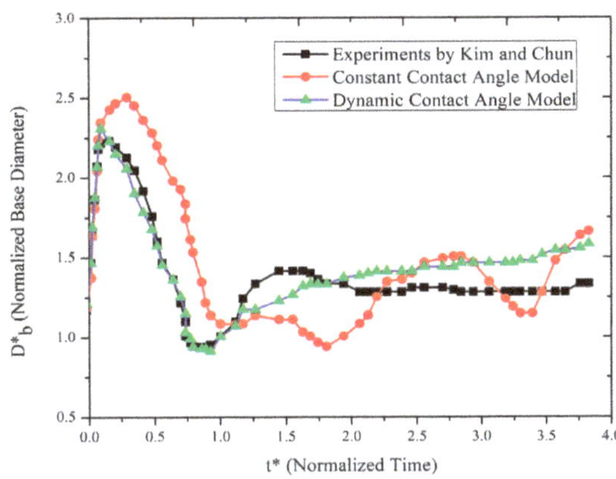

Figure 4. Comparison of experimental studies [25] and numerical results for a water droplet impinging on a polycarbonate surface. Quantities with superscript * denote the normalized quantities, as defined earlier.

The constant contact angle model provides a reasonable prediction of the initial spreading behavior of the droplet; however, it tends to overpredict the recoiling behavior. In contrast, the dynamic contact angle model better captures the initial spreading and recoiling behaviors. Additionally, it more accurately represents the post-recoiling oscillations compared to the constant contact angle model.

The computational model, while slightly overpredicting the equilibrium base diameter, as observed from the qualitative shape evolution analysis, effectively captures the overall dynamics of the droplet. The predicted dynamics include the intricate details of the spreading and recoiling phases. The dynamic contact angle model's ability to better capture the transient behaviors and oscillations highlights its robustness and accuracy in simulating droplet dynamics.

Overall, the computational results provide a comprehensive and effective prediction of the spreading and recoiling behavior of a water droplet impinging on a polycarbonate surface. This validation underscores the model's capability to accurately simulate complex fluid–solid interactions, which is crucial for applications in various engineering and scientific studies.

3.1.2. Method Validation: Impact of a Water Droplet on a Hydrophobic Surface

After examining the droplet dynamics on a hydrophilic surface in the previous section, this study extends the analysis to a hydrophobic surface. The computational model is tested for a scenario where a water droplet impacts a hydrophobic surface. Azimi et al. [29] investigated the nanostructured modification of material surfaces and the induced hydrophobicity of rare-earth oxide ceramics. They fabricated ceramic nanostructures, consisting of a thin layer on cubical silicon micro-posts, which exhibited significant hydrophobicity with contact angles of $160 \pm 2^{\circ}$, as illustrated in Figure 5.

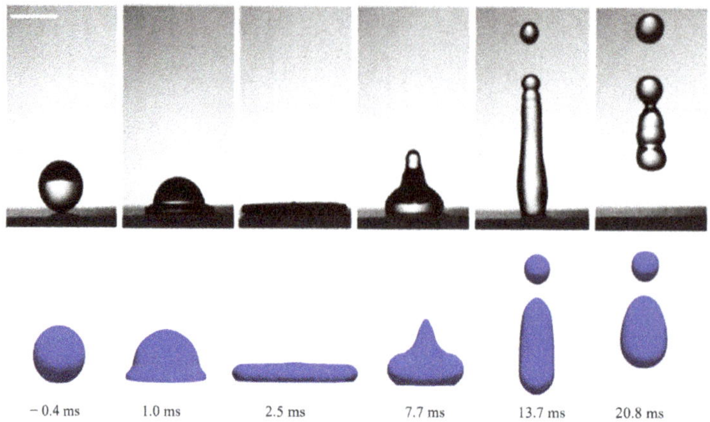

Figure 5. Comparison of experiments and the computational results. (**Top**) High-speed camera images of a water droplet impinging on the nanograss-covered micro-posts sputtered with ceria (impact velocity ~1.6 m s^{-1}); scale bar, 2.5 mm [29]. (**Bottom**) Shape evolution of the water droplet emulating the experimental study using a multiphase CFD solver. Experimental results are reproduced with permission from Azimi et al. [29]. Copyright 2013 Nature Publishing Group.

In the computational domain, a water droplet with a diameter of 2.5 mm is made to fall on the hydrophobic surface with an initial velocity of 1.6 m/s. The bottom wall is modeled as a hydrophobic surface with a contact angle of 160°, while the other boundary conditions remain consistent with those described in Section 2.1. Due to the hydrophobic nature of the nanostructured surface, the water droplet bounces off upon impact. The computational model accurately captures this behavior, as shown in Figure 5.

The study demonstrates the model's capability to simulate the complex interactions between droplets and hydrophobic surfaces. The results provide valuable insights into the influence of surface properties on droplet dynamics, which are crucial for applications in surface engineering, coating technologies, and fluid dynamics research.

The comparison of the base diameter variation with time between the experimental data and the simulations is illustrated in Figure 6. The results demonstrate a high degree of correlation, with the simulations accurately capturing the temporal evolution of the base diameter, thereby validating the computational model's effectiveness in predicting droplet dynamics on hydrophobic surfaces.

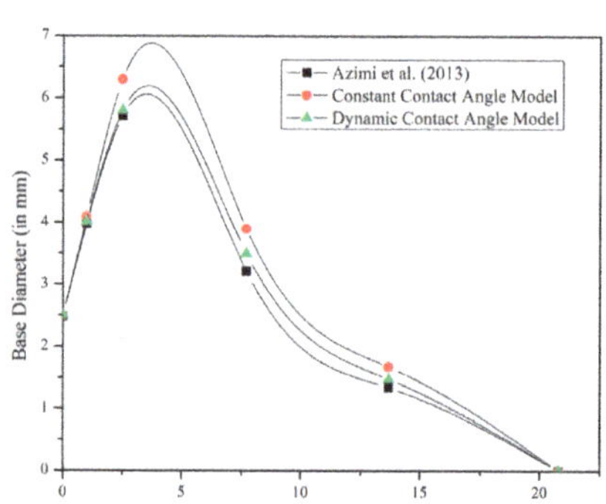

Figure 6. Comparison of experimental data reported by Azimi et al. [29] for a water droplet impinging on a hydrophobic surface and the computational results using a multiphase CFD solver for the same case study.

Based on the experimental data reported by Azimi et al. [29], the base diameter of the droplet increases post-impact, reaching a maximum of 5.71 mm at 2.48 ms before decreasing as the droplet bounces off the surface at 20.8 ms. The experiments did not document or report the subsequent rebound and oscillatory behavior stages. However, the computational model replicates the observed trend, with the dynamic contact angle model providing more accurate results than the constant contact angle model. The constant contact angle model tends to overpredict the base diameter. In contrast, the dynamic contact angle model offers better predictions due to its more sophisticated approximation of the contact line velocity at the solid–liquid interface.

Overall, the multiphase CFD solver effectively simulates a water droplet's flow and spreading behavior impinging on a polycarbonate surface. These findings are crucial for understanding the variation in base diameter across different fluid–solid interactions and for conducting parametric studies to optimize process parameters. The subsequent section will explore the effects of non-dimensional numbers, such as Reynolds, Weber, and Ohnesorge numbers, to investigate the interplay of various forces associated with this phenomenon.

3.2. Effect of Non-Dimensional Reynolds (Re), Weber (We), and Ohnesorge (Oh) Numbers

The numerical study has examined droplet dynamics in relation to spreading and recoiling behavior and fluid–solid interactions (contact angle). The computational models have demonstrated good agreement with experimental results. This section investigates and discusses the effect of relevant non-dimensional numbers on droplet dynamics. These dimensionless numbers are crucial in fluid dynamics as they elucidate the interplay of different forces associated with flow phenomena.

Key non-dimensional numbers pertinent to droplet dynamics include the Reynolds (Re), Weber (We), and Ohnesorge (Oh) numbers. These numbers are defined as ratios of different forces and are dimensionless. The Reynolds number (Re) is the ratio of inertial to viscous forces and is essential for understanding flow profiles and fall velocity in droplet dynamics. The Weber number (We) represents the ratio of inertial to surface tension forces and characterizes the droplet shape influenced by inertia (impact velocity). The Ohnesorge number (Oh) is the ratio of viscous forces to inertial and surface tension forces. The definitions of Re, We, and Oh numbers are provided below:

$$Re = \frac{inertia}{viscous\ force} = \frac{\rho u D}{\mu}$$

$$We = \frac{inertia}{surface\ tension\ force} = \frac{\rho u^2 D}{\sigma}$$

$$Oh = \frac{viscous\ force}{\sqrt{inertial.surface\ tension\ force}} = \frac{\mu}{\rho \sigma D}$$

where ρ is the fluid density, U is the characteristic velocity, L is the characteristic length, μ is the dynamic viscosity, and σ is the surface tension. These dimensionless numbers play a pivotal role in characterizing the droplet dynamics and understanding the influence of various forces on the behavior of droplets upon impact.

The case study presented in Section 3.1 analyzes droplet dynamics for different Reynolds numbers (Re) ranging from 1400 to 3000. The focus is primarily on the initial spreading and recoiling behavior, influenced by the interplay of inertial and viscous forces. As depicted in Figure 7, the initial spreading is governed by the Reynolds number, with the spreading increasing as the Reynolds number rises. The peak spread diameter increases with higher Reynolds numbers, indicating that the base diameter spread is directly proportional to the Reynolds number. This behavior is attributed to the increased inertial forces and the more significant impact of the collision force associated with higher Reynolds numbers. However, it is important to note that further increases in the Reynolds number can lead to splash generation, which is undesirable for printing applications.

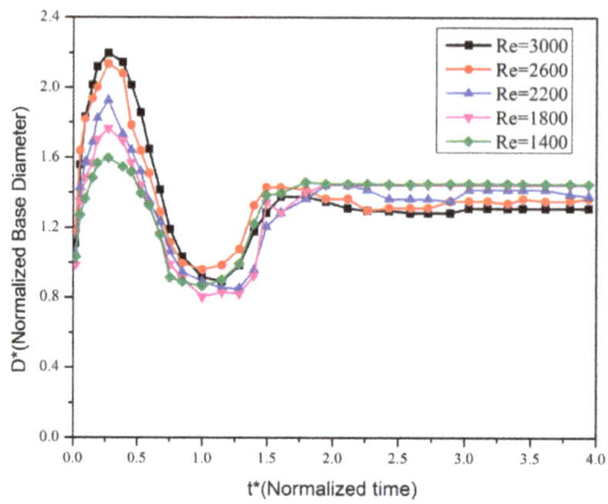

Figure 7. Droplet evolution (base diameter) comparison for different Reynolds numbers. Quantities with superscript * denote the normalized quantities, as defined earlier.

Interestingly, there is minimal variation in the equilibrium base diameter across different Reynolds numbers, with convergence to a similar equilibrium base diameter despite the varying peaks. These results are consistent with the literature and experimental data, confirming the Reynolds number's significant role in dictating the droplet's initial spread. The recoiling and equilibrium stages do not exhibit significant variations with changes in the Reynolds number.

We subsequently study the effect of the Weber number (We) and Ohnesorge number (Oh) on droplet dynamics to understand the interplay of inertial, surface tension, and viscous forces, as shown in Figure 8.

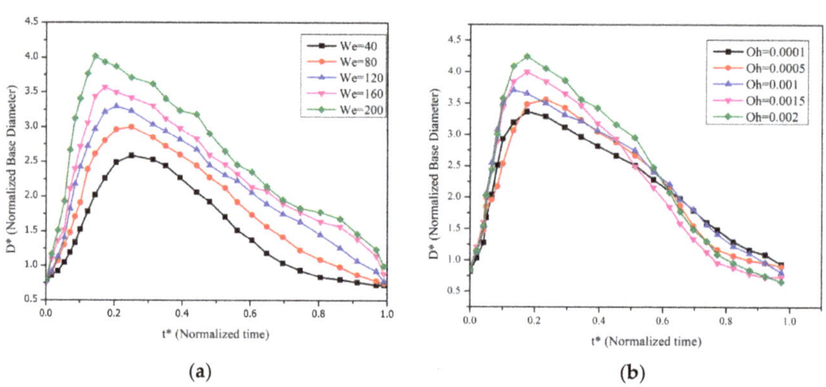

Figure 8. Droplet evolution (base diameter) comparison for different (**a**) Weber and (**b**) Ohnesorge numbers. Quantities with superscript * denote the normalized quantities, as defined earlier.

The Weber number (We), which ranges from 40 to 200, is examined for its impact on the droplet's initial spreading and recoiling phases. The Weber number, the ratio of inertial to surface tension forces, is examined primarily for its impact on the initial spreading and recoiling phases. The Weber number primarily influences the initial spreading phase.

As shown in Figure 8a, the normalized base diameter is plotted against the normalized time for different Weber numbers. The peak spread diameter increases as the Weber number rises due to the increased inertial effect. However, the recoiling phase remains relatively unaffected, with recoiling time almost constant across different Weber numbers. Further investigation is required to fully understand the role of the Weber number in determining recoiling behavior. The Weber number is particularly significant in dictating droplet generation at microscales, where the interplay of inertial and surface tension forces is critical.

The Ohnesorge number (Oh) is used to predict the nature of the recoiling phase and analyze the spreading and recoiling behavior of droplet impingement on a solid surface. The Ohnesorge number (Oh) is defined as the ratio of viscous forces to inertial and surface tension forces, and the Ohnesorge number quantifies the relative importance of these forces for a given inertia. Figure 8b compares the base diameter evolution for different Ohnesorge numbers, with all cases having the same inertia, to distinctly compare the effects of viscous and surface tension forces.

Figure 8b shows that the spreading phase exhibits minimal variation across different Ohnesorge numbers, indicating that the Ohnesorge number does not significantly influence the initial spreading when the inertial force is constant. However, the recoiling phase is shorter for lower Ohnesorge numbers as the viscous and surface tension forces directly scale in this phase. The Ohnesorge number determines the resistance force to the recoiling behavior in the droplet impingement phenomenon.

One potential explanation for the counterintuitive behavior observed is the relationship between the Ohnesorge number and surface tension. A higher Oh value indicates stronger viscous forces relative to surface tension forces. In droplet spreading, the dominant forces are viscous and surface tension forces. A higher Ohnesorge number implies that viscous forces play a more significant role, potentially weakening the influence of surface tension in restraining the droplet's spreading. Additionally, a higher Oh value may correspond to a decrease in the capillary length scale, further diminishing the importance of surface tension relative to viscous forces and promoting spreading.

It is crucial to note that the interaction between the Ohnesorge number and surface tension is complex and can be influenced by factors such as the Weber number. The combined influence of the Ohnesorge and Weber numbers on droplet behavior warrants comprehensive analysis to accurately elucidate the underlying physics. Further investigation and detailed analysis are necessary to understand why a higher *Oh* value, indicative of more dominant viscous forces, enhances droplet spreading rather than inhibiting it. Exploring the interplay between the Ohnesorge number, Weber number, and surface tension can elucidate the intricate dynamics governing droplet behavior in various scenarios.

In summary, these non-dimensional numbers are significant in determining the interplay of forces associated with droplet dynamics, primarily inertia, surface tension, and viscous forces. The Reynolds and Weber numbers directly affect the spreading phase due to the dominance of inertial forces, while the Ohnesorge number governs the recoiling phase.

3.3. Fluid–Fluid Interaction (Droplet Coalescence) and Prediction of Final Print Resolution in Droplet-Based Printing

During continuous printing, the print nozzle traverses in a direction dictated by the print speed, depositing subsequent droplets accordingly. The print setup and traverse speed determine the precise positioning of the droplets on the substrate. It is crucial to investigate the dynamics of droplets during printing, particularly their interactions with the substrate and other droplets. The interactions between droplets and the dynamics of coalescence are pivotal in determining the final print resolution in continuous droplet-based printing.

The coalescence mechanism, which encompasses rupture, deformation, and interface merging of droplets, warrants further investigation within the context of droplet-based printing. Extensive studies have been conducted on the dynamics of coalescence, focusing on droplet–droplet interactions [30–32], interactions with solid surfaces [33–36], and interactions with liquid films [32,35,37,38]. Various computational approaches have been employed to capture the interfaces during coalescence, including the solution of Navier–Stokes equations and computational fluid dynamics (CFDs) techniques [39–41], the lattice Boltzmann method, and molecular dynamics (MDs) simulations [42–45].

The present work extends these computational approaches by utilizing the volume of fluid (VOF) method to solve flow equations and capture the merging of interfaces driven by surface tension forces. The methodology involves the addition of a droplet, with its center location determined by the print speed or traverse speed. Initially, a sessile droplet is deposited and allowed to impinge upon the substrate, followed by the deposition of a satellite droplet, whose position is also governed by the print speed.

A schematic diagram illustrating the positions of the droplets and the center-to-center distance of the subsequent droplet is presented in Figure 9 below.

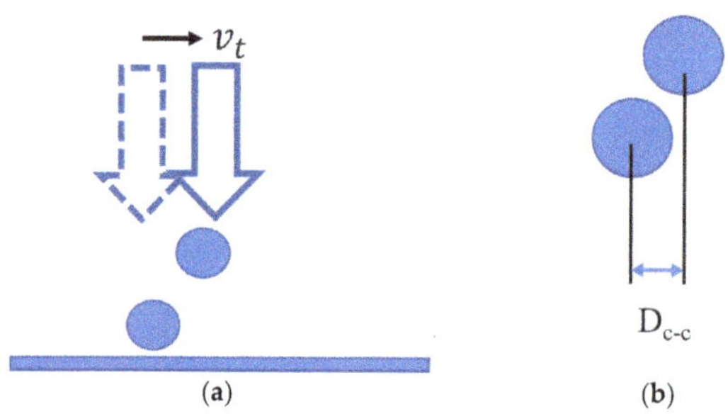

Figure 9. Sketching of the in-print conditions. (**a**) The sessile droplet position with print speed v_t. (**b**) Center-to-center distance for two consecutive impinged droplets.

3.3.1. Flow and Spreading Behavior of Two Coalescing Droplets

Extending the current studies, the computational model rigorously simulates the dynamics and coalescence of two water droplets impinging upon a polycarbonate substrate to gain insights into the flow and spreading behavior. The computational domain and boundary conditions employed are consistent with those listed in Section 3.

The traverse speed and the corresponding lateral migration of the print nozzle along the X-axis govern the positional determination of the subsequent droplet. Observations indicate that the impinging droplets exhibit characteristic spreading and recoiling phases analogous to those observed in the case of a solitary uniform droplet. However, the offset distance between the two droplets notably influences the overall dynamics.

Figure 10 provides a comprehensive overview of the overall spreading and coalescence mechanisms within the two-dimensional computational domain, highlighting the intricate interplay of forces and the resultant flow patterns and morphological changes.

Although the droplet dynamics adhere to the aforementioned spreading and recoiling phases, the overall behavior exhibits notable asymmetry due to the addition of a secondary droplet. The subsequent impingement of the satellite droplet significantly influences the initial spreading of the sessile droplet. This collision introduces a higher degree of dynamism to the flow, with initial observations indicating air entrapment at the interface. However, as the flow progresses, the entrapped air dissipates, culminating in the formation of a crown which represents the apex of phase accumulation.

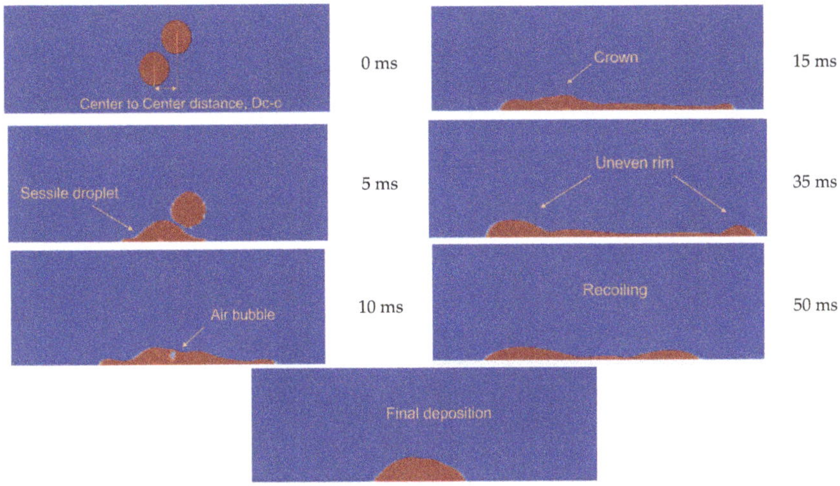

Figure 10. Selected frames from the simulation depicting salient features for the coalescence of two droplets.

The crown formation predominantly occurs at the first droplet's impingement site, indicating a localized accumulation of phases at the collision point. This phase is followed by the formation of an unequal rim at the conclusion of the spreading phase. The recoiling behavior is characterized by the inward movement of this unequal rim towards the center. Notably, the recoiling phase is prolonged compared to a solitary uniform droplet, attributable to the flow's increased complexity and dynamic nature.

The system exhibits multiple oscillations during the subsequent spreading and recoiling phases, resulting in a longer final deposition time than a single droplet. The present study highlights the intricate spreading and recoiling phases and the asymmetrical flow behavior of coalescing droplets. These findings underscore the necessity for extending the analysis to a three-dimensional framework to fully capture the complexities of the phenomenon.

3.3.2. Prediction of the Flow Dynamics with Relevance to Droplet-Based Printing

The computational model is extended to a three-dimensional domain, maintaining the same boundary conditions as those utilized in the previous section. A more viscous fluid (similar to printing ink) is considered to study the flow behavior and predict the final print resolution, emulating the inkjet printing process. A droplet with a diameter of 3.6 mm is considered to impinge on a polycarbonate surface with an initial velocity of 0.68 m/s. The fluid properties include a viscosity of 4.98×10^{-5} Pa.s and a surface tension of 0.0798 N/m. The equilibrium contact angle between the printing ink and the substrate is set at 70° to mimic a hydrophilic surface.

The flow behavior is analyzed through the evolution of the droplets during coalescence, as depicted in Figure 11. The three-dimensional analysis reveals that the droplet follows a single uniform droplet's spreading and recoiling behavior. However, the higher viscosity of the fluid resists the spreading and recoiling phases, underscoring the significant role of viscosity in droplet dynamics. The flow behavior is further examined using velocity streamlines, which elucidate the material flow patterns during droplet impact and coalescence.

As shown in Figure 11, the initial droplet begins to spread upon impact, and subsequently, the satellite droplet impinges on the sessile droplet, altering the overall dynamics. The coalescence process demonstrates the merging of the two droplets, with the interfaces disappearing as mixing occurs. The spreading phase is prominent, while the recoiling phase is less dominant due to the higher viscosity of the fluid. The final print resolution is determined by the coalescence and merging of the interfaces of these two droplets.

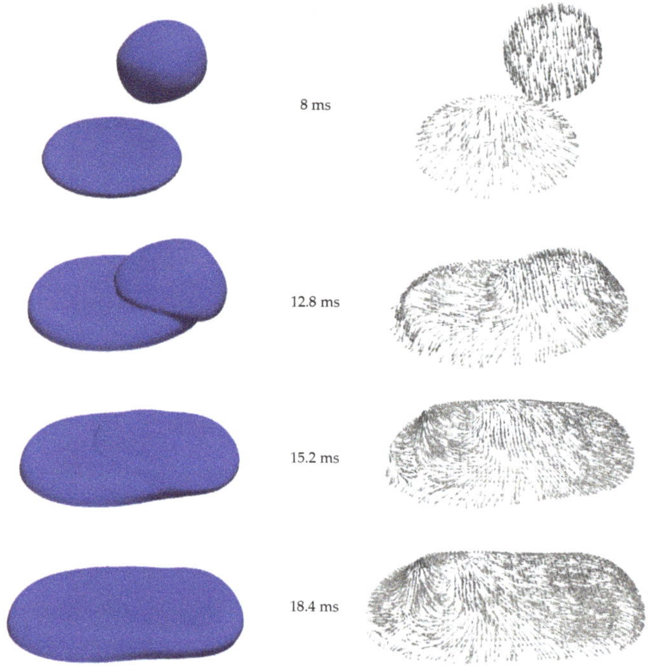

Figure 11. The flow behavior of two droplets during impact and coalescence, along with corresponding velocity streamlines.

The velocity streamlines provide insights into the flow patterns of the print material throughout the spreading and recoiling phases. Initially, the velocity streamlines are directed outwards for the sessile droplet and downwards for the satellite droplet, as observed at 8 ms in Figure 11. Upon the satellite droplet's impact on the sessile droplet, the flow pattern remains consistent instantaneously. However, as the sessile droplet continues to spread, the flow patterns for the satellite droplet exhibit both outward and downward directions. This behavior is followed by a strong mixing of the velocity streamlines at the interface where the two droplets merge. This information is crucial for determining the material flow in different regions, aiding in the prediction of the surface characteristics of the final printed part. The flow pattern reverses during the recoiling phase, although it is less dominant than the spreading phase. The final print resolution and surface finish result from this dynamic flow pattern of the printing material.

Furthermore, the effect of print speed on the print resolution is analyzed at various print speeds. The print speed directly influences the center-to-center distance between two droplets. Different print speeds ranging from 0.5 m/s to 5 m/s were modeled, representing varying center-to-center distances between the droplets, as illustrated in Figure 12. This analysis is crucial for predicting the final resolution and surface characteristics at different printing speeds.

Three distinct cases are examined, ranging from drop-on-drop deposition to higher print speeds. The comparison of the final resolution for these cases is presented in Figure 12. This comparative analysis provides valuable insights into how print speed affects the coalescence dynamics, droplet interaction, and, ultimately, the surface finish of the printed part.

This study aims to optimize the printing process to achieve desired resolutions and surface qualities by understanding the relationship between print speed and droplet spacing. The findings highlight the importance of precise control over print speed to ensure high-quality outcomes in droplet-based printing applications.

The print resolution increases with the increase in print speed, which has significant implications for manufacturing time and surface finish. Higher printing speeds enable faster manufacturing processes, thereby reducing production times. However, this acceleration often comes at the expense of print quality. Extreme print speeds can lead to undesirable outcomes such as puddle formation, poor surface finish, and wastage of print material.

Therefore, it is essential to strike a balance between printing speed and achievable print resolution. The parametric study conducted in this research helps identify the optimal printing speeds that yield the desired print resolutions while maintaining acceptable surface quality. By analyzing different print speeds and their effects on droplet coalescence and interaction, this study provides valuable insights into optimizing the printing process for various applications.

The findings underscore the importance of carefully selecting print speeds to ensure high-quality outcomes in droplet-based printing. This balance is crucial for achieving efficient manufacturing processes without compromising the integrity and precision of the printed structures.

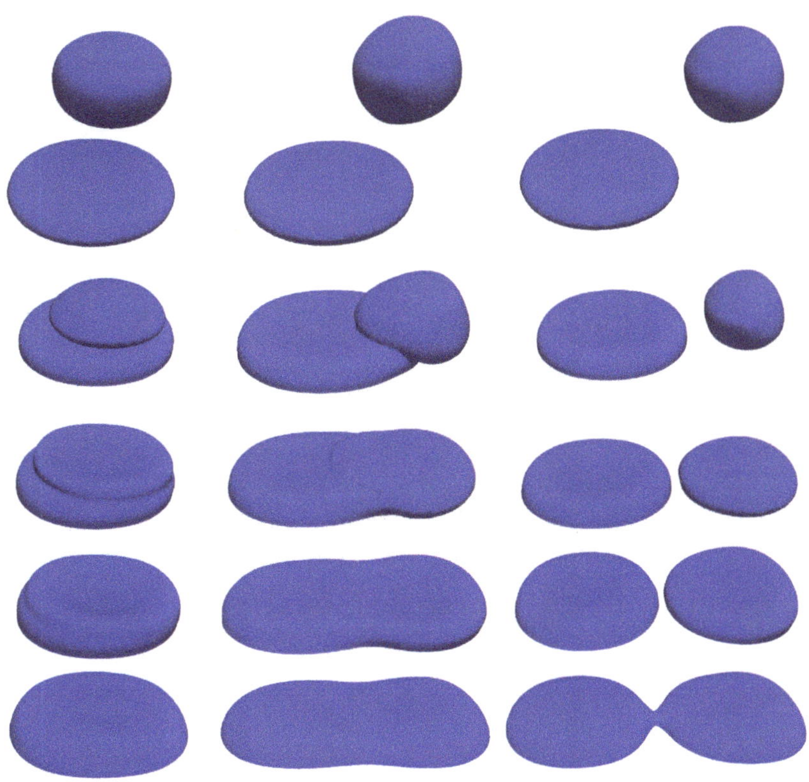

Figure 12. Comparison of droplet coalescence and shape evolution for different print speeds (0.5 m/s, 3 m/s, and 5 m/s from left to right), snapshots taken at 2.4 ms, 8 ms, 15.2 ms, 18.2 ms, and 40 ms (from top to bottom).

4. Discussion

The numerical model extends the qualitative analysis to a three-dimensional study. This quantitative study agrees well with the experimental studies conducted by Kim and Chum [28] for hydrophilic surfaces and by Azimi et al. [29] for hydrophobic surfaces. Notably, the computational model captures the drop evolution with impingement on nanostructured hydrophobic surfaces, as reported by Azimi et al. [29]. Additionally, an in-depth analysis of dimensionless numbers (Re, We, and Oh) quantifies the role of different forces associated with printing. This non-dimensional number-based analysis sheds light on the mechanics of droplet dynamics and the printability of materials specific to additive manufacturing applications.

The numerical model and modeling framework agree with the experimental data by following the shape evolution of a droplet on a polycarbonate surface and capturing the mechanisms in droplet-based printing. The present work illustrated different stages of spreading, recoiling, lamella formation, and wetting for droplet dynamics in printing, as reported in other experimental and numerical studies.

A three-dimensional model provided an effective modeling framework and improved the quantitative analysis of droplet-based printing in additive manufacturing. The numerical model extended the qualitative analysis reported by Shah and Mohan [15] to a three-dimensional study, showing good agreement with the experimental studies. The computational model even captured the various stages of the interaction of droplets with the polycarbonate surface in agreement with the experimental study. The accuracy of the VOF method heavily relies on the resolution of the computational grid. When dealing with small droplets or complex geometries, capturing the precise shape and position of the droplet interface becomes challenging. Insufficient grid resolution may result in inaccuracies, especially near the contact line between the droplet and the solid substrate.

Furthermore, the study predicted the final printing resolution depending on the process parameters and material properties. In addition, a dimensionless analysis of the effect of characteristic numbers (Re, We, and Oh) revealed the role of different forces associated with printing. The inertial forces mainly govern the initial spreading phenomenon, while the recoiling is characterized by the interplay of inertia, viscous, and surface tension forces. This study gives an insight into the mechanics of droplet dynamics, in general, and the printability of materials specific to additive manufacturing applications.

The numerical method captures the flow and spreading behavior of two coalescing droplets and shows the asymmetric flow pattern with the crown formation and uneven rim around the center. The flow behavior and the velocity streamlines reveal the intermixing of the material during the printing process. Higher printing speeds lead to less manufacturing time, but with poor resolution and surface finish. The parametric studies are conducted to achieve the desired printing resolution with optimal printing speeds. This information is much needed and helps decipher the conditions for optimizing the printing process. Furthermore, the study helps predict the final printing resolution depending on the process parameters and material properties.

Future studies could consider understanding the spreading behavior with multiple droplets, incorporating the material phase change and solidification, and the surface variation effects when the materials are printed on the subsequent layers. The inclusion of the phase-change model into the present work can extend this study to analyze the photopolymerization of polymers in 3D printing and its effect on print surfaces and overall print quality. The effect of temperatures can be studied further, and their implications on the final printing quality can be examined. Many industrial materials like polymers, bio-inks, and other commercial printing inks can also be analyzed with parametric studies for better print quality and fidelity.

The current modeling framework with complex computational models, phenomenological models for surface characteristics effects, and higher computing times is required for such analyses.

5. Conclusions

The present investigations on multiphase modeling of flow and spreading behavior in droplet-based printing yield several key insights. The numerical model and modeling framework demonstrate strong agreement with experimental data, accurately capturing the shape evolution of droplets on both hydrophilic and hydrophobic surfaces. The model effectively captures the mechanisms involved in droplet-based printing, including different stages of spreading, recoiling, lamella formation, and wetting, as reported in other experimental and numerical studies.

The extension to a three-dimensional model provides an effective framework for improving the quantitative analysis of droplet-based printing in additive manufacturing. The computational model successfully captures the interaction of droplets with nanostructured hydrophobic surfaces, providing essential information for optimizing the 3D printing process.

The study predicts the final printing resolution based on process parameters and material properties. A dimensionless analysis of characteristic numbers (Reynolds number, Weber number, and Ohnesorge number) reveals the roles of different forces associated with printing. Inertial forces primarily govern the initial spreading phenomenon, while the interplay of inertial, viscous, and surface tension forces characterizes the recoiling phase. This analysis provides valuable insights into the mechanics of droplet dynamics and the printability of materials specific to additive manufacturing applications.

The study of droplet coalescence in the context of 3D printing reveals asymmetric flow behavior around the center. Flow and velocity streamlines indicate the mixing and dynamics of interface merging and droplet coalescence. Higher printing speeds result in shorter manufacturing times but poorer print quality. The effect of print speeds on final resolution and surface finish is analyzed, emphasizing the need to optimize the printing process for different speeds.

While our current study provides valuable insights into droplet coalescence, a comprehensive quantitative analysis of the surface tension-driven coalescence phenomenon is necessary. The studies can be extended to perform process optimization by conducting a sensitivity analysis for various process parameters, including traverse speed, droplet size, and impinging velocities. This future scope will enhance our understanding of droplet coalescence and its implications for print quality and resolution. Future studies should investigate the effects of temperature on final printing quality. Incorporating a phase-change model into the present work could extend the analysis of the photopolymerization of polymers in 3D printing and its impact on print surfaces and overall print quality. Many industrial materials, such as polymers, bio-inks, and other commercial printing inks, can be analyzed through parametric studies to achieve better print quality and fidelity. This study can also be extended to bio-ink materials and biological cells to predict and characterize printability and print quality in bioprinting.

Future work should incorporate non-Newtonian models to simulate compound droplets and their applications in droplet-based printing. It could also focus on understanding the spreading behavior with multiple droplets, incorporating material phase change and solidification, and examining surface variation effects when materials are printed on subsequent layers.

Author Contributions: Conceptualization, R.V.M.; Methodology, R.S. and R.V.M.; Software, R.S.; Formal analysis, R.S.; Investigation, R.S.; Resources, R.V.M.; Writing—original draft, R.S.; Writing—review & editing, R.V.M. All authors have read and agreed to the published version of the manuscript.

Funding: This research received no external funding. The APC was waived by the publisher.

Data Availability Statement: The raw data supporting the conclusions of this article will be made available by the authors on request. The data are not publicly available due to Researcher Intent.

Acknowledgments: The authors acknowledge the support provided for this work at the Joint School of Nanoscience and Nanoengineering, a member of the National Nanotechnology Coordinated Infrastructure (NNCI), which is supported by the National Science Foundation. We thank the University of North Carolina at Chapel Hill and the Research Computing group for providing computational resources and support that have contributed to these research results. Computing resource support from the University of North Carolina at Charlotte is also acknowledged.

Conflicts of Interest: The authors declare no conflicts of interest.

Abbreviations

The following abbreviations are used in this manuscript:

VOF	Volume of Fluid
CFD	Computational Fluid Dynamics
CFL	Courant–Friedrichs–Lewy
PISO	Pressure Implicit with Splitting of Operators
SIMPLE	Semi-Implicit Method for Pressure-Linked Equations
MULES	Multidimensional Universal Limiter for Explicit Solution
Re	Reynolds number
We	Weber number
Oh	Ohnesorge number

Appendix A. Grid Independence Study Results

Convergence Analysis: We conducted a grid independence study using multiple grid resolutions to ensure that the results are independent of the mesh size. The study involved testing grid sizes of $50 \times 50 \times 35$, $80 \times 80 \times 45$, $100 \times 100 \times 60$, $125 \times 125 \times 75$, and $150 \times 150 \times 90$ cells. The results showed that the key parameters, such as droplet spread diameter and velocity profiles, converged with increasing grid resolution. Specifically, the differences in these parameters between the $125 \times 125 \times 75$ and $150 \times 150 \times 90$ grids were less than 1%, indicating that further refinement does not significantly affect the simulation results.

Computational Time Analysis: The parallel computations were executed on the Longleaf high-performance computing (HPC) cluster at the University of North Carolina at Chapel Hill, NC, USA. Each simulation utilizes 8 nodes with 36 processing cores to efficiently distribute the computational workload. The above-mentioned grid sizes were employed to analyze the computational time, providing insights into the scalability and performance of the simulations across different mesh resolutions. The $100 \times 100 \times 60$ grid required approximately two hours of computational time, while the $125 \times 125 \times 75$ grid required 3.5 hours, and the $150 \times 150 \times 90$ grid required 5 hours. The $125 \times 125 \times 75$ grid was chosen as the optimal resolution, balancing accuracy and computational efficiency.

Table A1. Grid Independence Study.

Mesh Resolution	Cells	Computational Time (Approx Hours)	Key Parameter (Droplet Spread Diameter in mm)
Coarse	50 × 50 × 35	1.0	6.532
Medium	80 × 80 × 45	1.5	6.010
Fine	100 × 100 × 60	2.0	5.791
Finer	125 × 125 × 75	3.5	5.767
Finest	150 × 150 × 90	5.0	5.720

References

1. Yarin, A.L. DROP IMPACT DYNAMICS: Splashing, Spreading, Receding, Bouncing. *Annu. Rev. Fluid Mech.* **2006**, *38*, 159–192. [CrossRef]
2. Elliot, T.A.; Ford, D.M. Dynamic contact angles. *J. Chem. Soc.* **1972**, *68*, 1814–1823. [CrossRef]
3. Fukai, J.; Shiiba, Y.; Yamamoto, T.; Miyatake, O.; Poulikakos, D.; Megaridis, C.M.; Zhao, Z. Wetting effects on the spreading of a liquid droplet colliding with a flat surface: Experiment and modeling. *Phys. Fluids* **1995**, *7*, 236–247. [CrossRef]
4. Asai, A.; Shioya, M.; Hirasawa, S.; Okazaki, T. Impact of an Ink Drop on Paper. *J. Imaging Sci. Technol.* **1993**, *37*, 205.
5. Bayer, I.S.; Megaridis, C.M. Contact angle dynamics in droplets impacting on flat surfaces with different wetting characteristics. *J. Fluid Mech.* **2006**, *558*, 415–449. [CrossRef]
6. Scheller, B.; Bousfield, W. Newtonian Drop Impact with a Solid Surface. *AIChE J.* **1995**, *41*, 1357–1367. [CrossRef]
7. Roisman, I. Inertia Dominated Drop Collisions. II. An Analytical Solution of the Navier–Stokes Equations for a Spreading Viscous Film. *Phys. Fluids* **2009**, *21*, 052104. [CrossRef]
8. Toivakka, M. Numerical Investigation of Droplet Impact Spreading in Spray Coating of Paper. In Proceedings of the 2003 TAPPI 8th Advanced Coating Fundamentals Symposium, Chicago, IL, USA, 8–10 May 2003.
9. Pasandideh-Fard, M.; Qiao, Y.M.; Chandra, S.; Mostaghimi, J. Capillary Effects During Droplet Impact on a Solid Surface. *Phys. Fluids* **1996**, *8*, 650–659. [CrossRef]
10. Feng, J.Q. A Computational Study of High-Speed Microdroplet Impact onto a Smooth Solid Surface. *J. Appl. Fluid Mech.* **2017**, *10*, 243–256. [CrossRef]
11. Adeniyi, A.A.; Morvan, H.P.; Simmons, K.A. A coupled Euler-Lagrange CFD modeling of droplets-to-film. *Aeronaut. J.* **2017**, *121*, 1897–1918. [CrossRef]
12. Yokoi, K.; Vadillo, D.; Hinch, J.; Hutchings, I. Numerical Studies of the Influence of the Dynamic Contact Angle on a Droplet Impacting on a Dry Surface. *Phys. Fluids* **2009**, *21*, 072102. [CrossRef]
13. Gujjula, S. Numerical Investigation of Droplet Spread: Effect of Contact Angle Models. Master's Thesis, University of Cincinnati, Cincinnati, OH, USA, 2015.
14. Wang, L.S.; Krull, B.; Lucas, D.; Meller, R.; Schlegel, F.; Tekavcic, M.; Xu, J.Y. Simulation of droplet entrainment in annular flow with a morphology adaptive multifield two-fluid model. *Phys. Fluids* **2023**, *35*, 103312. [CrossRef]
15. Shah, R.G.; Mohan, R.V. Multiphase Modeling of Droplet-Based 3D Printing: Predicting Printability, Resolution and Shape Fidelity in Additive Manufacturing Processes. In Proceedings of the ASME 2023 International Mechanical Engineering Congress and Exposition, New Orleans, LA, USA, 29 October–2 November 2023.
16. Vontas, K. Enhanced VOF-based CFD simulations of droplet impact. Bachelor's Thesis, Piraeus University of Applied Sciences, University of Brighton, Brighton, UK, 2016.
17. Teng, T.; Joralmon, D.; Anyigbo, T.; Li, X. Acoustic Levitation-Assisted Contactless Printing of Microdroplets for Biomedical Applications. *J. Manuf. Sci. Eng.* **2024**, *146*, 011002. [CrossRef]
18. Wu, D.; Xu, C. Predictive Modeling of Droplet Formation Processes in Inkjet-Based Bioprinting. *Trans. ASME.* **2018**, *140*, 101007. [CrossRef]
19. Aladese, A.; Jeong, H. Recent Developments in 3D Printing of Droplet-Based Microfluidics. *BioChip J.* **2021**, *15*, 313–333. [CrossRef]
20. Britel, A.; Tomagra, G.; Apra, P.; Varzi, V.; Sturari, S.; Amine, N.; Olivero, P.; Picollo, F. 3D printing in microfluidics: Experimental optimization of droplet size and generation time through flow focusing, phase, and geometry variation. *RSC Adv.* **2024**, *14*, 7770–7778. [CrossRef]
21. Jiang, J.; Chen, X.; Mei, Z.; Chen, H.; Chen, J.; Wang, X.; Li, S.; Zhang, R.; Zheng, G.; Li, W. Review of droplet printing technologies for flexible electronic devices: Materials, control, and applications. *Micromachines* **2024**, *15*, 333. [CrossRef]

22. Abdolmaleki, H.; Kidmose, P.; Agarwala, S. Droplet-based techniques for printing of functional inks for flexible physical sensors. *Adv. Mater.* **2021**, *33*, 2006792. [CrossRef]
23. Zhang, H.; Cui, L.; Lee, P.-H.; Kim, Y.; Moon, S.K.; Choi, J.P. Data-Driven autonomous printing process optimization and real-time abnormality identification in aerosol jet-deposited droplet morphology. *Virtual Phys. Prototyp.* **2024**, *19*, e2429530. [CrossRef]
24. Tofan, T.; Borodinas, S.; Kačianauskas, R.; Jasevičius, R. Modeling 3D droplet movement using a drop-on-demand inkjet printhead model. *Processes* **2022**, *10*, 1467. [CrossRef]
25. Jia, Y.; Cheng, X.; Song, W.; Zhou, Y.; Zhao, H. A droplet state prediction method for inkjet printing under small sample conditions based on the two-stage TrAdaBoost.R2 algorithm. *AIP Adv.* **2025**, *15*, 015120. [CrossRef]
26. The OpenFOAM Foundation. OpenFOAM v2021 User Guide. Available online: https://cfd.direct/openfoam/user-guide (accessed on 10 January 2023).
27. Fujimoto, H.; Shiotani, Y.; Tong, A.Y.; Hama, T.; Takuda, H. Three-Dimensional Numerical Analysis of the Deformation Behavior of Droplets Impinging onto a Solid Substrate. *Int. J. Multiph. Flow* **2007**, *33*, 317–332. [CrossRef]
28. Kim, H.-Y.; Chun, J.-H. The recoiling of liquid droplets upon collision with solid surfaces. *Phys. Fluids* **2001**, *13*, 643–659. [CrossRef]
29. Azimi, G.; Dhiman, R.; Kwon, H.-M.; Paxson, A.T.; Varanasi, K.K. Hydrophobicity of rare-earth oxide ceramics. *Nat. Mater.* **2013**, *12*, 315–320. [CrossRef] [PubMed]
30. Hsu, A.S.; Roy, A.; Leal, L.G. Drop-size effects on coalescence of two equal-sized drops in a head-on collision. *J. Rheol.* **2008**, *52*, 1291–1310. [CrossRef]
31. Acevedo-Malavé, A.; García-Sucre, M. Coalescence collision of liquid drops I: Off-center collisions of equal-size drops. *AIP Adv.* **2011**, *1*, 032117. [CrossRef]
32. Tang, C.; Zhang, P.; Law, C.K. Bouncing, coalescence, and separation in head-on collision of unequal-size droplets. *Phys. Fluids* **2012**, *24*, 022101. [CrossRef]
33. Josserand, C.; Thoroddsen, S.T. Drop impact on a solid surface. *Annu. Rev. Fluid Mech.* **2016**, *48*, 365–391. [CrossRef]
34. Antonini, C.; Amirfazli, A.; Marengo, M. Drop impact and wettability: From hydrophilic to superhydrophobic surfaces. *Phys. Fluids* **2012**, *24*, 102104. [CrossRef]
35. Rein, M. The transitional regime between coalescing and splashing drops. *J. Fluid Mech.* **1996**, *306*, 145–165. [CrossRef]
36. Thoroddsen, S.T.; Etoh, T.G.; Takehara, K. Air entrapment under an impacting drop. *J. Fluid Mech.* **2003**, *478*, 125–134. [CrossRef]
37. Thoroddsen, S.T.; Takehara, K. The coalescence cascade of a drop. *Phys. Fluids* **2000**, *12*, 1265–1267. [CrossRef]
38. Mohamed-Kassim, Z.; Longmire, E.K. Drop coalescence through a liquid/liquid interface. *Phys. Fluids* **2004**, *16*, 2170–2181. [CrossRef]
39. Chen, Y.; Shen, C.; Peterson, G.P. Hydrodynamics and morphologies of droplet coalescence. *Ind. Eng. Chem. Res.* **2015**, *54*, 9257–9262. [CrossRef]
40. Shen, C.; Chen, Y.; Yu, C.; Liu, X. Numerical study on the liquid-liquid interface evolution during droplet coalescence. *Microgravity Sci. Technol.* **2020**, *32*, 737–748. [CrossRef]
41. Zimmermann, P.; Mawbey, A.; Zeiner, T. Calculation of droplet coalescence in binary liquid-liquid systems: An incompressible Cahn–Hilliard/Navier–Stokes approach using the non-random two-liquid model. *J. Chem. Eng.* **2020**, *65*, 1083–1094. [CrossRef]
42. Baroudi, L.; Kawaji, M.; Lee, T. Effects of initial conditions on the simulation of inertial coalescence of two drops. *Comput. Math. Appl.* **2014**, *67*, 282–289. [CrossRef]
43. Lim, S.J.; Choi, M.C.; Weon, B.M.; Gim, B. Lattice Boltzmann simulations for water coalescence. *Appl. Phys. Lett.* **2017**, *111*, 101602. [CrossRef]
44. Gross, M.; Steinbach, I.; Raabe, D.; Varnik, F. Viscous coalescence of droplets: A lattice Boltzmann study. *Phys. Fluids.* **2013**, *25*, 052101. [CrossRef]
45. Perumanath, S.; Borg, M.K.; Chubynsky, M.V.; Sprittles, J.E.; Reese, J.M. Droplet coalescence is initiated by thermal motion. *Phys. Rev. Lett.* **2019**, *122*, 104501. [CrossRef]

fluids

MDPI

Article

Impact of Viscous Droplets on Dry and Wet Substrates for Spray Painting Processes †

Qiaoyan Ye [1,*], Oliver Tiedje [1], Bo Shen [2] and Joachim Domnick [2]

[1] Fraunhofer Institute for Manufacturing Engineering and Automation, 70569 Stuttgart, Germany; oliver.tiedje@ipa.fraunhofer.de

[2] Faculty of Science, Energy and Building Services, University of Applied Sciences, 73728 Esslingen, Germany; joachim.domnick@hs-esslingen.de (J.D.)

* Correspondence: qiaoyan.ye@ipa.fraunhofer.de

† This paper is an extended version of our paper published in Ye, Q.; Dreher, M.; Tiedje, O.; Shen, B.; Domnick, J. Numerical simulation of flake orientation during droplet impact on substrates in spray painting processes. In Proceedings of the ICLASS, Edinburgh, UK, 29 August–2 September 2021.

Abstract: This paper presents numerical studies of the viscous droplet impact on dry and wetted solid walls for spray painting applications, focusing on air entrapment, film structure, and flake (flat pigment) orientation. The results were compared with experimental observations using various high-speed camera arrangements. For paint droplet impact on dry substrates, a dynamic contact angle model was developed and used in numerical simulations. This contact angle model was verified with experimental observations. For the droplet impact on wet surfaces, characteristic crater sizes (diameter and depth) were defined considering also the effect of the film thickness. A strong correlation with the droplet impact Reynolds number was observed. In addition, a user-defined 6DOF (6-degrees-of-freedom) solver was implemented in a CFD program to perform calculations of rigid body motions within the impacting droplet, technically relevant for the resulting effect of flakes in metallic effect paints. The developed models were applied in parameter studies to further clarify the existing dependencies on application and fluid parameters more quantitatively. The simulation results are helpful to understand and to improve painting processes with respect to the final quality parameters.

Keywords: droplet impact; liquid film; air entrapment; flake orientation

1. Introduction

The collision of liquid drops with solid surfaces occurs in many engineering applications such as spray painting, spray cooling, and ink-jet printing. This paper presents a summary of the research results of a study of spray painting using model liquids and real paint. The motivation for the investigation lies in the huge industrial demand of understanding the occurrence of air entrapment during drop impact, which may lead to so-called pinholes after drying and baking, the final wavy surface structure (also called orange peel), and the behavior of pigment during droplet impact, enabling, e.g., metallic effects of coatings, which are very popular in the automotive industry.

A lot of research has already been carried out on droplet impact dynamics. Experimental observations of the impact of liquid drops on dry solid surfaces at room temperature, including an analysis of air entrapment, have been reported extensively [1–7]. When using flash photographic methods and high-speed cameras in conjunction with different light settings, such as back or oblique lighting with and without light diffusers, the authors observed bubble formation at the stagnation point and assumed bubble formation because of a dimple created on the drop surface at the impact point [1,6]. In investigations using viscous drops, Thoroddsen et al. [4] found in increasing number of bubble entrapment events during the drop spreading process, resulting from the localized contact of the levitated lamella with the solid substrate, especially for intermediate values of the impact Reynolds number (Re~250–350). Similar research has also been carried out by Palacios et al. [5]. Besides myriads of air bubbles at the interface between liquid and solid, they also observed two rings of microbubbles under impacting glycerol/water drops on a dry glass surface at Reynolds and Weber numbers around the splashing/deposition threshold. They further analyzed the dependency of the behavior of the two rings of microbubbles on the drop impact velocity and the relevant dimensionless numbers.

The maximal droplet-spreading diameter that is interesting for technical applications was correlated (see [8,9]) with non-dimensional numbers, i.e., Weber and Reynolds numbers, by using experimental results. In general, large droplets, e.g., D > 500 μm, have to be used in the experiment. For smaller droplets (50–300 μm), especially of opaque liquids, which are relevant for spray painting processes, it is very difficult to obtain experimentally qualified time-resolved images of the entrapment of air bubbles by drop impingement. However, the smaller droplet size range has been considered in various numerical studies, focusing also on the air entrapment during droplet impact [10–12].

The impact dynamics of non-Newtonian droplets, namely yield-stress fluid droplets, have been studied experimentally [13,14] and numerically [11,12], in which the influence of the rheological parameters on the droplet spreading and recalling processes was evaluated. Furthermore, contact angle models that have to be applied in numerical simulations were studied and discussed in detail [15,16].

With regard to the droplet impact on wet solid surfaces, there are some experimental and numerical investigations [17–21]. Weiss and Yarin [17] have simulated the impact of droplets on liquid films using a boundary-integral method for different impingement velocity and surface tension values. Different outcomes of droplet impact on substrates were analyzed. However, in their study, the effect of viscosity was neglected. Thoroddsen and Thoraval [22] examined the dynamics of a droplet on a liquid film both experimentally and numerically using relatively large droplets (D = 4.64 mm). They also investigated the air disc formation and visualized the evolution of smaller bubbles. Microbubble entrainment on thin liquid films under droplet impacts on an inclined surface was experimentally investigated by Tran et al. [21]. They found that the formation of large-area microbubbles is driven by a thin intervening air gap and contact line instability, analogous to classic coating instabilities.

In spray coating processes with metallic-effect paints, pigments (flat, 5–40 μm broad flakes) are widely used. From practical experiences, it is already well-known that the initial pigment/flake orientation in the paint layer, primarily at the early stage of film formation and before the subsequent solvent evaporation and baking, influences the final metallic effect significantly. Therefore, it is strongly suspected that processes during viscous droplet impact have a decisive influence on pigment orientation. For a strong metallic effect, usually, a pigment orientation parallel to the substrate is desired. Droplets with high impact velocities, namely high kinetic energies, were considered to be helpful for a surface-parallel flake orientation. Studies focusing on the flake movement inside the droplet and in the film are still quite limited. Although a few predictions of flake orientation were carried out based on mathematical analysis [23] and the flow field of two-dimensional droplet impact calculation [24], the corresponding results should be further verified. Recently, a numerical study [25] of the flake orientation during droplet impact on solid surfaces was carried out. Multiple flake particles inside the droplet were applied. The effects of the initial positions of flakes on the final orientation and of the interaction between flakes were investigated.

In our previous study [26–28], numerical simulations of the flake orientation in viscous droplets and the film structure due to droplet impact on wet substrates were performed. However, the parameter investigations were limited. The present paper offers detailed numerical studies of viscous droplet impact on dry and wetted solid walls, focusing on air entrapment, film structure, and flake orientation. In that way, a contact angle model necessary in numerical simulations was developed for the paint droplet impact process. In addition, a user-defined 6DOF (6-degrees-of-freedom) solver was implemented in a CFD program to enable rigid body (flake) motion calculations within the impacting droplet. The developed models were applied in parameter studies to further clarify the existing dependencies on application and fluid parameters more quantitatively. Experimental studies were carried out with a high-speed camera to produce experimental observations that validates the simulation results.

The paper is structured as follows: The experimental setup and measuring technique are explained in Section 2. Section 3 provides the basic numerical methods. The simulation results of droplet impact on a dry solid wall, including validation of the dynamic contact angle model, are shown in Section 4. In Section 5, we discuss the film structure as well as the entrapped air bubbles during droplet impact on dry/wet solid walls. Section 6 provides the simulation results of the flake orientation during droplet impact on dry/wet substrates. Section 7 summarizes the present work.

2. Experimental Setup and Measuring Techniques

In this section, basic experimental setups and measuring techniques are described. The experimental studies carried out provided useful information for the validation of the following simulation results.

2.1. Investigations of Single Droplet Impacts on Dry Solid Walls

Viscous droplets were generated using an industrial Drop-on-Demand device, in which a piezo-valve system that generates a short dosing impulse in the range of 1 ms was applied. Details of the device can be found in [29]. When using a 200 µm nozzle diameter, well-defined droplets (D = 280–450 µm) could be generated, depending on the liquid properties and the operating conditions. Table 1 shows characteristic droplet properties that were used in our numerical study. The experimental setup is shown in Figure 1. A high-speed camera (FASTCAM SA-X, Photron, Buckinghamshire, UK, Version: Photron Fastcam Viewer 4) was applied to obtain images of the evolution of the droplet contours and to determine the impact velocities (Figure 1a). The observation of the air entrapment on the substrate (in these cases, a transparent glass plate) was carried out using the up-view setup (observation from the glass plate backside).

Table 1. Drop properties.

Liquid	D (µm)	ρ (kg/m^3)	σ (N/m)	η (Pa·s)	U (m/s)
Glycerol/water	280–450	1200	0.063	0.02	1–3
Waterborne paint	280–450	1025	0.025	Shear thinning	1–3

2.2. Investigations of Droplet Impact and Air Entrapment on Wetted Solid Walls Using an Atomizer

Visualization of droplet impact events on a glass plate using an industrial air atomizer was carried out with the setup shown in Figure 2. An air spray gun (SATA LPS RB 2000RP, SATA, Kornwestheim, Germany) and a solvent basecoat were applied. In the first step, a wet paint layer with a well-defined thickness was applied on the glass plate. Subsequently, the resulting droplet impact processes on the wet layer were recorded by the high-speed camera.

Snapshots with an image size of approx. 1.8 mm × 1.8 mm and a 1024 × 1024 pixel resolution were analyzed. In general, many air bubbles could be observed in the paint layer. In Section 5, these visualization results will be compared to the simulation results.

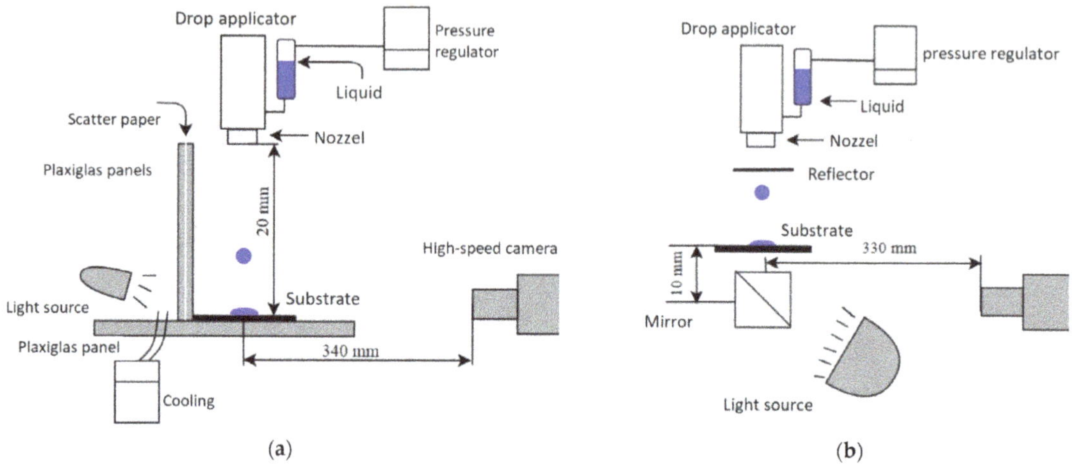

Figure 1. Experimental single droplet setup: (**a**) side observation, (**b**) upward observation.

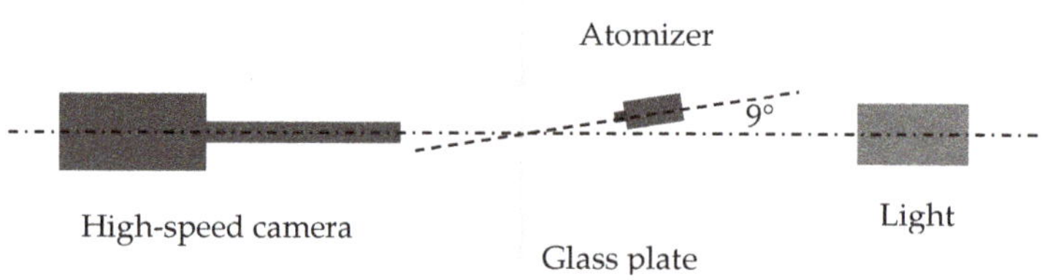

Figure 2. Experimental setup for droplet impact on wetted solid wall.

2.3. Rheological Properties of the Paint Used for Droplet Impacts on Wetted Solid Walls

Three model paints consisting of varying mass fractions of n-butyl acetate in a basic paint formulation were used in the present study. For the determination of the rheological properties, the paint material was collected from the solid wall directly after spray painting, thereby revealing the effect of solvent evaporation on the viscosity during the droplet movement between atomizer and target. The shear viscosity was experimentally measured using a rotational rheometer in the low shear rate range up to 3000 s^{-1} and a capillary rheometer in the high shear rate range up to 10^6 s^{-1}. The measured viscosities as a function of the shear rate are illustrated in Figure 3. Like most of the water-based paint liquids used in industrial applications, the model paints show a non-Newtonian behavior, namely a shear-thinning behavior, i.e., the viscosity decreases with an increasing shear rate.

In the numerical simulations, the observed non-Newtonian behavior was considered by applying Cross models, which are also plotted in Figure 3 as solid lines.

$$\eta = \eta_\infty + \frac{\eta_0 - \eta_\infty}{1 + (t \cdot \dot{\gamma})^m} \tag{1}$$

In the above equation, η and $\dot{\gamma}$ are the shear viscosity and the shear rate, respectively. η_∞ represents the shear viscosity when the shear rate approaches infinity, and η_0 the viscosity when the shear rate approaches zero, the so-called zero viscosity. t and m are the consistency index and flow index, respectively. Each model paint has been applied in two variants with a static surface tension of either 0.024 N/m or 0.067 N/m.

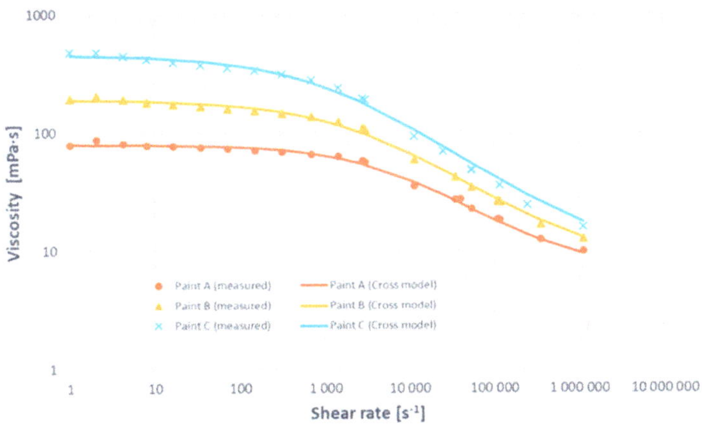

Figure 3. Flow curves of the model paints, measured and fitted by the Cross model.

3. Basic Numerical Methods

The droplet impact and spreading process on the surface can be treated by the Volume of Fluid (VOF) method for interfacial flow problems. Hereby, two or more immiscible fluids can be modeled by solving a single set of momentum and mass equations, including the tracking of the volume fractions of each of the fluids throughout the domain. The numerical simulations in this work were carried out with the commercial CFD code ANSYS-FLUENT (Version 24) based on the finite-volume approach.

In the present application, time-dependent VOF calculations were performed using an explicit scheme. A geometric reconstruction scheme for the volume fraction discretization was used, ensuring a sharp and low-diffusive interface discretization. To accommodate surface tension effects, the CSF model with wall adhesion modeling was used. The PRESTO scheme was applied for the pressure discretization.

Cartesian grids with local mesh refinement in the relevant regions were created. As it is well-known that the VOF method is quite sensitive to the grid resolution, grid independence was verified by comparing measured and calculated spread and height factors of selected droplet impact events. $D/\Delta x = 150$ was obtained as a reasonable cell size and therefore applied around the droplet and in the liquid film on the substrate. Here, D denotes the droplet diameter and Δx the grid size. In the far-field, coarser hexahedral meshes were used to reduce the total number of cells and the connected computational costs. For the simulation of droplet impact processes on wet solid surfaces, a relatively large computational domain had to be used. To reduce computational cells effectively, a dynamic mesh adaptation was applied, which was found to be especially necessary for large droplets. Consequently, the computational domains in the present study varied between 20 and 150 million cells. According to different simulation tasks, two-dimensional axisymmetric and 3-dimensional computational domains were applied. Further detailed important boundary conditions and models are described as follows.

3.1. Simulation of Viscous Droplet Impact on Dry Solid Surfaces

A computational domain with an extension of $2D \times 2D \times 1.5D$ was created for the simulation of droplet impacts on the dry solid surface with the flake pigment. Only a quarter of the spherical droplet was calculated using symmetric boundary conditions to reduce the total number of cells. A variable flow field calculation time step was set, considering the CFL (Courant–Friedrichs–Lewy) number to be less than 1. The resulting time steps ranged from 5×10^{-9} to 1×10^{-7} s. Based on common spray painting conditions, droplet impact velocities between 0.5 and 10 m/s were investigated. The initial droplet position was chosen to be a few micrometers above the wall surface to be able to calculate the gas flow field in the vicinity of the droplet. The pressure inside the droplet, induced by the surface tension of the liquid, was calculated and properly initialized. The remaining domain was set to ambient pressure. A dry smooth wall with no-slip boundary conditions was used.

It is well-known that the numerical description of a dynamically moving contact line or contact angle during droplet impact on dry surfaces is not trivial and therefore quite challenging. Nevertheless, the concept of dynamic contact angles is widely used in VOF simulations to accommodate the adhesion behavior. However, the applicability of the corresponding models is quite problem-dependent, including variations in liquid and substrate properties as well as changing operating parameters. Hysteresis effects of the contact line and contact angle make the model development even more difficult. For further information, refer to [11,16]. Figure 4 shows the definition of the contact angle and contact line. The static (equilibrium) contact angle θ_E is defined by Young's equation $\sigma_{LG}\cos\theta_E = \sigma_{SG} - \sigma_{SL}$, which is related to the surface tension of solid/vapor σ_{SG} and solid/liquid σ_{SL} interfaces. The contact angle observed experimentally at a moving contact line is defined as the dynamic contact angle θ_D.

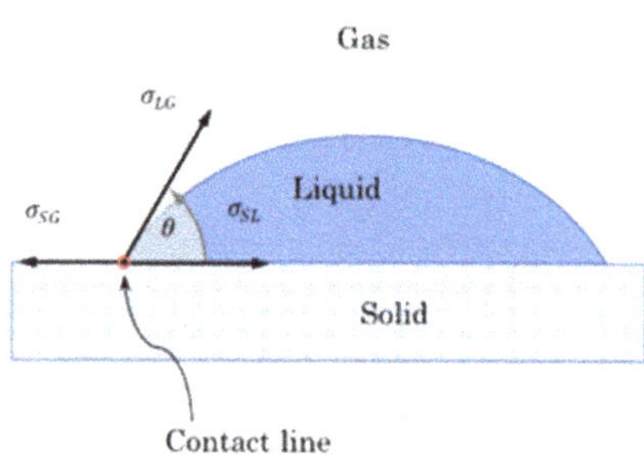

Figure 4. Definition of contact angle.

We have carried out experimental observations of viscous droplet impact on dry solid surfaces, focusing on the contour evolution during the spreading and receding process phases. Based on these results, a model for the dynamic contact angle was developed, which is further used in the simulations. Equation (2) describes this model in a mathematical manner. The main problem in the simulation of droplet movement on a dry substrate is the switching mechanism from the propagation contact angle to the static contact angle. The contact line oscillates strongly in the hysteresis region, resulting in a negative velocity while the droplet is not yet in the retraction phase. With the following treatment, smooth and sensible switching can be achieved.

$$\theta_D = \begin{cases} \theta_A, & \text{if } v_{cl} > 0 \\ \theta_{int} = 90^\circ, & \text{if } \overline{v_{cl}} < |\delta| \cap v_{cl} < \delta, \quad \text{where } \delta = -0.1 \\ \theta_{\nabla\phi}, & \text{if } \theta_{\nabla\phi} < \theta_A \\ \theta_E, & \text{if } \theta_{\nabla\phi} \leq \theta_E \end{cases} \quad (2)$$

During the spreading phase, where the contact line velocity (v_{cl}) is significantly positive, a constant advancing contact angle (θ_A) is deployed. This angle is the largest contact angle that is observed before the wetting line starts to move in the direction of the gas phase. At maximum wetting, when the contact line movement stops and the droplet usually starts to recede, an intermediate contact angle (θ_{int}) is set. This helps to suppress contact line movements, which are in accordance with experimental observations. $\overline{v_{cl}} < |\delta| \cap v_{cl} < \delta$ is the mathematic treatment to determine the hysteresis situation. An average contact line velocity $\overline{v_{cl}}$ based on three iteration time steps was used. A δ value was defined around zero velocity. If the angle between the fluid interface and the wall, calculated by the gradient of the VOF field ($\nabla\phi$), is smaller than the advancing contact angle, the interphase angle is set as the contact angle $(\theta_{\nabla\phi})$. It is secured that this angle decreases monotonically to the static or equilibrium contact angle (θ_E). On the wall cells far away from the contact line, the equilibrium contact angle (θ_E) is used, which ensures a reasonable wetting behavior on the wall.

3.2. Simulation of Viscous Droplet Impact on Wet Solid Surfaces

A thin film with a height between 30 and 60 μm and an infinitive width was applied on the substrate. As a first step, droplet impact processes were calculated using a two-dimensional axisymmetric geometry to obtain a general insight into the behavior of the gas–liquid interface as a function of relevant dimensionless numbers, namely, Re-, Oh-, and We-numbers. As a second step, the computational domain of 8D × 8D × 3D was created for more detailed simulations of the impact of droplets that contain a flake. Clearly, the number of computational cells increased tremendously. Local grid adaption and dynamic mesh adaption were carefully performed, ensuring the necessary grid resolution. Especially in case of large droplets, e.g., 300 μm, the inhomogeneous grid region enlarged significantly due to the necessary inclusion of the crown formation and wavy structures around the impact location. A typical mesh model based on dynamic adaptation is shown in Figure 5. In this case, the mesh size around the interface between liquid and gas is set to 1 μm, while being 16 μm at a larger distance from the interface.

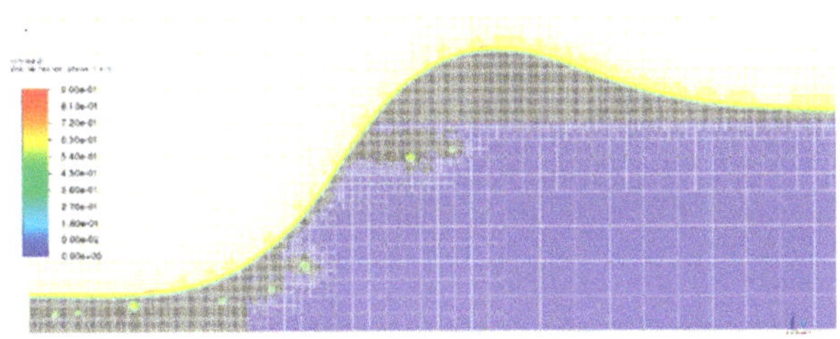

Figure 5. Mesh size distribution determined using dynamic mesh adaption model.

3.3. Simulation of Flake Orientation

For the simulation of the flake movement within an impacting droplet, a three-dimensional computational domain with a quarter of the spherical droplet was applied. A dynamic mesh model based on the overset meshing method included in ANSYS Fluent was applied. Hereby, a 'component' mesh that is linked to the flake and a 'background' mesh, representing the domain, are overlaid and connected via an interpolated mesh interface (Figure 6). In the present study, only a single rectangular-shaped flake in the computational domain was applied. The size of the considered flake was chosen to be 1 × 16 × 16 μm^3, reasonably representing a typical aluminum flake geometry inside a metallic-effect paint. However, this small size results in corresponding moments of inertia of 1.6×10^{-23} [kg·m^2], which are too marginal for the actual 6DOF motion solver by ANSYS. As a consequence, a custom motion equation solver was implemented via a user-defined function (UDF), assuming rigid body motion. Therefore, the flake motion may be simplified to the movement of its center of mass and the rotation around it. Consequently, the momentum conservation (3) is calculated in the global inertial coordinate system, and the angular momentum (4) is determined in body coordinates:

$$\dot{v} = \frac{1}{m}\sum F \tag{3}$$

$$\dot{\omega} = I^{-1}\left(\sum M - \omega \times I\omega\right) \tag{4}$$

Here, m denotes the mass of the flake, I the inertia tensor, and F, M the flow-induced forces and moments, respectively. $\dot{v}$ and $\dot{\omega}$ represent the time derivatives of the velocity and the angular velocity. In Equation (3), the sum includes pressure, viscosity, and gravitational forces. An Adams–Moulton algorithm of the fourth order with a rather complex variable time-step formulation was derived and implemented to integrate the above given ordinary differential equations (ODEs) in time. This enabled stable simulations of flake movement and orientation using a reasonable time-step size, such as dt = 1×10^{-8} to 1×10^{-7} s, which was solely adjusted by the flow solver. Table 2 shows the typical parameters applied in the present simulations. Shear thinning non-Newtonian viscosity was applied for model paints, as shown in Figure 3.

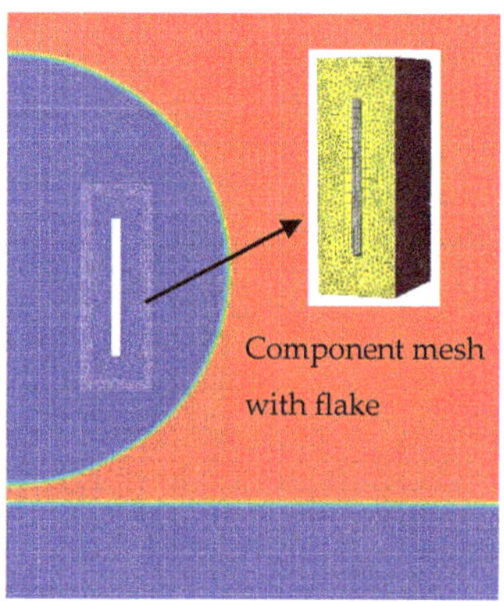

Figure 6. Contours of VOF overlaid with mesh at t = 0 s, blue: liquid, red: air.

Table 2. Parameters used in simulations.

Droplet diameter D (μm)	24, 50, 100, 300	Flake size (μm^3)	$1 \times 16 \times 16$
Droplet velocity U (m/s)	0.5–20	Flake density (kg/m^3)	3200
Newtonian viscosity η (Pa·s)	0.01–0.08	Ratio of flake to droplet l/D	0.053–0.32
Non-Newtonian viscosity	Shear thinning	Re-number, Re = ρUD/η	1–100
Surface tension σ (N/m)	0.025, 0.063	We-number, We = $\rho U^2 D/\sigma$	3–1200
Liquid density ρ (kg/m^3)	920, 1020, 1200	Oh-number, Oh = $\eta/\sqrt{(\sigma D \rho)}$	0.1–2.2
Static contact angle (°)	30, 50, 60		
Height of film (μm)	30–65		

4. Experimental Validation of the Numerical Models for the Droplet Impact on Dry Solid Surfaces

Validations of the numerical models for the impact process of viscous droplets on a dry solid surface were carried out using the single droplet generator and the visualization technique, as described in Figure 1a. Glycerol/water droplets (D = 400 μm, η = 20 mPa·s, σ = 0.063 N/m, θ_E = 55°) and model paint droplets (D = 300 μm, η = 17 mPa·s at

$\gamma = 10^5$ (1/s), $\sigma = 0.025$ N/m, $\theta_E = 53°$) were created. Simulations were performed using the two-dimensional axisymmetric domain. Figure 7 shows the comparison between the high-speed-camera recording and the simulation of a glycerol/water droplet impact. The obtained videos of the impact processes for both droplets were further analyzed with respect to the evolution of the droplet spread factors d/D_0 and the height ratios h/D_0. In the simulations, advancing angles θ_A of 120° for glycerol/water droplets and 95° for paint droplets were used, estimated from the high-speed videos. In Figure 8, experimental and simulative spread factors are compared. Quite good agreement can be observed, which indicates that the proposed dynamic contact angle model delivers a good performance and is suitable for more advanced applications.

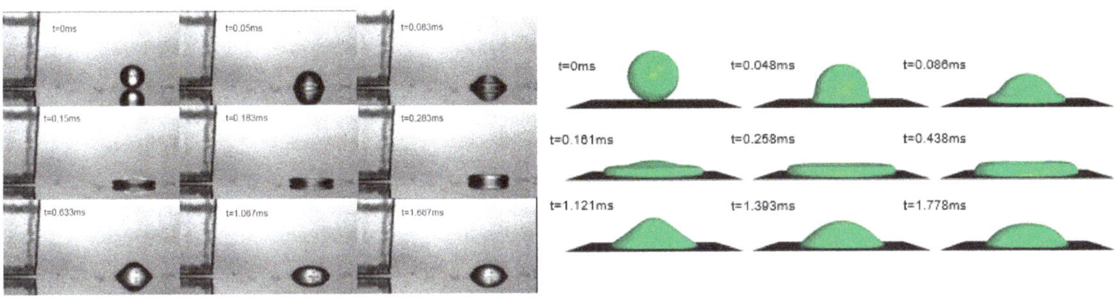

(**a**) (**b**)

Figure 7. Glycerol/water droplet impact process, experiment (**a**) vs. simulation (**b**), droplet velocity U = 2.6 m/s.

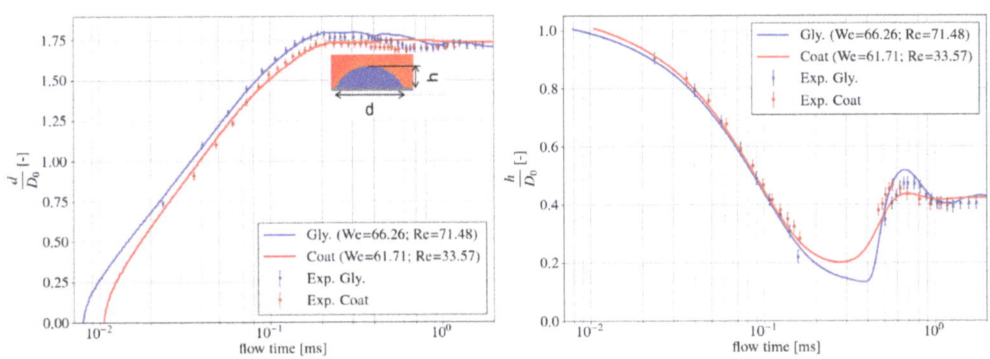

Figure 8. Comparison of droplet spread factors between experiment and simulation during the impact process, Gly. is Glycerol/water droplet, Coat is paint drop.

5. Simulation of Droplet Impact on Dry/Wet Solid Surface

5.1. Air Entrapment in Droplet Impact Processes

It is well-known that air entrapment at droplet impact may occur independently of the state of the surface, i.e., dry or wet. In the case of painting applications, relevant droplets are rather small, with medium volume sizes between 20 and 50 µm. Moreover, paint liquids are very often opaque, resulting in significant difficulties for useful experimental investigations of bubble formations. Therefore, numerical simulations provide to some extent the only appropriate tool to further analyze the mechanism of the air entrapment during droplet impact. Here, numerical simulations were carried out using a quarter of the spherical droplet with a symmetric boundary.

Figure 9 shows simulation results of the droplet contour evolution during the impact process of a 300 μm droplet at an impact velocity of 4 m/s on a dry surface. A thin layer of air bubbles (further also called an air disc) is formed directly at the contact area between droplet and substrate. This air disc is continuously enlarging during the drop spreading until a fully wetted contact line is created and the maximum air disc diameter is observed, which is smaller than the maximum droplet spreading diameter, as shown in Figure 8 at t = 0.127 ms. The air disc then contracts into larger bubbles during the receding process, in which some bubbles can also be released, as shown in the figure at t = 5.8 ms. At the quasi-static state, there are still large air bubbles on the substrate, as the release of the remaining bubbles slows further in the quasi-static state. The total amount of remaining air bubbles in a quasi-static state depends on the properties of the liquid and substrate, as well as the operating conditions. For a water droplet with an identical size and impact conditions, the maximum extension of the air disc and the final number of bubbles on the substrate in the quasi-static state are significantly reduced [12]. Some influence factors on the air bubble formation are indicated by Figure 10, where the quasi-static state of the impact of a viscous 50 μm droplet is analyzed. More bubbles are observed in the case of a higher viscosity (Figure 10a). In addition, a higher impact velocity leads to a larger air disc during the spreading process and ultimately to more bubbles (Figure 10b). With a smaller equilibrium contact angle $\theta_E = 30°$, i.e., improved wettability, bubbles tend to reduce their wall contact area, leading to larger distances between the bubble centers and the wall (Figure 10c). This should also ease any additional bubble release in a quasi-static state. Since Figure 10 was created based on the quarter of the spherical droplet with a symmetric boundary, the results shown only demonstrate the dependency between the air entrapment and the different operating conditions, but not the exact distribution of air bubbles inside a droplet.

Air can also be trapped during droplet impacting on a wetted substrate. In this case, characteristic rings around the impact center are formed, consisting of several air bubbles, which can be observed both in the simulation (Figure 11b) and experimentally (Figure 11a). In the following simulation results, bubble formation can be observed continuously.

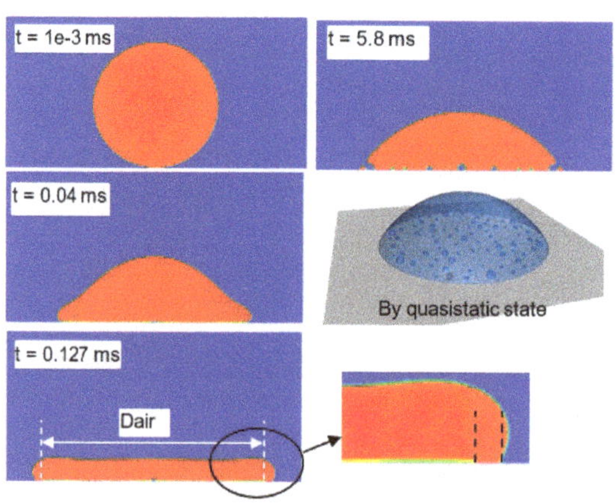

Figure 9. Contours of liquid fraction (red: liquid, blue: air) in impact of a viscous drop (D = 300 μm, U = 4 m/s, η = 20 mPa·s, σ = 24 mN/m, $\theta_E = 60°$.

5.2. *Evolution of Gas–Liquid Interface at Droplet Impact on Wet Solid Surfaces*

As stated in the introduction, there are many studies considering droplet impact on wet substrates. Depending on impact velocity and liquid viscosity, droplet impact on a thin liquid film can be essentially divided into spreading/deposition, cratering/receding, crown formation, and crown with splashing. The latter will not be studied in the present investigation because it is very seldom in spray coating applications. In spray coating applications, typical mean diameters of the droplet range between 30 and 50 μm. Considering existing correlations between droplet impact velocities and droplet diameters for different atomizers in spray painting processes [30], droplets with diameters of less than 100 μm and impact velocities up to 20 m/s were used in our investigations of the interface evolution at droplet impact on wet surfaces. Here, the two-dimensional axisymmetric computational domain and model paints were applied in simulations.

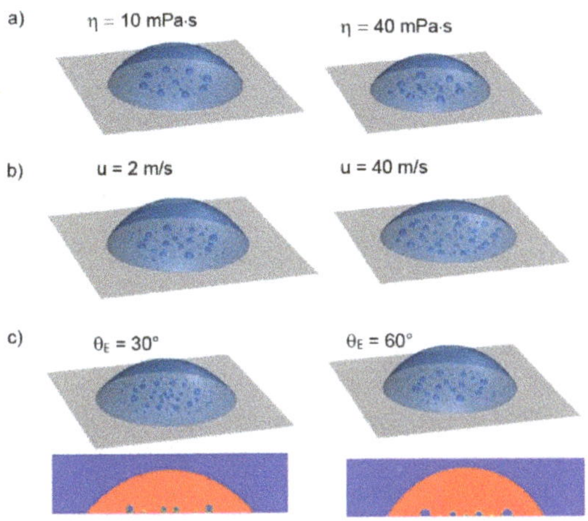

Figure 10. Air entrapment at droplet impact on a dry surface in the quasi-static state (D = 50 μm, σ = 24 mN/m). (**a**) Effect of the viscosity, (**b**) effect of the impact velocity, (**c**) effect of the static contact angle, here cross-sections are also depicted to show the adhesion state of air bubbles on the substrate.

(**a**) (**b**)

Figure 11. Air entrapment at droplet impact on a wet substrate. (**a**) Experimental observation, (**b**) simulation results (H_f = 60 μm, D = 280 μm, U = 2.2 m/s, η = 30 mPa·s, σ = 28 mN/m).

Figure 12 shows typical results of different paint droplet impacting on a thin liquid film. A small droplet at a low impact speed deposits smoothly on the liquid film (Figure 12a). With an increasing impact energy of the droplet, namely, either with a larger diameter or with a higher velocity, a small crater can be formed in the target liquid layer (see Figure 12b). If the impact energy is high enough, the so-called crater rim or crown rises at the circumference of the crater above the original surface of the liquid layer on the target (Figure 12c). Air bubbles (blue color) can be observed clearly in the liquid film, which is air entrapment, a typical phenomenon of the droplet impact process.

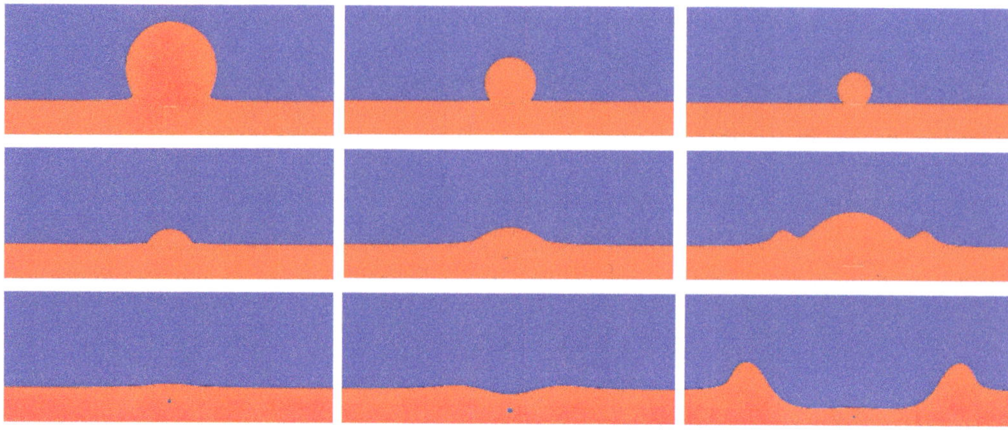

Figure 12. Droplet impact on a wet solid surface (red: liquid, blue: air). Film H_f = 30 µm. (**a**) D = 30 µm, U = 4 m/s, (**b**) D = 45 µm, U = 4 m/s, (**c**) D = 80 µm, U = 8 m/s.

For paint liquids with a shear thinning behavior, the variation in the apparent viscosity during the impact process must be considered. Figure 13 shows the evolution of the viscosity inside the droplet and in the liquid film during the cratering process. In this case, a droplet with a diameter of 65 µm impacts at a 6 m/s velocity on a wet solid surface with a film thickness of 65 µm. The evolution of the viscosity in the whole relevant fluid region from the early stage of droplet impact until the end phase of the impact can be obtained. For a better analysis, the characteristic viscosity η^* of the droplet is determined to be 14.86 mPa·s (η^* is defined as the viscosity at the characteristic shear rate $\dot{\gamma}^*$ of droplets in the impact process, $\dot{\gamma}^* = 2{\cdot}U/D$). The corresponding Reynolds number ($Re^* = \rho UD/\eta^*$) and the Ohnesorge number ($Oh^* = \eta^*/\sqrt{\rho\sigma D}$) are 24.2 and 0.39, respectively. At the early stage of drop impact, the viscosity is quite low in the collision region due to the very high shear rate in this region (larger than 500,000 s^{-1}). The viscosity in the collision region is approx. 10 mPa·s. Far away from the impact center, the viscosity is about 85 mPa·s, corresponding to the value at the shear rate of zero. At the end phase of the impact, the liquid viscosity is increased to 70–75 mPa·s. In this quasi-static state, a ring-shaped region of low viscosity inside the film under the remaining depression can be observed, where the air bubbles are located.

The parameter study of the paint droplet impact on the wet film was carried out using different droplet diameters, impact velocities, as well as film thicknesses, as shown in Table 2, with a focus on the result of the gas–liquid interface. The obtained dimensionless crater sizes, namely, the ratios between the maximal crater diameter *Dc* (see Figure 13), the droplet diameter *D*, and the maximal crater depth h_c to *D* are plotted as a function of the Reynolds number Re^* in Figure 14. Considering the effect of the film thickness on the crater size, we introduced the characteristic crater diameter and depth as follows:

$$D^* = \left(\frac{D_c}{D}\right)^{\frac{1}{Re^*}\cdot\left(\frac{H}{D}\right)^{-n}} \tag{5}$$

$$h^* = \left(\frac{h_c}{D}\right)^{\frac{1}{Re^*}\cdot\left(\frac{H}{D}\right)^{-n}} \tag{6}$$

where Re^* is the characteristic Reynolds number and H is the film thickness. The exponent n indicates the effect of the dimensionless film thickness on the crater size with $n = 0.1$. In

Figure 14, the characteristic crater sizes are plotted as a function of the Reynolds number Re^*. A strong correlation between both dimensionless values is observed and can be described using an exponential function:

$$D^* = A_0 + A_1 e^{-A_2 \cdot Re^*} \tag{7}$$

$$h^* = B_0 + B_1 e^{-B_2 \cdot Re^*} \tag{8}$$

where $A_0 = 1.0755$, $A_1 = 0.62629$, $A_2 = 0.19858$, $B_0 = 0.99827$, $B_1 = -0.63523$, and $B_2 = 0.13853$. Finally, the dependency of the different droplet impact processes on the Reynolds and Ohnesorge numbers is analyzed and depicted in Figure 15, with a blue dot for deposition and a red circle for a crater/crown. In general, the simulation results shown in this section will help to analyze and understand all further investigations, such as the flake orientation during droplet impact on wet solid surfaces, entrainment of air bubbles in the impact process, and the final film leveling.

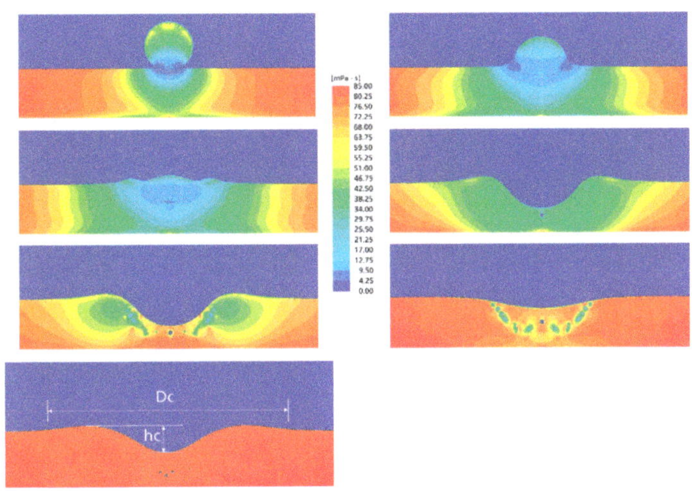

Figure 13. Evolution of the viscosity contours (model paint A, D = 65 µm, H_f = 65 µm, U = 6 m/s, $Re^* = 24.2$, $Oh^* = 0.39$).

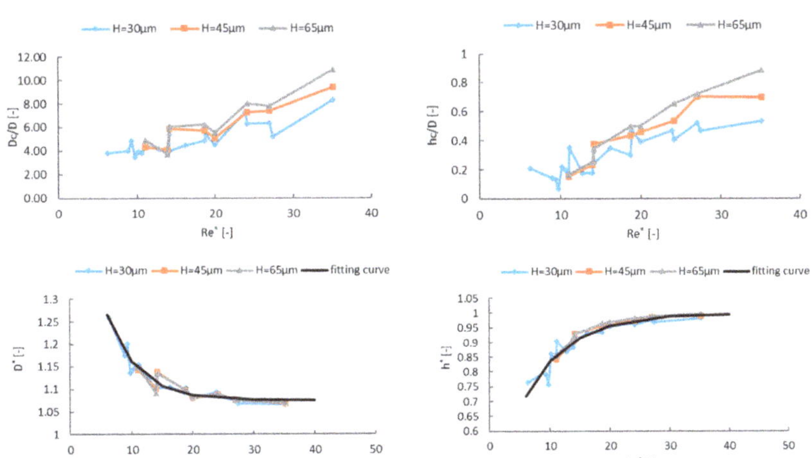

Figure 14. Crater sizes as a function of the Reynolds number (Re^*).

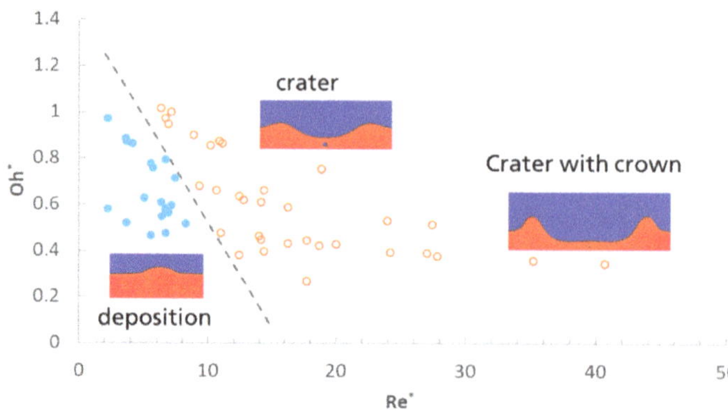

Figure 15. Droplet impact formations at different Reynolds and Ohnesorge numbers (ratio between film thickness and drop diameter is between 0.375 and 2.16).

6. Simulation of Flake Orientation at Droplet Impact

Effect pigments (such as metallic pigments) are an essential component of paints with a metallic (or mica) effect. Figure 16 shows a typical metallic pigment, that is, a thin plate with a thickness about 1 μm and a diameter of 8–16 μm. Essentially, flat metallic pigments or flakes act like small mirrors reflecting incident light preferentially at the incident angle. As shown in Figure 16b, the final intensity distribution depends mainly on the orientation distribution of the flakes in the film. An optimum surface-parallel orientation of the flake results in an optimum metallic effect. In simulations, a simple geometry of the flake, namely, a rectangular plate of $1 \times 16 \times 16\ \mu m^3$, was applied.

(a) 

(b) 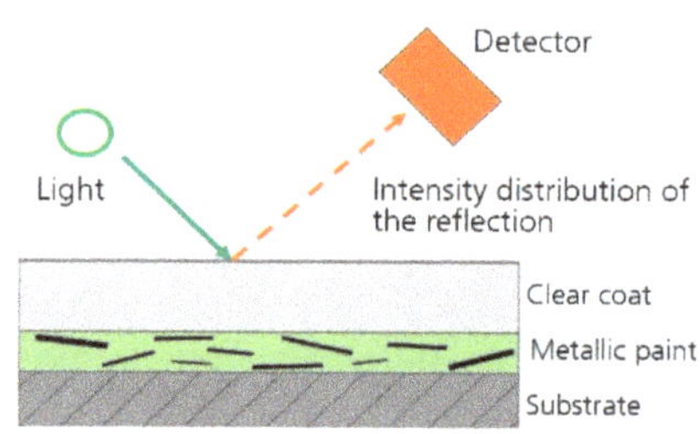

Figure 16. Metallic pigments and their effect in the paint layer. (**a**) Metallic pigment Alushine® [31], (**b**) reflection of a real metallic coating.

Due to the small sizes of the flakes, the short time scales of the process, and the opaqueness of the paints, it is almost impossible to observe the flake movement during the impact process experimentally. Therefore, numerical simulations provide probably the only alternative for the investigation of flake orientation in droplet impact processes.

In the following parameter study of flake orientation, Newtonian materials with constant viscosity were investigated at first, and later we will also show the results of using non-Newtonian liquid. Based on the characteristic shear rate proposed above, a reasonable range of Newtonian viscosities, i.e., 0.01–0.08 Pa·s, was chosen, representing a good approximation of the viscosities of typical paint materials at the high shear rate during the impact processes. Moreover, a surface tension of 0.025 N/m was used in the simulations for paint liquids. The detailed parameters used in the simulations of flake orientation are given in Table 2.

6.1. Dry Solid Surfaces

The initial position and orientation of effect pigments inside an impacting droplet may be randomly distributed. For the simulations of droplet impact on dry substrates, the flake has initially been located at the horizontal mid plane and at a 50% droplet radius in the vertical direction. The initial orientation of the flake is perpendicular to the surface, which can be considered the worst case for the final desired surface-parallel orientation.

Figure 17 shows the droplet contour evolution and the flake orientation during the droplet impact process. The color bar shows the contact angle. The outer contact angle changes from 120° at the beginning of the spreading process to 60° in the final quasi-static situation. The initial state of the flake is depicted in Figure 17a. During the inertia-driven spreading phase, as shown in Figure 17b,c, the flake is rotating significantly. In Figure 17d, the maximum spreading is reached and surface tension effects become dominant. During this stage, despite droplet contour movements, the above-mentioned contact line hysteresis is observed, which is identical to our experimental findings. Here, the flake is increasing its orientation angle with respect to the surface again. Some changes of flake orientation after t = 0.032 ms (Figure 17d) can be observed (Figure 17e,f,g), until the final location and orientation of the flake at a final quasi-steady state, as shown in Figure 17h. Air entrapment at droplet impact and air bubble formations (blue bubbles) on the wall can also be observed in Figure 17.

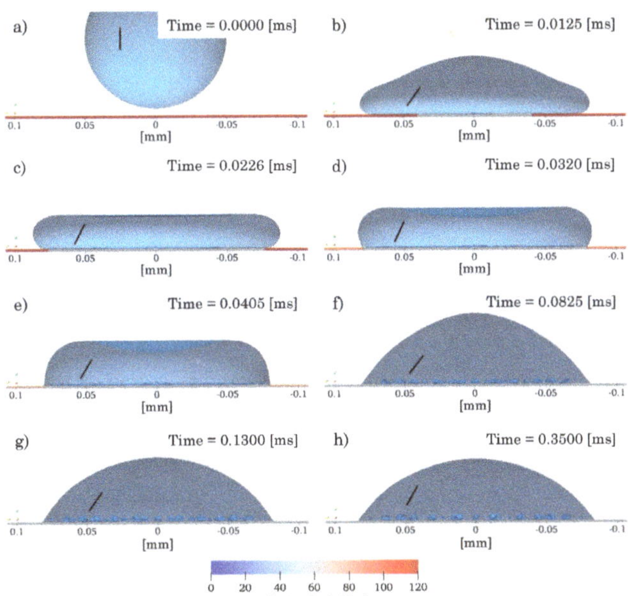

Figure 17. Evolution of the glycerol/water droplet contour and flake orientation during dry wall droplet impact (D = 100 μm, η = 20 mPa·s, σ = 0.063 N/m, U = 6 m/s, θ_E = 55°, We = 68.6, Re = 36). (**a**–**h**) correspond to the situation of time sequence from 0.0–0.35 ms.

The effects of the droplet liquid properties and the impact parameters on the temporal evolution of the flake orientation are shown in Figure 18. Glycerol/water and paint droplets with η = 20 mPa·s were applied. For small droplets, in the considered case of 50 μm, the quasi-static state is reached at t = 0.3 ms. Since the size ratio between the flake and droplet diameter is relatively large, i.e., l/D = 0.32 for the 50 μm droplet, the movement of the flake is quite limited. In contrast, larger variations in the flake orientation angle can be observed for 300 μm diameter droplets. In addition, the time to reach a quasi-static state significantly increases with increasing We- or Re-numbers. The final flake angle in the quasi-static state depends, presumably, mainly on the Re-number for a given low viscosity. In general, it seems that the final flake orientation improves with an increasing Re-number, at least in the case of droplet impact on dry surfaces.

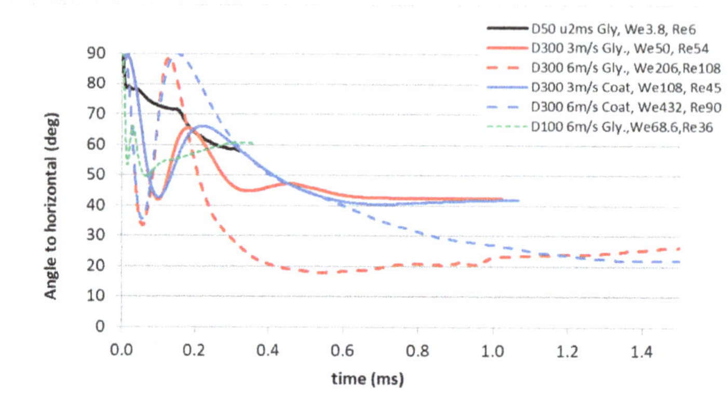

Figure 18. Comparison of flake orientation with different viscous droplets.

6.2. Impact on Wet Solid Surfaces

The flake orientation at droplet impact on wet surfaces appears to be technically more relevant, as only the very first droplets after the start of application impact on a dry surface. At a later stage, a wet paint film has already been formed on the surface, and the resulting flake orientation is mainly dependent on the complete flow field at and close to the impact location. Since simulations of flake orientation during droplet impact are quite time consuming due to the need for large computational domains, our parameter studies are so far limited to droplets with a diameter of 50 μm only. Also, it should be noted that the droplet diameter must be greater than the longest extension of the flake, i.e., very small droplets do not contain any flakes.

Typical droplet deposition and cratering processes at different impact velocities are shown in Figures 19–21, respectively. The wet film thickness is 60 μm. The final flake angle of the droplet deposition in Figure 19 is about 30° with the small impact velocity (U = 0.5 m/s). A typical crater occurs when increasing the impact velocity up to 10 m/s. During the deepening of the crater, as shown in Figure 20, the flake angle decreases and even tends to almost 0 (flake parallel to the surface) at t = 0.03 s. However, during crater retreat, the flake angle increases again significantly and reaches 49.3° in the final quasi-static state. By further increasing the impact velocity to 20 m/s, as shown in Figure 21, the crater size increases heavily, and crown formation can be observed. A larger amount of liquid and the flake are pushed outward from the crater center. The crater retreat begins when the crown disappears. Compared to the case with U = 10 m/s, the process of the crater retreat is much slower due to the thin film at the bottom and the resulting lower surface force. This leads to a lower flake angle (38°) in the final quasi-static state.

The effect of a surface tension variation on the flake orientation at different impact velocities is shown in Figure 22, using a glycerol/water mixture and paint, both with a constant viscosity of 20 mPa·s. When comparing glycerol/water and paint droplets at a 10 m/s impact velocity, the surface tension shows only a weak effect on the final flake orientation in a quasi-static state, approaching 50° for both cases; however, the time to reach the quasi-static state is significantly extended in the case of the paint droplet with a lower surface tension. At lower impact velocities between 0.2 m/s and 5 m/s, the final flake angles are constantly smaller (around 30°), independent of the surface tension. When further increasing the impact velocity to 20 m/s, the final flake angle seems to decrease. In this case, at t = 0.618 ms the film lever shown in Figure 21 is almost horizontal, corresponding to a quasi-steady state.

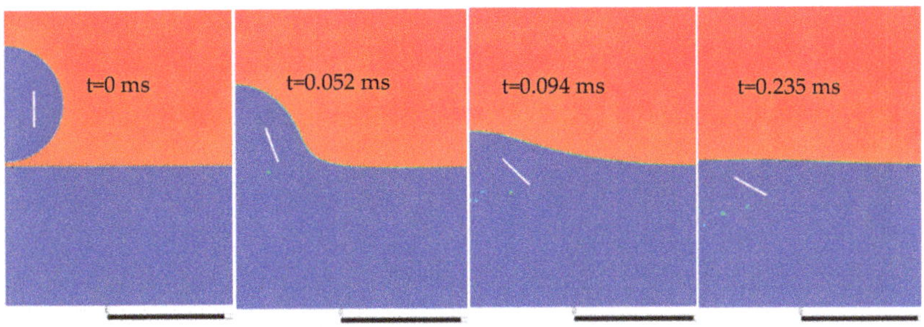

Figure 19. Evolution of droplet contour, film contour, and flake orientation (black ruler corresponds to 50 μm), D = 50 μm, η = 20mPa·s, U = 0.5 m/s, σ = 0.025 N/m, We = 0.51, Re = 1.28, Oh = 0.56.

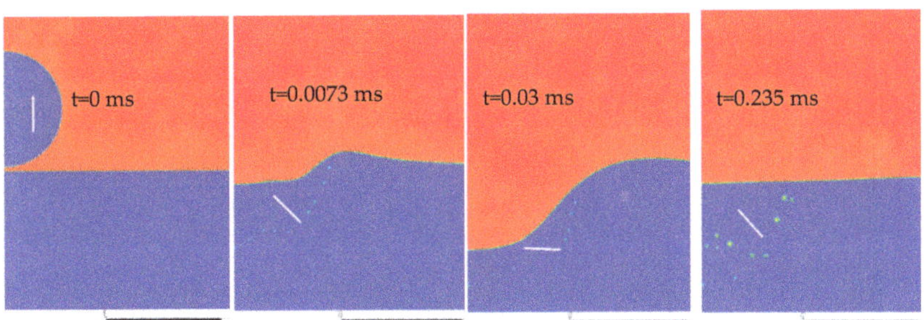

Figure 20. Evolution of droplet contour, film contour, and flake orientation, D = 50 μm, η = 20 mPa·s, U = 10 m/s, σ = 0.025 N/m, We = 204, Re = 25.5, Oh = 0.56.

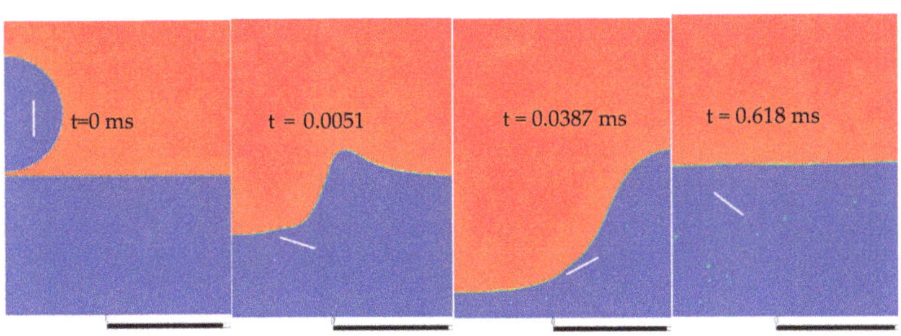

Figure 21. Evolution of droplet contour, film contour, and flake orientation, D = 50 μm, η = 20 mPa·s, U = 20 m/s, σ = 0.025 N/m, We = 816, Re = 51, Oh = 0.56.

The effect of the viscosity on the flake angle is shown in Figure 23 for a droplet size D of 50 μm, a film thickness H of 60 μm, an impact velocity U of 5 m/s, and a surface tension σ of 0.025 N/m. Clearly, the final flake angle in a quasi-static state increases significantly with an increasing viscosity, especially in case of the 40 mPa·s and the 80 mPa·s liquids. Figure 23 also includes the simulation results of a non-Newtonian, shear-thinning model paint. The corresponding viscosity curve (paint A) is given in Figure 3. For a further explanation of the result, Figure 23 needs to be analyzed, showing the distribution of the apparent fluid viscosity during different stages of droplet impact. As expected, the apparent viscosity diminishes in fluid regions with high shear rates due to the presence of strong velocity gradients. Since the flake is mainly located in a 20 mPa·s fluid zone, especially at a later stage of the impact process (Figure 24c), the flake orientation and the final flake angle correspond nicely to the results of the Newtonian 20 mPa·s case. Although the present investigations focus on a wet file thickness of 60 μm, only a weak effect of the film thickness was found in the range between 30 μm and 60 μm, as shown in Figure 25, which are typical wet film thicknesses of metallic paint layers.

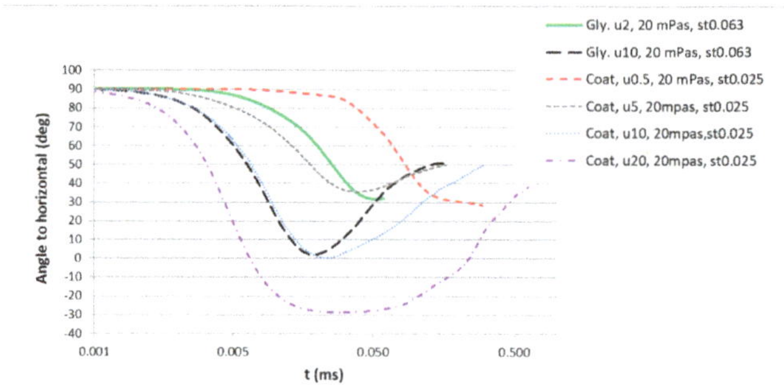

Figure 22. Effect of surface tension on flake orientation, H_f = 60 μm, D = 50 μm (Gly. is glycerol/water with surface tension st = 0.063 N/m, Coat is the droplet with constant viscosity but similar surface tension to water paint st = 0.025 N/m).

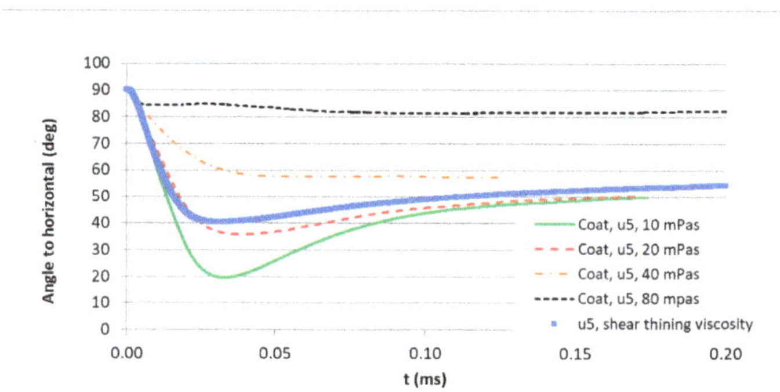

Figure 23. Effect of viscosity on the flake angle, U = 5 m/s, H_f = 60 μm, D = 50 μm.

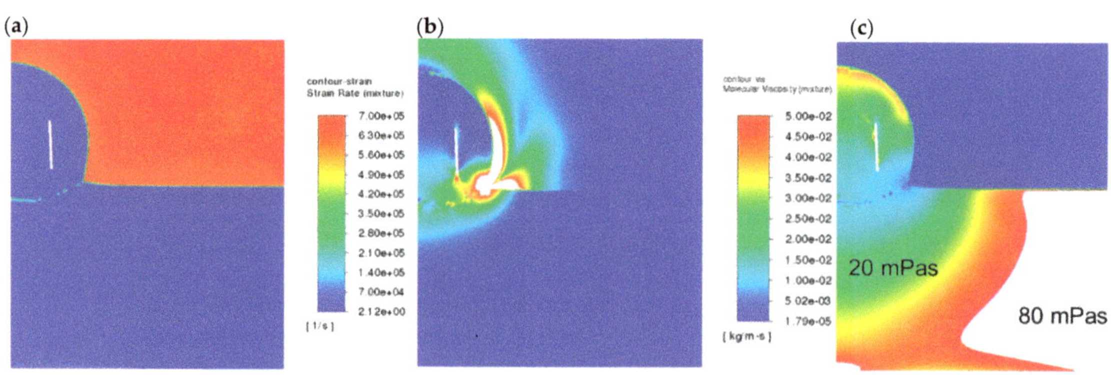

Figure 24. Simulation results of droplet impact using a shear thinning viscosity (U = 5 m/s, D = 50 μm): (**a**) VOF contours (red: air, blue: liquid); (**b**) strain rate (1/s) distribution (values that are larger than 7e5 are blanked out); (**c**) viscosity distribution (value that is larger than 50 mPa·s is blanked out).

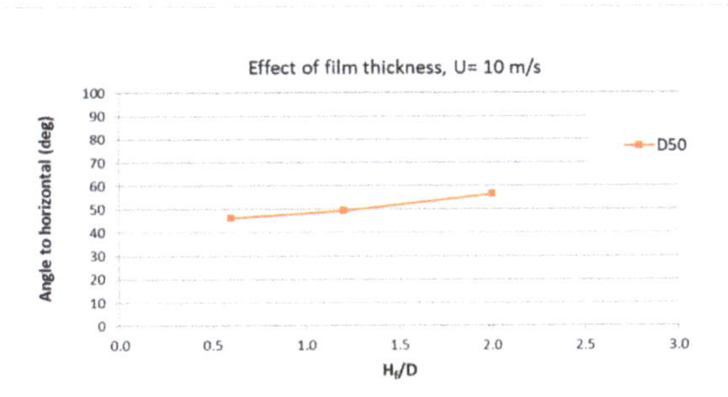

Figure 25. Effect of film thickness on the flake orientation (η = 20 mPa·s, H_f = 30–100 μm, D = 50 μm, U = 10 m/s).

In summary, the important influencing factors on the evolution of the gas–liquid interface and subsequently on the final flake orientation are the droplet diameter, droplet impact velocity, and liquid viscosity. The correlation between the characteristic Re-number Re^* and the final flake angle for paint droplets is depicted in Figure 26. For the desired minimum flake angle, the Re^* should be in the range of 40–60. In connection with the subsequent reduction in the flake angle during paint layer drying and baking, these low values should be an appropriate starting condition for an excellent final metallic effect of the paint layer. Low flake angles can also be observed in the case of $0.3 < Re^* < 1.5$, corresponding to small droplets and/or very low impact velocities. However, technically, this region appears to have only a little practical importance, as smaller droplets have a lower probability of containing a flake. Moreover, a larger number of very small droplets are subject to unwanted evaporation effects.

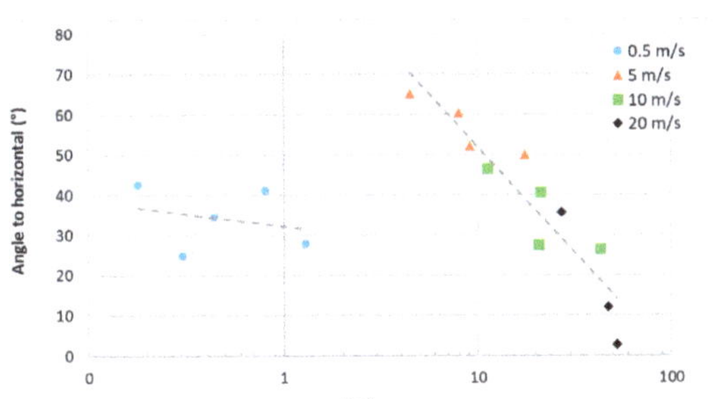

Figure 26. Correlation of Re* number and final flake orientation H_f = 60 μm.

The region of very high Re-numbers over 100 was not considered in the present investigations, as the impact of very large droplets and high impact velocities may lead to splashing and formation of ripple structures on the film surface, often resulting in a so-called orange-peel structure of the final paint layer. Paint surfaces with visible orange peel do not correspond to the desired quality. In the present numerical study, the flake is initially oriented vertically to the solid surface, which is the worst case. For additional simulation results of the effect of the initial position of the flake inside droplets on the final flake angle, refer to [26].

7. Summary and Conclusions

The present paper summarizes the results of numerical investigations of viscous droplet impact on dry and wetted solid walls for spray coating applications, focusing on air entrapment, film structure, and flake (pigment) orientation. Parameter studies were performed including Newtonian liquids as well as non-Newtonian, shear-thinning liquids typical for water-based paint materials. Operating parameters typical for high-quality spray painting applications were applied. Also, a dynamic contact angle model that is suitable to be used in the spray painting processes was proposed. Where possible, the obtained results could be validated by experimental observations using a high-speed camera.

It was found that there is always the formation of air entrapment during droplet impact on a dry or wet substrate, almost independent of the surface wettability. Thin air layers (air disc) result from the direct contact between the droplet surface and the substrate. The size of the air disc, the contraction of the air disc to bubbles, and the release of air bubbles depend on material and target properties and application parameters. A reduction in the static contact angle of the liquid enhances the bubble release from the target wall as well as from the liquid film. Of course, higher liquid viscosities delay bubble release from both the wall and the liquid film.

At the considered conditions, when focusing on the spray painting applications, craters are formed at Re-numbers greater than 10 in the case of impact on wet surfaces. Characteristic crater diameters and crater depths could be defined and correlated with the Re-number and the film thickness. Smooth droplet deposition and leveling were found at lower Re-numbers in connection with small Oh-numbers.

Particularly for the analysis of the metallic effect of paints, a detailed numerical study of the flake orientation during droplet impact on dry and wet solid surfaces was carried out. Therefore, a rigid body motion solver for calculating pigment movement was implemented in ANSYS-Fluent, including specific numerical approaches to ensure stable solutions of the flake motion equations and enable simulations with practical relevance. In the case of droplet impact on dry surfaces, relatively large final flake angles with respect to the surface are obtained in the quasi-steady state after the receding process, together with a significant dependency on the Re-number. In the case of droplet impact on wet surfaces, it was found that higher liquid viscosities reduce the flake movement, resulting in smaller differences between the initial and the final flake angles. The dependency of the final flake angle on the characteristic Re-number was analyzed, yielding suitable Re^*-numbers for a final optimal flake angle in practical applications.

The presented set of simulation results delivers useful information for future investigations with enhanced practical relevance, including, e.g., the distribution of flake orientation depending on the droplet size distribution and the size/velocity correlation at droplet impact. Also, film drying and baking effects on air release and flake orientation must be considered to allow an estimation of the final paint film quality.

Author Contributions: Validation, O.T. and J.D.; Investigation, Q.Y. and B.S.; Writing—original draft, Q.Y.; Writing—review & editing, O.T. and J.D.; Supervision, Q.Y.; Project administration, O.T. All authors have read and agreed to the published version of the manuscript.

Funding: The present investigations were supported by the German Federal Ministry for Economic Affairs and Energy through the Arbeitsgemeinschaft industrieller Forschungsvereinigungen (AIF). The work was also supported by the High-Performance Computing Center (HLRS) of the University of Stuttgart.

Data Availability Statement: Dataset available on request from the authors.

Acknowledgments: We would like to thank our colleagues at the Fraunhofer Institute for Manufacturing Engineering and Automation in Stuttgart, who provided us the experimental results, which were very helpful for the present work.

Conflicts of Interest: The authors declare no conflict of interest.

References

1. Chandra, S.; Avedisian, C.T. On the collision of a droplet with a solid surface. *Proc. R. Soc. Lond. Ser. A* **1991**, *432*, 13–41. [CrossRef]
2. Fujimoto, H.; Shiraishi, H.; Hatta, N. Evolution of liquid/solid contact area of a drop impinging on a solid surface. *Int. J. Heat Mass Transf.* **2000**, *43*, 1673–1677. [CrossRef]
3. Thoroddsen, S.T.; Etoh, T.G.; Takehara, K.; Ootsuka, N.; Hatsuki, Y. The air bubble entrapped under a drop impacting on a solid surface. *J. Fluid Mech.* **2005**, *545*, 203–212. [CrossRef]
4. Thoroddsen, S.T.; Takehara, K.; Etoh, T.G. Bubble entrapment through topological change. *Phys. Fluids* **2010**, *22*, 051701. [CrossRef]
5. Palacios, J.; Hernández, J.; Gómez, P.; Zanzi, C.; López, J. On the impact of viscous drops onto dry smooth surfaces. *Exp. Fluids* **2012**, *52*, 1449–1463. [CrossRef]
6. Thoroddsen, S.T.; Sakakibara, J. Evolution of the fingering pattern of an impacting drop. *Phys. Fluids* **1998**, *10*, 1359–1374. [CrossRef]
7. Li, E.Q.; Thoroddsen, S.T. Time-resolved imaging of a compressible air disc under a drop impacting on a solid surface. *J. Fluid Mech.* **2015**, *780*, 636–648. [CrossRef]
8. Arogeti, M.; Sher, E.; Bar-Kohany, T. A single spherical drop impact on a flat, dry surface—A unified correlation. *At. Sprays* **2017**, *27*, 759–770. [CrossRef]
9. Wörner, M. Maximum spreading of an impacting drop. *Int. J. Multiph. Flow* **2023**, *167*, 104528. [CrossRef]
10. Mehdi-Nejad, V.; Mostaghimi, J.; Chandra, S. Air bubble entrapment under an impacting droplet. *Phys. Fluids* **2003**, *15*, 173–183. [CrossRef]
11. Kim, E.; Baek, J. Numerical study of the parameters governing the impact dynamics of yield-stress fluid droplets on a solid surface. *J. Non-Newton. Fluid Mech.* **2012**, *173–174*, 62–71. [CrossRef]
12. Ye, Q.; Tiedje, O. Investigation on Air Entrapment in Paint Drops under Impact onto Dry Solid Surfaces. *High Perform. Comput. Sci. Eng.* **2016**, *16*, 355–374.
13. German, G.; Bertola, V. Impact of shear-thinning and yield-stress drops on solid substrates. *J. Phys. Condens. Matter* **2009**, *21*, 375111. [CrossRef]
14. Nigen, S. Experimental investigation of the impact of an (apparent) yield-stress material. *At. Sprays* **2005**, *15*, 103–118. [CrossRef]
15. Šikalo, Š.; Wilhelm, H.-D.; Roisman, I.V.; Jakirlić, S.; Tropea, C. Dynamic contact angle of spreading droplets: Experiments and simulations. *Phys. Fluids* **2005**, *17*, 062103. [CrossRef]
16. Linder, N.; Criscione, A.; Roisman, I.V.; Marschall, H.; Tropea, C. 3D computation of an incipient motion of a sessile drop on a rigid surface with contact angle hysteresis. *Theor. Comput. Fluid Dyn.* **2015**, *29*, 373–390. [CrossRef]
17. Weiss, D.A.; Yarin, A.L. Single drop impact onto liquid films: Neck distortion, jetting, tiny bubble entrainment, and crown formation. *J. Fluid Mech.* **1999**, *385*, 229–254. [CrossRef]
18. Rein, M. Phenomena of liquid drop impact on solid and liquid surfaces. *Fluid Dyn. Res.* **1993**, *12*, 61–93. [CrossRef]
19. Wang, A.-B.; Chen, C.-C. Splashing impact of a single drop onto very thin liquid films. *Phys. Fluids* **2000**, *12*, 2155–2158. [CrossRef]
20. Banks, D.; Ajawara, C.; Sanchez, R.; Surti, H.; Aguilar, G. Effects of drop and film viscosity on drop impacts onto thin films. *At. Sprays* **2013**, *23*, 525–540. [CrossRef]
21. Tran, H.; He, Z.; Pack, M. Microbubble entrainment on thin liquid films under drop impacts. *J. Colloid Interface Sci.* **2025**, *682*, 915–923. [CrossRef]
22. Thoroddsen, S.T.; Thoraval, M.-J. Using drop impacts to study the dynamics of sub-micron liquid structures. In Proceedings of the 15th International Symposium on Flow Visualization, Minsk, Belarus, 25–28 June 2012.
23. Kirchner, E. Flow-induced orientation of flakes in metallic coatings: II. The orientation mechanism. *Prog. Org. Coat.* **2018**, *124*, 104–109. [CrossRef]
24. Schlüsener, T. Untersuchungen zum Einfluss der thermo- und hydrodynamischen Vorgänge bei der Lackapplikation und -trocknung auf die Farbtonausbildung wasserbasierter Metallic-Lacke. Ph.D. Thesis, Technical University of Darmstadt, Darmstadt, Germany, 2000. (In German)
25. Saito, Y.; Hashiguchi, N.; Cheng, Z.; Okabe, T.; Miyagawa, T.; Shirota, M.; Matsukawa, Y.; Aoki, H.; Daikoku, M.; Fukuno, J. Numerical analysis of droplet impact behavior and changes in flake particle orientation within droplets on a solid surface. *Chem. Eng. Sci.* **2025**, *306*, 121299. [CrossRef]
26. Ye, Q.; Dreher, M.; Tiedje, O.; Shen, B.; Domnick, J. Numerical simulation of flake orientation during droplet impact on substrates in spray painting processes. In Proceedings of the ICLASS, Edinburgh, UK, 29 August–2 September 2021. [CrossRef]
27. Shen, B.; Ye, Q.; Tiedje, O.; Domnick, J. On the impact of viscous droplets on wet solid surfaces for spray painting processes. In Proceedings of the ICLASS 2021, 15th Triennial International Conference on Liquid Atomization and Spray Systems, Edinburgh, UK, 29 August–2 September 2021; Volume 1. [CrossRef]
28. Ye, Q.; Dreher, M.; Shen, B. Numerical simulation of flake orientation during droplet impact on substrates in spray painting processes. In *High Performance Computing in Science and Engineering '21*; Nagel, W.E., Kröner, D.H., Resch, M.M., Eds.; Springer International Publishing: Cham, Switzerland, 2023; pp. 321–337. [CrossRef]

29. Ye, Q.; Rhein, S.; Hess, T.; Niemeier, W.; Tiedje, O. Investigation on viscous droplet generation for selective coating processes. In Proceedings of the 10th International Conference on Multiphase Flow, ICMF 2019, Rio de Janeiro, Brazil, 19–24 May 2019.
30. Ye, Q.; Domnick, J. Analysis of droplet impingement of different atomizers used in spray coating processes. *J. Coat. Technol. Res.* **2017**, *14*, 467–476. [CrossRef]
31. Maile, F.J.; Pfaff, G.; Reynders, P. Effect pigments—Past, present and future. *Prog. Org. Coat.* **2005**, *54*, 150–163. [CrossRef]

MDPI AG
Grosspeteranlage 5
4052 Basel
Switzerland
Tel.: +41 61 683 77 34

Fluids Editorial Office
E-mail: fluids@mdpi.com
www.mdpi.com/journal/fluids

www.ingramcontent.com/pod-product-compliance
Lightning Source LLC
Chambersburg PA
CBHW041945260926
51850CB00018B/841
* 9 7 8 3 7 2 5 8 5 0 8 7 7 *